Structures and Functions of Retinal Proteins

Structures et fonctions des rétino-protéines

Colloques **INSERM**
ISSN 0768-3154

Other *Colloques* published as co-editions by John Libbey Eurotext and INSERM

133 Cardiovascular and Respiratory Physiology in the Fetus and Neonate. *Physiologie Cardiovasculaire et Respiratoire du Fœtus et du Nouveau-né.*
Scientific Committee : P. Karlberg,
A. Minkowski, W. Oh and L. Stern;
Managing Editor : M. Monset-Couchard.
ISBN : John Libbey Eurotext 0 86196 086 6
INSERM 2 85598 282 0

134 Porphyrins and Porphyrias. *Porphyrines et Porphyries.*
Edited by Y. Nordmann.
ISBN : John Libbey Eurotext 0 86196 087 4
INSERM 2 85598 281 2

137 Neo-Adjuvant Chemotherapy. *Chimiothérapie Néo-Adjuvante.*
Edited by C. Jacquillat, M. Weil and D. Khayat.
ISBN : John Libbey Eurotext 0 86196 077 7
INSERM 2 85598 283 7

139 Hormones and Cell Regulation (10th European Symposium). *Hormones et Régulation Cellulaire (10ᵉ Symposium Européen).*
Edited by J. Nunez, J.E. Dumont and R.J.B. King.
ISBN : John Libbey Eurotext 0 86196 084 X
INSERM 2 85598 284 7

147 Modern Trends in Aging Research. *Nouvelles Perspectives de la Recherche sur le Vieillissement.*
Edited by Y. Courtois, B. Faucheux, B. Forette, D.L. Knook and J.A. Tréton.
ISBN : John Libbey Eurotext 0 86196 103 X
INSERM 2 85598 309 6

149 Binding Proteins of Steroid Hormones. *Protéines de liaison des Hormones Stéroïdes.*
Edited by M.G. Forest and M. Pugeat.
ISBN : John Libbey Eurotext 0 86196 125 0
INSERM 2 85598 310 X

151 Control and Management of Parturition. *La Maîtrise de la Parturition.*
Edited by C. Sureau, P. Blot, D. Cabrol, F. Cavaillé and G. Germain.
ISBN : John Libbey Eurotext 0 86196 096 3
INSERM 2 85598 311 8

Suite page 433
(Continued p. 433)

Structures and Functions of Retinal Proteins

Structures et fonctions des rétino-protéines

Proceedings of the Vth International Conference on Retinal Proteins
held in Dourdan (France)
June 28 - July 3, 1992

Edited by

Jean-Louis Rigaud

LES EDITIONS
INSERM

British Library Cataloguing in Publication Data
A catalogue record for this book
is available from the British Library

ISBN 0 86196 355 5
ISSN 0768-3154

First published in 1992 by

Editions John Libbey Eurotext
6 rue Blanche, 92120 Montrouge, France. (33) (1) 47 35 85 52
ISBN 0 86196 355 5

John Libbey and Company Ltd
13 Smiths Yard, Summerley Street, London SW18 4HR,
England.
(44) (81) 947 27 77

Institut National de la Santé et de la Recherche Médicale
101 rue de Tolbiac, 75654 Paris Cedex 13, France.
(33) (1) 44 23 60 00
ISBN 2 85598 509 9

ISSN 0768-3154

© 1992 Colloques INSERM/John Libbey Eurotext Ltd,
All rights reserved
Unauthorized publication contravenes applicable laws

Foreword

The Retinal Conferences are organized every two years (Germany, 1984 ; Russia, 1986 ; Japan, 1988 ; USA, 1990) and regroup experts working on retinal chromoproteins providing the opportunity for exchanging opinions on new findings on photoreception and energy transduction in visual cells and bacteria.

Four retinal proteins have so far been identified in the halobacterial branch of archaebacteria. Two of them (the proton pump bacteriorhodopsin and the chloride pump halorhodopsin) convert light energy into chemical energy and two (sensory rhodopsins I and II) act as photoreceptors sensing the light and allowing the halobacterium cells to move.

On the other hand, rhodopsin is the protein responsible for generating an optic nerve impulse on the visual receptors of the three phyla that possess image-resolving eyes : molluscs, arthropods and vertebrates.

All these proteins form a ring of seven transmembrane helices and have a lysine on the middle of the 7th helix to bind retinal and form a protonated Schiff base. The molecular processes intervening between absorption of the photon by the retinal proteins and the ensuing biological responses can be divided into two sets of events : (1) the sets of events at the level of the photoreceptor molecule leading to production of its activated form ; (2) the subsequent sets of events triggered by these photoactivated forms which lead to ion translocation or formation of signaling state inducing cellular responses such as conductance changes, flagellar motility or generation of electrochemical gradients.

Thus the "Vth International Conference on Retinal Proteins" organized in Dourdan worked to the finality to give general account of the latest achievements in these interdisciplinary fields where photochemical, biophysical, biochemical, molecular biological and physiological aspects are intimately linked.

I wish to express my indebtedness to those 190 "eye and bacterium" scientists who have contributed to the enrichment of the program by their high standard oral and poster presentations. All of them have provided a constructive opportunity for formal and informal but always useful exchange of ideas, suggestions in a cordial atmosphere warmed up by a nice weather and French wine and champagne.

Finally I would like to express the sincere thanks to the academic and research institutions which made possible the organization of the congress through their valuable support and sponsorship.

Jean-Louis Rigaud

Organizer
Organisateur

Jean-Louis Rigaud *(CEA, Saclay)*

Local Committee
Comité local

M. Chabre *(Nice)*
E. Padrós *(Barcelona)*
J.L. Popot *(Paris)*
M. Seigneuret *(Paris)*
G. Zaccaï *(Grenoble)*

International Advisory Committee
Comité international

N. Abdulaev *(Russia)*
R. Bogomolni *(USA)*
T. Hara *(Japan)*
R. Henderson *(England)*
M. Heyn *(Germany)*
K.P. Hofmann *(Germany)*
L. Keszthelyi *(Hungary)*
G. Khorana *(USA)*
J.K. Lanyi *(USA)*
D. Oesterhelt *(Germany)*
V.P. Skulachev *(Russia)*
T. Yoshizawa *(Japan)*

Avant-propos

Ces conférences ont lieu tous les deux ans, les précédentes ayant été organisées en Allemagne (1984), en Russie (1986), au Japon (1988) et aux États-Unis (1990). Elles rassemblent les plus grands spécialistes travaillant sur les protéines liant le rétinal, protéines impliquées dans les processus de photoréception et de transduction d'énergie dans les cellules visuelles (rhodopsine) et les bactéries (bactériorhodopsine, halorhodopsine, rhodopsines sensorielles). Ces conférences sont donc caractérisées par la rencontre de deux grands domaines qui, bien que différents au regard de leurs fonctions biologiques, présentent de grandes similarités au regard des mécanismes de photoréception, similarités dues à la structure commune rétinal plus opsine des protéines impliquées.

Si la rhodopsine et la bactériorhodopsine ont des rôles biologiques très différents, elles présentent cependant de grandes similarités au regard de la première série d'événements, c'est-à-dire au regard des mécanismes de photoréception qui ont amené naturellement les chercheurs à des études expérimentales et théoriques comparatives. D'autre part, la frontière entre les mondes de "l'oeil" et de "la bactérie" s'est largement estompée depuis la découverte récente de rhodopsines sensorielles dans les archaebactéries. Deux rétino-protéines, SRI et SRII, assurent une vision de la couleur qui permet aux halobactéries de se déplacer vers des micro-environnements favorables. Ces protéines agissent comme les yeux des halobactéries et remplissent la même fonction que les pigments visuels de la rétine des vertébrés (plusieurs composants de la chaîne de transduction du signal ont pu être caractérisés : protéines G, C.GMP. ...). Plus récemment, il a pu être mis en évidence chez *Chlamydomonas* un système visuel contenant la rhodopsine qui agit comme photorécepteur fonctionnel permettant l'orientation de ce micro-organisme à la lumière.

Toutes ces protéines membranaires (rhodopsines, bactériorhodopsine, halorhodopsine, rhodopsines sensorielles) appartiennent à la famille des protéines à sept hélices avec, en leur milieu, un chromophore, le rétinal (tout-trans dans les bactéries, 11-cis dans les pigments visuels) lié de façon covalente à une lysine de la protéine par l'intermédiaire d'une base de Schiff protonée. Dans tous les cas, la lumière provoque une isomérisation du rétinal (*trans* → 13 *cis* dans les bactéries, 11 → *cis* tout *trans* dans les pigments visuels) suivie d'un changement conformationnel qui conduit soit à la translocation d'ions, soit à l'interaction avec des protéines G pour démarrer la cascade visuelle (transducine C GMP phosphodiestérase).

La 5e Conférence Internationale sur les rétino-protéines organisée à Dourdan a eu pour objectif de présenter les derniers travaux dans ces domaines interdisciplinaires où photochimie, biophysique, biochimie, biologie moléculaire et physiologie sont intimement interconnectées.

Je tiens à remercier les 190 collègues de nombreux pays qui ont accepté de participer à cette conférence, de présenter oralement et par voie d'affiches leurs derniers travaux et de prendre part activement à toutes les discussions.

Enfin, je voudrais exprimer mes remerciements à toutes les institutions et organismes qui, par leur parrainage et leur soutien, ont permis à cette Conférence de se dérouler dans des conditions idéales.

Jean-Louis Rigaud

With the sponsorship of
Avec le soutien de :

Centre National de la Recherche Scientifique (CNRS)
Commissariat à l'Energie Atomique (CEA)
Commission of the European Community (CEC)
Institut National de la Santé et de la Recherche Médicale (INSERM)
International Union of Biochemistry and Molecular Biology (IUBMB)
Ministère de la Recherche et de l'Espace (MRE)
Organisation des Nations Unies pour l'Education, la Science et la Culture (UNESCO)
Société Française de Biophysique
Société Française de Photobiologie
Université de Paris VII (Laboratoire de Biologie Cellulaire)

Under the auspices of
Avec le patronage de :

Société Française de Biochimie et de Biologie Moléculaire

We also thank, for their interest and participation, the following firms :
Bruker
Jobin Yvon
Vegatec

Contents
Sommaire

V   Foreword
VII *Avant-propos*

I. MOLECULAR STRUCTURE OF RETINAL PROTEINS
I. *STRUCTURE MOLÉCULAIRE DES RÉTINO-PROTÉINES*

3 T.W. Kahn, D.M. Engelman
The roles of extramembranous bilayer loops and of retinal in the folding and stability of bacteriorhodopsin
Rôle des boucles extramembranaires et du rétinal dans le repliement et la stabilité de la bactériorhodopsine

9 F.A. Samatey, J.L. Popot, C. Etchebest, G. Zaccaï
Rotational orientation of transmembrane α-helices in bacteriorhodopsin studied by neutron diffraction
Orientation rotationnelle des α-hélices transmembranaires de la bactériorhodopsine étudiée par diffraction de neutrons

13 P. Tufféry, J.L. Popot, R. Lavery
Modelling bundles of transmembrane α-helices : a test study on bacteriorhodopsin
Modélisation de l'organisation des α-hélices transmembranaires : application à l'étude de la bactériorhodopsine

17 M. Nina, B. Roux, J.C. Smith
Ground state potential surface calculations for butadiene and retinal
Calcul des potentiels de surface à l'état fondamental du butadiène et du rétinal

21 S.L. Helgerson, E. Dratz, R.D. Renthal
Evidence for beta-turns in the surface loops and α-I/α-II helices in the transmembrane regions of bacteriorhodopsin
Mise en évidence de structures bêta dans les boucles de surface et d'hélices αI-αII dans les régions transmembranaires de la bactériorhodopsine

25 J.Y. Cassim, J.E. Draheim, N.J. Gibson
The *in situ* molecular dynamics of the transmembrane protein bacteriorhodopsin : static or kinetic ?
Dynamiques moléculaires in situ *de la bactériorhodopsine : statique ou cinétique ?*

29 J. Torres, E. Padrós
Fourier-transform infrared estimation of the secondary structure of bacterioopsin integrated into lipid vesicles
Estimation de la structure secondaire de la bactériorhodopsine reconstituée en liposomes par IRTF

33 J. Cladera, E. Padrós
Fourier-transform infrared study of the secondary structure of heat-denatured bacteriorhodopsin. Comparison with the native form
Etude par IRTF de la structure secondaire de la bactériorhodopsine dénaturée par la chaleur. Comparaison avec la forme native

37 T. Hamanaka, Y. Kito, M. Seidou, K. Wakabayashi, Y. Amemiya
Structure analysis of the live squid photoreceptor by X-ray diffraction
Analyse structurale par diffraction de rayons X du photorécepteur de calmar

41 M. Seigneuret, D. Lévy, J.M. Neumann
One-dimensional and multidimensional NMR of bacteriorhodopsin
Etudes de la structure de la bactériorhodopsine par RMN à une et plusieurs dimensions

45 L. Pardo, M. Duñach
A preliminar three-dimensional model of rhodopsin
Un modèle préliminaire de la structure tridimensionnelle de la rhodopsine

II. MOLECULAR ANALYSIS OF PHOTORECEPTORS
II. ANALYSE MOLÉCULAIRE DES PHOTORÉCEPTEURS

51 P.A. Hargrave
Structure and function of vertebrate rhodopsin
Structure et fonction de la rhodopsine des vertébrés

55 C. Venien-Bryan, A. Davies, J.R. Wilkinson, K. Langmack, J. Baverstock, C. Nobes, H.R. Saibil
Structure and function of squid rhodopsin
Structure et fonction de la rhodopsine de calmar

59 W.J. DeGrip, L.L.J. DeCaluwé, R.G. Foster, J.J.M. Janssen, H.W. Korf, O. Bousché, K.J. Rothschild
Identification and molecular analysis of vertebrate visual and non-visual photoreceptor proteins
Identification et analyse moléculaire des photorécepteurs visuels des vertébrés

63 L.L.J. DeCaluwé, D.M.F. VanAalten, J. VanOostrum, W.J. DeGrip
Studies towards the molecular mechanism of rhodopsin : protein engineering and molecular modeling
Etudes du mécanisme moléculaire de la rhodopsine : modélisation moléculaire et ingénierie des protéines

67 T.P. Sakmar, K. Fahmy, T. Chan, M. Lee
Mutagenesis studies of rhodopsin phototransduction
Etude par mutagenèse de la phototransduction par la rhodopsine

71 T. Yoshizawa
Cone visual pigments : structure, evolution and function
Pigments visuels des cônes : structure, évolution et fonction

75 A. Carne, S. Conway, J.N. Keen, L. Shu-Hua, R.A. McGregor, P.D. Monk, J.S. Lott, J.D.D. Pottinger, N.J.P. Ryba, A. Sinclair, J.B.C. Findlay
Dissecting the squid phototransduction membrane
Analyse détaillée des composants membranaires impliqués dans la transduction de la vision chez le calmar

79 P. Towner, W. Gärtner
The opsin sequence of the mantid *Sphodromantis* sp.
Séquence de l'opsine de l'espèce Sphodromantis

83 **P. Röhlich, A. Szél, V.I. Govardovskii, T. van Veen**
Two cone types expressing different visual pigments in rodents
Deux types de cônes contenant des pigments différents chez les rongeurs

III. MUTAGENESIS STUDIES OF BACTERIORHODOPSIN
III. ÉTUDES PAR MUTAGENÈSE DE LA BACTÉRIORHO-DOPSINE

89 **M. Dyall-Smith, M. Holmes, M. Kamekura, F. Doolittle**
Halobacterial vector development and the opportunities for gene expression and analysis
Vecteur de développement chez les bactéries halophiles : opportunités pour l'expression et l'analyse génétique

93 **M. Chang, X. Fan, R. Needleman**
Expression in *H. halobium* of bacteriorhodopsin mutants obtained by site-directed mutagenesis
Expression dans Halobacterium halobium *de mutants de la bactériorhodopsine obtenus par mutagenèse dirigée*

97 **V. Hildebrandt, U. Bauer, N.A. Dencher, G. Büldt, P. Wrede**
Bacteriorhodopsin precursor is partially processed at the N-terminal end in hetereologous *in vitro* and *in vivo* expression systems
Le précurseur de la bactériorhodopsine est partiellement «processé» à l'extrémité N-terminale dans des systèmes d'expression in vivo *et* in vitro

101 **Y. Mukohata, K. Ihara, Y. Miyashita, T. Amemiya, T. Taguchi, M. Tateno, Y. Sugiyama**
Comparative studies on bacteriorhodopsin-type pumps : basic physicochemical data for proton pumps
Etudes comparatives de pompes de type bactériorhodopsine : données physicochimiques de base pour l'étude des pompes à protons

105 **J. Otomo, H. Tomioka, Y. Urabe, H. Sasabe**
Properties and the primary structures of new bacterial rhodopsins
Propriétés et structures primaires de nouvelles rhodopsines bactériennes

IV. PHOTOCYCLE
IV. *CYCLE PHOTOCHIMIQUE*

111 **S.P. Balashov, R. Govindjee, M. Kono, E. Lukashov, T.G. Ebrey, Y. Feng, R. K. Crouch, D.R. Menick**
Arg82Ala mutant of bacteriorhodopsin expressed in *H. halobium* : drastic decrease in the rate of proton release and effect on dark adaptation
Mutant Arg82-Ala de la bactériorhodopsine exprimée dans H.halobium : *diminution importante de la vitesse de relargage du proton et effet sur l'adaptation à la lumière*

115 **R. Govindjee, E. Lukashev, M. Kono, S.P. Balashov, T.G. Ebrey, J. Soppa, J. Tittor, D. Oesterhelt**
Tyr57Asn mutant of bacteriorhodopsin : M formation and spectral transformations at high pH
Mutant Tyr57-Asn de la bactériorhodopsine : formation de l'intermédiaire M et transformations spectrales à haut pH

119 **S.P. Balashov, E.S. Imasheva, N.V. Karneyeva, F.F. Litvin, T.G. Ebrey**
Conformers and multiple primary photoproducts of bacteriorhodopsin and N intermediate at low temperature
Conformères et photoproduits primaires dans le cycle de la bactériorhodopsine. Etude des intermédiaires N à basse température

123 **H. Okazaki, C.W. Chang, T. Akaike, O. Oshida, T. Yasukawa**
Biological significance of the *trans-cis* isomerization of retinal in proton transfer processes of bacteriorhodopsin
Signification biologique de l'isomérisation trans-cis *du rétinal dans les processus de transfert du proton dans la bactériorhodopsine*

127 **J. Le Coutre, K. Gerwert**
The azide-effect in D96→N/G mutated bacteriorhodopsins monitored by timeresolved FTIR-difference-spectroscopy
Etude par spectroscopie IRTF résolue dans le temps des effets de l'azide sur la bactériorhodopsine mutée D96→N/G

131 **A. Maeda**
Fourier transform infrared studies on light energy transfer process of bacteriorhodopsin
Etudes par spectroscopie IRTF des processus de transfert de l'énergie lumineuse par la bactériorhodopsine

135 P. Ormos, K. Chu, J. Mourant
Infrared spectroscopy of the reactions of the M form of bacteriorhodopsin : conclusions about the mechanism of proton pumping
Etude par spectroscopie infra-rouge des réactions de l'état M de la bactériorhodopsine : conclusions sur le mécanisme de pompage de protons

139 W. Eisfeld, M. Stockburger
Optical transient studies on the photochemical cycle of bacteriorhodopsin (0.5 µs - 500 ms, pH 4-10)
Etudes par spectroscopie optique transitoire du cycle photochimique de la bactériorhodopsine (0,5 µs-500 ms ; pH 4-10)

143 C. Pusch, R. Diller, W. Eisfeld, R. Lohrmann, M. Stockburger
The light-induced proton-pump of bacteriorhodopsin studied by resonance Raman and optical transient spectroscopy
Etudes par spectroscopie optique transitoire et par résonance Raman du pompage de protons par la bactériorhodopsine

147 R. Lohrmann, M. Stockburger
Evidences for structural changes at the chromophoric site of bacteriorhodopsin during the K-to-L transition
Mise en évidence de changements structuraux au site du chromophore de la bactériorhodopsine lors de la transition K-L

151 M. Rohr, P. Schulenberg, W. Gärtner, S.E. Braslavsky
Detection of conformational changes during the photocycle of bacteriorhodopsin by laser-induced optoacoustic spectroscopy (LIOAS)
Détection de changements conformationnels lors du photocycle de la bactériorhodopsine : analyse par spectroscopie opto-acoustique induite au laser (LIOAS)

155 B. Hessling, G. Souvignier, K. Gerwert
A new approach to analyse kinetic data of bacteriorhodopsin, factor analysis and decomposition
Une nouvelle approche de l'analyse des études cinétiques de la bactériorhodopsine : analyse des facteurs et décomposition

159 L.A. Drachev, A.D. Kaulen, A.Y. Komrakov
Relationship of M-intermediates in bacteriorhodopsin photocycle
Relation entre les intermédiaires M dans le cycle photochimique de la bactériorhodopsine

163 **A.D. Kaulen, L.A. Drachev, S.V. Dracheva**
M-type intermediate formation during 13-*cis* bacteriorhodopsin photocycle and light-dark adaptation
Formation d'un intermédiaire de type M lors du cycle photochimique et de l'adaptation à la lumière

167 **L.V. Khitrina, L.A. Drachev, S.V. Eremin, A.D. Kaulen, A.A. Khodonov**
M-intermediate in the 13-*cis*-cycle of bacteriorhodopsin analogs
Formation d'un intermédiaire de type M dans le cycle 13 cis d'analogues de la bactériorhodopsine

171 **I.V. Chizhov, M. Engelhard, A.V. Sharkov, B. Hess**
Two quantum absorption of ultrashort laser pulses by the bacteriorhodopsin chromophore
Absorptions de deux quanta par le chromophore de la bactériorhodopsine lors d'impulsions laser ultracourtes

175 **Z. Tokaji, Z. Dancsházy**
Light density controls the mechanism of the photocycle of bacteriorhodopsin
La densité de la lumière contrôle les mécanismes du cycle photochimique de la bactériorhodopsine

179 **N.M. Kozhevnikov**
Dynamic holograms in bacteriorhodopsin : theory and application for phase-modulated optical beams detecting
Hologrammes dynamiques : théorie et application de la bactériorhodopsine pour un détecteur à faisceaux optiques modulés en phase

183 **S. Scherling, H. Sigrist**
Engineering of a light-sensitive molecular device
Ingénierie d'un système moléculaire sensible à la lumière

V. BACTERIORHODOPSIN, HALORHODOPSIN :
ION TRANSLOCATION, CHARGE MOVEMENT
V. *BACTÉRIORHODOPSINE, HALORHODOPSINE :
TRANSLOCATION D'IONS ET MOUVEMENTS DE
CHARGES*

189 **M. Ikonen, A. Sharonov, N. Tkachenko, H.Lemmetyinen**
The photovoltage of bacteriorhodopsin in X- and Z-type monolayers and Z-type multilayer Langmuir-Blodgett film
Le potentiel induit par la lumière de la bactériorhodopsine incorporée dans des couches monomoléculaires de type X ou Z, et dans des films multicouches Langmuir-Blodgett de type Z

193 **C. Gergely, G. Váró**
Charge motions in the D85N and D212N mutants of bacteriorhodopsin
Mouvements de charges dans les mutants D85N et D212N de la bactériorhodopsine

197 **A. Dér, R. Tóth-Boconádi, S. Száraz**
Electric signals and the photocycle of bacteriorhodopsin
Signaux électriques et cycle photochimique de la bactériorhodopsine

201 **S. Moltke, M.P. Heyn, M.P. Krebs, R. Mollaaghababa, H.G. Khorana**
Low pH photovoltage kinetics of bacteriorhodopsin with replacements of Asp-96, -85, -212 and Arg-82
Potentiels induits par la lumière à bas pH : études sur la bactériorhodopsine mutée sur Asp96, Asp85, Asp212 et Arg82

205 **P. Scherrer, U. Alexiev, H. Otto, M.P. Heyn, T. Marti, H.G. Khorana**
Proton movement and surface charge in bacteriorhodopsin detected by selectively attached pH-indicators
Mouvements de protons et charges de surface dans la bactériorhodopsine étudiée par des indicateurs de pH sélectifs

213 **N.A. Dencher, J. Heberle, G. Büldt, H.D. Höltje, M. Höltje**
Active and passive proton transfer steps through bacteriorhodopsin are controlled by a light-triggered hydrophobic gate
Les étapes dans les transferts actifs et passifs de protons au travers de la bactériorhodopsine sont contrôlés par une barrière hydrophobe déclenchée par la lumière

217 **G. Thiedemann, J. Heberle, N.A. Dencher**
Bacteriorhodopsin pump activity at reduced humidity
Activité de pompe de la bactériorhodopsine dans une humidité réduite

221 **J. Heberle, N.A. Dencher**
The surface of the purple membrane : a transient pool for protons ?
La surface de la membrane pourpre : un réservoir transitoire de protons

225 **F. Sepulcre, E. Padrós**
Fluorescence studies on the surface potential of deionized forms of purple and bleached membranes
Etudes par fluorescence du potentiel de surface des formes désionisées de la membrane pourpre et de la membrane blanchie

229 **V.P. Skulachev**
Rhodopsins : from ion pumps to specialized photoreceptors
Les rhodopsines : des pompes ioniques aux récepteurs photosensibles spécialisés

233 **T.J. Walter, M.S. Braiman**
FT-IR difference spectroscopy of halorhodopsin in the presence of different anions
Etude par IRTF de l'halorhodopsine en présence de différents anions

237 **H. Tomioka, N. Kamo, K. Fujikawa, II. Sasabc**
Absorbance change of a C-50 carotenoid, bacterioruberin related to Cl^- translocation in a Cl^- pump, halorhodopsin
Relations entre le changement d'absorbance d'un caroténoïde C50 (bactériorubérine) et le transport de chlore dans une pompe à chlore (halorhodopsine)

VI. STRUCTURE AND FUNCTION OF RETINAL
VI. STRUCTURE ET FONCTION DU RÉTINAL

243 **A.H. Chen, F. Derguini, P. Franklin, S. Hu, K. Nakanishi, B.R. Silvo, J. Wang**
The triggering process of visual transduction
Les processus de déclenchement de la transduction de la vision

247 A.S. Ulrich, I. Wallat, M.P. Heyn, A. Watts
Evidence for a curved retinal in BR from solid-state ^{2}H-NMR
Mise en évidence par RMN du solide d'un rétinal incurvé dans la bactériorhodopsine

251 R.S.H. Liu, L.U. Colmenares
^{19}F-NMR in studies of fluorinated visual pigment analogs. A method for detecting neighbouring groups or empty space in a binding site
RMN du fluor 19 d'analogues de pigments visuels fluorés : une méthode pour détecter des groupes voisins ou un espace vide dans un site de liaison

255 T. Hara, R. Hara, I. Hara-Nishimura, M. Nishimura, A. Terakita, K. Ozaki
The rhodopsin-retinochrome system in the squid visual cell
Le système rhodopsine-rétinochrome dans les cellules visuelles de calmar

259 I. Hara-Nishimura, M. Kondo, M. Nishimura, R. Hara, T. Hara
Structural comparison of retinal photopigments in cephalopods
Comparaison de la structure des pigments rétiniens photosensibles des céphalopodes

263 N. Sekiya, A. Kishigami, F. Tokunaga, T. Takahashi, K. Yoshihara
A structural study of retinochrome using fluorinated retinal analogues
Une étude structurale du rétinochrome à l'aide d'analogues fluorés du rétinal

267 I.M. Pepe, C. Cugnoli, J.N. Keen, J.B.C. Findlay
Retinal photoisomerase from honeybee compound eye : physiological role
Rôle physiologique de la photo-isomérase du rétinal de l'oeil d'abeille

271 C. Cugnoli, R. Fioravanti, O. Golisano, I.M. Pepe
Retinal photoisomerase from honeybee compound eye : enzymic mechanism
Mécanismes enzymatiques de la photo-isomérase du rétinal de l'oeil d'abeille

275 **I.B. Fedorovich, K.M. Grant, A.M.J. Brennan, M.A. Ostrovsky, C.A. Converse**
Photodamage to interphotoreceptor retinoid-binding protein
Dommage induit par la lumière sur l'IRBP (interphotoreceptor retinoid-binding protein)

277 **J. Schwemer, F. Spengler**
Opsin synthesis in blowfly photoreceptors is controlled by an 11-cis retinoid
La synthèse de l'opsine dans les récepteurs à la lumière de la mouche dorée est contrôlée par un 11 cis-rétinoïde

VII. RHODOPSINS : STRUCTURE-FUNCTION
VII. RHODOPSINE : STRUCTURE-FONCTION

283 **R.R. Birge, R.B. Barlow Jr, J.R. Tallent**
On the molecular origins of thermal noise in vertebrate and invertebrate photoreceptors
Origines moléculaires du bruit thermique dans les photorécepteurs de vertébrés et d'invertébrés

287 **M. Tsuda, M. Nakagawa, T. Iwasa, S. Kikkawa**
Light induced conformational changes of octopus rhodopsin
Changements conformationnels induits par la lumière dans la rhodopsine de pieuvre

291 **Y. Shichida**
Changes in chromophore-opsin interaction in the photobleaching processes of visual pigments
Changements dans l'interaction opsine-chromophore lors du blanchiement par la lumière des pigments visuels

295 **D. Garcia-Quintana, P. Garriga, M. Duñach, J. Manyosa**
The role of the Cys110 -Cys187 disulfide bond in rhodopsin investigated spectrophotometrically
Le rôle du pont disulfure Cys110-Cys187 dans la rhodopsine. Etude spectrophotométrique

299 **T. Iwasa, N.G. Abdulaev, M. Nakagawa, S. Kikkawa, M. Tsuda**
Cysteine residues in rhodopsins
Les résidus cystéines dans les rhodopsines

303 N. Virmaux, L. Menguy, N. Boukra, G. Nullans, A. van Dorsselaer, P.F. Urban
Phospholipids associated with rhodopsin purified by concanavalin-A sepharose
Purification par chromatographie sur sépharose-concanavaline A des phospholipides liés à la rhodopsine

VIII. BACTERIAL SENSORY RHODOPSINS : PHOTOTAXIS
VIII. RHODOPSINES SENSORIELLES DES BACTÉRIES : PHOTOTAXISME

309 J.L. Spudich
Phototransduction by sensory rhodopsin I
Phototransduction par la rhodopsine sensorielle (SRI)

313 W. Marwan, M. Montrone, D. Oesterhelt
Signal transduction in *Halobacterium halobium* mediated by the switch factor fumarate
La transduction du signal lumineux dans Halobacterium halobium *est régulée par le fumarate*

317 B. Scharf, M. Engelhard, F. Siebert
A carboxyl group is protonated during the photocycle of the photophobic receptor psR-II from *Natronobacterium pharaonis*
Protonation d'un groupe carboxylique au cours du photocycle du récepteur psR-II de Natronobacterium pharaonis

321 Y. Imamoto, Y. Shichida, J. Hirayama, H. Tomioka, N. Kamo, T. Yoshizawa
Chromophore configuration and photoreaction cycle of phoborhodopsin from *Natronobacterium pharaonis*
Configuration du chromophore et cycle photochimique de la phoborhodopsine de Natronobacterium pharaonis

325 E. Ferrando, D. Oesterhelt
A molecular genetic approach for studying the halobacterial photoreceptor sensory rhodopsin I
*Une approche par génétique moléculaire de l'étude de la rhodopsine sensorielle (SRI) d'*Halobacterium halobium

329 **J.A.W. Heymann, W.A. Havelka, D. Oesterhelt**
Overexpression of halorhodopsin : new perspectives for structure-function studies
Surexpression de l'halorhodopsine : nouvelles perspectives pour des études de relation structure-fonction

333 **S.I. Bibikov, R.N. Grishanin, A.D. Kaulen, W. Marwan, D. Oesterhelt, V.P. Skulachev**
Direct evidence for the involvement of membrane potential changes in the photosensory transduction of halobacteria
Mise en évidence directe de changements de potentiels membranaires dans les processus de transduction du signal photosensible chez les halobactéries

IX. RHODOPSIN-LIKE PIGMENTS IN *CHLAMYDOMONAS*
IX. PIGMENTS DE TYPE RHODOPSINE CHEZ CHLAMYDOMONAS

339 **L. Keszthelyi**
Light excited electric signals from *Chlamydomonas*
Signaux électriques induits par la lumière chez Chlamydomonas

343 **O.A. Sineshchekov, E.G. Govorunova, A. Dér, L. Keszthelyi**
Retinal-induced photoelectric responses in *Chlamydomonas reinhardtii* "blind" mutants
Réponses photo-électriques induites par le rétinal dans des mutants "aveugles" de Chlamydomonas reinhardtii

347 **I. Dumler, S. Korolkov, M. Garnovskaya**
Rhodopsin-like pigment and G-proteins in the eyespot of *Chlamydomonas reinhardtii*
Pigments de type rhodopsine et protéines G dans l'ocelle de Chlamydomonas reinhardtii

351 **A. Mirshahi, A. Nato, D. Lavergne, G. Ducreux, J.P. Faure, M. Mirshahi**
Chloroplastic membrane-bound arrestin-like immunoreactive proteins in tobacco and *Chlamydomonas* cells
Protéines de la membrane chloroplastique des cellules de tabac et de Chlamydomonas *présentant des analogies immunoréactives avec l'arrestine*

355 A. Razaghi, F. Borgese, B. Fiévet, R. Motais, A. Nato, J. Oliver, A. Vandewalle, M. Mirshahi, J.P. Faure
Proteins related to retinal S-antigen (arrestin) in non photosensitive cells
Protéines présentant des analogies avec l'arrestine (rétinal S-antigène) dans des cellules non photosensibles

X. SIGNAL TRANSDUCTION
X. TRANSDUCTION DU SIGNAL LUMINEUX

361 H.E. Hamm, N.O. Artemyev, J.S. Mills, N.P. Skiba, H.M. Rarick, C. Lambert, E.A. Dratz
Sites and mechanisms of interaction of rod G protein with rhodopsin and cGMP phosphodiesterase
Sites et mécanismes d'interaction de la protéine G rod avec la rhodopsine et la phosphodiestérase cGMP

365 R.R. Rando
Posttranslational modifications of retinal G proteins
Changements post-translationnels des protéines G de la rétine

369 A. Plangger, R. Paulsen
Phosphorylation of rhodopsin in fly photoreceptor membranes is controlled by the light-dependent binding of an arrestin homolog
La phosphorylation de la rhodopsine des membranes photoréceptrices de la mouche est contrôlée par une liaison d'un homologue de l'arrestine

375 J. Kibelbek, D.C. Mitchell, B.J. Litman
The effect of rhodopsin phosphorylation on the formation and decay of metarhodopsin and rhodopsin–G_t interactions
L'influence de la phosphorylation de la rhodopsine sur la formation et la disparition de la métarhodopsine et sur les interactions rhodopsine-protéine G

379 F. Bruckert, C. Pfister
Binding of Ca^{2+}, Mg^{2+} and Tb^{3+} to bovine retinal arrestin
Liaison du calcium, du magnésium et du terbium à l'arrestine de la rétine bovine

383 F. Pages, P. Deterre, C. Pfister
The fast GTP hydrolysis by transducin bound to phosphodiesterase
L'hydrolyse rapide du GTP par le complexe transducine-phosphodiestérase

387 P. Catty, C. Pfister, F. Bruckert, P. Deterre
The retinal phosphodiesterase-transducin complex : membrane binding, subunits interactions and activity
Le complexe phosphodiestérase-transducine dans la rétine : liaison à la membrane, interaction des sous-unités et activité

391 K.W. Koch
Recovery of the photoresponse in vertebrate photoreceptors : role of Ca^{2+}-dependent regulation of guanylate cyclase
Récupération de la réponse à la lumière dans les photorécepteurs des vertébrés : rôle de la régulation dépendante du calcium de la guanylate cyclase

395 O. Goureau, M. Lepoivre, F. Mascarelli, Y. Courtois
Nitric oxide synthase activity in bovine retina
Activité de la syntase d'oxyde nitrique dans la rétine bovine

399 J.A. Clausen, L. Kelly, M. Brown, E. O'Gara, A. Delaney, A.D. Blest
A 23 kDa Ca^{2+}-binding putative cysteine protease in arthropod rhabdomeres
Une protéine de poids moléculaire 23000 liant le calcium chez Arthropod rhabdomeres

403 R. Hardie, B. Minke
Light-activated channels in *Drosophila* are coded by the *trp* gene
Les canaux ioniques activés par la lumière sont codés par le gène trp *chez la drosophile*

407 R.H. Lee, B. Lieberman, H. Yamane, D. Bok, B.D.K. Fung
Purification and identification of the βΓ-transducin complex of cone photoreceptor cells
Purification et identification du complexe transducine βΓ des cellules photoréceptrices des cônes

411 Y. Kito, K. Narita, M. Seidou, M. Michinomae, K. Yoshihara, J.C. Partridge, P.J. Herring
A blue sensitive visual pigment based on 4-hydroxyretinal is found widely in mesopelagic cephalopods
Un pigment visuel rendu sensible au bleu par le 4-hydroxyrétinal existe chez les céphalopodes mésopélagiques

415 G. Wolbring, W. Haase, N.J. Cook
Immunolocalization of protein kinase A in bovine retina
Localisation par immunologie de la protéine kinase A de la rétine bovine

419 Author index
Index des auteurs

421 List and address of participants
Liste et adresse des participants

I. Molecular structure of retinal proteins

I. Structure moléculaire des rétino-protéines

The roles of extramembranous bilayer loops and of retinal in the folding and stability of bacteriorhodopsin

Theodore W. Kahn and Donald M. Engelman

Yale University, Department of Molecular Biophysics and Biochemistry, 260 Whitney Avenue, New Haven, CT 06511, USA

INTRODUCTION

Because parts of integral membrane proteins must fold in the hydrophobic region of lipid bilayers, the balance of interactions stabilizing the proteins' structures is likely to be both quantitatively and qualitatively different from that found in soluble proteins. Factors contributing to the stability of proteins that contain several transbilayer α-helices can be separated conceptually into those stabilizing the helices themselves and those stabilizing the interactions between the helices in the tertiary structure. This perspective is embodied in the two-stage model proposed by Popot and Engelman (1990). In this concept the stability of individual helices in bilayers is thought to arise from main chain hydrogen bonding and from the hydrophobic nature of the sidechains (Engelman et al, 1986), and the association of helices is driven by a different set of energies. In this brief paper, we summarize work showing that bacteriorhodopsin (BR) can be refolded from two independently stable transmembrane helices and the complementary five-helix fragment, and also examine the contributions to thermodynamic stability made by the covalent continuity of links between helices and by the binding of retinal to the protein.

Of the factors stabilizing helix-helix association in BR, four major categories can be identified. First, the extramembranous loops connecting the helices may constrain their positions. Second, polar amino acid side chains within the membrane may cause the helices to associate in such a way as to allow the formation of hydrogen bonds and/or ion pairs (Engelman, 1982). Third, the favorable packing of helices with each other as opposed to with lipids may drive the helices together in order to maximize van der Waals contacts and minimize the cavities that might result if the helices were each surrounded by lipids. Fourth, the binding interactions between the retinal moiety and the protein may stabilize the structure of the protein. In the following, we specifically address the contributions of the links between helices and retinal binding.

NOMENCLATURE

We have adopted a new nomenclature designed to show in a clear way which helices are present in a sample and whether any covalent connections between the helices have been cut. The helices are lettered according to Engelman et al. (1980). A dot (·) between two letters indicates that no covalent bond connects those two helices. If the sample contains retinal, the name of the sample is prefixed with the letter R. Subscripts indicate the state of the sample: S indicates a sample in the form of membrane sheets as obtained from *Halobacterium halobium*; V a sample reconstituted into lipid vesicles. If the sample is being referred to in an abstract sense, no subscripted letter will be used. Some examples:

$R(ABCDEFG)_s$ Purple membrane.
$(ABCDEFG)_v$ Bleached BR that has been reconstituted into vesicles.
$R(A \cdot B \cdot CDEFG)_v$ BR reconstituted in vesicles from fragments containing the first helix, the second helix, and the last five helices, with retinal added.

RECONSTITUTION OF BR FROM (A) + (B) + (CDEFG) + R

In a forthcoming publication (Kahn and Engelman, 1992) we show that BR can be reformed from peptides containing the first helix, the second helix and the remaining five helices, which have been independently refolded in lipid bilayers and then allowed to interact with each other. The use of single helices is particularly important as it is a test of the two-stage model. The work follows the original experimental design and reconstitution protocols of Popot et al. (1987) in which the reformation of BR from a two helix fragment and a five helix fragment was demonstrated. A most important idea is that the separate helices will be stable in lipid bilayers. This aspect is not part of the current discussion, but has been demonstrated in other work (Hunt et al 1991; same authors in preparation 1992), in which it was shown by Fourier transform infrared dichroism and circular dichroism that (A) and (B) are largely helical when independently reconstituted in vesicles.

To test the hypothesis that BR is formed by the association of independently stable helices, two peptides were chemically synthesized, one containing the helix A sequence, and the other the helix B sequence. A reconstitution using (A) and (B) is a direct test of whether these two individual peptides are separately capable of folding and associating with (CDEFG) to form the complete BR structure. As synthesized, (A) extends from residue 6 to 42, and (B) from 36 to 71. (A) and (B) overlap by 7 residues, the overlap being entirely within the polar region of the sequence between the first and second helices. The overlap was included so as to give each of these extremely hydrophobic peptides a significant number of polar residues so as to promote solubility in mixed organic/aqueous solvent systems and in aqueous detergent solutions.

(A), (B), and (CDEFG) were independently reconstituted in native vesicles at a 2:1 lipid to protein mass ratio using the method of Popot et al (1987). This lipid to protein ratio was used because vesicle fusion becomes more efficient as the lipid to protein is raised. However, the ratio could not be much higher since at a 10:1 ratio BR reconstituted from the three fragments mixed together did not regenerate the native spectrum. The three populations of vesicles were then fused so as to yield a 2:2:1 molar ratio of $(A)_v:(B)_v:(CDEFG)_v$. Retinal was added either before or after fusion, and a slow appearance of the characteristic absorption spectrum of BR occurred over many hours, being complete after two days. If either (A) or (B) was omitted, no purple chromophore recovery was observed. When the fused vesicles were dried at low humidity, a crystal lattice with x-ray diffraction intensities and reflection positions like those of native purple membrane was observed. The pattern was of much lower quality than that of native purple membrane, but its presence shows that some portion of the material regained the structure of native purple membrane.

CONTRIBUTIONS OF LOOPS AND OF RETINAL TO BR STABILITY

While it has been demonstrated that the BR molecule can fold with one or two of its helix-connecting loops cleaved, and there is some evidence (Popot et al, 1987) that fragments can interact in the absence of retinal, these observations only show that the links and retinal are not absolutely required for successful refolding, but do not allow the assessment of their contributions to the total stability of BR. To address the issue of stability we have used differential scanning calorimetry (DSC) and thermal denaturation observed by ultraviolet circular dichroism and absorption spectroscopy to compare native BR, cleaved BR, and BR from which retinal has been removed. Retinal binding and the loop connections were each found to make a small contribution to stability, and even a sample that was cleaved twice and bleached to remove retinal showed a denaturation temperature well above room temperature. Removal of retinal destabilized the protein more than cleaving one loop, and about as much as cleaving two loops. Retinal binding and the connections in the loops were found to stabilize BR in independent ways, since their effects proved additive. Cleavage of the molecule into fragments did not reduce the intermolecular cooperativity of the denaturation, nor did dilution of the protein by excess lipid in order to eliminate the purple membrane crystal lattice. The details of these experiments are to be found in a manuscript by Kahn, Sturtevant, and Engelman (Biochemistry, in press, 1992).

Table 1. Thermodynamic values from DSC of BR.

Sample	T_m (°C)	ΔH_{cal} (kcal/mol)	ΔH_{vH} (kcal/mol)	$\Delta H_{vH}/\Delta H_{cal}$
$R(ABCDEFG)_S$	101	100	148	1.5
$R(AB \cdot CDEFG)_S$	95	56	110	2.0
$(ABCDEFG)_S$	85	23	76	3.3
$(AB \cdot CDEFG)_S$	80	16	84	5.4
$R(ABCDEFG)_S$ pH 9	83	148	111	0.75
$R(AB \cdot CDEFG)_S$ pH 9	78	62	127	2.1
$(ABCDEFG)_S$ pH 9	65	56	71	1.3
$(AB \cdot CDEFG)_S$ pH 9	57	22	80	3.7
$R(ABCDEFG)_V$	95	121	191	1.6
$R(AB \cdot CDEFG)_V$	83	88	94	1.1
$(ABCDEFG)_V$	79	20	122	6.1
$(AB \cdot CDEFG)_V$	71	20	56	2.8
$R(A \cdot B \cdot CDEFG)_V$	71	39	67	1.7

Table 1 summarizes thermodynamic data obtained from DSC of a range of samples Thermodynamic parameters were extracted as described in Kahn, Sturtevant and Engelman (1992). Transition temperatures (T_m) and van't Hoff enthalpies were also obtained from changes in circular dichroism or absorption spectra (not shown). Cleavage of a polypeptide bond in the loop between helices reduces the T_m by about 5-10°, and the removal of retinal reduces it by 15°. In soluble proteins it has been shown that the change in T_m varies linearly with $\Delta\Delta G$ over small ranges of temperature, suggesting that the melting behavior can be roughly interpreted in terms of free energy changes (Becktel and Schellman, 1987; Hu et al., 1992). It appears that the effects of cleaving links and of removing retinal are roughly additive in most comparisons. This suggests that they act independently to destabilize the protein.

DISCUSSION

In this paper we have summarized experimental tests of the hypothesis that BR can fold by the association of independently stable transmembrane helices. The fact that peptides containing the first and second helical segments of BR can be separately reconstituted across membranes and can subsequently reassociate with the rest of the molecule supports the two-stage hypothesis of membrane protein folding, and suggests that similar concepts may be useful in thinking about the oligomerization of proteins as well, since this model system is, in effect, a heterotrimer of membrane spanning polypeptides. From this set of observations, it seems that the first two helices of BR can be considered as independent folding domains, and that covalent connections in the loops are not essential for the appropriate association of helices.

In the introduction it was proposed that the factors involved in holding BR helices together can be assigned to the categories of interhelical loops, polar interactions, packing effects, and retinal binding. The results from our thermodynamic studies indicate that each connection in a loop or the binding of retinal makes a small contribution to the stability of BR. The stability of the protein is reduced by cleavage or bleaching, but not to the point where the protein unfolds or dissociates into separate fragments at room temperature. Work with another membrane protein system, the dimerization of the transmembrane region of glycophorin, has suggested that highly specific packing interactions can be a dominant factor in oligomerization and, by inference, in folding (Lemmon et al, 1992). Further experimentation will be needed to establish the relative role of this category of interactions in BR folding. It appears that side-by-side interaction within the bilayer play a major role in stabilizing the association of membrane-spanning helices.

REFERENCES

Becktel, W.J. and Schellman, J.A. (1987) *Biopolymers* 26, 18-59-1877.
Engelman, D.M. et al (1980) *P.N.A.S. U.S.A.*, 77, 2023.
Engelman, D.M. (1982) *Biophys. J.* 37, 187.
Engelman, D.M. et al. (1986) *Ann. Rev. Biophys. Biophys. Chem.* 15, 321-353.
Hu, C.-Q. et al. (1992) *Biochemistry* 31, 4876-4882.
Hunt, J. et al., (1991) *Biophys J.* 59, 400a.
Hunt, J. et al., (1992) manuscript in preparation.
Kahn, T.W. and Engelman, D.M. (1992) *Biochemistry*, accepted.
Kahn, T.W., Sturtevant, J.M. and Engelman, D.M. (1992) submitted, *Biochemistry*.
Lemmon, M.A., Flanagan, J.M., Hunt, J.F., Adair, B.D., Bormann, B.J., Dempsey, C.E., and Engelman, D.M. *J. Biol. Chem.* 267 (1992)7683-7689.
Popot, J.-L. and Engelman, D.M., (1990) *Biochemistry*, 29, 4031-4037.
Popot, J.-L. et al, (1987) *J. Mol. Biol.* 198, 655-676.

Rotational orientation of transmembrane α-helices in bacteriorhodopsin studied by neutron diffraction

Fadel A. Samatey[1,2], Jean-Luc Popot[2], Catherine Etchebest[2] and Giuseppe Zaccaï[1]

[1] Institut de Biologie Structurale and Institut Laue Langevin, BP 156, 38042 Grenoble Cedex, France.
[2] Institut de Biologie Physico-Chimique, 13, rue Pierre et Marie Curie, 75005 Paris, France

1. Introduction

Bacteriorhodopsin (BR) is the only protein in the purple membrane (PM) of *Halobacterium halobium*, in which it is organised with lipids in a highly ordered two-dimensional lattice. It functions as a light-driven proton pump. The colour of the membrane is due to a retinal chromophore bound via a Schiff base to a lysine residue in the protein. In 1975, the low resolution structure of BR was determined by a combination of electron microscopy (EM) and electron diffraction, showing that its transmembrane region is made up of 7 α-helices[1]. A few years later, when the sequence was determined, it was found to contain 7 stretches of mainly hydrophobic amino acid residues, thought to correspond to the transmembrane α-helices[2].

Many experiments have been performed on the crystalline membrane patches by X-ray diffraction, electron diffraction and microscopy, and neutron diffraction, with the aim of characterising the structure of BR to as high resolution as possible. The best model so far is the EM of Henderson et al.[3]. The 7 α-helices (ABCDEFG) of the protein sequence correspond to the seven helices of the low resolution structure (1234567) in the order AGFEDCB. The study was done by a combination of electron diffraction and microscopy. The resolution of the electron density map is 3.5 Å in the in-plane projection and about 10 Å perpendicular to it. Bulky features such as the aromatic side chains and parts of the retinal molecule are visible in this map. The model is in perfect agreement with the partial structure placing helices A and B in positions 1 and 7, respectively[4], and with the position of the retinal obtained in previous neutron diffraction experiments using specifically deuterated purple membrane samples.

For a detailed analysis of the functional mechanism of BR, however, it would be useful to have a structure of higher resolution, and further refinement of the model is under way by different approaches. The complementarity of the neutron methods to EM is based on their respective sensitivity to completely different aspects of the structure: the bulky side chains for EM and the deuterium labels for neutrons. In the present paper, a neutron diffraction approach has been developed to test to what extent the helix rotational positions in the model could be varied and still remain in agreement with the experimental data. There are two justifications for this study. On the one hand, helix D does not contain aromatic side groups, and is, therefore, less constrained by the electron density map. On the other, because of the existence of different conformers for amino-acid side-chains[5], different rotational positions for the helices might still be compatible with the electron density map.

2. Materials and Methods

The preparation of PM with BR specifically deuterated in all its valine residues, the collection of neutron data and other experimental protocols are described elsewhere[6]. Data collected previously by Popot et al.[4] on reconstituted PM samples containing deuterated leucine and tryptophane residues were also used in the analysis. Data from three purple membrane samples were analysed. In the first (sample VAL), all the valines of the BR molecule were deuterated. The second and third samples had the leucines and tryptophanes deuterated either in helices A and B (sample LW1) or in helices CDEFG (sample LW2).

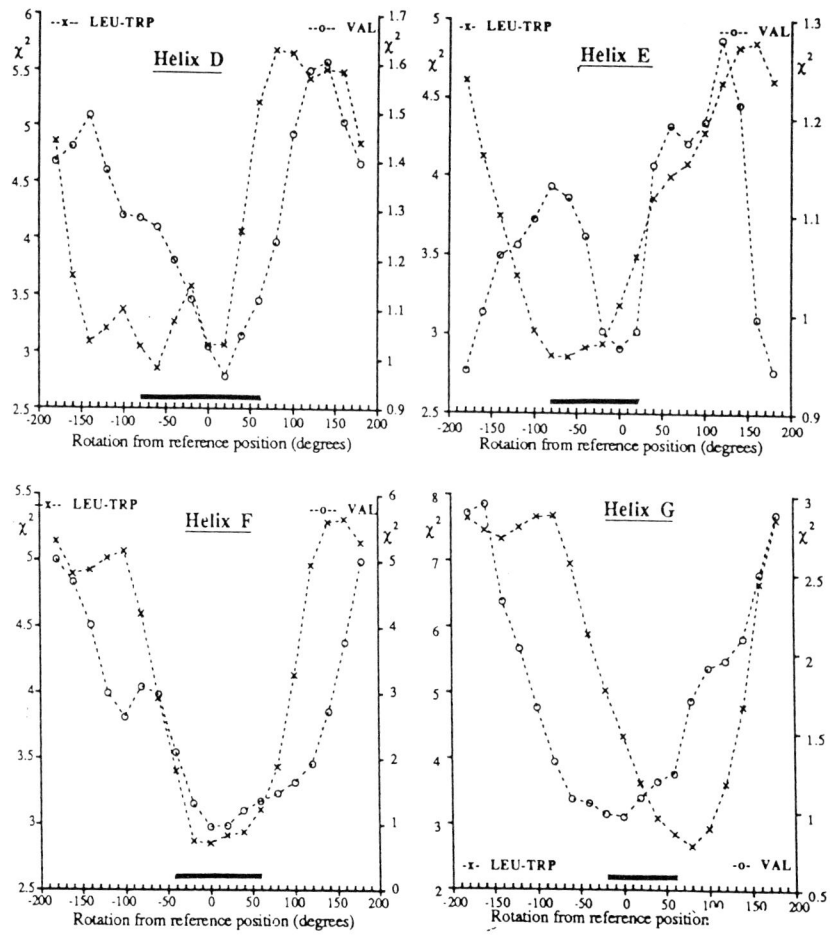

Figure 1. An example of rotational search. In each panel, the helix under study explores the rotational space by 20° increments while the remaining helices are left free to take their optimal position within a ±60° range around the reference position. --o--: fit to the VAL data set; --x--: fit to the LW2 data set.

The analysis was done by comparing the experimental diffraction intensities of the different samples with intensities calculated from models. Structure factors for each reflexion were calculated as the sum of the structure factor of the native (unlabeled) BR structure, F_N, and the structure factor corresponding to the difference between the native and labelled parts, F_{D-H}:

$$F = F_N + F_{D-H}$$

The F_N values are experimental, obtained from neutron experiments on native PM and the phases from EM, as described previously[4]. A set of algorithms was developed to generate coordinates of the labelled parts, for different models, as a function of the rotational orientation of each helix and the conformation of the labelled side-chains. For each of the three labelled samples, the F_{D-H} and predicted neutron diffraction intensities were then calculated for each model. Calculated and observed intensities were compared using reduced χ^2 statistics:

$$\chi_r^2 = (1/n) \Sigma (I_c - I_o)^2 / (\sigma_c^2 + \sigma_o^2)$$

where n is the number of reflections and I_c and I_o are calculated and observed intensities, respectively, with corresponding variances σ_c^2 and σ_o^2.

3. Results and Discussion

In the following analysis, the model of Henderson et al.[3] is defined as the reference model. The extent to which helices are rotated away from their position in the EM model is measured clockwise when looking from the helix N-terminus. The method was first tested on the LW1 data. The positions and rotational orientations of helices A and B have been determined previously quite precisely from these data by a simpler model building approach and difference Fourier analysis[4], and confirmed by the EM structure[3].

In the present search, helices A and B were rotated in steps of 20°; 25 different positions for the deuterated leucine residue in the loop connecting helices B and C and 6 different conformations for the other leucine side chains[5] were explored to test their effect on the χ^2 values. However, because of the prohibitively large number of combinations, the following protocol was applied. First, all combinations of helix rotations were explored with fixed side-chain conformations and a fixed loop position. Secondly, a restricted rotational search was performed around the model that gave the best fit, while the leucine in the loop was moved over each of its 25 positions. Thirdly, a restricted rotational search was performed around the best model, with the loop leucine in each of the three best positions (from the previous analysis), and for all combinations of 6 different side chain conformations for the helix leucines. From the last analysis, a set of 9 models which are structurally very similar is obtained, of which the one with the lowest χ^2 is the model found previously by Popot et al.[4]. This model was already the best one in the first analysis, in which the loop leucine and side-chain conformations had been kept fixed.

For the analysis of the other five helices, the data from the VAL and the LW2 samples were used. A protocol, similar to the one described above, was applied separately to each sample. The results were then combined in order to define for each helix a restricted set of rotational positions outside which no good fit to the experimental data could be obtained (indicated by the horizontal bars in the panels of Fig.1). The combination of all rotational positions allowed by the restricted domains generates 7200 rotational models. Among those, only 15 fit the data as well as or better than the reference model. After elimination of redundant models (models that differ from another one by rotating a single helix by 20°, thereby giving a lower fit to the data), 9 best models were obtained (including that of Henderson et al.[3]). The positions taken by each helix in these models are listed in the following table:

Helix	Deviation from 1990 EM model
C	$-20° \pm 20°$
D	$+20° \pm 20°$
E	$-20° \pm 20°$
F	$0°$ to $+20°$
G	$0°$ to $-20°$

The good agreement of the positions found for helices C and E-G with those established by EM confirms the reliability of the neutron diffraction approach and indicates that its accuracy is in the range of ± 20°. The position we find for D, at 20°±20° from that of the EM model, is compatible with the EM data, with the crosslinking data of Ding et al.[8], and with the results of the model-buiding examination of helix rotational orientations by Tufféry et al.[9]. It confirms, in particular, that the C^{α}-C^{β} bond of Asp^{115}, whose replacement by Asn affects both chromophore regeneration[7] and proton release[10], points towards neighbouring helix C. Asp^{115} is thought not to change its protonation state throughout the photocycle of BR[11]. Its location may permit it to establish strong interactions with helix C. Helix C contains two other aspartate residues, Asp^{85} and Asp^{96}, whose protonation and deprotonation, respectively, play a key role in proton pumping by BR (see refs 3, 7, 10, 11, and references therein).

4. References

1. Henderson, R., & Unwin, P.N.T. *Nature* **257**, 28-32 (1975)
2. Engelman, D.M., Henderson, R., McLachlan, A.D., & Wallace, B.A. *Proc. Natl. Acad. Sci. USA* **77**, 2023-2027 (1980)
3. Henderson, R., Baldwin, J.M., Ceska, T.A., Zemlin, F., Beckmann, E., & Downing, K.H. *J. Mol. Biol.* **213**, 899-929 (1990)
4. Popot, J.-L., Engelman, D.M., Gurel, O., & Zaccaï, G. *J. Mol. Biol.* **210**, 829-847 (1989)
5. Tufféry, P., Etchebest, C., Hazout, S., & Lavery, R. *J. Bio. Struc. & Dyn.* **8**, 1267-1289 (1991)
6. Samatey, F.A. Etude structurale de la bactériorhodopsine par diffraction des neutrons. (Thèse de Doctorat, Université Joseph Fourier, Grenoble, 1992)
7. Mogi, T., Stern, L.J., Marti, T., Chao, B.H., & Khorana, H.G. *Proc. Natl. Acad. Sci. USA* **85**, 4148-4152 (1988)
8. Ding, W.-D., Tsipouras, A., Ok, H., Yamamoto, T., Gawinowicz, M.A., & Nakanishi, K. *Biochemistry* **29**, 4898-4904 (1990)
9. Tufféry, P., Popot, J.-L., & Lavery, R. Modeling bundles of transmembrane α-helices: a test study on bacteriorhodopsin (see these proceedings)
10. Marinetti, T., Subramaniam, S., Mogi, T., Marti, T., & Khorana, H.G. *Proc. Natl. Acad. Sci. USA* **86**, 529-533 (1989)
11. Braiman, M.S., Mogi, T., Marti, T., Stern, L.J., Khorana, H.G., & Rothschild, K.J. *Biochemistry* **27**, 8516-8520 (1988)

Modelling bundles of transmembrane α-helices : a test study on bacteriorhodopsin

Pierre Tufféry[1, 2], Jean-Luc Popot[2] and Richard Lavery[1]

[1] CNRS URA 77 ; [2] CNRS URA 1187. Institut de Biologie Physico-Chimique, 13 rue Pierre et Marie Curie, 75005 Paris, France

1. Introduction

High-resolution crystallographic data on integral membrane proteins are notoriously difficult to obtain. On the other hand, the structure of the transmembrane region of integral proteins may be more accessible to ab initio structural prediction than that of soluble proteins, mainly because of the strong constraints that apply to the secondary structure of polypeptide segments traversing the hydrophobic core of a membrane :

(i) Increasing evidence suggests that the transmembrane region of many integral membrane proteins can be considered as resulting from the assembly of pre-formed transmembrane α-helices (Popot & Engelman, 1990).

(ii) Putative transmembrane helices can often be identified by inspecting the distribution of hydrophobic residues in the sequence of a protein. Examination of the lateral amphipathicity and the differential mutability of helix faces often gives indications as to which side of a given helix faces the inside of the transmembrane helix bundle.

(iii) The conformation of the polypeptide backbone in a transmembrane helix is constrained by hydrogen bonding.

At present, however, there is no general method to predict the way a collection of transmembrane helices packs into a three-dimensional bundle. We have recently developed fast approaches to predicting the side-chain conformations in a protein, starting from a knowledge of the sequence and the backbone conformation (Tuffery et al., 1991). In the present work, we have examined the feasibility of predicting the rotational orientation of α-helices in a transmembrane bundle by examining the quality of side chain packing in each possible combination of orientations.

We have chosen bacteriorhodopsin (BR) as a test protein. The current model of BR, based on high resolution electron microscopy, defines the arrangement and axis positions of its 7 transmembrane helices (Henderson et al., 1990). Helix rotational orientations have been studied by electron microscopy (Henderson et al., 1990) and neutron diffraction (Popot et al., 1989 ; Samatey et al., 1992), as well as the position of the retinal prosthetic group (reviewed in Henderson et al., 1990). Because atomic resolution data are not available yet, approximate side-chain positions are known only for the largest residues. The structure of the extramembrane regions is not known in detail. It has been shown however that several of the extramembrane loops connecting the helices, which are not taken into account in the present analysis, can be cut without preventing correct folding. This suggests that the loops, while they contribute to BR stability, do not direct the way transmembrane helices interacts with their neighbours (reviewed in Popot et al., 1992).

2. Methods

2.1. *Sampling backbone orientations.*

The BR model is comprised of the 7 transmembrane helices and the retinal, extramembrane loops being omitted. The backbone of each helix is considered a rigid block (in the conformation given by Henderson et al., 1990). Its orientation is varied by rotations around and displacements along the crystallographic helix axis. We have initially focused on determining rotational positions.

2.2. *Packing side chains.*

Combinations of side chain conformations giving a good packing are identified using the SMD (Sparse Matrix Driven) algorithm (Tuffery et al., 1991). SMD is based on the use of a library of 110 discrete conformations (rotamers) that are most frequently observed in high-resolution X-ray structures.

The conformational energy of a model with a given set of side chain conformations is given by :

$$E = \Sigma_{a,b} E_{ab} + \Sigma_{b,i} E_{bi}^{k} + \Sigma_{i,j} E_{ij}^{kl}$$

where a,b denote backbone atoms and i,j the side chains with the associated rotamers k,l. The first term in the sum represents backbone/backbone interactions, the second term side chain/backbone interactions and the third term side chain/side chain interactions.

A matrix of all the interaction energies defined above is constructed. The SMD algorithm profits from the fact that most side chains only interact with a limited number of other side chains. Thus, the matrix of their mutual interaction energies is sparse and can be rearranged into a set of sub-matrices of side chain interactions (corresponding to clusters of neighbouring side chains) which contain most of the dominant contributions. An iterative combinatorial search is performed on each cluster, including weighings that represent the extra cluster energy components, to determine the best conformation describing each side chain.

2.3. *Cutting the combinatorial problem short : approximation of the system of 7 helices by 11 helix pairs.*

The combinatorial problem of simultaneously positioning 7 helices using even a single degree of freedom is far too difficult to allow an exhaustive search. For example, sampling the rotation of helices around their axes in 20° steps generates a total of 18^7 possible models; finding the lowest energy set of rotamers in one full BR model currently requires 4.5 minutes of CPU time. Fortunately, examining subsets of helices shows that the conformation adopted by most side chains is affected by the vicinity of only one of the neighbouring helices. Thus, the full bundle may be reasonably approximated by pairs of helices. The geometry of the bundle suggests that 11 pairs are most important (fig. 1).

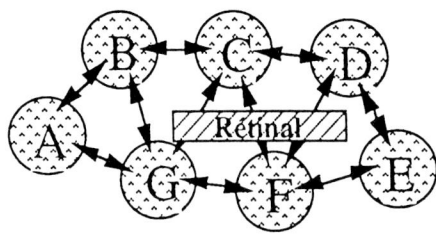

Fig. 1: Decomposition of the 7 helices bundle as a sum of 11 contributing pairs.

The search for an optimal combination of rotational orientations can therefore be divided into two steps:

1) For each pair, an energy map is calculated of the lowest energy of interaction obtainable for each of the 18x18 combinations resulting from rotating the two helices by 20° increments (Fig.2). This step size was shown to preserve the features of energy maps computed with 10° increments.

2) The maps are merged and the lowest energy combination of the 7 helices is retained as the solution.

3. Results

Fig. 2 shows examples of energy maps obtained for helix pairs AB and AG. The following conclusions can be drawn:

i) The combination of rotational positions that corresponds to crystallographic positions (0°,0°) does yield a low energy of interaction, as expected.

ii) This solution is not unique; in its current state, our approach does not permit to unambiguously determine the rotational arrangement of two helices whose axes are known.

iii) Because each helix is in contact with several neighbours, most ambiguities can be resolved by merging the maps. For instance, merging the 2 interaction maps between helices A, B and G leads to a unique solution for the A helix (0°, i.e. the correct arrangement).

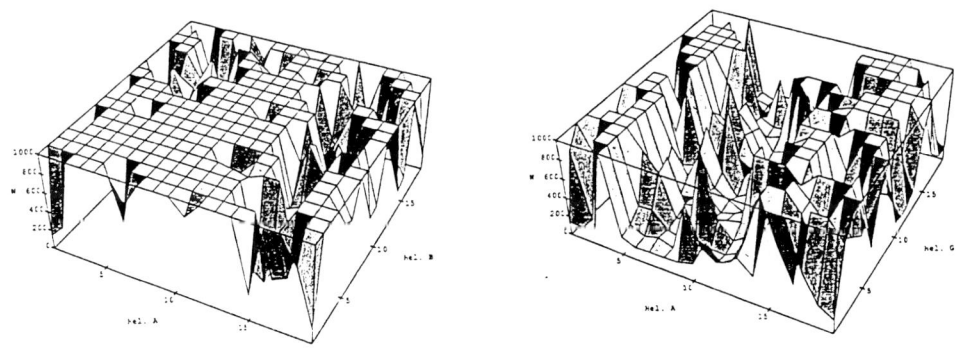

Fig2: Example of energy maps for he pairs AB and AG (20° steps).

The results obtained when merging the 11 maps covering the whole transmembrane region of BR are illustrated by the data in Table 1. When all 7 helices are left free to rotate over 360°, the search correctly identifies the rotational positions of only helices A, B and C (line 1). Examination of the data indicates that the aberrant positioning of the other helices results from helix D being allowed access to a position very remote from the correct one. When helix D is denied access to a 60° sector (from +100° to +140° from the crystallographic position), the rotational positioning of the full bundle is achieved with only minor deviations (line 2). Finally, considering the opportunity of displacing the D helix of 2.5 Å along its axis (R. Henderson, personal communication) leads to the same results without any constraint on helix D (line 3).

Constraints on helix rotation	Helix A	B	C	D	E	F	G
No constraints	0	0	0	120	160	-40	-140
Constraint on helix D	0	0	0	20	0	0	-20
Refined helix D (no constraints)	0	0	0	-20	0	0	-20

Table 1. Deviation (degrees) between the rotational orientation of the 7 transmembrane α-helices in BR obtained model building and that determined by electron and neutron diffraction.

4. Conclusion

Steps that could lead to ab initio prediction of the atomic structure of a transmembrane helix bundle can be enumerated as follows :

1) identifying transmembrane helices in the sequence
2) deciding which side of each helix faces the core of the bundle
3) determining the arrangement of helices in the bundle
4) determining interaxis angles and distances
5) determining backbone coordinates : rotational orientation, register, backbone distortions
6) determining side chain conformations

Steps 1 and 2 can often be achieved by combining sequence analysis and experimental data. Step 6 can be completed with fair reliability (Tuffery et al., 1991). We show here that, knowing axis positions and backbone conformations, rotational positions of the helices can be established (part of step 5). We are currently investigating the applicability of the same methodology to determining longitudinal positions and interaxis distances and angles, and examining the effect and predictability of backbone distortions from ideal α-helix coordinates (steps 4 & 5). Simultaneously taking into account all degrees of freedom and bridging the gap from step 2 to step 6 remains a redoubtable challenge. As it stands, our approach should nevertheless be useful to examine other proteins of the BR family as well as the impact of mutations on BR, and it has the potential to help in the interpretation of the electron density maps obtained by high resolution electron microscopy.

References

Henderson R., Baldwin J. M., Ceska T. A., Zemlin F., Beckmann E., Downing K. H. (1990) J. Mol. Biol. 213,899-929.
Popot J.L., Engelman D. M., Gurel O., Zaccaï G.(1989) J. Mol. Biol. 210,829-847.
Popot J.L., Engelman D. M. (1990) Biochemistry 29,4031-4037.
Popot J.L., deVitry C., Atteia A., in "Membrane protein structure: experimental approaches", (S. H. White, ed.), Oxford University Press (1992), in Press.
Samatey F., Popot J.L., Etchebest C., Zaccaï G. (1992) (these proceedings).
Tuffery P., Etchebest C., Hazout S., Lavery R. (1991) J. Biomol. Struct. Dynam. 6,1267-1289.

Acknowledgements

The authors would like to thank Organibio and the French Ministry of Research and Technology for their support of this work through the research program CM^2AO "Computer Aided Conception of Macromolecules".

Ground state potential surface calculations for butadiene and retinal

Mafalda Nina[1], Benoît Roux[2] and Jeremy C. Smith[2]

[1] Section de Biophysique des Protéines et des Membranes. Département de Biologie Cellulaire et Moléculaire. C.E.-Saclay, 91191 Gif-sur-Yvette Cedex, France. [2] Département de Physique, Université de Montréal, Montréal, Québec, Canada

INTRODUCTION. The determination of a three-dimensional structure of bacteriorhodopsin at close-to-atomic resolution (Henderson *et al.* 1990) renews the possibility that further understanding of the function of this molecule can be obtained employing theoretical methods. A full description of the photocycle requires the determination of excited state and ground state retinal potential surfaces. Due partly to the conjugated nature of the molecule, this may be nontrivial (Tavan *et al.* 1985). The ground state problem can be approached using two complementary techniques; molecular mechanics and quantum mechanics. *Ab initio* quantum mechanical calculations can provide conformational energetics and charge distributions for small molecules. This data can be used to parameterize a molecular mechanics force field for subsequent larger scale simulations. In this preliminary report we use *ab initio* calculations to examine the charge distribution and rotational potential of the C_2-C_3 bond of the retinal fragment, butadiene. The molecular mechanics force field is parameterized to reproduce the *ab initio* results. The resulting potential is transferred to retinal and used to investigate the *6-s-cis* ---> *6-s-trans* rotational potential of the isolated retinal molecule.

METHODS. The butadiene *ab initio* calculations were performed with the Gaussian 90 program (Frisch *et al.* 1990) using 6-31G, 6-31G* and 6-31G** basis sets at the Hartree-Fock (HF) level and with second-order Moller-Plesset perturbation (MP2) correction. The energy as a function of rotation around the C_2-C_3 single bond dihedral was obtained at the RHF/6-31G* level with full geometry optimisation subject to a dihedral constraint. Similarly, the molecular mechanics rotational curves were obtained by constraining the required dihedrals and minimizing the potential energy due to the other terms. Our *ab initio* results for the geometries and torsional energetics are in accord with similar calculations recently published (Guo & Karplus, 1991).

The molecular mechanics program used is CHARMM (Brooks *et al.* 1983). The form of the potential energy, V is as follows;

$$V = \sum_{bonds} k_b(b-b_0)^2 + \sum_{angles} k_\theta(\theta-\theta_0)^2 + \sum_{dihedrals} k_\phi(1+\cos(n\phi-\delta)) + \sum_{impropers} k_\omega(\omega-\omega_0)^2$$

$$+ \sum_{\substack{i,j \\ nonbonded}} 4\varepsilon_{i,j}\left[\left(\frac{\sigma_{i,j}}{r_{i,j}}\right)^{12} - \left(\frac{\sigma_{i,j}}{r_{i,j}}\right)^{6}\right] + \sum_{\substack{i,j \\ nonbonded}} \frac{q_i q_j}{r_{i,j}} \quad (1)$$

The terms b, θ, ϕ and ω are the bond lengths, angles, dihedrals and improper torsions for which b_0, θ_0, ϕ_0 and ω_0 are the reference values and k_b, k_θ, k_ϕ and k_ω the force constants. For the dihedrals the periodicity is n and δ is the phase angle. The nonbonded interactions, for atom pairs i and j separated by distance $r_{i,j}$ consist of a 12-6 Lennard-Jones van der Waals term (well-depth $\varepsilon_{i,j}$, diameter $\sigma_{i,j}$) and a Coulombic term between atomic partial charges, q_i and q_j. Initial parameters were taken from an existing force field for unsaturated hydrocarbons (Smith & Karplus, 1992). The charge distributions were obtained using Mulliken population analysis of the *ab initio* calculations performed at MP2 / 6-31G* level, on the *trans* conformer of butadiene. We found that the charges were basis-set dependent but that their relative distribution was constant i.e., the charges obtained using any basis set could be derived from those of any other basis set by multiplying all charges by the same constant. This constant was taken as the single electrostatic variable in the molecular mechanics calculations. Its optimised value was 0.9 of the MP2 / 6-31G* charges. The bond length and angle terms were parameterized to reproduce the *ab initio trans* butadiene geometry. The dihedral terms were also optimised to reproduce the C_2-C_3 potential. It was not possible to reproduce the *ab initio* potential with a combination of Van der Waals, electrostatic and a single 2-fold torsional terms. It was found necessary to include 1, 2, 3 and 4-fold dihedral terms for the C_2-C_3 dihedral angle i.e.,

$$V_{\phi,C_2,C_3} = 0.16(1+\cos\phi) + 2.23(1+\cos(2\phi-\pi)) + 1.13(1+\cos(3\phi)) + 0.30(1+\cos 4\phi) \quad (2)$$

where ϕ is the C_2-C_3 dihedral angle, in radians and V_ϕ is in $kcal.(mol.rad)^{-1}$.

RESULTS.

Butadiene. The butadiene rotational potential curves as obtained by the *ab initio* and molecular mechanical calculations are shown in Fig.2. The potential has two minima, one *trans* and one *gauche*. The form of the *ab initio* curve is closely reproduced in the molecular mechanics potential. The form of the molecular mechanical curve is strongly influenced by the torsional terms (Eq. 2) and repulsive non-bonded effects. The *cis-gauche* energy difference of 1.14 $kcal.mol^{-1}$ constains significant contribution from torsional terms (0.47 $kcal.mol^{-1}$), Van der Waals terms (0.27 $kcal.mol^{-1}$) and electrostatic terms (0.32 $kcal.mol^{-1}$). The empirical force

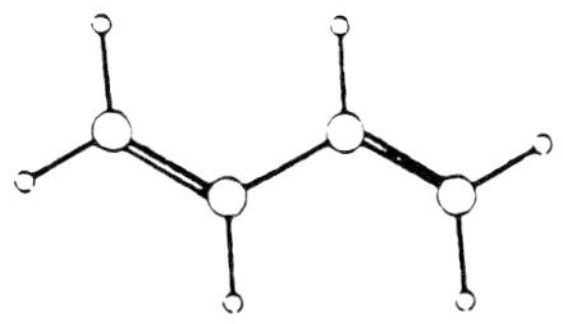

Fig. 1 *Trans* butadiene

field *cis*, *gauche* and *trans* geometries reproduce all bond lengths and angles to within 0.01 Å and 2° of the *ab initio* values, respectively. The CCC angles of the molecular mechanics values are 124.3°, 124.7° and 125.5° in the *trans*, *gauche* and *cis* geometries respectively compared to corresponding *ab initio* values at RHF/6-31G* level of 124.14°, 125.30° and 127.16°. More subtle geometrical effects, such as the slight lengthening of the double C-C bond at the 100° geometry are not reproduced by the molecular mechanics force field.

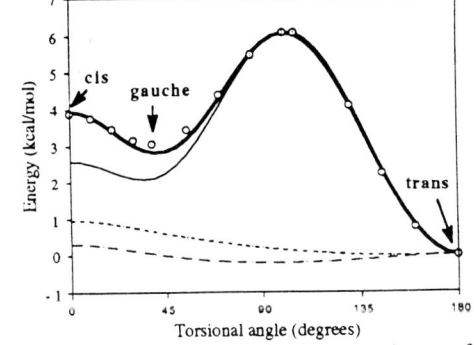

Fig 2. *Ab initio* and molecular mechanics energies as a function of C_2-C_3 dihedral angle in butadiene.
Ab initio RHF/6-31G* energy; o o o o.
Molecular mechanics energies; —— total; – – – Van der Waals; - - - - electrostatic; ——— dihedral.

Molecular Mechanics Calculations on Isolated Retinal. The molecular mechanics force field derived for butadiene was transferred to the polyene chain of retinal. Energy minimization of *all-trans* retinal gave a geometry with an RMS deviation of 0.31Å from the corresponding crystal structure (Hamanaka *et al.*, 1972).

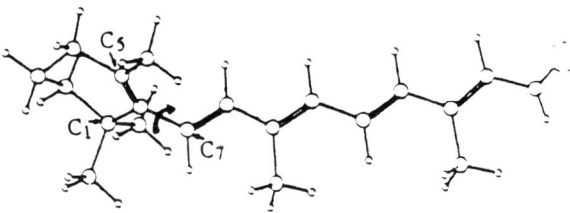

Fig.3 Minimized-geometry for 6-*s*-*cis* retinal. The arrow indicates the C_6-C_7 single bond.

The rotational potential around the C_6-C_7 bond, the single bond adjacent to the ionone ring, is of particular interest in this molecule. In bacteriorhodopsin, there is evidence that retinal adopts an *all-trans 6-s-trans* approximately planar geometry (Van der Steen *et al.* 1986).The C_6-C_7 angle (defined here as the angle between the plane defined by atoms C_1, C_5 and C_6 and that defined by C_6, C_7 and C_8) is -58° (6-*s*-*cis*) in the X-ray structure. This angle is -51° in our energy minimized geometry (Fig. 3). The calculated potential energy as a function of C_6-C_7 torsion is plotted in Fig.4. The lowest energy conformation is at -51° (6-*s*-*cis*) and the second lowest energy conformation is at +51° with a barrier of 5.0 $kcal.mol^{-1}$ between the two conformers. The energy difference between the +51° and -51° forms is almost zero (0.05 $kcal.mol^{-1}$).

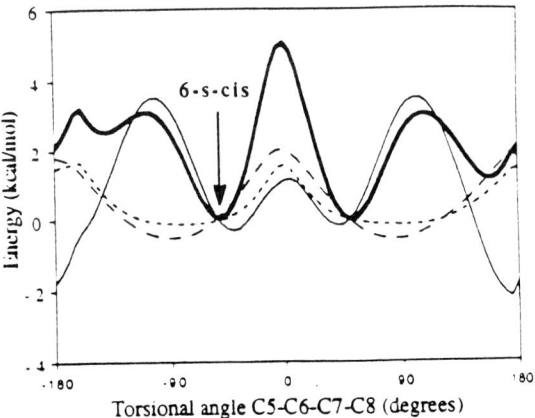

Fig. 4. Molecular mechanics potential energy as a function of the C_6-C_7 rotation in retinal.
—— total energy; – – – Van der Waals; ——— dihedral; - - - - - angles.

DISCUSSION. The work presented here represents preliminary calculations towards the development of parameterization of a molecular mechanics potential function for retinal. The characteristics of the *ab initio* rotational potential of butadiene are well reproduced by the molecular mechanics curve. The molecular mechanics force field was transferred to retinal to calculate the ring-plane - chain rotational potential. A more complete report on the present parameterization will be given elsewhere. Other aspects of the retinal ground state potential and interactions with surrounding protein groups in bacteriorhodopsin will be investigated.

REFERENCES

Brooks, B.R., Bruccoleri R.E., Olafson B.D., States D.J., Swaminathan S. and Karplus M.(1983): CHARMM: a program for macromolecular energy minimization and dynamics calculations. *J. Comp. Chem.* **4**, 187-217.

Frisch J.M.J., Head-Gordon M., Trucks G.W., Foresman J.B., Schlegel H.B., Raghavachari K., Robb M., Binkley J.S., Gonzalez C., Defrees D.J., Fox D.J., Whiteside R.A., Seeger R., Melius C.F., Baker J., Martin R.L., Kahn L.R., Stewart J.J.P., Topiol S., and Pople J.A. Gaussian 90, Revision, Gaussian, Inc., Pittsburgh PA, 1990.

Guo H. and Karplus M. (1991): *Ab initio* studies of polyenes. I. 1,3-butadiene. *J. Chem. Phys.* **94 (5)**, 3679-3699

Hamanaka T., Mitsui T., Ashida T., and Kakudo M.(1972): The crystal structure of *all-trans* retinal. *Acta Crys.* **B28**, 214-222.

Harbison, G.D., Smith S.O., Pardoen J.A., Courtin J.M.L., Lugtenburg J., Herzfeld J., Mathies R.A. and Griffin R.G.(1985): Solid-state ^{13}C NMR detection of a perturbed *6-s-trans* chromophore in bacteriorhodopsin. *Biochemistry* **24**, 6955-6962.

Henderson, R., Baldwin J.M., Ceska T.A., Zemlin F., Beckmann E. and Downing K.H. (1990): Model for the structure of bacteriorhodopsin based on high-resolution electron cryo-microscopy. *J. Mol. Biol.* **213**, 899-929.

Smith J.C and Karplus M. (1992): Empirical force field study of geometries and conformational transitions of some organic molecules. *J. Am. Chem. Soc.* **114**, 801-812.

Tavan P., Schulten K. and Oesterhelt D. (1985): The effect of protonation and electrical interactions on the stereochemistry of retinal schiff bases. *Biophys. J.* **47**, 415-430.

Van der Steen R., Biesheuvel P.L., Mathies R.A., Lugtenberg J. (1986): Retinal Analogues with Locked 6-7 conformations show that bacteriorhodopsin requires the *6-s-trans* conformation of the chromophore. *J. Am. Chem. Soc.* **108**, 6410-6411.

Evidence for beta-turns in the surface loops and α-I/α-II helices in the transmembrane regions of bacteriorhodopsin

Sam L. Helgerson[1], Edward Dratz[2] and Robert D. Renthal[3]

[1] Baxter Biotech Group, Hyland Division, Duarte, CA 91010, USA; [2] Department of Chemistry and Biochemistry, Montana State University, Bozeman, MT 95717, USA; and [3] Division of Earth and Physical Sciences, University of Texas, San Antonio, TX 78249, USA

Bacteriorhodopsin (bR) is an integral membrane protein produced by the archaebacterium *Halobacterium halobium*. Bacteriorhodopsin is found in the purple membrane fraction (pm) of the *H. halobium* plasma membrane and is the only protein in isolated pm. Engelman *et al.* (1980) proposed a structural model for bR with seven transmembrane α-helical segments connected by surface loops. This model was based on the primary sequence and a low-resolution map of the electron diffraction density obtained from purple membrane (Henderson & Unwin, 1975). Recently, a near-atomic resolution electron density map has provided a more detailed structural model for the seven transmembrane α-helices (Henderson *et al.*, 1990). However, the resolved structure does not define the type of helical conformations (α_I or α_{II}) present or the conformations of the surface loops.

The present work analyzes the secondary structure of bacteriorhodopsin using the techniques of isotopic enrichment and FT-IR difference spectroscopy. The bR protein samples were selectively labeled in the peptide backbone carbonyl groups by biosynthetic incorporation of [^{13}C-1]-enriched leucine, lysine or valine residues. Thus, the secondary structure was resolved for local regions of the protein around these residues. Our results are consistent with the structural model proposed by Henderson *et al.* (1990) for the transmembrane regions of bR. They extend the structural information on bR to provide evidence for a distribution of α_I-type and α_{II}-type helical structures within the transmembrane regions and for three beta-turn structures in two surface loops.

MATERIALS AND METHODS

Preparation and characterization of purple membrane samples. *Halobacterium halobium* strain JW-3 (Weber & Bogomolni, 1982) was obtained from Dr. Walther Stoeckenius (University of California, San Francisco). The bacteria were grown on a chemically defined synthetic growth medium (Helgerson *et al.*, 1992) supplemented with [^{13}C-1]-labeled amino acids. The media also contained a tracer amount of the appropriate [^{14}C-1]-labeled amino acid in each case. Native bR for control experiments was obtained from cells grown in a complex medium or in the synthetic medium without isotopically enriched amino acids. The purple membrane fraction from the *H. halobium* cell membrane was isolated by standard procedures (Oesterhelt & Stoeckenius, 1974). The pm fractions had OD_{280}/OD_{570} ratios of <1.6 indicating negligible contamination by residual red membrane proteins. The bR content in light-adapted pm samples was determined using a molar extinction coefficient of 63,000 $M^{-1}cm^{-1}$ at 570 nm and a molecular weight of 26,000 kD.

Isotopically enriched bR was isolated from *H. halobium* grown on synthetic media containing either [^{13}C-1]-leucine (50% sp. act.), [^{13}C-1]-valine (100% sp. act.), or [^{13}C-1]-lysine (100% sp. act.). Enrichment levels in each bR sample were determined by HPLC analysis of the amino acids in hydrolyzed pm samples and quantitated from the incorporation of the [^{14}C-1]-labeled amino acid tracer (Helgerson et al., 1992). The measured isotopic enrichment levels were equal within experimental error to the original specific activities of the [^{13}C]-labeled amino acids added to the growth media, i.e., 55% for [^{13}C-1]-LEU, 100% for [^{13}C-1]-VAL, or 90% for [^{13}C-1]-LYS in the three bR samples.

FT-IR difference spectroscopy. The FT-IR spectra were recorded using a Mattson Polaris spectrometer (Mattson Instruments, Inc., Madison, WI, USA). Samples were prepared by drying pm suspensions onto a silica window under a stream of nitrogen. FT-IR spectra were recorded by averaging 128 scans at a resolution of 2 cm^{-1}. The spectrum of each [^{13}C-1]-amino acid labeled sample was subtracted from the spectrum of native bR. The C-H stretching region (2700-2800 cm^{-1}) was used to scale the spectral intensities before subtraction. The amide I region (1560-1690 cm^{-1}) in each difference spectrum was analyzed by fitting with the sum of multiple Lorentzian-shaped bands. The center frequency for each band was fixed during the fitting procedure while the bandwidth and intensity were allowed to vary. Curve fitting analysis was performed with the data analysis program SPECTRA-CALC (Galactic Industries, Salem, NH, USA).

RESULTS AND DISCUSSION

The [^{13}C-1]-amino acid labeled bR samples were studied by FT-IR spectroscopy to determine the types of helical and nonhelical secondary structures present in bacteriorhodopsin. Correlations between the regular secondary structures found in proteins and the vibrational frequencies for the associated amide I modes have been established (Krimm & Bandekar, 1986; Byler & Susi, 1986). The amide I mode is primarily a stretching vibration of the -C=O group in the peptide bond. The specific isotopic labeling of peptide backbone carbonyl groups with ^{13}C allows for the determination of local secondary structure at the labeled residues. The substitution of ^{12}C by ^{13}C produces a shift of ca. 40 cm^{-1} in vibrational frequencies associated with -C=O groups. Dollinger et al. (1986) have measured a shift of this magnitude in the -C=O stretching vibration of side chain carboxyl groups in IR spectra of [^{13}C-4]-aspartate labeled bR.

FT-IR difference spectra were obtained by subtracting the spectrum of [^{13}C-1]-LEU bR, [^{13}C-1]-VAL bR, or [^{13}C-1]-LYS bR from the spectrum of native bR. The amide I region from 1560-1690 cm^{-1} in each difference spectrum was analyzed to determine the types of secondary structures present. Curve fitting analysis was done on the positive portion of each difference spectrum, i.e., vibrational modes associated with ^{12}C-carbonyl groups present in the native bR structure that were shifted in the ^{13}C-carbonyl labeled samples. Each difference spectrum was fitted with six Lorentzian-shaped bands centered at 1672, 1662, 1653, 1641 and 1633/1680 cm^{-1} and assigned to beta-turn, α_{II}-helix, α_{I}-helix, random coil, and beta-sheet secondary structures, respectively (Krimm & Bandekar, 1986; Dunach et al., 1989). The 1633 cm^{-1} and 1680 cm^{-1} bands for the beta-sheet secondary structures are due to the low and high frequency components of the -C=O stretching vibrational modes, respectively (Byler & Susi, 1986).

Table 1 gives the results of the secondary structure analysis for the three [^{13}C-1]-amino acid labeled bR samples. The percentage of total area contributed by each FT-IR band is shown along with the number of residues in bacteriorhodopsin corresponding to that percentage. The primary sequence of bR has a total of 248 amino acid residues including 36 leucine, 21 valine, and 7 lysine residues. The number of residues represented in each band has been rounded off to the nearest integer.

TABLE 1

ANALYSIS OF FT-IR DIFFERENCE SPECTRA FOR NATIVE BR MINUS [^{13}C-1]-AMINO ACID LABELED BR

SAMPLE	BAND (cm^{-1})	AREA (%)	RESIDUES (#)
[^{13}C-VAL]-bR	1633+1680	5	1
	1641	21	4
	1653	13	3
	1662	50	11
	1672	10	2
[^{13}C-LEU]-bR	1633+1680	5	2
	1641	12	4
	1653	26	9
	1662	55	20
	1672	2	1
[^{13}C-LYS]-bR	1633+1680	4	--
	1641	10	1
	1653	27	2
	1662	54	4
	1672	5	--

The results indicate that one leucine and two valine residues are in beta-turn secondary structures. Beta-turns are formed by four residues and are stabilized by a hydrogen bond between the backbone carbonyl and amide groups of the first and fourth residues, respectively. Prediction of secondary structures (Chou & Fasman, 1978) in the interhelical surface loop sequences of bR gave high probabilities for beta-turn formation starting at nine positions (not shown). Three of these turns are predicted to start at residues VAL-34, LEU-62, and VAL-69, respectively. The other predicted turns start at ASP-36, ASN-76, ASP-102, ASP-104, TYR-131, and GLY-195. They do not include the three residues investigated in the present work.

Both electron cyro-microscopy (Henderson et al., 1990) and circular dichroism spectroscopy (Gibson & Cassim, 1989) have shown that ca. 80% of the overall structure of bR is in α-helical conformation. The results of Henderson et al. (1990) have conclusively proven that the transmembrane portions of bR are all α-helical structures. These transmembrane helices contain 75% of the total amino acids in bR. The results shown in Table 1 are consistent with the proposed structural model for bR. The bands corresponding to labeled residues in α_I-helices (1653 cm^{-1}) and α_{II}-helices (1662 cm^{-1}) can be added together to give the total helical content in each bR sample. This indicates that 65% of the valine residues, 80% of the leucine residues, and 80% of the lysine residues are in helical secondary structures in bR. The average amount of α-helical conformation in bR is then 75% as measured by the combined sites where these three types of residues occur.

The conformations of α_I- and α_{II}-helices differ in the orientation of the peptide bonds relative to the helical axis. The amide groups are parallel to the axis in an α_I-helix but are tilted away from the axis in an α_{II}-helix (Krimm & Bandekar, 1986). A pure α_{II}-helix would not form hydrogen bonds along the intrahelix peptide backbone. Thus, an α_I-helix is energetically favored unless the amide groups of the α_{II}-helix can form hydrogen bonds with other acceptor groups in order to stabilize the structure. This suggests that specific interactions may occur between residues found in α_{II}-helices and extra-helical acceptor groups either within the protein or in the surrounding environment.

A specific distribution of α_I-type and α_{II}-type helical conformations may be important for both the structure and function of native bR (Krimm & Dwivedi, 1982). The overall α_{II}-type helical content of bR decreases in favor of the α_I-type conformation when native purple membrane is converted to the nonfunctional blue membrane form (Dunach et al., 1989). The structure and function of bR in blue membrane is disrupted in that the retinylidene chromophore displays an altered photochemical cycle and the protein does not function as a light-driven proton pump. Also, the types of helices found in bR may have general implications for the transmembrane structures of other proteins (Gibson & Cassim, 1989). The structural motif of the seven transmembrane α-helices in bR serves as a model for the putative structures of several families of integral membrane proteins including the G protein-linked receptors (Dohlman et al., 1987). The techniques reported here will be generally applicable to studying structures of both membrane and cytoplasmic proteins isolated from natural sources or expressed in recombinant systems.

REFERENCES

Byler, D.M. & Susi, H. (1986): Examination of the secondary structure of proteins by deconvolution FTIR spectra. *Biopolymers* 25, 469-487.

Chou, P.Y. & Fasman, G.D. (1978): Prediction of the secondary structure of proteins from their amino acid sequence. *Adv. Enzymol.* 47, 45-148.

Dohlman, H.G., Caron, M.G. & Lefkowitz, R.J. (1987): A family of receptors coupled to guanine nucleotide regulatory proteins. *Biochemistry* 26, 2657-2664.

Dollinger, G., Eisenstein, L., Lin, S-L., Nakanishi, K., Odashima, K. & Termini, J. (1986): Bacteriorhodopsin: Fourier transform infrared methods for studies of protonation of carboxyl groups. *Meth. Enzymol.* 127, 649-662.

Dunach, M., Padros, E., Muga, A. & Arrondo, J.L.R. (1989) Fourier-transform infrared studies on cation binding to native and modified purple membranes. *Biochemistry* 28, 8940-8945.

Engelman, D.M., Henderson, R., McLachlan, A.D. & Wallace, B.A. (1980): Path of the polypeptide in bacteriorhodopsin. *Proc. Natl. Acad. Sci., USA* 77, 2023-2027.

Gibson, N.J. & Cassim, J.Y. (1989): Evidence for an α_{II}-type helical conformation for bacteriorhodopsin in the purple membrane. *Biochemistry* 28, 2134-2139.

Helgerson, S.L., Siemsen, S.L. & Dratz, E.A. (1992): Enrichment of bacteriorhodopsin with isotopically labeled amino acids by biosynthetic incorporation in *H. halobium*. *Can. J. Microbiol.* (in press).

Henderson, R. & Unwin, P.N.T. (1975): Three-dimensional model of purple membrane obtained by electron microscopy. *Nature (Lond.)* 257, 28-32.

Henderson, R., Baldwin, J.M., Caska, T.A., Zemlin, F., Beckmann, E. & Downing, K.H. (1990): Model for the structure of bacteriorhodopsin based on high-resolution electron cyromicroscopy. *J. Mol. Biol.* 213, 899-929.

Krimm, S. & Bandekar, J. (1986): Vibrational spectroscopy and conformation of peptides, polypeptides, and proteins. *Adv. Protein Chem.* 38, 183-364.

Krimm, S. & Dwivedi, A.M. (1982): Infrared spectrum of the purple membrane: clue to proton conduction mechanism? *Science (Wash. DC)* 216, 407-408.

Oesterhelt, D. & Stoeckenius, W. (1974): Isolation of the cell membrane of *H. halobium* and its fractionation into red and purple membrane. *Meth. Enzymol.* 31, 667-678.

The *in situ* molecular dynamics of the transmembrane protein bacteriorhodopsin: static or kinetic?

Joseph Y. Cassim, James E. Draheim[1] and Nicholas J. Gibson[2]

Department of Microbiology and Program in Biophysics, College of Biological Sciences, The Ohio State University, Columbus, Ohio 43210; [1] *Department of Chemistry, Adrian College, Adrian, Michigan, 49221* and [2] *Department of Chemistry, University of Arizona, Tucson, Arizona 15721 USA*

There are two opposing views of the *in situ* molecular dynamics of the transmembrane protein bacteriorhodopsin (bR) of the purple membrane (PM): (*1*) a static model based on electron cryo-microscopy and diffraction (EMD) studies of Henderson and co-workers (Unwin & Henderson, 1975; Henderson et al., 1990) and (*2*) a kinetic model based on the far-UV oriented circular dichroism (OCD) and midinfrared linear dichroism (IRLD) studies of Cassim and co-workers (Gibson & Cassim, 1989 a,b; Draheim et al., 1991). The rationale for the static model is the crystalline nature of the PM. Nearly two decades of EMD studies have firmly established the PM as a two-dimensional crystal. There is an *a priori* expectation that a protein in a crystal lattice structure would exhibit very restricted molecular dynamics due to structural constraints imposed upon it by such an ordered environment. This should limit the reversible conformational changes of the bR to only subtle localized tertiary structure changes. However, evidence from OCD studies supported by IRLD studies contradicts this static view. This evidence is consistent with a more kinetic view of the dynamics in which the bR possess inherent potential to undergo dramatic reversible global tertiary structure changes without any secondary structure changes. Although this kinetic model, as it will be shown, is well supported experimentally and theoretically, it remains controversial due to the fact that many still consider it to be a radical view.

The analysis of the OCD is based on the exciton theory of α-helical polypeptides (Woody, 1985; and references therein). The far-UV spectra of α-helical proteins arise from the amide π-π^*(NV$_1$) and n-π^* transitions of the polypeptide. The degenerate n-π^* transitions, which are polarized somewhat perpendicular to the helix axis, generate a negative CD band at *ca* 225 nm. The degenerate π-π^*(NV$_1$) transitions undergo strong excitonic coupling and split into three excitonic transitions. One, which is polarized parallel to the helix axis, generates a negative CD band at *ca* 207 nm. The other two, which are double degenerate and polarized perpendicular to the helix axis generate a positive CD band at *ca* 190 nm. Therefore, when the incident light is parallel to the helix axis the negative CD band at *ca* 207 nm cannot be generated since the incident light and 207-nm transition dipole moment will be co-parallel. Deviations from this parallelism will result in the emergence of this band (Bazzi & Woody, 1985; Gibson & Cassim, 1989a; Wu et al., 1990). The magnitude of the emerging band will depend on the angle, θ_α, between the helix axis and the incident light. Due to the cylindrical symmetry of the α-helix, θ_α is given by the relationship $\theta_\alpha = \sin^{-1} (2R_\alpha/3R_0)^{\frac{1}{2}}$ where R_α and R_0 are the magnitudes of the 207-nm band at any θ_α and the θ_α at which the helix axis is randomly oriented with respect to the incident light, respectively. The theoretical expected value of θ_α for the random orientation case is given by $\theta_\alpha = \sin^{-1} (2/3)^{\frac{1}{2}} = 54.735°$. In the OCD studies of PM films, the incident light is oriented parallel to the film normal. Since PM films used for such studies are relatively very thin, consisting of no more than 6 to 12 monolayers, the mosaic spread angle θ_m (the angle between the film normal and the membrane normal) is estimated to be zero (Gibson & Cassim, 1989b). Therefore, in OCD studies the incident light is essentially oriented perpendicular to the membrane plane.

An independent spectral method for determining θ_α is the IRLD. The theory for determining θ_α from the measured dichroic ratio at the amide frequencies is straightforward and is based on the molecular geometry of the bR in the PM film (Draheim et al., 1991). θ_α is determined from the following relationship for PM films oriented at 45° to the incident light.

$$D = 1 + \frac{2}{3}\left[\frac{f_1 S_1 + f_{11} S_{11}}{f_1 (1-S_1) + f_{11}(1-S_{11}) + f_u}\right] \quad \text{where } S_1 = S_m S_\alpha S_{ml} \text{ and } S_{11} = S_m S_\alpha S_{mll}$$

$S = \frac{3\cos^2\theta - 1}{2}$ where S is S_m, S_α, S_{ml}, or S_{mll} and θ is θ_m, θ_α, θ_{ml} or θ_{mll}, respectively.

D is the observed dichroic ratio at the amide A, I and II frequencies, f_1, f_{11} and f_u are the fraction of residue of the bR in the α_1-helix, α_{11}-helix and aperiodic secondary structure, respectively (consistency between θ_α calculated from OCD and IRLD occurs when f_1, f_{11} and f_u are 0.5, 0.3 and 0.2, respectively), and θ_{ml} and θ_{mll} are the amide transition dipole moment angles (the angle between the moment and the helix axis) for α_1 and α_{11} helices, respectively (for which the best values from literature are 17-25°, 22-29°, 82-88° and 40-50°, 42-53°, and 86-91° for the amide A, I, and II transitions, respectively). θ_m is assumed to be 10° due to the relatively thick films necessary for these studies (consisting of 100-200 monolayers).

The OCD spectrum of the native PM is free of any contributions from the 207-nm band, which according to theory is indicative of all the helical segments of the bR being oriented nearly parallel to the PM normal; that is, θ_α is most likely 0°. This was confirmed independently by IRLD studies. On the other hand, the spectrum of an ethanol-treated PM film indicates a very large contribution from this band. The difference spectrum between the ethanol-treated and the native PM films result in a nearly Gaussian band centered at ca 207 nm as predicted by theory. In the ethanol-treated PM film the helical segments of the bR are conformationally identical to those in the native PM as far as secondary structure is concerned. However, the segments are completely randomly oriented in respect to the membrane normal. The unequivocal proof for this is obtained from IRLD spectral analyses (Draheim et al., 1991). An experimental value of 54.735 ± 0.001° is obtained for θ_α which is very close to the theoretical expected value of 54.736°. A similar spectral change, but of lesser magnitude, is observed when the native PM film is perturbed by a number of diversified factors at ambient temperature. For example, (1), light bleaching in the presence of hydroxylamine, (2) exhaustive high-vacuum dehydration, (3) dry-glycerol impregnation and, (4) glucose-embedding (Draheim et al., 1988; Gibson, 1988; Gibson and Cassim, 1989b). Unlike the ethanol-treatment these perturbations cause reversible changes. The secondary structures of the bRs in these perturbed membranes are also not altered from that of the native membranes since perturbation resulted only in the emergence of the 207-nm band as expected from theory and no significant changes in the other CD bands. The magnitudes of the 207-nm bands in the difference spectra of these cases are roughly one fifth of that of the ethanol-induced one. The changes in θ_α are easily calculated by obtaining R_0 from the magnitude of the ethanol-induced 207-nm band and R_α from the magnitudes of this band induced by the other perturbations. Surprisingly, the θ_α values, thus calculated, do not vary significantly in spite of the diversity of the perturbing factors used. Within an experimental uncertainty of ±2°, the θ_α change, $\Delta\theta_\alpha$, is essentially limited to 22° for all these perturbed PM cases. In the case of the bleaching perturbation, this value was also verified by IRLD (Draheim et al., 1991). In addition, it was observed that a combination of two perturbing factors does not affect this seemingly limiting value of $\Delta\theta_\alpha$. These findings seem to suggest the presence of an inherent instability in the native PM supramolecular structure which regulates the magnitude of $\Delta\theta_\alpha$ to a constant limiting value independent of the perturbing factor.

To test this hypothesis, means were found which would enhance the inherent structural instability without causing any change in the native θ_α (Gibson, 1988; Gibson & Cassim, 1989b). The bR in the PM was irreversibly chemically altered by a variety of different agents. (1) cross-linking with dimethyl adipimidate, (2) extensive papain digestion and (3) reduction with sodium borohydride in the presence of light. In all cases, θ_α remained about 0° as determined independently by the two spectral methods in films of very different thicknesses (Draheim et al., 1991). Also the secondary structure of the bR was unchanged. However, when these altered PMs were subjected to light bleaching, the magnitudes of the 207-nm bands in the difference spectra of the bleached PMs was now increased to about one half of the magnitude of the ethanol-induced one. Therefore, the $\Delta\theta_\alpha$ of 22±2° noted in the perturbation of the native PM is increased to 36±2.5° in the chemically altered PMs at ambient temperature. As before, the findings were verified by the two independent spectral methods. Also the secondary structure remained invariant to the bleaching perturbation as before. Clearly, these findings are in accord with the inherent instability hypothesis of the bR structure, since induced enhancement of this instability results in the limiting value of $\Delta\theta_\alpha$ of native PM being increased to a higher value independent of the nature of the destabilizing agent used.

One exception to these results is that when the ground-state bR_{568} is converted to the photointermediate state M_{412}, $\Delta\theta_\alpha$ is 10° not 22° (Draheim & Cassim, 1985). However, these studies are necessarily made at very low temperatures and not ambient in order to trap this intermediate state. Notably, the Henderson EMD method failed to detect this change (Glaeser et al., 1986). This method requires among other things glucose embedding and very low temperatures. According to the Henderson EMD model of the bR, the helical segments are tilted away from the membrane normal resulting in a θ_α of about 10° (Unwin & Henderson, 1975; Henderson et al., 1990). As mentioned before, both OCD and IRLD spectral studies have indicated a θ_α of about 0° for native PM films at ambient temperature. A reasonable explanation that can accommodate these conflicting findings would be that lower temperatures constrain the dynamics of bR resulting in the value of $\Delta\theta_\alpha$ being reduced to about 10° from 22°. Reduction of molecular dynamics at very low temperatures is a highly expected possibility. It follows, therefore, that if the glucose effect results in a $\Delta\theta_\alpha$ of 22° at ambient temperature, it may expectedly be 10° at lower temperatures. This would bring the EMD results in consistency with the spectral ones. Furthermore, since the $\Delta\theta_\alpha$ due to M_{412} formation is about 10° at lower temperatures, it could be 22° at ambient. This can provide an explanation for why the EMD analyses failed to detect this change if 10° is the true limiting value of $\Delta\theta_\alpha$ at lower temperatures as 22° seems to be at ambient temperature.

The prospect of such conspicuous dynamic capabilities inherent in the bR may be antithetical to most expectations. Therefore, it might be argued that there are alternative explanations for the induced spectral changes other than the change in θ_α given in the foregoing paragraphs. One such alternative explanation might be that θ_m is changing instead of θ_α. Changes in θ_m would result in identical changes of the 207-nm band. Since the magnitude of this band depends on the angle between the helical segments and the incident light, it does not matter whether the orientation of the segments have changed with respect to the membrane normal or the orientation of the membrane normal has changed with respect to the film normal. However, if the following observations are considered concerning the induced spectral changes, it can be easily seen that this alternative explanation is not a plausible one. (1) The results are independent of the film thicknesses. (It is expected θ_m would be greatly influenced by film thickness whereas θ_α would not be.) (2) PM bleached in suspension or in situ in films resulted in the same changes. (3) When PM films are treated with an overlying buffer solution no changes are observed even after several days. (4) When the PM films are treated with a hydroxylamine solution in the dark there are no changes until the film is radiated with intense light. (5) When the PM films are impregnated with glycerol there are no changes until the film is subjected to dry-nitrogen flushing and this change is very rapidly reversed when the film is re-exposed to ambient (average 50% relative humidity) humidity. (6) When the exhaustively high-vacuum dehydrated PM film is re-exposed to ambient humidity, the induced change is very rapidly reversed. (7) When glucose solutions are layered on PM films, the glucose solution forms a thick syrup which becomes glasslike with further drying, with no changes appearing until the dried film is incubated in a 90% relative humidity environment for a time. (8) Sodium borohydride reduced PM films have identical spectra as the native ones, with changes appearing only when reduced films are radiated with intensive white light for a time. It is clear that liquids added to the films do not disturb its θ_m. However, there are agents which do, light radiation and critical removal of water molecules. It is difficult to see how these agents can alter the physical ordering of the PM discs in the film and alter θ_m when physical disturbances resulting from solution addition which would be more reasonably expected to do so, do not. The close packed nature of the PM film would not allow for any major three-dimensional reordering without a large input of energy to overcome the electrostatic interactions between PM discs.

Another possible alternative explanation may be that these spectral changes are experimentally caused artifacts resulting from changes in optical distortions and not changes in the structure of the bR. However, this possibility is very remote since previously studied effects of optical distortions on CD spectra indicated that all bands should be altered, although disproportionally (Urry & Long, 1978). It is clear from the difference spectra that in these perturbation studies only the 207-nm band is significantly altered. In addition, many of the OCD results have been substantiated by IRLD studies. Since optical artifacts inherent in far UV spectral measurements are expected to be very different than those present in mid-IR measurements, it is not possible for such experimentally induced artifacts to be correlated in these two very different spectral techniques. Therefore, changes in optical distortion are also not a viable alternative.

Recently, it was suggested that the emergence of the 207-nm band in these perturbation studies of the PM films by OCD spectra may be due to changes in the dihedral angles of the bR helix (Glaeser et al., 1991). However, changes in dihedral angles are expected to have a very small effect on the 207-nm band since this is an exciton band, which mainly depends on the orientation of the helical axes in the respect to the incident light. On the other hand, such changes are expected to alter the 225-nm and 197-nm bands much more significantly. However, no large changes have been observed in these bands comparable to those observed in the 207-nm band.

It is apparent that the only viable explanation for these induced spectral changes is the change in θ_α. All other alternative explanations possible within the framework of current scientific knowledge is readily eliminated by experimental evidence. A $\Delta\theta_\alpha$ of 22° for native PM and 36° for chemically altered PM indicates that the bR is inherently capable of very significant reversible molecular dynamics. However, a most important question remains for future resolution. Is this inherent *in situ* dynamics capabilities of bR unique to bR or is it a general characteristic of many other transmembranes consisting of two or more helical segments?

REFERENCES

Bazzi, M.D. & Woody, R.W. (1985): Oriented secondary structure in integral membrane proteins. *Biophys. J.* 48, 957-966.
Draheim, J.E. & Cassim, J.Y. (1985): Large scale global structural changes of the purple membrane during the photocycle. *Biophys. J.* 47, 497-507.
Draheim, J.E., Gibson, N.J. & Cassim, J.Y. (1988): Dehydration-induced structural changes of the purple membrane of *Halobacterium halobium*. *Biophys. J.* 54, 931-944.
Draheim, J.E., Gibson, N.J. & Cassim, J.Y. (1991): Dramatic *in situ* conformational dynamics of the transmembrane protein bacteriorhodopsin. *Biophys. J.* 60, 89-100.
Gibson, N.J, (1988): An analysis of the secondary and tertiary structure of the membrane protein bacteriorhodopsin and changes induced by glucose or reduction of the retinal-lysine linkage. *Ph.D. thesis*. pp 1-258, Columbus, Ohio: The Ohio State University.
Gibson, N.J & Cassim, J.Y. (1989a): Evidence for an α_{II}-type helical conformation for bacteriorhodopsin in the purple membrane. *Biochemistry* 28, 2134-2139.
Gibson, N.J & Cassim, J.Y. (1989b): Nature of forces stabilizing the transmembrane protein bacteriorhodopsin in purple membrane. *Biophys. J.* 56, 769-780.
Glaeser, R.M., Baldwin, J., Ceska, T.A. & Henderson, R. (1986): Electron diffraction analysis of the M_{412} intermediate of bacteriorhodopsin. *Biophys. J.* 50, 913-920.
Glaeser, R.M., Downing, K.H. & Jap, B.K. (1991): What spectroscopy can still tell us about the secondary structure of bacteriorhodopsin. *Biophys. J.* 59, 934-938.
Henderson, R., Baldwin, T.A., Zemlin, F., Beckman, E. & Downing, K.H. (1990): Model for the structure of bacteriorhodopsin based on high-resolution electron cryo-microscopy. *J. Mol. Biol.* 213, 899-929.
Stoeckenius, W., Lozier, R.H. & Bogomolni, R.A. (1979): Bacteriorhodopsin and the purple membrane of halobacteria. *Biochem. Biophys. Acta* 505, 215-278.
Unwin, P.N.T. & Henderson, R. (1975): Molecular structure determination by electron microscopy of unstained crystalline specimens. *J. Mol. Biol.* 94, 425-440.
Urry, R.W. & Long, M.M. (1978): Ultraviolet absorption, circular dichroism, and optical rotatory dispersion in biomembrane studies. In *Physiology of membrane disorder*, eds. T.E. Andreoli, J.F. Hoffman & D.D. Fanestil, pp. 107-124. New York: Plenum Medical Book Company.
Woody, R.W. (1985): Circular dichroism of peptides. In *The peptides*. Vol. 7, eds. S. Udenfriend & J. Meienhofler, pp. 15-114. New York: Academic Press.
Wu, Y., Huang, H.W. & Olah, G.A. (1990): Method of oriented circular dichroism. *Biophys. J.* 57, 797-806.

Fourier-transform infrared estimation of the secondary structure of bacterioopsin integrated into lipid vesicles*

Jaume Torres and Esteve Padrós

Unitat de Biofísica, Departament de Bioquímica i de Biologia Molecular, Facultat de Medicina, Universitat Autònoma de Barcelona, 08193 Bellaterra, Barcelona, Spain

INTRODUCTION

The renaturation of BR [1] in vesicle systems has experienced a great success in recent years and has made possible the structural and functional study of folding intermediates. Popot *et al.* (1987) extended the pioneering studies of Huang *et al.* (1981) and Liao *et al.* (1983) by developing conditions for the renaturation of intact and chymotryptic BR fragments under its natural (HL) bilayer environment.

In the present work, we have applied FTIR spectroscopy to the study of bacterioopsin inserted into HL vesicles, the main goal being the obtention of information on the effects of BR/BR contacts on the BR secondary structure. Since the vesicles were made at a 10:1 lipid-to-protein ratio, our results should correspond to a case where only BR monomers can be present, and no BR/BR interactions exist. On the other hand, since BR reconstituted in lipid vesicles is a widely used system for the study of the BR function (Khorana, 1988), it was also of interest to gain further insight into its conformational characteristics, as compared to the native BR.

MATERIALS AND METHODS

Sample Preparation

Purple membrane was isolated from *Halobacterium halobium* strain S9 as described

---

*This work was supported by DGICYT (PB89-0301)

[1] Abbreviations: BR, bacteriorhodopsin; BO, bacterioopsin; FTIR, Fourier-transform infrared; FWHH, full width at half height; HL, *Halobacterium halobium* lipids; IR, infrared; SDS, sodium dodecyl sulfate.

(Oesterhelt & Stoeckenius, 1974). *Halobacterium* lipids were prepared following Popot *et al.* (1987). Apomembranes were prepared as described by Gerber *et al.* (1979). Transfer of bacterioopsin to SDS solution and integration into HL vesicles was achieved following the method of Popot *et al.* (1987). Integration into lipid vesicles was performed in a lipid/protein ratio of 10:1 (by weight). Taurocholate was also used (Popot *et al.*, 1987) due to the difficult separation of the protein from the PDS precipitate. Its final concentration was 0.9 mM, well below its critical micellar concentration in these conditions (Popot *et al.*, 1987). Suspensions in D_2O were prepared by lyophilizing a small volume (200 μl) of a concentrated suspension in K buffer (Popot *et al.*, 1987) and resuspending it with an equal amount of D_2O in order to obtain the same salt concentration.

IR spectroscopy

Samples were placed in 6 μm pathlength CaF_2 IR cells with tin spacers in the H_2O case, and 25 μm teflon spacers in the D_2O case. IR spectra were recorded on a Mattson Polaris FTIR spectrometer equipped with a MCT detector, working at an instrumental resolution of 1 cm^{-1}. 1000 scans were averaged using a sample shuttle, apodized with a triangle function and Fourier transformed. The spectrometer was continuously purged with dry air (dew point lower than -60 °C). The sample temperature (20 °C) was set using a home-made cell jacket of circulating water. Spectra of solvent were collected under identical conditions and subtracted from the experimental one until a flat line between 1800 and 2000 cm^{-1} was obtained. Contribution of the lipids to the amide I region was eliminated by subtracting a lipid reference. The third-order derivative was performed using a breakpoint of 0.1. Deconvolutions were done using a FWHH of 14 cm^{-1}, and K factors of 2.2 in H_2O and 2.5 in D_2O. Curve-fitting of the deconvoluted spectra were performed using the number and positions of the peaks in the deconvoluted and derived spectra. The bandwith and intensities were also estimated from deconvoluted and derived curves. The methodological errors arising from deconvolution and curve-fitting were estimated as described by Cladera *et al.* (1992). The experimental mean and standard deviation were obtained by averaging four samples in H_2O and three samples in D_2O.

RESULTS AND DISCUSSION

Figure 1A shows a representative deconvoluted spectrum of BO integrated into vesicles made from native HL in H_2O, along with the best curve-fitted bands. The three bands in the 1671-1685 cm^{-1} region are attributed to reverse turns, although some weak ß-sheet contribution cannot be completely discarded. These bands are centered at 1685, 1677 and 1671 cm^{-1} in H_2O, giving a total amount of 12.7% ± 2.7% reverse turns. In D_2O (Fig. 1B), the total amount of reverse turns is 7.8% ± 1.3% ; the difference arises from a shift of some of these components to lower wavenumbers.

Four bands centered at 1664, 1657, 1651 and 1644 cm^{-1} in the H_2O spectra are assigned mostly to a-helical structures. As in native BR, the bands at 1664 and 1657 can be attributed to $a_{\parallel}$ and $a_{\perp}$ helices (Krimm & Dwivedi, 1982; Cladera *et al.*,

1992). The band at 1651 cm^{-1} in H$_2$O could be attributed to two different structures: a-helices and unordered structure (Byler & Susi, 1986; Cladera *et al.*, 1992), although from the curve-fitted bands in H$_2$O alone we cannot determine the amount of unordered structure. However, if we assume that the band at 1642 cm^{-1} in D$_2$O (14% ± 4.1%) results from the 1644 cm^{-1} band (11.3% ± 2.3%) and the shift of part of the 1651 cm^{-1} band (corresponding to unordered structure) in H$_2$O, the contribution of unordered structure can be estimated as being about 3 %. Finally, the band around 1644 cm^{-1} in H$_2$O has been proposed recently to correspond to 3$_{10}$ helices in native BR (Cladera *et al.*, 1992).

The region between 1637 and 1619 cm^{-1} in H$_2$O can be ascribed to different types of ß structure (Byler & Susi, 1986), the total area corresponding to this region being 29.3 % ± 6.0%. In D$_2$O we find three bands which give a similar amount of 27.4% ± 3.8%. The overall quantification of the secondary structure of bacterioopsin included in liposomes at a lipid-to-protein ratio of 10:1 can be summarized as being 55.0% ± 10.0% a-helices, 12.7% ± 2.7% reverse turns, 29.3 % ± 6.0% ß-sheet and about 3% unordered structure. Comparing the total a-helix content with that of the native purple membrane (about 62%, Cladera *et al.*, 1992) which is similar to that of the bleached membrane (Cladera *et al.*, unpublished results), a decrease of about 7 % helix is obtained. This decrease can almost be totally assigned to the $a_{||}$ type, and it is worth noting that this reduction is accompanied by an increase of about 12% in ß-sheet content. Part of the increased ß-sheet content seems to arise from intermolecular ß-sheet resulting from aggregation, which has been reported to yield bands in the 1620-1625 cm^{-1} region (Jackson & Mantsch, 1992). In fact, an increase of about 8% is observed in this region with respect to native BR (Cladera *et al.*, 1992), although the presence of intramolecular ß-sheet between helices cannot be excluded. In view of these results, the appearance of these ß-sheet interactions would affect mostly the regions containing $a_{||}$ helices.

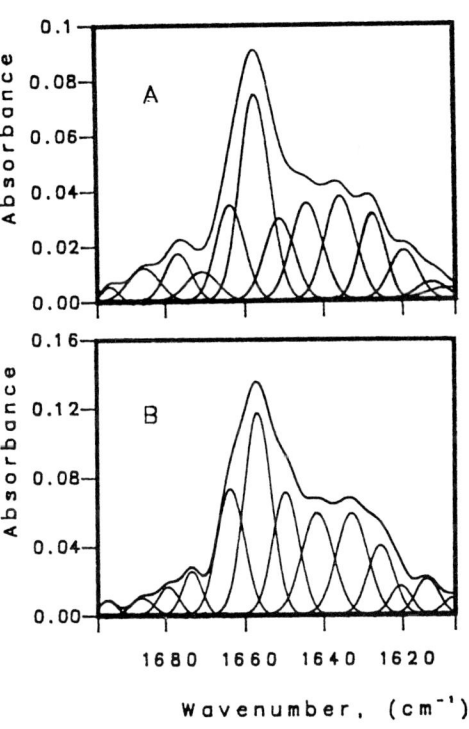

Fig.1 Deconvoluted spectra and best-fitted component bands of BO inserted in HL vesicles at a 10:1 lipid-to-protein ratio in H$_2$O K buffer (FWHH = 14, K = 2.2)(A), and D$_2$O K buffer (FWHH = 14, K = 2.5) (B).

The alterations in the BO secondary structure can be related to the decrease of the regeneration ability of reconstituted samples with respect to bleached bacteriorhodopsin upon retinal addition described by Popot *et al.* (1987), this ability being reduced as the lipid-to-protein ratio was increased. The existence of improperly renatured material detected by these authors in samples reconstituted at a 1:1 lipid-to-protein ratio seems to be responsible of this reduced regeneration ability. Thus, our results suggest that these incorrectly-folded BO molecules possess an increased proportion of ß-sheets, with little or no $a_{\parallel}$ helices. The properly-folded molecules are expected to exhibit a secondary structure very similar to that of the native form, but this aspect could only be ascertained if an homogeneous sample were available.

REFERENCES

Byler, M. & Susi, H. (1986): Examination of the secondary structure of proteins by deconvolved FTIR spectra. *Biopolymers* 25, 469-487.

Cladera, J., Sabés, M. & Padrós, E. (1992): Fourier-transform infrared analysis of bacteriorhodopsin secondary structure. (Submitted to *Biochemistry*)

Gerber, G.E., Anderegg, R.J., Herlihy, W.C, Gray, C.P., Biemann, K. & Khorana, H.G. (1979): Partial primary structure of bacteriorhodopsin: sequencing methods for membrane proteins. *Proc. Natl. Acad. Sci., USA* 76, 277-231.

Huang, K.S., Bayley, H., Liao, M.J., London, E. & Khorana, H.G. (1981): Refolding of an internal membrane protein. Denaturation, renaturation and reconstitution of intact bacteriorhodopsin and two proteolytic fragments. *J. Biol. Chem.* 226, 3802-3809.

Jackson, M. & Mantsch, H.H. (1992): Halogenated alcohols as solvents for proteins: FTIR spectroscopic studies. *Biochim. Biophys. Acta*, 1118, 139-143.

Kates, M., Kushawa, S.D. & Sprott, G.D. (1982): Lipids of purple membrane from extreme halophiles and of methanogenic bacteria. *Methods Enzymol.* 88, 98-111.

Khorana, H.G. (1988): Bacteriorhodopsin, a membrane protein that uses light to translocate protons *J. Biol. Chem.* 263, 7539-7422.

Krimm, S. & Dwivedi, A.M. (1982): Infrared spectrum of the purple membrane: Clue to a proton conduction mechanism?. *Science* 216, 407-408.

Liao, M.J., London, E. & Khorana, H.G. (1983): Regeneration of the native bacteriorhodopsin structure from two chymotryptic fragments. *J. Biol. Chem.* 258, 9949-9955.

Oesterhelt, D. & Stoeckenius, W. (1974): Isolation of the cell membrane of *Halobacterium halobium* and its fractionation into red and purple membrane. *Methods Enzymol.* 31, 667-678.

Popot, J.L., Gerchman, S.E. & Engelman, D.M. (1987): Refolding of bacteriorhodopsin in lipid bilayers. *J. Mol. Biol.* 198, 655-676.

Fourier-transform infrared study of the structure of heat-denatured bacteriorhodopsin. Comparison with the native form*

Josep Cladera and Esteve Padrós

Unitat de Biofísica, Departament de Bioquímica i de Biologia Molecular, Facultat de Medicina, Universitat Autònoma de Barcelona, 08193 Bellaterra Barcelona, Spain

INTRODUCTION

DSC studies of the thermal denaturation of bacteriorhodopsin (BR), have shown that the enthalpy of the main thermal transition (at about 96°C in H_2O) is 3 to 4 times lower than that of the globular proteins (Sánchez-Ruiz *et al.*, 1987). This seems to indicate that thermal denaturation of BR might be a process in which the resulting protein conformation retains several characteristics of its native state. In order to gain further insight into these conformational changes, we have applied Fourier-transform infrared spectroscopy (FTIR) to the study of the thermally-denatured BR. FTIR spectroscopy is a useful tool to obtain information about the secondary structure of proteins from the amide I region analysis. This can be achieved by means of mathematical techniques which allow to resolve the original overlapped component bands. One of these methods combines deconvolution and curve-fitting techniques, being possible to obtain a quantification of the different types of secondary structure (Byler & Susi, 1986). Using these methods we performed a previous study of the bacteriorhodopsin secondary structure from 1 cm^{-1}-resolution FTIR spectra. The results indicated, after estimation of methodological errors, that native BR contains 51-72% α-helices, 13-19% reverse turns, 11-17% ß-sheets and 3-7% unordered segments (Cladera *et al.*, 1992b).

MATERIALS AND METHODS

Sample preparation

Purple membrane was isolated from *Halobacterium halobium* strain S9 as described (Oesterhelt & Stoeckenius, 1974). Membrane samples in H_2O were thermally denatured at 99°C (Cladera *et al.*, 1992a) and allowed to cool at 20°C. Native and denatured purple membrane suspensions in D_2O (pD 6.5) were prepared by washing the

---

*This work was supported by DGICYT (PB89-0301)

membrane three times with D_2O and keeping the final suspension overnight before data collection. Samples in D_2O at a protein concentration of about 20 mg/ml were placed in 50 µm pathlength CaF_2 IR cells with teflon spacers.

IR data acquisition

IR spectra were acquired on a Mattson Polaris spectrometer equipped with a MCT detector, at a resolution of 2 cm^{-1}. 500 scans were averaged using a sample shuttle, apodized with a triangle function and Fourier-transformed. The spectrometer was continuously purged with dry air. To obtain the pure spectrum of samples in D_2O the solvent spectrum was subtracted. The criteria for a good subtraction was the obtention of a flat line between 1800 and 2000 cm^{-1}.

Resolution enhancement and curve-fitting

Quantification of the secondary structure of native and thermally-denatured BR was done by fitting the deconvoluted spectra (Byler & Susi, 1986). A quantitative assessement of the errors involved in deconvolution and curve-fitting was obtained by the following procedure:

(1) Absorbance experimental spectra were Fourier-self deconvoluted using the programs developped by Moffatt *et al.* (1986), using a lorentzian band shape, a full width at half height (FWHH) of 14 cm^{-1} and a resolution enhancement factor (k) of 2.7. This allowed the obtention of the number and position of the component bands.

(2) A least-square iterative curve-fitting was performed over the deconvoluted spectrum, using gaussian bandshape and allowing the peak positions, heights and bandwidths to vary simultaneously until a good fit was achieved. The band areas obtained were used to quantify the different types of secondary structure.

(3) The bands obtained in (2) were transformed to 100% lorentzian bandshape with the bandwidth multiplied by the k factor (2.7), and added again to give a spectrum which we called synthetic spectrum, very similar in shape to the experimental one.

(4) The synthetic spectrum was Fourier-self deconvoluted as in (1). The resulting spectrum was called synthetic deconvoluted spectrum.

(5) The synthetic deconvoluted spectrum was curve-fitted as in (2) and the resulting areas were compared with the known component areas of the synthetic spectrum. The methodological errors thus estimated were: (a) taking each band individually, the respective integrated areas differed by 1-40%; (b) taking the curve-fitted bands in groups of two consecutive bands, and comparing the sum of the two integrated areas with the corresponding original summed areas, the errors were less than 15%; (c) taking groups of three consecutive bands, the error was less than 10%.

RESULTS AND DISCUSSION

Figure 1A shows the deconvoluted spectrum of native BR and the corresponding fitted bands. The structural assignments to these bands are indicated in Table 1. The relative percentages corresponding to these bands are in good agreement with those presented in Cladera *et al.* (1992b) at 1 cm^{-1} resolution, where FTIR spectra of BR in H_2O and D_2O suspensions were compared. From this comparison we concluded that in D_2O about 2% of the band at 1665 cm^{-1} corresponds to reverse turns and the band at 1638 cm^{-1} has two components, one third corresponding to ß-structure and two thirds to 3_{10} helix. Taking these facts into account, as well as the contribution of the aminoacid side chains, we can conclude from the percentages presented in Table 1 that native BR contains about 62% a-helix, 16% reverse turns, 16% ß-sheets and 4% unordered segments. Figure 1B shows the deconvoluted spectrum and fitted bands of thermally denatured BR. Band assignments are indicated in Table 1. The main differences between native and denatured states are a clear decrease of the band at 1665 cm^{-1} corresponding to $a_{\parallel}$ helices and a clear increase of the 1624 cm^{-1} ß-structure band. In the case of the denatured BR the band at 1641 cm^{-1} could now correspond entirely to ß-structure or to 3_{10} helices. If this band is assigned to 3_{10} helices and taking into account the methodological errors mentioned before, the thermally denatured BR will contain

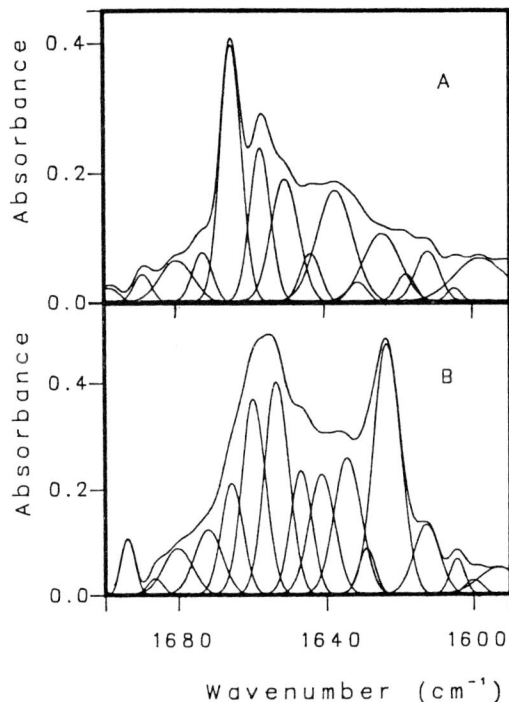

Fig.1. Deconvoluted spectra (FWHH = 14, k = 2.7) with the best-fitted individual components bands: (A) native bacteriorhodopsin; (B) denatured bacteriorhodopsin.

45.5%±4.6% a-helix (five bands assigned to a-helix) 36.2%±3.6% ß-structure (three bands) 12.0%±1.2% reverse turns (three bands) and 8.2%±3.3% unordered structure (one band). If the 1641 cm^{-1} band is assigned to ß-structure, the total a-helical content will decrease to about 36% and the ß-structure would increase to about 46%.

In any case, the structural quantification shows that thermally-denatured BR retains important conformational features of the native form. Comparison of the amide II absorbance spectra of native and denatured forms in D_2O (data not shown), allows us to conclude that the H→D exchange level is very similar in both cases, being the denatured BR amide II intensity just slightly lower than that of native BR. This means that after thermal denaturation the protein remains highly buried in the bilayer.

The secondary structure percentages obtained indicate that an important amount of a-helical structure remains after denaturation. Moreover, the amount of reverse turns is only slightly decreased in comparison with the native form, and just a little increase of unordered segments is observed. The most important secondary structural change is the appearance of the 1624 cm^{-1} ß-structure band, characteristic of thermal-, solvent- or pressure-induced protein denaturation. This band is usually assigned to intermolecular ß-structure (Jackson & Mantsch, 1992). In our case this structure seems to arise mainly from the $a_{\parallel}$-helix of the native form, a structure which is thought to be highly buried in the membrane and whose energetic stability is lower

Table 1: Position, fractional areas and assignments of the amide I bands of native and denatured bacteriorhodopsin in D_2O, pD 6.5.

NATIVE BR			DENATURED BR		
freq.[a] (cm^{-1})	area[b] (%)	assignment	freq.[a] (cm^{-1})	area[b] (%)	assignment
1689	2.2	turns	1694	2.4	ß
1680	6.0	turns	1687	0.8	turns
1673	4.3	turns	1680	4.1	turns
1665	23.5	α_{II} + turns	1672	5.6	turns
1659	13.8	α_{II} + α_{I}	1665	7.4	α_{II} + turns
1651	14.3	α_{I}	1659	13.7	α_{II} + α_{I}
1645	4.2	unordered	1653	15.0	α_{I}
1639	17.4	3_{10} helix + ß	1647	7.9	unordered
1632	1.6	ß	1641	9.4	3_{10} helix or ß
1625	10.9	ß + C=N	1635	11.2	ß
1618	2.3	ß	1629	2.3	ß
			1624	20.3	ß

[a]Frequency positions are expressed in cm^{-1}, and rounded off to the nearest integer. [b]The relative area values are affected by the methodological error, as described in the text.

than that of α_I-helix. Upon denaturation, α_{II}-helix in the bilayer could experience a conformational change which can be interpreted as partial unfolding of the protein due to local disruptions of intermolecular hydrogen bonds and the formation of new stronger ones. These new ß-sheets could form either from different transmembrane segments of the same molecule or from transmembrane segments of different protein monomers.

REFERENCES

Byler, D.M. and Susi, H. (1986): Examination of the secondary structure of proteins by deconvolved FTIR spectra. *Biopolymers* 25, 469-487.

Cladera, J., Galisteo, M.L., Sabés, M., Mateo, P.L. and Padrós, E. (1992a): The role of retinal in the thermal stability of the purple membrane. *Eur. J. Biochem. (in press)*.

Cladera, J., Sabés, M. and Padrós, E. (1992b): Fourier-transform infrared analysis of bacteriorhodopsin secondary structure. Submitted to *Biochemistry*.

Jackson, M. and Mantsch, H.H. (1992): Halogenated alcohols as solvents for proteins: FTIR spectroscopic studies. *Biochim. Biophys. Acta* 1118, 139-143.

Moffatt, D.J., Kauppinen, J.K., Cameron, D.G., Mantsch, H.H. and Jones, R.N. (1986): Computer programs for infrared spectrophotometry. *NRCC Bulletin 18, National Research Council of Canada, Ottawa, Canada*.

Oesterhelt, D. and Stoeckenius, W. (1974): Isolation of the cell membrane of *Halobacterium halobium* and its fractionation into red and purple membranes. *Methods Enzymol.* 31, 667-678.

Sanchez-Ruiz, J.M. and Mateo, P.L. (1987): Differential scanning calorimetry of membrane proteins. *Cell. Biol. Rev.* 11, 15-45.

Structures and Functions of Retinal Proteins. Ed. J.L. Rigaud. Colloque INSERM / John Libbey Eurotext Ltd.
© 1992, Vol. 221, pp. 37-40

Structure analysis of the live squid photoreceptor by X-ray diffraction

Toshiaki Hamanaka, Yuji Kito[1], Masatsugu Seidou[1], Katsuzo Wakabayashi and Yoshiyuki Amemiya[2]

Department of Biophysical Engineering, Faculty of Engineering Science, Osaka University, Toyonaka, Osaka 560, Japan, [1] *Department of Biology, Faculty of Science, Osaka University, Toyonaka, Osaka 560, Japan and* [2] *Photon Factory, Laboratory for High Energy Physics, Tsukuba, Ibaraki 305, Japan*

INTRODUCTION

The initial step of the visual process is the absorption of light by the visual pigment in the photoreceptor cell. The squid visual pigment is located in microvilli which are cylindrical extension of the cell membrane, arranged hexagonally within the rhabdomes. Until now, the retina fixed by glutaraldehyde was used on the structural study of invertebrate rhabdomes by x-ray diffraction, because this tissue disintegrated within few hours of dissection and more than ten hours exposure time was needed to record a diffraction pattern with the conventional x-ray apparatus (Worthington et al., 1976; Saibil, 1982; Saibil & Hewat, 1987). In the present study, we could succeed in recording the x-ray diffraction pattern from unfixed squid retina by the use of the synchrotron radiation and a storage phosphor screen, the imaging plate, as a detector. By the recent advances in x-ray experiment, it become possible to investigate the structure of squid photoreceptors in the physiological conditions and the structural response to the light illumination. The structure of the photoreceptors was analyzed by the model calculation, using reflections up to 1/8 nm^{-1}.

MATERIALS AND METHODS

Living, active specimens of the firefly squid, Watasenia Scintillans were captured at Toyama Bay of the Japan Sea and brought to Tsukuba within several hours. For the x-ray experiment, a 1-mm thick slice of retina was kept in a sample chamber with Myler windows, containing an artificial seawater at 4°C. Schematic diagram of a vertical slice of squid retina is shown in Fig.1. X-ray experiments have been performed with a mirror-monochromator optics at the beam line 15A1 of Photon Factory (Amemiya et al., 1983). The wavelength of the radiation was 0.15 nm. The sample to detector distance was 2196 mm. X-ray scattering intensity was recorded on the imaging plate and stored on magnetic tape after converting to the digital signals with the image reader and the image processor (Amemiya et al., 1988). The exposure time was 2-10 minutes and each recording finished within 30 minutes after the decapitation. The sample preparation and the x-ray measurement were done under dim red light at the experimental hatch.

For the structure analysis, the integrated intensity of each reflection was obtained by summing up the intensities at pixels within a diffraction spot after the suitable subtraction of background scattering and being corrected by a Lorentz factor. The structure of the cross section of rhabdome has been sought by model building (Fig.3). The electron density profiles of a model structure were approximated by step functions. To evaluate the models, the calculated intensities and the Patterson maps, calculated as direct Fourier transforms of diffraction patterns, were compared with the observed ones. The data processing and the model calculations were carried out on the ACOS 2000 computer (NEC) at Institute for protein Research, Osaka University and PDP 11/34 computer (DEC).

RESULTS AND DISCUSSION

Figure 2 shows the x-ray diffraction pattern from the live squid photoreceptor. The low angle x-ray diffraction spots are observed due to the 60 nm hexagonal lattice of microvilli. The six diffuse maxima, which might originate from the bilayer structure of microvillar membrane, are also observed around $1/4$ nm^{-1}. The intensities are different among equivalent Bragg reflections and stronger near the vertical axis of the diffraction pattern, suggesting the random orientation of rhabdomes around the vertical axis of retina in the whole sample. This is supported by the light microscopic observation that the regularity of the rhabdome arrangement is local in the firefly squid retina.

For analysis of the structure, the Patterson functions were calculated based on the models, in which the microvillar membrane was assumed to have the bilayer electron density distribution. The radii and electron density of microvilli and the microvillar cytoskelton were varied in the model calculation (Fig.3). Figure 4 shows the Patterson function calculated from the observed x-ray diffraction intensity up to 8 nm^{-1}. Figure 5A shows the initial model which was suggested by the electron micrograph of a vertical section of the glutaraldehyde-fixed and negatively stained retina (Michinomae et al., unpublished data). However, there is no resemblance between the observed Patterson map shown in Fig. 5B and the calculated one (Fig.4). The highest peaks at positions (1/3,2/3) and (2/3,1/3) in the observed Patterson map are not seen in the calculated one. The peaks at

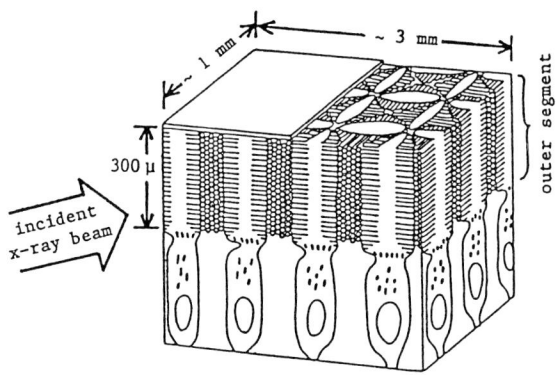

Fig. 1. Schematic diagram of a vertical slice of squid retina. The retina consist almost entirely of photoreceptor cells. The photoreceptive outer segments are cylindrical extensions of the cell membrane, packed hexagonally in the rhabdomes. The microvilli are ~50 nm in diameter and ~1 μm long.

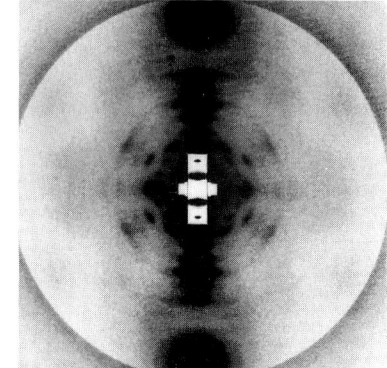

Fig. 2. The x-ray diffraction pattern from a live squid retina in artificial seawater. The storage ring was operated at 2.5 GeV with a beam current of 188 mA. The exposure time was 5 minutes. The background scattering was radially subtracted. The diffraction spots were indexed by a hexagonal lattice of 60 nm. The diffuse maxima at higher diffraction angle are centered at 4 nm^{-1}.

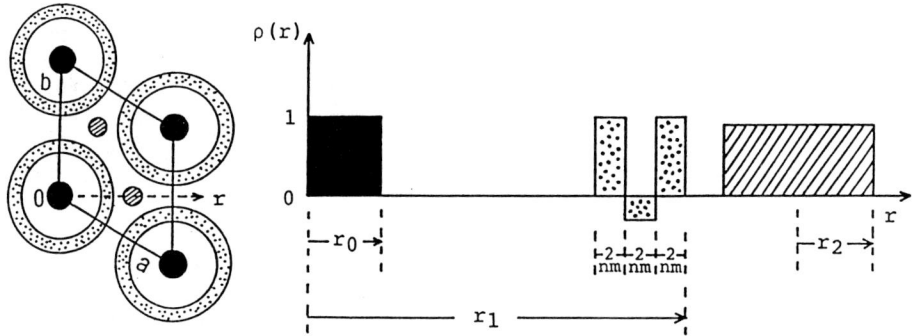

Fig. 3. Model construction of the photoreceptor outer segments. (a) The cross section of the rhabdome, in which the microvillar membrane (doted ring), cytoskelton core (filled circle) and inter-microvillus material (shaded circle) are positioned in a hexagonal lattice of 60 nm. (b) The electron density profile approximated by step function, along a dashed line in a. The bilayer profile is assumed in the microvillar membrane. The relative electron density is normalized to that of the hydrophilic region of the bilayer profile.

(1/3,0), (1/3,1/3) and (0,1/3) seem to be positioned by the 30° rotation in the calculated Patterson map. The highest peaks suggested the existence of inter-microvillous materials at (1/3,2/3) and (2/3,1/3). As shown in Fig. 6, a fairly good agreement was obtained between the observed and the model Patterson maps except the region around a lattice corner, by introducing the inter-microvillus materials. The significant improvement was found by considering the rather complicated structure as a model of the inter-microvillus materials, which is shown in Fig. 7. In this model, the radius of microvilli is 24 nm and the relative electron density of hydrophobic region is close to 0, referred to the aqueous medium. The radius and the electron density of the cytoskelton are 4.8 nm and 1.2, respectively. The parameters of the inter-microvillous material are $r_2=6$ nm (1), $r_3=3$ nm (0.9) and $r_4=3$ nm (0.6), where figures in parentheses represent relative electron densities and r_3 and r_4 stand for the radii of a small dashed circle and another one, respectively.

Fig. 4. The Patterson map obtained by the direct Fourier transform of the observed x-ray diffraction intensity up to 8 nm^{-1}. Negative contours are shown as dashed lines.

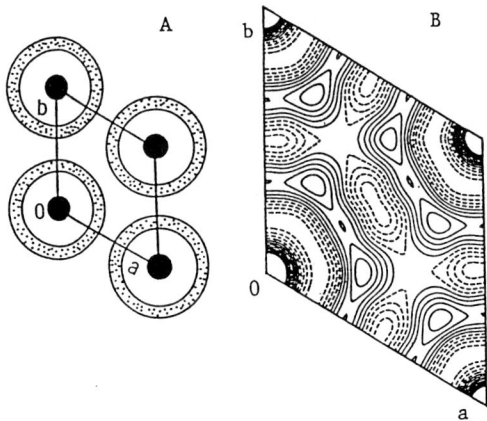

Fig. 5. The Patterson map (B) based on the model shown in A, suggested by the electron micrograph of a vertical section of the glutaraldehyde-fixed and negative-stained retina.

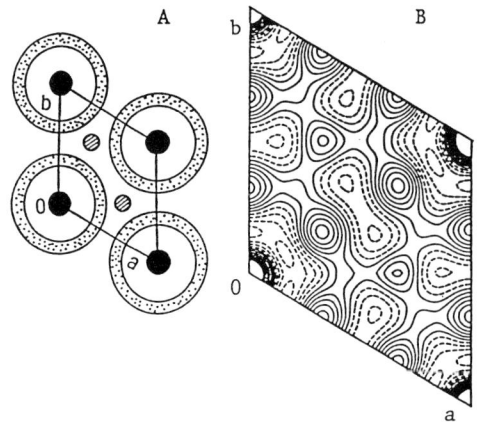
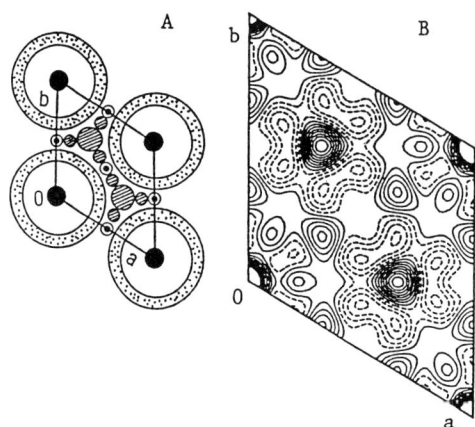

Fig. 6. The Patterson map based on the model in which the inter-microvillous materials are positioned.

Fig. 7. The Patterson map of the best model, having a rather complicated structure as the inter-microvillus material.

The present x-ray study has clearly shown the existence of massive inter-microvillous materials which could be scarecely observed in the electron micrograph of negatively stained retina. The inter-microvillous materials may correspond to the membrane junction pointed out by Saibil & Hewat (1987). The best model showed the large gap between adjacent microvillar membranes (~12 nm), in contrast to the electron microscope section of fixed retina. Therefore, the inter-microvillous materials may play a role in aligning the microvilli. The hydrophobic region in the bilayer profile of microvillar membrane have a electron density close to the aqueous medium and higher than that of the disk membrane of vertebrate photoreceptor. For the detailed discussion of the bilayer membrane profile, the intensities up to reflections at $1/4$ nm^{-1} are needed to be taken into the calculation. Unfortunately, the diffraction around $1/4$ nm^{-1} was diffuse and the quantatative estimation of intensity was difficult, due to the disorientation of rhabdomes, which made the diffraction spot to the arc, and the variation in size and shape of microvilli, giving a broad background peak under Bragg reflection spots.

This work has been performed under the approval of the Photon Factory Advisory Committe (Proposal Nos. 87-040 and 89-062).

REFERENCES

Amemiya, Y., Wakabayashi, K., Hamanaka, T., Wakabayashi, T., Matsushita, T., and Hashizume, H. (1983): Design of small-angle x-ray diffractometer using synchrotron radiation at the Photon Factory. Nucl. Instr. and Meth. 208, 471-477.

Amemiya, Y., Matsushita, T., Nakagawa, A., Satow, Y., Miyahara, J., and Chikawa, T. (1988): Design and performance of an imaging plate system for x-ray diffraction study. Nucl. Instr. and Meth. A266, 645-653.

Saibil, H.R. (1982): An ordered membrane-cytoskeleton network in squid photo-receptor microvilli. J. Mol. Biol. 158, 435-456.

Saibil, H.R., and Hewat, E. (1987): Ordered transmembrane and extracellular structure in squid photoreceptor microvilli. J. Cell Biol. 105, 19-28.

Worthington, C.R., Wang, S.K., and Folzer, C.M. (1976): Low angle x-ray diffraction patterns of squid retina. Nature 262, 626-628.

One-dimensional and multidimensional NMR of bacteriorhodopsin

Michel Seigneuret, Daniel Lévy and Jean-Michel Neumann*

*Laboratoire de Biophysique Cellulaire, URA 526-Equipe ATIPE, Université Paris VII, 2, place Jussieu, 75251 Paris Cedex 05 and * Département de Biologie Cellulaire et Moléculaire, URA 1290, CE Saclay, 91191 Gif-sur-Yvette Cedex, France*

1. INTRODUCTION

In the case of soluble proteins, high-resolution solution NMR has emerged in the past ten years as a technique able to provide *ab initio* complete 3D structure with medium to high resolution. It is possible to use solution NMR on membrane proteins after solubilization in detergent in the native state as is illustrated here in the case of bacteriorhodopsin.

Bacteriorhodopsin, the 26kD light-driven proton pump of *Halobacterium Halobium* has several times been used as a model system for the development of particular structural approaches to membrane proteins. Recent improvements in electron microscopy of purple membrane (Henderson et al 1990) and mutagenesis (Mogi et al 1989) have provided a medium resolution model of the protein 3D structure which has to be ascertained by other approaches. Furthermore many aspects of the bacteriorhodopsin structure remain unknown such as the exact limits and conformation of aqueous regions and the orientations and conformation of most aminoacid sidechains. A better structural knowledge of the bacteriorhodopsin structure may also increase our understanding of the proton transport mechanism. Potential technical advantages of bacteriorhodopsin for NMR include bacterial origin (i.e. feasability of biosynthetic labeling), availability in large quantities and stability. It must nevertheless be emphasized that bacteriorhodopsin is already a rather large protein for NMR. Indeed, for multidimensional NMR, molecular weights above 20kD presently represent " the state of the art" even for soluble proteins. Due to the additional problems encountered with membrane proteins that will be discussed below, structure determination of bacteriorhodopsin has to rely on a combination of well established 1D NMR and of newly develloping multidimensional NMR techniques.

2. HIGH-RESOLUTION SOLUTION NMR OF BACTERIORHODOPSIN.

2.1. Requirements for high-resolution

A prerequisite for solution NMR studies of bacteriorhodopsin is that the native or active conformation be maintained in the detergent-solubilized state. We have found that the absorption spectrum and the ability of dark-light adaptation are largely unchanged after solubilization in a number of detergents. Recent studies (Milder et al. 1991) indicate that the photocycle of bacteriorhodopsin is very close to that of purple membrane in the detergent dodecylmaltoside that we have selected for NMR studies.

High S/N ratio in NMR demands high concentrations (typically millimolar) of the purified protein. We have designed a method for delipidation and solubilization of bacteriorhodopsin into highly concentrated monomers in several detergents (Seigneuret et al. 1991a). Recent improvement in the method have allowed us to obtain bacteriorhodopsin concentrations up to 1.5-2 mM.

We found that the attainable resolution in NMR experiments with solubilized bacteriorhodopsin depends strongly on the chosen detergent (Seigneuret et al. 1991a). Indeed such resolution is directly related to the rotational correlation time of the protein in solution .The later itself depends upon 3 parameters influenced by the detergent: effective size of the protein, viscosity and temperature. By combining hydrodynamic and stability measurements, we showed that such parameters widely differ from one detergent to another, promoting differences of one order of magnitude in the rotational correlation time of the protein. The effect of detergent on resolution was confirmed by recording ^{13}C NMR spectra of bacteriorhodopsin containing $(1,^{13}C)$ phenylalanine (i.e. carbonyl labeled) in varios detergents. These data indicated that the best choice for solubilization of bacteriorhodopsin for NMR is dodecylmaltoside, in which the protein rotates relatively rapidly and is stable for days at temperatures up to 50°C. The resolution achieved is illustrated in Fig.1 for bacteriorhodopsins labeled with various carbonyl ^{13}C aminoacids.

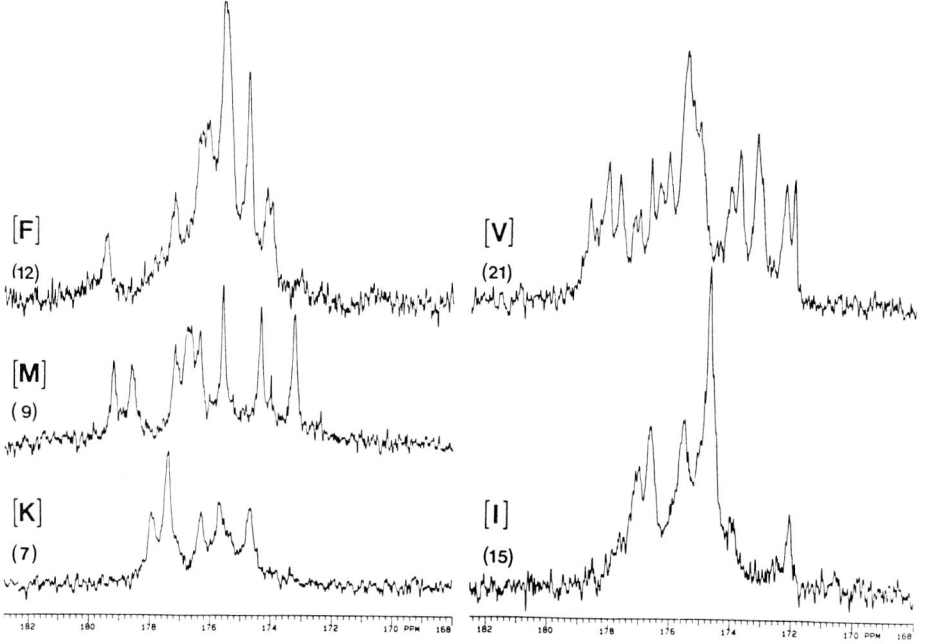

Fig. 1. 13C NMR spectra of bacteriorhodopsin in dodecylmaltoside at 50°c after labeling with several carbonyl 13C aminoacids. The aminoacid type as well as its occurence in the sequence is indicated near each spectrum.

2.2. One-dimensional NMR of bacteriorhodopsin.

Structural information in 1D NMR is mainly obtained through the effect of "site-directed" extrinsic agents on NMR spectra. We have used this approach for the determination of the topography of methionine residues inthe bacteriorhodopsin structure (Seigneuret et al. 1991b, Seigneuret and Kainosho, submitted). For this purpose, bacteriorhodopsins containing ^{13}C methionine labeled either on the carbonyl carbon or on the terminal methyl were used. Extrinsic agents whose effects were assessed included the methionine oxidizer H_2O_2, paramagnetic probes aimed at the hydrophilic or hydrophobic surfaces and ^2H_2O in order to assess exchangeable amide protons.

The following results were obtained with regard to the location of the 9 methionine residues of bacteriorhodopsin. Two residues, namely Met 68 and 163 appear to be on the hydrophilic surfaces of the protein, one being close to the membrane. Three residues are located on the hydrophobic surfaces among which Met 20 is close to the aqueous interface. All 4 remaining methionine residues are inside the protein. This later result allows us to evaluate current models of the bacteriorhodopsin folding and tertiary structure. Our data are in complete agreement with the folding model of Henderson et al. (1990). The tertiary structure model of Henderson et al. (1990), from electron microscopy, is largely similar to that of Mogi et al. (1989), from mutagenesis, except for an ambiguity on the rotational orientation of helix B and D. Only one combination of the two proposed orientations is in agreement with the occurence of 3 methionine residues on the hydrophobic surface.

2.3 Multidimensional NMR of bacteriorhodopsin.

The 1D NMR methods described above can be used to obtain a particular topographic orientation or to test a structural model. However, it is likely that NMR will establish itself as a major method for structure determination of membrane proteins only if it can provide global structural data in a model independant manner. We have thus recently attempted to apply multidimensional NMR to bacteriorhodopsin with the secondary structure as a first goal. For soluble proteins above 20kD the now established procedure relies on ^1H-^{15}N 3D experiments (Clore and Gronenborn, 1991).

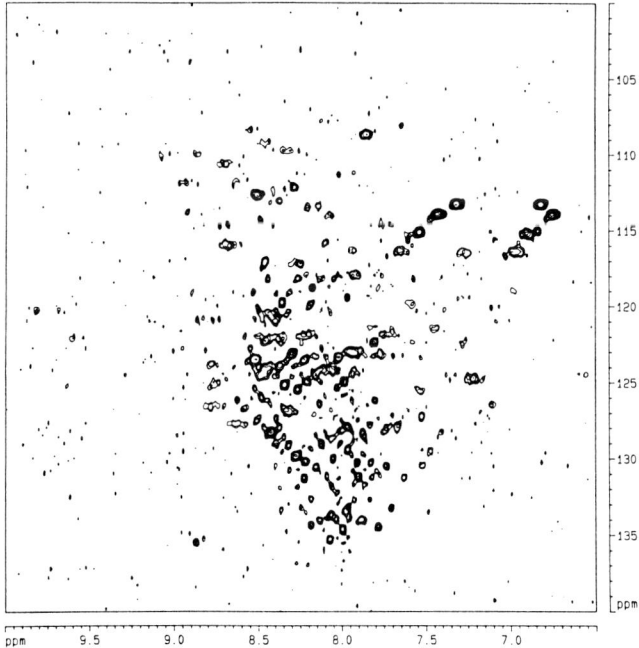

Fig. 2. ^1H-^{15}N HSQC spectrum of uniformly 15N-labeled bacteriorhodopsin in dodecylmaltoside at 47°C.

We have uniformly labeled bacteriorhodopsin with ^{15}N by growing H. Halobium in a culture medium containing hydrolysed labeled algae as major nutrient. Fig.5 shows the amide region of the ^1H-^{15}N HSQC spectrum at 50°C of the labeled protein solubilized in dodecylmaltoside at 1mM. About 140 resolved resonances as well as many unresolved resonances can be observed. This indicate that such isotope-assisted multidimensional methods are indeed applicable to relatively large membrane protein such as bacteriorhodopsin. Such HSQC or HMQC experiments still have to be combined with NOESY experiments in order to yield structural information. A limitation is the increase in rotational correlation time due to the detergent that leads to much shorter T_2 values than for a soluble protein of similar molecular weight. We have found that, due to this effect, methods based on ^1H-^1H scalar correlation such as HOHAHA and COSY are unapplicable. We are thus turning to other methods for residue-specific assignments, namely HSQC of bacteriorhodopsins selectively labeled with ^{15}N aminoacids. We have also found that, even for NOESY, usually less sensitive to size, only a limited number of correlations can be obtained. Our present procedure for overcoming this difficulty relies on fractional deuteration of the protein (together with ^{15}N labeling) in order to lenghten the T_2 sufficiently for HMQC-NOESY to be performed. The data obtained from such experiments should allow a relatively precise determination of the secondary structure of bacteriorhodopsin including aqueous portions

3. CONCLUSION

This work indicates that it is possible to perform high-resolution solution NMR experiments on a relatively large membrane protein such as bacteriorhodopsin solubilized in detergent. Relatively simple 1D NMR experiments can be used to obtain topographic data on a membrane protein and should be applicable to larger molecular weights. In the case of multidimensional NMR experiments, bacteriorhodopsin probably represents the upper limit of protein sizes that can be investigated. In any case it is likely that NMR has a significant role to play in membrane protein structural biology due to its ability to provide relatively precise sequence-related structural data and its potentiality in reporting on dynamic and functional aspects.

REFERENCES

Clore, G.M. and Gronenborn, A.M. (1991) Prog. NMR Spectrosc. 23.43
Henderson, R., Baldwin, J.M., Ceska, T.A., Zemlin, F., Beckman, E. and Downing, K.H. (1990) J. Mol.Biol. 213. 899
Milder, S.J., Thorgeirsson,T.E., Stroud,R.M. and Klinger, D.S. (1991) Biochemistry 30. 1751
Mogi, T., Marti, T.and Khorana, H.G.(1989) J. Biol. Chem. 264. 14197
Seigneuret, M., Neumann, J.M. and Rigaud, J.L. (1991a) J. Biol. Chem. 266. 10066
Seigneuret, M., Neumann, J.M., Levy, D. and Rigaud, J.L. (1991b) Biochemistry 30. 3885

A preliminar three-dimensional model of rhodopsin

Leonardo Pardo and Mireia Duñach*

*Laboratory of Computational Medicine, Departament de Bioestadística, and *Unitat de Biofísica, Departamento de Bioquímica i Biologia Molecular, Facultat de Medicina, Universitat Autònoma de Barcelona, 08193 Bellaterra, Spain*

The tertiary structure of bacteriorhodopsin has been used as a structural template to construct a three-dimensional model of rhodopsin. This model is proposed in light of *i)* the functional and structural similarities between rhodopsin and bacteriorhodopsin, *ii)* the sequence homology between rhodopsin, the neurotransmitter G-protein coupled receptors, and bacteriorhodopsin, and *iii)* the information on chimeric receptors and site-directed mutagenesis studies.

INTRODUCTION

The 3-D structure of G-protein coupled receptors (GPCR) has not yet been characterized by X-ray cristallography. However, in analogy with bacteriorhodopsin (BR), a common secondary structural model consisting of seven transmembrane helical (TMH) segments has been predicted for the large family of GPCRs. These include rhodopsin and other visual opsins, neurotransmitter receptors, peptide hormone receptors, among others. Some intramembrane helices of these receptors interact to form a hydrophobic pocket into which the ligand binds. Like the cis - trans retinal isomerization in rhodopsin, the interaction of the extracellular ligand with the receptor, may initiate charge movements and conformational changes across the membrane that propagate to the cytoplasmic surface and result in G-protein activation.

TERTIARY STRUCTURE

The lack of detailed structural information continues to hamper the identification of those structural features of individual GPCRs that are responsible for their specific functions. In order to get insight into these processes, the molecular architecture of BR (Henderson *et al.*, 1990) is commonly regarded as a structural template to construct 3-D models for the GPCR (Findlay & Pappin, 1986; Dahl *et al.*, 1991; Hibert *et al.*, 1991). However, these models do not resolve the difficulty caused by the apparent lack of significant homology between the seven TMH of BR and GPCR, or even between BR and the mammalian opsins. In a very recent work (Pardo *et al.*, 1992), an appreciable homology between some hydrophobic domains of BR and

Table 1. Sequence homology betwen BR and GPCR: shaded areas indicate sequence identity between BR and GPCR; bordered areas indicate sequence identity among GPCR. RHO: bovine rhodopsin.

										HELIX 3 IN GPCR												
BR III	78	I	Y	W	A	R	Y	A	D	W	L	F	T	T	P	L	L	L	D	L	A	L	L	V	D	A	D
BR VII	205	T	L	L	F	M	V	L	D	V	S	A	K	V	G	F	G	L	I	L	L	R	S	R	A	I	F	G
RHO	110	C	N	L	E	G	F	F	A	T	L	G	G	E	I	A	L	W	S	L	V	V	L	A	I	E	R	Y
5HT$_{1A}$	109	C	D	L	F	I	A	L	D	V	L	C	C	T	S	S	I	L	H	L	C	A	I	A	L	D	R	Y
α$_1$	113	C	D	I	W	A	A	V	D	V	L	C	C	T	A	S	I	L	S	L	C	A	I	S	L	D	R	Y
β$_3$	110	C	E	L	W	T	S	V	D	V	L	C	V	T	A	S	I	E	T	L	C	A	L	A	V	D	R	Y
H$_2$	91	C	N	I	Y	T	S	L	D	V	M	L	C	T	A	S	I	L	N	L	F	M	I	S	L	D	R	Y

										HELIX 5 IN GPCR											
BR V	135	F	V	W	W	A	I	S	T	A	A	M	L	Y	I	L	Y	V	L	F	F	G	F	T	S	K
BR III	77	P	I	Y	W	A	R	Y	A	D	W	L	F	T	T	P	L	L	L	D	L	A	L	L	V
RHO	201	E	S	F	V	I	Y	M	F	V	V	H	F	T	I	P	M	I	I	I	F	F	C	Y	G	Q
5HT$_{1A}$	193	H	G	Y	T	I	Y	S	T	F	G	A	F	Y	I	P	L	L	L	M	L	V	L	Y	G	R
α$_1$	104	P	F	Y	A	L	F	S	S	L	G	S	F	Y	I	P	L	A	V	I	L	V	M	Y	C	R
β$_3$	202	M	P	Y	V	L	L	S	S	S	V	S	F	Y	L	P	L	L	V	M	L	F	V	Y	A	R
H$_2$	180	L	V	Y	G	L	V	D	G	L	V	T	F	Y	L	P	L	L	V	M	C	I	T	Y	Y	R

										HELIX 7 IN GPCR						
BR VII	210	V	L	D	V	S	A	K	V	G	F	G	L	I	L	L	R	S	R	A	I	F	G
BR I	9	E	W	I	W	L	A	L	G	T	A	L	M	G	L	G	T	L	Y	F	L	V	K
RHO	290	I	P	A	F	F	A	K	S	A	A	I	Y	N	P	V	I	Y	I	M	M	N	K
5HT$_{1A}$	384	I	I	N	W	L	G	Y	S	N	S	L	L	N	P	V	I	Y	A	Y	F	N	K
α$_1$	235	V	V	F	W	L	G	Y	F	N	S	C	I	N	P	I	I	Y	P	C	S	S	K
β$_3$	330	A	L	N	W	L	G	Y	A	N	S	A	F	N	P	L	I	Y	C	R	S	P	D
H$_2$	272	V	V	L	W	L	G	Y	A	N	S	A	L	N	P	I	L	Y	A	T	L	N	R

those of the neurotransmitter receptors has been found if the sequential order of the helices is ignored. These findings prompted us to investigate the proposed sequence homology between rhodopsin, the neurotransmitter receptors, and BR. Table 1 summarizes the helices related with higher homology: helices 3, 5, and 7 of GPCR with helices VII, III and I of BR, respectively. The most remarkable features in helix-3 of GPCR are the conserved residues that define the helix borders: Cys-110 (of bovine rhodopsin) is involved in an essential disulfide bond at the intradiscal surface (Karnik & Khorana, 1990) and the charged pair Glu-134/Arg-135 is required for the binding of the G-protein at the cytoplasmic end (Franke *et al.*, 1990). In the same helix, Glu-113 has been identified as the stabilizing counterion for the protonated retinylidene Schiff base in rhodopsin (Sakmar *et al.*, 1989). An aspartate residue is conserved one turn above the α-helix in all GPCR that bind protonated amine ligands, and has been postulated to form an ion pair with the cationic amine (Strader *et al.*, 1988). In rhodopsin, Glu-113 which has a longer side chain, occupies an equivalent position as the acidic residue involved in the binding of the cationic ligand in GPCR.

As shown in Table 1, no homology is observed between helix-VII in BR and helix-7 in rhodopsin. In this comparison, Lys-296 in rhodopsin has been superimposed to Lys-216 in BR (residues to which the retinal binds). However, a more detectable homology is found between helix-I in BR and helix-7 in GPCR. These findings suggest a modification of the BR template to account for the new positioning of the helices in the tertiary structures of GPCR. Therefore, considering the helices related with a higher homology (Table I), we propose the tertiary organization shown in Fig. 1. In this preliminary 3-D model of rhodopsin, only helices 3 to 7 are matched to helices VII, VI, III, II and I in BR, respectively. Moreover, experimental results on chimeric $α_2$- and $ß_2$-receptors (Kobilka, 1988) support the conclusion that helix-7 lies adjacent to helices 3 and 4. This requirement can be fullfiled with the proposed organization of the helix bundle of GPCR.

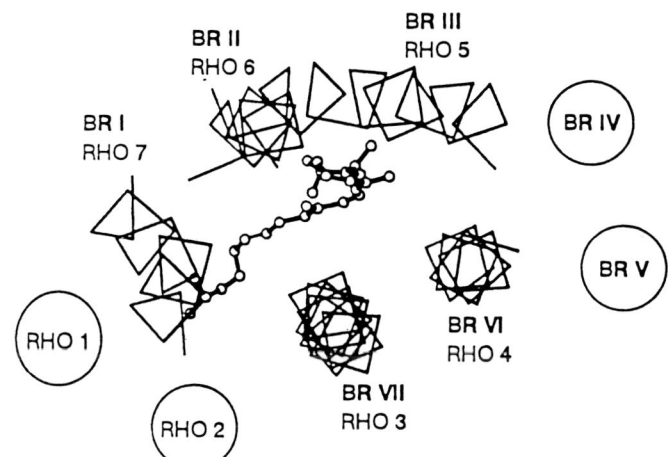

Fig. 1. Schematic representation of the helix bundle organization of Bacteriorhodopsin (BR), and the proposed arrangement for the helices in rhodopsin (RHO) that reflect the homology suggested in the text. The tertiary structure is the cytoplasmic view.

The proposed 3-D model of rhodopsin can explain experimental results from site-directed mutagenesis. The Schiff base counterion, Glu-113, is located inside the membrane bilayer near the intradiscal surface, and is in a favorable position to interact with the protonated Schiff base in helix-7 (see Fig. 2), which is the site of attachment of retinal. Glu-122, which is specific for rhodopsin, is also found near the retinal ring (data not shown) and could be involved in wavelength regulation, as suggested by Sakmar *et al.* (1989). In this model, the side chain of Phe-116, which is specific for the opsin family and the conserved Tyr-268, would sandwich the retinylidene chain, impeding the chromophore to adopt a configuration different from 11-cis in the dark-adapted state of opsins. Three highly conserved aromatic residues define the hydrophobic pocket around the retinal ring. As illustrated in Fig. 2, Trp-265 and Phe-261 in helix 6 are located at one side of the ß-ionone, whereas Phe-221 in helix 5 lies at the opposite side of the retinal ring. Trp-126 in helix-3 is found at the same level as Phe-261 and could interact with Trp-265 (data not shown). All these hydrophobic residues provide a rigid environment around the retinal binding pocket. Some of these aromatic aminoacids have been modified by a photoactivable analog of 11-cis retinal (Nakayama & Khorana, 1990), indicating that the ß-ionone ring of retinal is oriented towards these residues. It is worth mentioning that Phe-293, Phe-276, and Phe-203 occupy the region below the protonated Schiff base close to the intradiscal surface. These hydrophobic residues could prevent water molecules from hydrolizing the Schiff base in the dark-adapted state of rhodopsin.

In this organisation, four conserved proline residues in helices 4, 5, 6 and 7 are found surrounding the retinal binding site at a similar depth within the membrane. They may introduce helix breakers capable of promoting movements in the helices when rhodopsin is activated. Upon absorbance of light, aromatic residues close to retinal may alter their orientation and affect adjacent helices at the level of proline residues. These helical motions could propagate to the surface and modify the accessibility of the cytoplasmic loops, facilitating receptor/G-protein interactions.

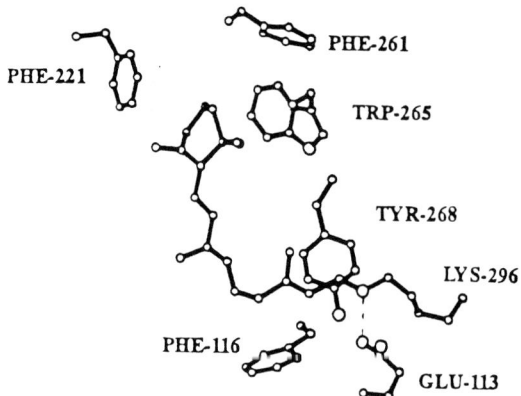

Fig. 2. A view of the retinal environment. Glu-113 (TMH 3) acts as the counterion for the protonated Schiff base. Phe-116 (TMH 3) and Tyr-268 (TMH 3) would sandwich the retinal. Phe-221 (TMH 5), Phe-261 (TMH 6) and Trp-265 (TMH 6) surround the retinal ring.

REFERENCES

Dahl, S., Edvardsen, O. & Sylte, I. (1991): Molecular dynamics of dopamine at the D receptor. *Proc. Natl. Acad. Sci. USA*, 88, 8111-8115.

Findley, J.B.C. & Pappin, D.J.C. (1986): The opsin family of proteins. *Biochem. J.* 238, 625-642.

Franke, R., König, B., Sakmar, T., Khorana, H.G. & Hoffmann, K. (1990): Rhodopsin mutants that fail to activate transducin. *Science*, 250, 123-125.

Henderson, R., Baldwin, J.M., Ceska, T.A., Zemlin, F., Beckmann, E. & Downing, K.H. (1990): Model for the structure of bacteriorhodopsin based on high-resolution electron cryo-microscopy. *J. Mol. Biol.* 213, 899-929.

Hibert, M.F., Trumpp-Kallmeyer, S., Bruinvels, A. & Hoflack, J. (1991): Three-dimensional models neurotransmitter G binding proteins-coupled receptors. *Mol. Pharmacol.* 40, 8-15.

Karnik, S.S. & Khorana, H.G. (1990): Assembly of functional rhodopsin requires a desulfide bond between cysteine residues 110 and 187. *J. Biol. Chem.* 265, 17520-17524.

Kobilka, B., Kobilka, T., Daniel, K., Regan, J., Caron, M. & Lefkovitz, R.: (1988): Chimeric α -, ß - adrenergic receptors: delineation of domains involved in effector coupling and ligand binding specificity. *Science*, 24, 1310-1316.

Nakayama, T.A. & Khorana, H.G. (1990): Orientation of retinal in bovine rhodopsin determined by cross-linking using a photoactivable analog of 11-cis retinal. *J. Biol. Chem.* 265, 15762-15769.

Pardo, L., Ballesteros, J.A., Osman, R. & Weinstein, H. (1992): On the use of the transmembranal domain of bacteriorhodopsin as a template for modeling the 3-D structure of G-protein coupled receptors. *Proc. Natl. Acad. Sci. USA*, 89, 4009-4012.

Sakmar, T.P., Franke, R.R. & Khorana, H.G. (1989): Glutamic acid-113 serves as the retinylidene Schiff base counterion in bovine rhodopsin. *Proc. Natl. Acad. Sci. USA*, 86, 8309-8313.

Strader, C., Sigal, I., Candelore, M., Rands, E., Hill, W. & Dixon, R.A. (1988): Conserved Asp-79 and Asp-113 of the ß-adrenergic receptor have different roles in receptor function. *J. Biol. Chem.* 263, 10267-10271.

II. Molecular analysis of photoreceptors

II. Analyse moléculaire des photorécepteurs

Structure and function of vertebrate rhodopsin

Paul A. Hargrave

Department of Ophthalmology and Department of Biochemistry and Molecular Biology, University of Florida, Gainesville, Florida 32610, USA

The outer segment of the vertebrate rod cell is composed of stacks of disk-shaped membranes, whose major constituent is the photoreceptor protein, rhodopsin (Fig. 1a). Rhodopsin consists of a $M_r \sim 40$ kDa protein, opsin, in combination with a light-sensitive ligand, 11-*cis* retinal. Upon reception of light, rhodopsin activates the G-protein, transducin, which in turn activates cGMP-phosphodiesterase. The resulting hydrolysis of cGMP leads to closing of cation channels in the plasma membrane and hyperpolarization of the rod cell membrane.

THE STRUCTURE OF RHODOPSIN

Amino acid sequences have been determined for more than two dozen opsins, from sources as diverse as *Drosophila* and man. Recent additions include sequences of *Limulus*, pig and goldfish, and chicken cone pigments (Wang et al., 1992; Okana et al., 1992). All opsins share basic structural features. Their polypeptide chain consists of stretches of hydrophilic amino acids punctuated by 7 stretches of predominantly hydrophobic amino acids, 21-28 amino acids in length. Most opsins contain one or two of the sequences Asn-Xxx-Thr/Ser to which mannose- and glucosamine-containing oligosaccharides are attached. All opsins contain a lysine near the middle of their 7th transmembrane helix that serves as a site of attachment of retinal, in Schiff-base linkage. Also, most opsins, as well as most other members of the family of G-protein-linked receptors contain one or two cysteines in their carboxyl-terminal region following its emergence from the 7th helix (reviewed in Hargrave, 1991). In bovine rhodopsin these cysteines have been shown to be covalently attached to palmitate, and are believed to serve as a site of membrane anchoring of the polypeptide chain (Ovchinnikov et al., 1988). Cysteines corresponding to residues 110 and 187 in the bovine sequence are invariant in rhodopsins and related receptors. They form a disulfide bridge important for the structural stability of the protein (Karnik and Khorana, 1990).

Topographic studies support a model in which rhodopsin's amino terminus faces the inside surface of the disk membrane, and the polypeptide chain traverses the lipid bilayer 7 times, exposing the carboxyl-terminus to the rod cell cytoplasm (see references in Hargrave and McDowell, 1992; Fig. 1b). The carboxyl-terminal contains a sequence rich in serine and threonine residues that serve as sites of phosphorylation by rhodopsin kinase. The 3rd cytoplasmic loop appears to contain site(s) phosphorylated by protein kinase C (Newton and Williams, 1991). Three cytoplasmic surface loops comprise a site for binding and activation of the G-protein, transducin. Portions of the cytoplasmic surface of phosphorylated photoactivated rhodopsin also participate in binding arrestin. Physical studies suggest that rhodopsin has an ovoid shape and that the seven helices comprise a bundle that forms a binding site for the retinal (Fig. 1c).

THE RETINAL POCKET

The counterion for the protonated retinyl Schiff base has been identified as Glu113 (Zhukovsky and Oprian, 1989; Sakmar et al., 1989; Nathans, 1990). When uncharged glutamine replaces Glu133, the rhodopsin spectrum shifts from 498 to 380 nm and can be modulated by the pH and ionic environment. Amino acids in the retinal pocket have been identified by reaction with a photoactivated retinal analog (Nakayama and Khorana, 1990). Trp265 in helix VI has been modified along with several amino acids in helix III (Nakayama and Khorana, 1991). Replacement of Trp265 and Tyr268 by site specific mutagenesis causes a large shift in the absorbance spectrum of retinal, implicating these positions in the retinal binding pocket.

Comparison of amino acid sequences of the homologous red and green cone pigments of primates shows changes in the wavelength of maximum absorption that correlate with specific amino acid substitutions. Substitution of a retinal pocket Phe with Tyr, or Ala with Thr or Ser, causes visual pigment spectra to be red-shifted. The amino acid positions in rhodopsin that are comparable are positions 164 (in helix IV) and 261 and 269 (in helix VI). These mutations have recently been shown to be present naturally in the human population, meaning that not everyone sees the color red exactly the same (Winderickx et al., 1992).

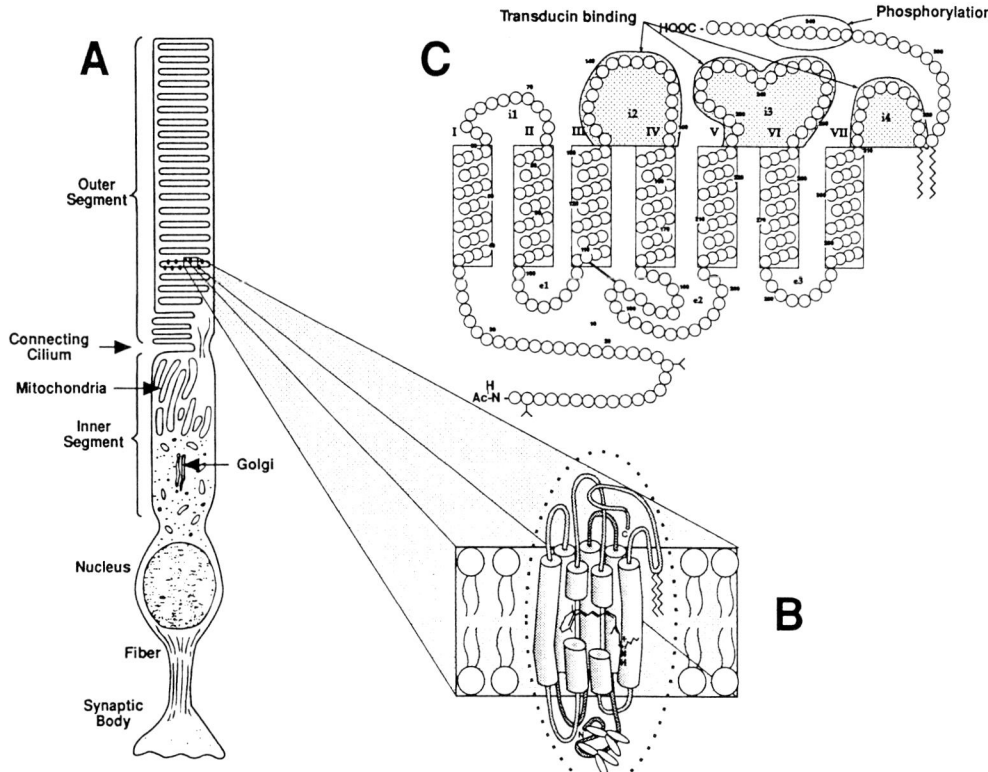

Fig. 1.   (A) Diagram of a vertebrate rod cell; (B) a helix-bundle model for rhodopsin; (C) a topographic model for rhodopsin in the disk membrane.

MUTATIONS IN RHODOPSIN CAN CAUSE EYE DISEASE

More than 40 different mutations in rhodopsin have been described that are responsible for the blinding human disease, retinitis pigmentosa (RP) (Dryja et al., 1991; Sung et al., 1991). For RP, the majority of the mutations are single base changes that have led to a single amino acid substitution (Fig. 2). These amino acid substitutions seem to affect different functions of rhodopsin, but in all cases lead to production of an abnormal protein that eventually leads to death of the cell (reviewed in Hargrave and O'Brien, 1991). Several mutant rhodopsins have been produced in cell culture and their properties studied in order to elucidate structure/function relationships (Doi et al., 1990; Sung et al., 1991). A mutation in rhodopsin has been found in a patient with congenital night blindness; a disease in which rod vision is completely absent (Sieving et al., 1992). Deficiencies in color vision are due to mutations in genes for the cone visual pigments (Nathans et al., 1989). Some of these mutations involve deletions or gene fusions.

In RP, two mutations, Thr4Lys and Thr17Met, both destroy a consensus sequence for glycosylation and lead to a protein that must fail to glycosylate and fold normally. There are many mutations that occur within the transmembrane regions of the protein. Each of these would be suspected to disrupt helical structure or helical packing due to the different properties of the new amino acid side chain. Burying of a charged amino acid in a predominantly hydrophobic environment, e.g., Thr58Arg, would undoubtedly be disruptive, as could be the substitution of amino acids of different size and properties, such as His211Pro and Phe220Cys. Substitutions of amino acids in surface-exposed loops e1 and e2 presumably interfere with formation of rhodopsin's structurally essential disulfide bridge. Future studies must deal with the question, "Why do these mutations in rhodopsin lead to rod cell death and retinal degeneration"?

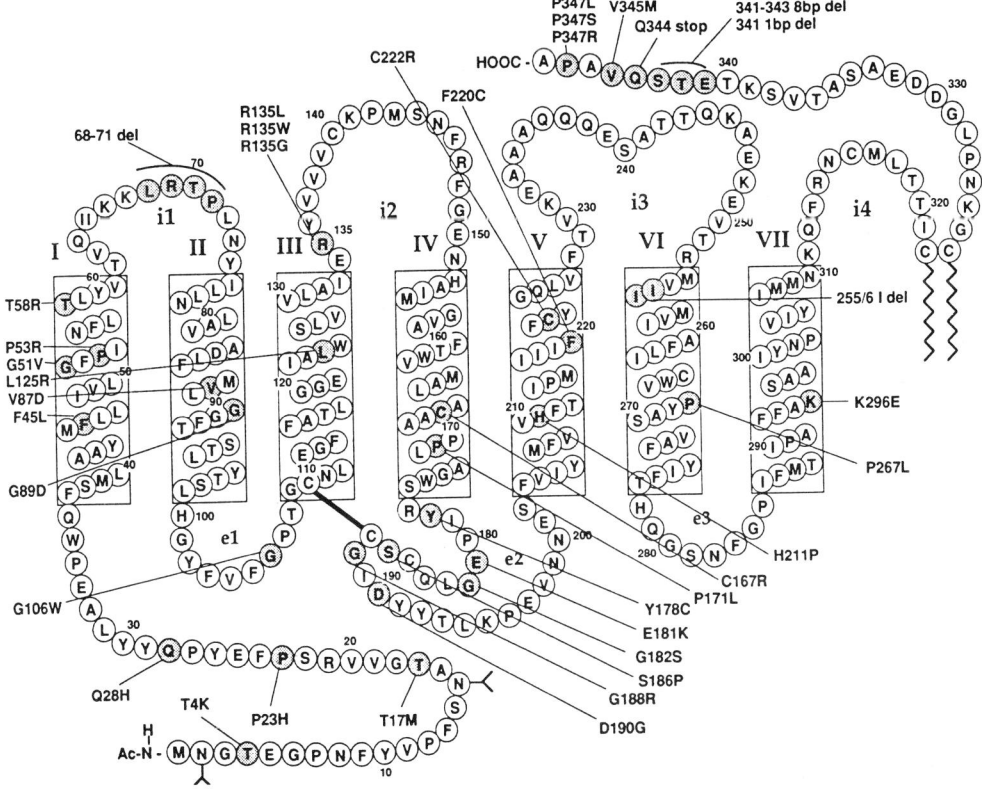

Fig. 2.   Mutations in human rhodopsin that cause retinitis pigmentosa.

REFERENCES

Doi, T., Molday, R.S. and Khorana, H.G. (1990): Role of the intradiscal domain in rhodopsin assembly and function. Proc. Natl. Acad. Sci. USA 87: 4991-4995.

Dryja, T.P., Hahn, L.B., Cowley, G.S., McGee, T.L. and Berson, E.L. (1991): Mutation spectrum of the rhodopsin gene among patients with autosomal dominant retinitis pigmentosa. Proc. Natl. Acad. Sci. USA 88: 9370-9374.

Hargrave, P.A. (1991): Seven-helix receptors. Current Opin. Struct. Biol. 1: 575-581.

Hargrave, P.A. and McDowell, J.H. (1992): Rhodopsin and phototransduction - A model system for G-protein-linked receptors. FASEB J. 6: 2323-2331.

Hargrave, P.A. and O'Brien, P.J. (1991): Speculations on the molecular basis of retinal degeneration in retinitis pigmentosa. In Retinal Degenerations, eds R.E. Anderson, J.G. Hollyfield and M.M. LaVail, Boca Raton, Florida, CRC Press.

Karnik, S.S. and Khorana, H.G. (1990): Assembly of functional rhodopsin requires a disulfide bond between cysteine residues 110 and 187. J. Biol. Chem. 265: 17520-17524.

Nakayama, T.A. and Khorana, H.G. (1990): Orientation of retinal in bovine rhodopsin determined by cross-linking using a photoactivatable analog of 11-cis-retinal. J. Biol. Chem. 265: 15762-15769.

Nakayama, T.A. and Khorana, H.G. (1991): Mapping of the amino acids in membrane-embedded helices that interact with the retinal chromophore in bovine rhodopsin. J. Biol. Chem. 266: 4269-4275.

Nathans, J. (1990): Determinants of visual pigment absorbance: Role of charged amino acids in the putative transmembrane segment. Biochemistry 29: 937-942.

Nathans, J., Davenport, C.M., Maumenee, I.H., Lewis, R.A., Hejtmancik, J.F., Litt, M., Lovrien, E., Weleber, R., Bachynski, B., Zwas, F., Klingaman R. and Fishman, G. (1989): Molecular genetics of human blue cone monochromacy. Science 245: 831-838.

Newton, A.C. and Williams, D.S. (1991): Involvement of protein kinase C in the phosphorylation of rhodopsin. J. Biol. Chem. 266: 17725-17728.

Okana, T., Kojima, D., Fukada, Y. Shichida, Y. and Yoshizawa, T. (1992) Proc. Natl. Acad. Sci. USA (In press).

Ovchinnikov, Y.A., Abdulaev, N.G. and Bogachuk, A.S. (1988): Two adjacent cysteine residues in the C-terminal cytoplasmic fragment of bovine rhodopsin are palmitylated. FEBS Letts. 230: 1-5.

Sakmar, T.P., Franke, R.R. and Khorana, H.G. (1989): Glutamic acid-113 serves as the retinylidine schiff base counterion in bovine rhodopsin. Proc. Natl. Acad. Sci. USA 86: 8309-8313.

Sieving, P.A., Richards, J.E., Bingham, E.L., Naarendorp, F. (1992): Dominant congenital complete nyctalopia & Gly90Asp rhodopsin mutation. Invest. Ophthalmol. & Vis. Sci. 33: 1397.

Sung, C.-H., Davenport, C.M., Hennessey, J.C., Maumenee, I.H., Jacobson, S.G., Heckenlively, J.R., Nowakowski, R., Fishman, G., Gouras, P. and Nathans, J. (1991): Rhodopsin mutations in autosomal dominant retinitis pigmentosa. Proc. Natl. Acad. Sci. USA 88: 6481-6485.

Sung, C.-H., Schneider, B.G., Agarwal, N., Papermaster, D.S. and Nathans, J. (1991): Functional heterogeneity of mutant rhodopsins responsible for autosomal dominant retinitis pigmentosa. Proc. Natl. Acad. Sci. USA 88: 8840-8844.

Wang, S.Z., Adler, R. and Nathans, J. (1992): A visual pigment from chicken that resembles rhodopsin - Amino acid sequence, gene structure, and functional expression. Biochemistry 31: 3309-3315.

Winderickx, J., Lindsey, D.T., Sanocki, E., Teller, D.Y., Motulsky, A.G. and Deeb, S.S. (1992): Polymorphism in red photopigment underlies variation in colour matching. Nature 356: 431-433.

Zhukovsky, E.A. and Oprian, D.D. (1989): Effect of carboxylic acid side chains on the absorption maximum of visual pigments. Science 245: 928-930.

Structure and function of squid rhodopsin

Catherine Venien-Bryan[1], Anthony Davies*, J. Richard Wilkinson*,
Keith Langmack[2], Jenny Baverstock[3], Catherine Nobes[4] and Helen R. Saibil*

* Department of Crystallography, Birkbeck College London, Malet St, London WC1E 7HX, UK.
Present addresses : [1] Department of Biochemistry, Oxford University. [2] Department of Medical Physics, Addenbrookes Hospital, Cambridge. [3] Sandoz Institute for Medical Research, Gower Place, London. [4] Department of Human Anatomy, Oxfort University, UK

Structure of squid photoreceptor microvilli

The photoreceptor membranes of the squid retina are hexagonally packed, 60 nm cylinders of membrane, or microvilli, with a thin cytoskeletal core (Saibil & Hewat, 1987; Saibil, 1990a). The membranes are highly ordered, with regular intermembrane contacts and membrane-cytoskeleton linkages (Fig. 1). The rhodopsin molecules, which form the bulk of the membrane protein, are constrained in orientation to produce a high linear dichroism, necessary to provide the high polarization sensitivity of the cephalopod (squid, octopus) retina (Saidel et al, 1983). The cytoskeletal filament consists of actin crosslinked to the membranes by a myosin I- like protein (*ninaC*; Montell & Rubin, 1988).

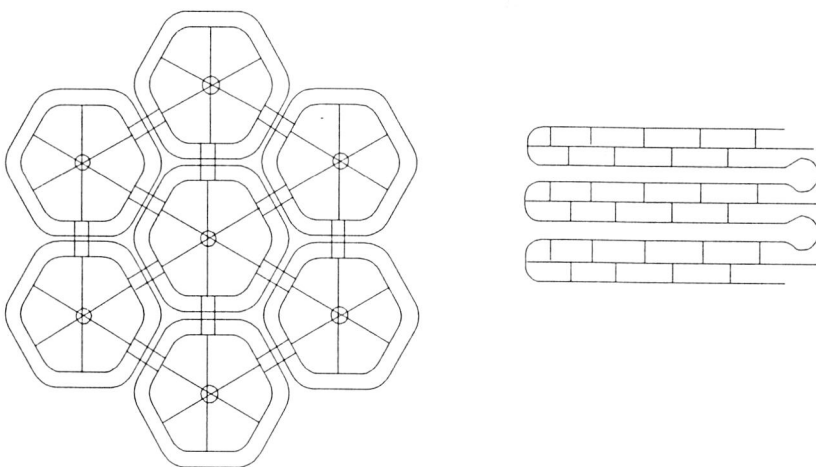

Fig. 1. Diagram of cross and longitudinal sections of squid photoreceptor microvilli, showing the cytoskeleton and membrane linkages.

Negative stain electron microscopy of isolated photoreceptor membranes

The microvillar membranes are readily isolated from squid retinas which have been rapidly frozen after dissection, and thawed in calcium-free buffer (Saibil, 1990b). Large quantities of pure photoreceptor membrane obtained by this method form sheets densely covered by small projecting structures (Fig. 2). These projections are also present in purified, reconstituted rhodopsin membranes. However, they are removed by treatment with *Staphyloccus aureus* endoproteinase glu-C, which removes the rhodopsin C-terminus. Cephalopod rhodopsins are unique in possessing a proline-rich C-terminal repeat (Ovchinnikov et al, 1988; Hall et al, 1991; see paper by Carne et al, this volume) with the consensus repeat sequence YPPQG. The images show that this repeat forms a structure extending from the membrane surface. Edge views of the projections in native membranes suggest that they extend about 100 Å from the membrane surface. The projections appear too large in diameter to represent single rhodopsin C-termini, and it seems most likely that the visible structures are formed of clusters of rhodopsin molecules, possibly with other proteins bound in the native membranes. This feature, and the low in plane mobility of squid rhodopsin shown by ESR studies (Ryba et al, in preparation), distinguish squid from vertebrate rhodopsins.

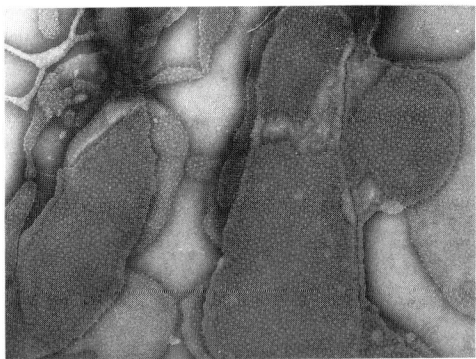

Fig 2. Negative stain electron micrograph of squid photoreceptor membranes. Magnification, x50,000.

Squid rhodopsin activation of the Gq - phospholipase C cascade

The major GTP-binding protein of squid photoreceptors has been identified by protein and cDNA sequencing (Pottinger et al, 1991) and by guanine nucleotide labelling and activity assays (Nobes et al, 1992). Previously identified G-proteins, revealed by light-dependent, cholera and pertussis toxin catalyzed labelling have recently been shown to be minor components. The major G-protein in the squid system is clearly in the Gq class, which is specific for activation of phospholipase C (Lee et al, 1990; Smrcka et al, 1991). It is not a substrate for modification catalyzed by bacterial toxins. The number of guanine nucleotide binding sites shows that the Gq:rhodopsin ratio is about 1:12 (Nobes et al, 1992), very similar to the transducin:bovine rhodopsin ratio. This is in accordance with the light-activated release of inositol trisphosphate in squid and other invertebrate photoreceptors. The C-terminal projection on squid rhodopsin does not appear to affect G-protein activation.

Conclusions

Squid photoreceptors consist of a highly ordered membrane-cytoskeleton network with many structural constraints not found in vertebrate photoreceptors. This may be related to the maintenance of microvillar shape and membrane turnover, and is also likely to be relevant to polarization sensitivity. Despite the immobility of squid rhodopsin, activation of Gq appears to proceed by a similar mechanism to that in the vertebrate rhodopsin-transducin system, a cyclic nucleotide cascade activated by a freely diffusing rhodopsin (Chabre & Deterre, 1989). However, the overall gain in invertebrate phototransduction is lower than that in rods, which may indicate a smaller gain at the rhodopsin-G-protein stage of the cascade.

Acknowledgements

We are grateful to the Marine Biological Association, Plymouth, for the supply of live squid, and the Wellcome Trust and the Science and Engineering Research Council for support.

References

Chabre, M. and Deterre, P. (1989) Molecular mechanism of visual transduction, Eur. J. Biochem. 179, 255-266.

Hall, M.D., Hoon, M.A., Ryba, N.J.P., Pottinger, J.D.D., Keen, J.N., Saibil, H.R. & Findlay, J.B.C. (1991) Molecular cloning and primary structure of squid *(Loligo forbesi)* rhodopsin, a phospholipase C-directed G-protein-linked receptor, Biochem. J. 274, 35-40.

Lee, Y.-J., Dobbs, M.B., Verardi, M.L. & Hyde, D.R. (1990) Neuron 5, 889-898.

Montell, C. & Rubin, G. (1988) The Drosophila *ninaC* locus encodes two photoreceptor cell specific protein with domains homologous to protein kinases and the myosin heavy chain head, Cell 52, 757-772.

Nobes, C, Baverstock, J. & Saibil, H.R. (1992) Activation of the GTP-binding protein Gq by rhodopsin in squid photoreceptors, Biochem. J. 287, in press.

Ovchinnikov, Yu. A., Abdulaev, N.G., Zolotarev, A.S., Artamonov, I.D., Bespalov, I.A., Dergachev, A.E. and Tsuda, M. (1988) Octopus rhodopsin. Amino acid sequence deduced from cDNA. FEBS Lett. 232, 69-72.

Pottinger, J.D.D., Ryba, N.J.P., Keen, J.N. & Findlay, J.B.C. (1991) The identification and purification of the heterotrimeric GTP-binding protein from squid *(Loligo forbesi)* photoreceptors, Biochem. J. 279, 323-326.

Saibil, H.R. (1990a) Cell and molecular biology of photoreceptors, Seminars in the Neurosciences 2, 15-23, Eds. Ashmore & Saibil, W.B.Saunders.

Saibil, H.R. (1990b) Structure and function of the squid eye, in Squid as Experimental Animals, Eds. W. Adelman, D. Gilbert & J. Arnold, Plenum Publishing Corp, pp. 371-397.

Saibil H.R. and Hewat E.A. (1987) Ordered transmembrane and extracellular structure in squid photoreceptor microvilli, J. Cell Biol. 105, 19-28.

Saidel, W.M., Lettvin, J.Y. and MacNicol, E.F., Jr. (1983) Processing of polarized light by squid photoreceptors. Nature 304, 534-536.

Smrcka, A.V., Hepler, J.R., Brown, K.O. & Sternweis, P.C. (1991) Regulation of polyphosphoinositide-specific phospholipase C activity by purified Gq, Science 251, 804-807.

Identification and molecular analysis of vertebrate visual and non-visual photoreceptor proteins

Willem J. DeGrip, Lieveke L.J. DeCaluwé, Russel G. Foster[1], Jacques J.M. Janssen, Horst-W. Korf[2], Olaf Bousché[3] and Kenneth J. Rothschild[3]

Department of Biochemistry, University of Nijmegen, P.O. Box 9101, 6500 HB Nijmegen, The Netherlands, [1] Department of Biology, University of Virginia, Charlottesville, VA 22901, USA, [2] Department of Neurobiology, J.W. Goethe University, Frankfurt D-6000, Germany and [3] Biophysics Program, Boston University, Boston, MA 02118, USA

INTRODUCTION

The photon energy is used by the vertebrate organism to signal various physiological processes. The most important of these are vision, entrainment of the circadian rhytm and seasonal timing of the reproductive system. In lower vertebrates the photoreceptor systems mediating these processes are located in different areas of the brain (Groos, 1982). Retinal photoreceptors mediate vision, while pineal photoreceptors are involved in circadian entrainment. The location of the "reproductive" photoreceptors is still uncertain. The present available evidence indicates, that in mammals these photoreceptor systems have been concentrated in the eye.

A major issue in this field is to establish whether an organism utilizes the same set of photoreceptor pigments in the different systems and whether in the mammalian eye the various processes are mediated by a single or by different photoreceptor systems. Using immunochemical approaches and retinaldehyde analysis we are attempting to identify and localize the various photoreceptor proteins. Furthermore, using molecular-biological approaches (sequence analysis, in vitro expression) we will analyze the relationship between the photoreceptor pigments of the various systems and study their structural and functional properties. In the next sections some recent progress will be presented.

THE LIZARD *ANOLIS CAROLINENSIS*: A TYPICAL LOWER VERTEBRATE?

Like in many other lower vertebrates (e.g. Groos, 1982), morphological evidence for the presence of photore-

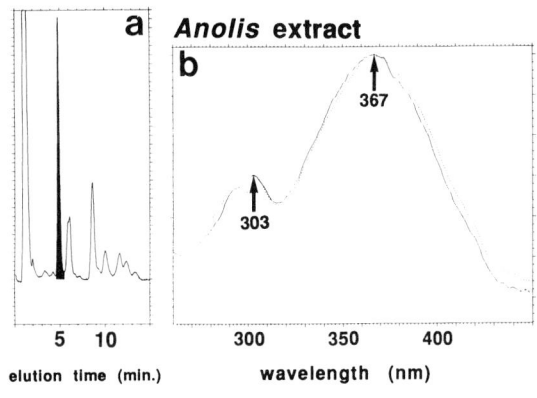

Fig. 1 Retinal analysis of ocular extract of *Anolis carolensis*. HPLC-trace in panel a and UV/Vis spectrum of main peak (filled in a) in panel b. The spectrum in b exactly corresponds to that of 11-cis retinal A_2.

ceptor cells in extra-ocular tissue (pineal organ, parietal eye) has been presented for the tree lizard *Anolis carolinensis* as well. We have been able to substantiate this finding by demonstrating immunoreactivity, with both anti-rod and anti-cone visual pigment antibodies, in eye and pineal organ of *Anolis*. This reactivity is directed against a 40 kD protein band. In addition, we have identified a small nucleus of CSF-contacting neurons in the hypothalamus which are immunopositive for anti-cone pigment antibodies, again directed against a 40 kD protein band. These observations are further substantiated by retinal analysis, showing that extracts of *Anolis* eye, pineal organ and brain sections, containing the hypothalamus, all contain sufficient amounts of retinal chromophore to allow for the presence of a photoreceptor protein of the visual pigment class. Remarkably, in all cases the chromophore is of the retinal A_2 type (Fig. 1). We consider this the first solid evidence for the location of the elusive "deep brain photoreceptor" in lower vertebrates, which probably mediates the photoperiodic reproductive timing.

THE MAMMALIAN CIRCADIAN PHOTORECEPTOR

Abundant evidence exists for the presence of photoreceptor cell proteins and corresponding photosensitivity in the pineal gland of lower vertebrates (e.g. Groos, 1982; Foster et al., 1989a; Sun et al., 1991 Araki et al., 1992; Lolley et al., 1992;). Recently, it has been demonstrated that visual pigment-like proteins (as well as other typical photoreceptor cell proteins like arrestin, rhodopsin kinase, IRBP and phosducin) are still expressed in the adult mammalian pineal gland, but no trace could be found of a corresponding retinal chromophore required to generate a photosensitive pigment (Korf et al., 1987; Foster et al., 1989b; Korf et al., 1991). We have extended these analyses to early postnatal developmental stages in rat and hamster, where the retina is still differentiating. Already at post-natal day 1, a substantial level of an opsin-like protein can be measured in the pineal gland which upon further development plateaus around day 20, but at no time point any retinal was detectable (Fig. 2). We have to conclude that in mammals, the pineal gland has indeed fully abandoned it original photosensitive function. The reason for continuing expression of a visual-pigment like apoprotein is presently fully unclear.

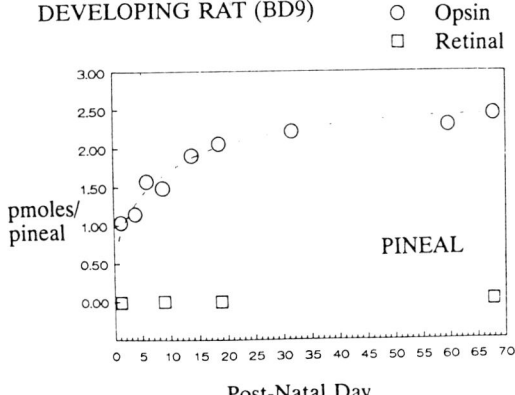

Fig. 2 Opsin and retinal levels in the pineal gland of the developing rat. This species reaches maturity in about 65 days. Opsin was measured by ELISA (Schalken & DeGrip, 1986), retinal by HPLC (Foster et al., 1989b).

Since the present evidence indicates, that during evolution of the mammal the circadian photoreceptor system has moved from the pineal gland to the eye, a major question is whether this function has been taken over by (a subclass of) the visual photoreceptor system or whether these processes are still mediated by two different sets of photoreceptors. Evidence for the latter option comes from recent studies on mice with retinal degeneration. At advanced stages of degeneration, where the visual photoreceptor cell population has been nearly entirely destroyed and vision fully abandoned, the photosensitivity to produce phase-shifts in the circadian rhythm is not affected at all (Foster et al., 1991)

IN VITRO EXPRESSION FOR MOLECULAR STUDIES

Due to the minute amounts available, structural and functional properties of non-visual photoreceptor proteins, and hence any relationship with their visual family, are only determined with great difficulty. Even the measurement of the spectral properties by microspectrophotometry is hardly feasible. This calls for an appropriate expression system to produce sufficient amounts of these proteins. We have developed an expression system for visual pigments, based on recombinant baculovirus (Janssen et al., 1988). This system is able to produce intact and correctly processed bovine opsin, which combines with 11-cis retinal into a photosensitive pigment. After purification and reconstitution into a suitable lipid environment, this pigment behaves identical to native bovine rhodopsin, according to analysis by UV/Vis and FTIR-spectroscopy (Janssen et al., 1991; DeCaluwé et al., 1992).

REFERENCES

Araki, M., Fukada, Y., Shichida, Y., Yoshizawa, T., & Tokunaga, F. (1992): Differentiation of Both Rod and Cone Types of Photoreceptors in the Invivo and Invitro Developing Pineal Glands of the Quail. *Dev. Brain Res.* 65, 85-92.

De Caluwé, G.L.J., Van Oostrum, J., Janssen J.J.M. & De Grip, W.J. (1992): In vitro synthesis of bovine rhodopsin using recombinant baculovirus. Meth. Neurosci. in press.

Foster, R.G., Provencio, I., Hudson, D., Fiske, S., DeGrip, W.J., & Menaker, M. (1991): Circadian photoreception in the retinally degenerate mouse (rd/rd). *J. Comp. Physiol. A.* 169, 39-50.

Foster, R.G., Schalken, J.J., Timmers, A.M.M., & DeGrip, W.J. (1989): A comparison of some photoreceptor characteristics in the pineal and retina: I. The Japanese quail (*Coturnix coturnix*). *J. Comp. Physiol. A.* 165, 553-565.

Foster, R.G., Timmers, A.M.M., Schalken, J.J., & DeGrip, W.J. (1989): A comparison of some photoreceptor characteristics in the pineal and retina: II. The Djungarian hamster (*Phodopus sungorus*). *J. Comp. Physiol. A.* 165, 565-572.

Groos, G. (1982): Topical review: The comparative physiology of extraocular photoreception. *Experientia* 38, 989-1128.

Janssen, J.J.M., DeCaluwé, G.L.J., & DeGrip, W.J. (1990): Asp-183, Glu-113 and Glu-134 are not specifically involved in Schiff base protonation or wavelength regulation in bovine rhodopsin. *FEBS Lett.* 260, 113-118.

Janssen, J.J.M., Mulder, W.R., DeCaluwé, G.L.J., Vlak, J.M., & DeGrip, W.J. (1991): Invitro Expression of Bovine Opsin Using Recombinant Baculovirus - The Role of Glutamic Acid (134) in Opsin Biosynthesis and Glycosylation. *Biochim. Biophys. Acta* 1089, 68-76.

Janssen, J.J.M., VanDeVen, W.J.M., VanGroningen-Luyben, W.A.H.M., Roosien, J., Vlak, J.M., & DeGrip, W.J. (1988): Synthesis of functional bovine opsin in insect cells under control of the baculovirus polyhedrin promotor. *Mol. Biol. Rep.* 13, 65-71.

Korf, H-W., Foster, R.G., Ekström, P. & Schalken, J.J. (1985): Opsin-like immunoreaction in the retinae and pineal organs of four mammalian species. Cel Tissue Res. 242, 645-648.

Korf, H-W., Kramm, C.M., & DeGrip, W.J. (1991): Further analysis of photoreceptor-specific proteins in the rodent pineal organ and retina. *Adv. Pineal Research*, 5, 115-122.

Korf, H-W. & Wicht, H. (1991): The Vertebrate Pineal Organ - Model for Receptor and Effector Mechanisms in Neuronal Systems. *Naturwissenschaften.* 78, 437-444.

Lolley, R.N., Craft, C.M., & Lee, R.H. (1992): Photoreceptors of the Retina and Pinealocytes of the Pineal Gland Share Common Components of Signal Transduction. *Neurochem. Res.* 17, 81-89.

Schalken, J.J. & DeGrip, W.J. (1986): Enzyme-linked immunosorbent assay for quantitative determination of the visual pigment rhodopsin in total eye-extracts. *Exp. Eye Res.* 43, 431-439.

Schalken, J.J., Janssen, J.J.M., Sanyal, S., Hawkins, R.K., & DeGrip, W.J. (1990): Development and degeneration in *rds* mutant mice: Immunoassay of the rod visual pigment rhodopsin. *Biochim. Biophys. Acta* 1033, 103-109.

Sun, J.H., Reiter, R.J., Mata, N.L., & Tsin, A.T.C. (1991): Identification of 11-cis-Retinal and Demonstration of Its Light-Induced Isomerization in the Chicken Pineal Gland. *Neurosci. Lett.* 133, 97-99.

Structures and Functions of Retinal Proteins. Ed. J.L. Rigaud. Colloque INSERM / John Libbey Eurotext Ltd.
© 1992, Vol. 221, pp. 63-66

Studies towards the molecular mechanism of rhodopsin : protein engineering and molecular modeling

L.L.J. DeCaluwé, D.M.F. VanAalten, J. VanOostrum and W.J. DeGrip

Department of Biochemistry, University of Nijmegen, P.O. Box 9101, 6500 HB Nijmegen, The Netherlands

Protein Engineering

Rhodopsin is the visual pigment of the rod photoreceptor cell in the vertebrate retina. It consists of an integral membrane protein, opsin, covalently linked to a chromophore, 11-*cis* retinal, via a protonated Schiff's base. Bovine rhodopsin has an absorbance band in the visible region with a maximum at 498 nm. Absorption of a photon triggers isomerization of the chromophore to all-trans. The resulting conformational changes (photolytic cascade) lead to signal site exposure, G-protein (Transducin) binding, and to subsequent desensitization by a specific rhodopsin kinase and a 45 kD protein (S-antigen or arrestin) (Stryer, 1991). In the photolytic cascade several intermediate steps can be distinguished. Metarhodopsin II is the active intermediate which triggers the signal transduction pathway. It slowly decays under release of the chromophore into metarhodopsin III. In order to study structure-function relationships on a molecular level we use recombinant AcNPV baculovirus and a host cell line of Spodoptera frugiperda (Sf9) for in vitro expression of bovine opsin. The strong promoter of the baculovirus matrix protein polyhedrin, which is produced in very large amounts in the late phase of infection, is used for the expression of heterologous protein (Luckow, 1991). Here we will describe the in vitro production of bovine rhodopsin and the effect of some amino acid substitutions on the biosynthesis and/or spectral properties.

After evaluating several factors in order to achieve maximal production of functional opsin, such as infection conditions and time, various culture media and cell lines, and variations in 5' noncoding region of the polyhedrin promoter and the opsin c-DNA, expression levels of 50 picomoles opsin/10^6 cells (ca 4 μg/mg protein; ca 4 mg opsin/liter culture) are now reproducibly achieved. Immunofluorescent analysis shows that opsin produced in Sf9 cells is targeted to the plasma membrane (Fig. 1A), and immunoblot analysis shows that v-ops migrates with the same apparent molecular weight (38 kD) as fully processed bovine opsin. This suggests that v-ops is glycosylated as well. This could be confirmed by adding the N-glycosylation inhibitor tunicamycin upon infection of Sf9 cells with recombinant virus. This results in the intact product now migrating with a lower apparent molecular weight (31 kD), corresponding to that of native non-glycosylated rhodopsin (Fig. 1B). The in vitro produced opsin (v-ops) can be regenerated with the chromophore 11-*cis* retinal into a photosensitive pigment (v-rho), which is spectrally identical to native rhodopsin. After purification and reconstitution into suitable lipids (egg PC or retina extract) v-rho display the normal pattern of late intermediates upon illumination (De Caluwé et al., 1992): Metarhodopsin II decays at room temperature with a half time of 4 ± 1 min into opsin and Metarhodopsin III. Our results so far lead to the conclusion that the functional properties of wildtype v-rho are identical to that of native rhodopsin.

The opsin sequence contains several highly conserved charged residues in critical positions (membrane domains, membrane surface) which could have important functional roles (translocation, folding, spectral tuning, signal propagation). Substitution of Asp83 slightly and of Glu113 strongly affected the spectral properties of the resulting mutant. These effects have been well documented elsewhere (Nathans, 1990; Janssen et al., 1990;

Sakmar et al., 1992) and will not be further discussed here. Mutations at these positions did not influence biosynthesis and targeting of the protein however. On the other hand, several positions were already found where mutation strongly interfered with the normal biosynthetic process (Arg69, Glu134, Lys248) either resulting in a glycosylated but truncated protein, or in a complete, but non-glycosylated and non-regenerable protein (Table 1). These residues are highly conserved, and meet the requirements for essential contribution to a start/stop transfer signal, alteration of which could lead to incorrect membrane translocation and/or protein folding. This will however require further study. Interestingly enough both the Glu134 and the Arg69 site are implicated in natural rhodopsin mutations correlated with autosomal dominant retinitis pigmentosa (Sung et al., 1991).

Fig.1

A: Immunohistochemical analysis of opsin biosynthesis in Sf9 cells. Sf9 cells were cultered on coverglass, infected with recombinant opsin virus and fixed at 2 dpi.
B: Immunoblot analysis of Sf9 cells infected with recombinant opsin virus in the absence (lane 1) or presence of increasing concentrations of tunicamycin (5 μg/ml (lane 2); 10 μg/ml (lane 3); 25 μg/ml (lane 4)). Opsin was identified by incubation with the polyclonal antiserum CERNJS858. Immunoreactivity was detected by reaction with fluorescein conjugated goat anti-rabbit IgG (A) or with HPO-labeled antibodies (B) (Janssen et al., 1991).

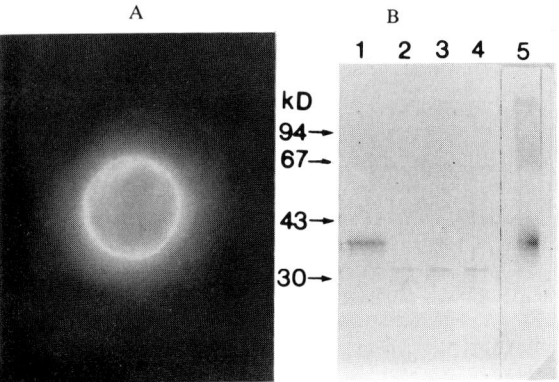

Table 1 Biosynthesis and spectral properties of native bovine rhodopsin and recombinant wildtype and selected mutant species.

species	expression level [1] pmol/10^6 cells	MW of product(s) [2]	$_{max}$ (nm) [3]
native bovine rhodopsin	-	38 kD	498
v-rho	40-50	38 kD	498
D83→N	30-40	38 kD	492
E113→D	15-25	38 kD	505
E134→D	50-60	38 kD ; 31 kD	498
E134→R	25-35	38 kD ; 31 kD	498
R69→H	30-40	23 kD	-
K248→L	60-70	38 kD ; 31 kD	498

[1] measured by inhibition-ELISA (De Caluwé et al., 1992)
[2] Apparent MW on SDS-PAGE gels. 38 kD represents the glycosylated and 31 kD the non-glycosylated complete protein; 23 kD represents a truncated, glycosylated protein.

Molecular Modeling

For the purpose of aiding structural studies as well as site-directed mutagenesis, a more reliable 3D model of bovine rhodopsin is being developed. First a 2D model was created for representing the various structural elements. A simple form of the model is depicted in Fig. 2. Several data were incorporated in this model: output from a program predicting the transmembrane propensities of a sequence (Klein et al., 1985), secondary structure predictions according to Chou & Fasman (1978) and Garnier et al. (1978), several site-directed mutagenesis studies (Janssen et al., 1991; Nakayama & Khorana, 1991), proteolytic experiments (Findlay & Pappin, 1986), antibody studies (Ovchinnikow, 1987) and FTIR data predicted (Pistorius & De Grip, 1992). This model was then used as a template for creating the 3D model. The helices in the 2D model were built using the Biosym modelling package Discover/Insight. Using the (low) homology with bacteriorhodopsin (bR), the helices of rhodopsin were aligned with the helices of bR (stage I). A program was written to calculate the hydrophobic faces ('sidedness') of the rhodopsin helices, using the hydropathy indices of Kyte & Doolittle (1982). This program was tested on the structure of bR (Henderson et al., 1990) and correctly predicted the 'sidedness' of the helices of bR within an experimental error of 10%. The program was applied to rhodopsin and the output was compared with the sidedness in stage I. The two methods (alignment and hydrophobic faces) predicted the same sidedness for helices 1, 6 and 7 (A, F and G), but helices 2, 3, 4 and 5 (B, C, D and E) needed additional rotation. This was done by making a new 3D superposition of these helices on their bR equivalents, in such a way that the hydrophilic face was pointing inward and that the 'ridges into grooves' pattern of bR was conserved. The resulting structure is pictured in Fig. 3. This stage II model was further refined by the use of minimization: in this way bad contacts between side chains were removed and possible stress in the backbone was relieved.

Up to now, the connecting loops, and the N- and C-terminus are not incorporated in the model, since almost no structural information concerning these parts is available. However, FTIR studies predict the C-terminus to have a certain amount of ß-sheet type structure, with a possible ß-turn (Pistorius & De Grip, 1992). This assignment is supported by secondary structure predictions using the methods of Chou & Fasman (1978) and Garnier et al. (1978). In order to evaluate this assignment, the residues 312-343 were modelled in a ß-strand motif, with a ß-turn at Asn326. During an initial minimization, the two ß-strands (312-325 and 330-343) were artificially pulled together using forcing constraints. The resulting structure was then minimized without the constraints, solvated in a 5 Å layer of water and submitted to a 5 ps molecular dynamics run at 900k. Structures were recorded at intervals of 0.5 ps. The structure after 1 ps is shown in Fig. 4. It is clear that some non-expected hydrogen bonds are formed and that the ß-turn has a strange fold, which might also cause the somewhat helical form of the ß-sheet. But the ß-sheet conformation is not unstable, since even after 5 ps dynamics at 900k an amount of ß-sheet hydrogen bonding can be observed. It is clear however, that the conformation of the ß-turn has to be determined more exactly, since this determines the hydrogen bonding further down the ß-sheet.

Fig. 2: Simple representation of the 2D model of bovine rhodopsin.

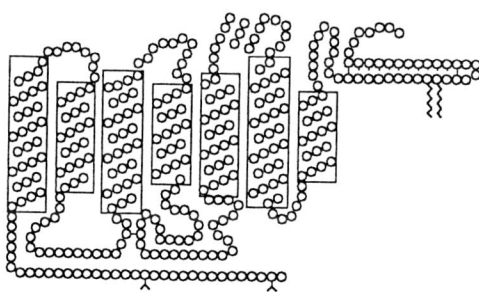

Fig. 3: 3D ribbon structure of the transmembrane helices of bovine rhodopsin.

Fig. 4: CA-carbon trace of residues 312-343 of bovine rhodopsin after several minimizations and 1 ps molecular dynamics at 900 K.

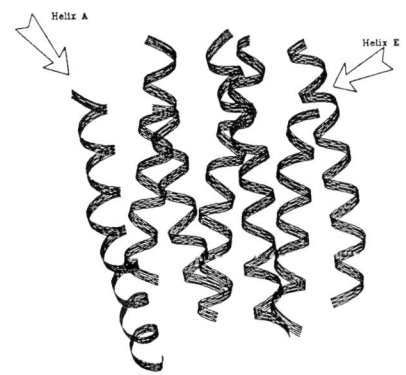
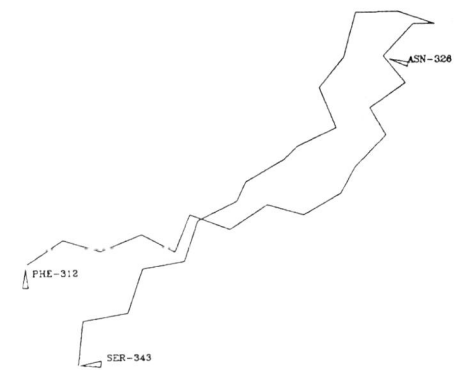

References

De Caluwé, G.L.J., Van Oostrum, J., Janssen, J.J.M. & De Grip, W.J. (1992): In vitro synthesis of bovine rhodopsin using recombinant baculovirus. *Meth. Neurosci.*, in press.

Chou, P.Y. & Fasman, G.D. (1978): Prediction of secondary structure of proteins from their amino acid sequence. *Adv. Enzymol.* 47, 45-148.

Garnier, J., Osguthorpe, D.J. & Robson, B. (1978): Analysis of the Accuracy and Implications of Simple Methods for Predicting the Secondary Structure of Globular Proteins. *J. Mol. Biol.* 120, 97-120.

Findlay, J.B.C. & Pappin, D.J.C (1986): The opsin family of proteins. *Biochem. J.* 238, 625-642.

Henderson, R., Baldwin, J.M., Ceska, T.A., Zemlin, F., Beckmann, E. & Downing, K.H. (1990): Model for the structure of bacteriorhodopsin based on high-resolution electron cryo-microscopy. *J. Mol. Biol.* 213, 899.

Janssen, J.J.M., De Caluwé, G.L.J. & De Grip, W.J. (1990): Asp83, Glu113 and Glu134 are not specifically involved in Schiff base protonation or wavelength regulation in bovine rhodopsin. *FEBS Lett.* 260, 113.

Janssen, J.J.M., Mulder, W.R., De Caluwé, G.L.J., Vlak, J.M. & De Grip, W.J. (1991): In vitro expression of bovine opsin using recombinant baculovirus: the role of glutamic acid (134) in opsin biosynthesis and glycosylation. *Biochim. Biophys. Acta* 1089, 68-76.

Klein, P., Kanehisa, M. & DeLisi, C. (1985): The detection and classification of membrane spanning proteins. *Biochim. Biophys. Acta* 815, 468-476.

Kyte, J. & Doolittle, R.F. (1982): A Simple Method for Displaying the Hydrpathic Character of a Protein. *J. Mol. Biol.* 157, 105-132.

Luckow, V.A. (1991): Cloning and Expression of Heterologous Genes in Insect Cells with Baculovirus Vectors. In *Recombinant DNA Technology and Applications*, ed. A. Prokop, R.K. Bajpai & C. Ho, pp. 97-152. New York: McGraw-Hill, Inc.

Nakayama, T.A. & Khorana, H.G. (1991): Mapping of the Amino Acids in Membrane-embedded Helices That Interact with the Retinal Chromophore in Bovine Rhodopsin. *J. Biol. Chem.* 266, 4269-4275.

Nathans, J. (1990): Determinants of Visual Pigment Absorbance: Identification of the Retinylidene Schiff's Base Counterion in Bovine Rhodopsin. *Biochemistry* 29, 9746.

Ovchinnikov, Y.A. (1987): Probing the folding of membrane proteins. *TIBS* 12, 434-438.

Pistorius, A. & De Grip, W.J. (1992): submitted for publication.

Sakmar, T.P., Franke R.R. & Khorana, H.G. (1989): Glutamic acid-113 serves as the retinylidene Schiff base counterion in bovine rhodopsin. *Proc. Natl. Acad. Sci. USA* 86, 8309.

Stryer, L. (1991): Visual Excitation and Recovery. *J. Biol. Chem.* 266, 10711.

Sung, C.H., Schneider, B.G., Agarwal, N., Papermaster, D.S. & Nathans, J. (1991): Functional Heterogeneity of Mutant Rhodopsins Responsible for Autosomal Dominant Retinitis-Pigmentosa. *Proc. Natl. Acad. Sci. USA* 88, 8840.

Mutagenesis studies of rhodopsin phototransduction

Thomas P. Sakmar, Karim Fahmy, Theresa Chan and Melissa Lee

Howard Hughes Medical Institute, Rockefeller University, NY 10021, USA

We employ the vertebrate visual proteins rhodopsin and transducin as a model system for structure-function studies on the molecular mechanisms of transmembrane signaling. These visual proteins are members of a super-family of related guanine nucleotide-binding regulatory proteins (G proteins) and G protein-coupled receptors. We are particularly interested in the structures of the retinal binding pockets of visual pigments and in identifying the structural domains of both rhodopsin and transducin involved in the G protein activation process.

AMINO ACID SUBSTITUIONS THAT CAUSE BATHOCHROMIC SPECTRAL SHIFTS IN BOVINE RHODOPSIN

Nearly all vertebrate visual pigments share a common chromophore, 11-*cis*-retinal. In humans, the differences in absorption maxima of the cone pigments that underlie human red-green color vision must result from differences in the amino acid sequences of the respective opsin proteins. Fifteen amino acid substitutions distinguish the human green pigment (λ_{max} = 530 nm) from the human red pigment (λ_{max} = 560 nm) (Nathans *et al.*, 1986). Three of these residues were suggested in a genetic analysis of eight primate visual pigments to produce this spectral difference of about 1000 cm^{-1} (Neitz *et al.*, 1991). The amino acid at each of these three positions in the rod pigment rhodopsin (λ_{max} = 500 nm), matches that of the green pigment (Table 1) (Nathans & Hogness, 1983; Nathans *et al.*, 1986). Therefore, it was postulated that the influence of these residues could be tested experimentally by substituting the amino acid residues of the red pigment into rhodopsin. A mutation resulting in a red shift in absorption maximum relative to rhodospin would indicate potential relevance in red-green spectral tuning.

We recently reported seven bovine rhodopsin mutants involving the three amino acid positions: three single substitutions, Ala 164 replaced by Ser (A164S), Phe 261 replaced by Tyr (F261Y), and Ala 269 replaced by Thr (A269T); three double substitutions; and one triple substitution (Chan *et al.*, 1992). Replacement of Ala 164 caused only a slight red shift effect (λ_{max} = 502 nm). However, replacement of Phe 261 or Ala 269 caused red-shifted λ_{max} values of 510 nm and 514 nm, respectively. The double replacement at both positions 261 and 269 simultaneously caused a red shift to 520 nm that was greater than either of the two substitutions alone but not strictly additive. Replacement at both positions 164 and 261 caused a red shift (λ_{max} = 512 nm) that was slightly greater than that of F261Y alone. Replacement at both positions 164 and 269 resulted in a λ_{max} value of 514 nm that was the same as that of the single substitution at position 269. Two of the three positions (261 and 269) in combination appear to account for the 775 cm^{-1} of the observed 1000 cm^{-1} difference between the human green and red pigments (Table 2). The triple mutant did not bind 11-*cis*-retinal to form a pigment. It is not known whether the triple mutant, if it could be induced to bind 11-*cis*-retinal, would display the full

1000 cm^{-1} red shift. However, the effects of all combinations of double replacements were qualitatively additive but not synergistic. For the triple mutant to account for the entire 1000 cm^{-1} shift, a synergistic effect would be required.

Table 1 Comparison of amino acids in various pigments at positions proposed to account for red-green spectral tuning*

bovine rhodopsin	human rhodopsin	human green	human red
Ala 164	Ala 164	Ala 180+	Ser 180
Phe 261	Phe 261	Phe 277	Tyr 277
Ala 269	Ala 269	Ala 285	Thr 285

*The numbering system shown is from previous reports of the deduced amino acid sequences of bovine rhodopsin (Nathans & Hogness, 1983), human rhodopsin (Nathans & Hogness, 1984), and human cone pigments (Nathans et al., 1986).
+a genetic polymorphism was reported at this position that could potentially result in Ser at this position as well (Nathans et al., 1986).

The most likely explanation for the observed red-shifted λ_{max} values is that a newly introduced hydroxyl-bearing amino acid residue can interact directly with the chromophore. However, it is possible that an individual amino acid replacement causes distant effects on the chromophore binding pocket. The effect of a mutation on absorption maximum may result from an indirect effect as well as a direct interaction. However, whereas blue-shifted λ_{max} values indicate a relative loss of chromophore-protein interactions, red-shifted λ_{max} values indicate an enhanced interaction. A mutant with a red-shifted λ_{max} value has a larger opsin shift than that normally observed in rhodopsin. Attributing an effect on absorption maximum to a specific amino acid-chromophore interaction is likely to be more valid in cases where a red shift rather than a blue shift is observed. A large number of rhodopsin mutants have been previously reported that cause blue-shifted absorption maxima (Nakayama & Khorana, 1990; Nathans, 1990a). No significant red-shifted mutants have been reported other than those involving the Schiff base counterion at position Glu 113 (Nathans, 1990b; Sakmar et al., 1989; Sakmar et al., 1991, Zhukovsky & Oprian, 1989).

Although residues in rhodopsin match those in the green pigment at the three positions tested, the rhodopsin and the green pigment are only about 70% homologous (Nathans et al., 1986). Obviously the retinal binding pocket in rhodopsin is significantly different from that of the green pigment as demonstrated by the 1,125 cm^{-1} difference between their spectral peaks. However, at the three positions proposed from primary structure comparisions to account for red-green pigment spectral tuning, rhodopsin and the green pigment share the same residues. In addition, the design of this experiment involves testing a hypothesis by correlating mutagenesis with the appearance of a red-shifted absorption maximum (increase in opsin shift), and not with the loss of an existing retinal-protein interaction as indicated by a blue-shifted absorption maximum (decrease in opsin shift).

Neitz et al. (1991) hypothesized that additive effects of changes at amino acid positions 180, 277, and 285 should account for all shifts in spectra among a set of primate visual pigments (see Table 1 for a comparison of numbering systems in rhodopsin versus cone pigments) (Neitz et al., 1991). They argued that the effects of changes at positions 180 and 285 were shifts of about 5 and 15.5 nm respectively and that the remaining 9- to 10-nm difference was produced by the substitution at position 277. We conclude that two of these residues (Tyr 277 and Thr 285) are primarily involved in spectral tuning that distinguishes red from green pigments, but that the effects of individual differences may not be strictly additive. For example, single substitutions in rhodopsin at position 261 (F261Y) and postion 269 (A269T) result in red shifts of 400 cm^{-1} and 550 cm^{-1}, respectively. However, in combination these two replacements cause a red shift of 775 cm^{-1}. Also, replacement at position 164 (A164S) results in a slight red shift (75 cm^{-1}). This effect was additive in combination with F261Y but not in combination with A269T (Table 2).

Table 2 Rhodopsin mutants designed to mimic naturally occuring substitutions in green and red pigments

Mutation(s)	λ_{max} (nm)*	Shift from rho (cm^{-1})+
A164S	502	75
F261Y	510	400
A269T	514	550
F261Y/A269T	520	775
A164S/F261Y	512	475
A164S/A269T	514	550
A164S/F261Y/A269T	n.d.	-

Site-directed mutagenesis was performed using restriction fragment replacement in a synthetic gene (Oprian et al., 1986). The altered genes were expressed in COS-1 cells and purified by an immunoaffinity procedure (Oprian et al., 1987; Sakmar et al., 1989).
* λ_{max} was determined from the peaks of photobleaching difference spectra. The precision is estimated to be +/- 2 nm. The λ_{max} of rhodopsin purified from COS cells was 500 nm.
+ λ_{max} shifts from that of rhodopsin are expressed in wavenumbers (cm^{-1}) to allow a direct comparison of energy differences. Values are rounded to the nearest 25 cm^{-1}. All shifts were to longer wavelengths (red shifts).
n.d. - the triple mutant did not bind 11-*cis*-retinal to form a pigment.

The red shift attributable to positions 277 and 285 in combination (775 cm^{-1}) is a significant fraction of the observed difference in absorption maxima between the human green and red pigments (1000 cm^{-1}). Other amino acid residues, including that at position 180, are likely to contribute to lesser degrees to account for the remaining 250 cm^{-1}. In rhodopsin, spectral tuning was shown not to be influenced by electrostatic interaction with carboxylates other that the counterion (Nathans, 1990a; Sakmar et al., 1989; Zhukovsky & Oprian, 1989). A neutral chromophore binding pocket model in which dipole and hydrogen bonding interactions predominate has been proposed (Birge et al., 1988; Sakmar et al., 1989; Zhukovsky & Oprian, 1989). Since the mutations described in this report account for more than three quarters of the difference in absorption maxima between green and red pigments, a similar neutral chromophore binding pocket model is likely to apply to the green and red color pigments as well. A complete understanding of spectral tuning in the visual pigments will require detailed spectroscopic studies, including resonance Raman spectroscopy of mutant rhodopsins (Lin et al., 1992) and cone pigments that have recently been expressed (Merbs & Nathans, 1992; Oprian et al., 1991).

FLUORESCENCE STUDIES OF RHODOPSIN-TRANSDUCIN INTERACTIONS

Light-activated rhodopsin catalyzes guanine nucleotide exchange by transducin. We are interested in identifying specific domains of rhodopsin and transducin involved in binding and activation. It has previously been shown by flash photolysis studies of site-directed rhodopsin mutants that loop CD and loop EF of rhodopsin are involved in activation of bound transducin (Franke et al., 1990). Other mutations prevent transducin binding. Recently, we have developed a spectrofluorimetric method designed to allow simultaneous illumination and excitation-emission fluorescence measurements of rhodopsin. Rhodopsin-catalyzed binding of GTP or a GTP analog to transducin results in a large increase in its intrinsic fluorescence. Mixtures of transducin and rhodopsin can be assayed by this method to determine the kinetic rate constants of their interaction and to evaluate the specific effects of mutations. Studies of a series of site-directed mutants of rhodopsin with alterations in their cytoplasmic domains are underway.

REFERENCES

Birge, R.R., Einterz, C.M., Knapp, H.M., and Murray, L.P. (1988): The nature of the primary photochemical events in rhodopsin and isorhodopsin. *Biophys. J.* 53, 367-385.

Chan, T., Lee, M., and Sakmar, T.P. (1992): Introduction of hydroxyl-bearing amino acids causes bathochromic spectral shifts in rhodopsin: Amino acid substitutions responsible for red-green color pigment spectral tuning. *J. Biol. Chem.* 267, 9478-9480.

Ferretti, L., Karnik, S.S., Khorana, H.G., Nassal, M. and Oprian, D.D. (1986): Total synthesis of a gene for bovine rhodopsin. *Proc. Natl. Acad. Sci. U.S.A.* 83, 599-603.

Franke, R.R., König, B., Sakmar, T.P., Khorana, H.G., and Hofmann, K.P. (1990): Rhodopsin mutants that bind but fail to activate transducin. *Science* 250, 123-125.

Karnik, S.S., Sakmar, T.P., Chen, H.-B., and Khorana, H.G. (1988): Cysteine residues 110 and 187 are essential for the formation of correct structure in bovine rhodopsin. *Proc. Natl. Acad. Sci. U.S.A.* 85, 8459-8463.

Lin, S.W., Sakmar, T.P., Franke, R.R., Khorana, H.G., and Mathies, R.A. (1992): Resonance Raman microprobe spectroscopy of rhodopsin mutants: Effect of substitutions in the third transmembrane helix. *Biochemistry* 31, 5105-5111.

Merbs, S.L., and Nathans, J. (1992): Absorption spectra of human cone pigments. *Nature* 356, 433-435.

Nakayama, T.A., and Khorana, H. G. (1990): Mapping of the amino acids in membrane-embedded helices that interact with the retinal chromophore in bovine rhodopsin. *J. Biol.Chem.* 266, 4269-4275.

Nathans, J., Thomas, D., and Hogness, D.S. (1986): Molecular genetics of human color vision: the genes encoding blue, green, and red pigments. *Science* 232, 193-202.

Nathans, J., and Hogness, D. S. (1984): Isolation and nucleotide sequence of the gene encoding human rhodopsin. *Proc. Natl. Acad. Sci. U.S.A.* 81, 4851-4855.

Nathans J., and Hogness, D. S. (1983): Isolation, sequence analysis, and intron-exon arrangement of the gene encoding bovine rhodopsin. *Cell* 34, 807-814.

Nathans, J. (1990): Determinants of visual pigment absorbance: role of charged amino acids in the putative transmembrane segments. *Biochemistry* 29, 937-942.

Nathans, J. (1990): Determinants of visual pigment absorbance: Identification of the retinylidene Schiff's base counterion in bovine rhodopsin. *Biochemistry* 29, 9746-9752.

Neitz, M., Neitz, J., and Jacobs, G.H. (1991): Spectral tuning of pigments underlying red-green color vision. *Science* 252, 971-973.

Oprian, D.D., Molday, R.S., Kaufman, R.J., and Khorana, H.G. (1987): Expression of a synthetic bovine rhodopsin gene in monkey kidney cells. *Proc. Natl. Acad. Sci. U.S.A.* 84, 8874-8878.

Oprian, D.D., Asenjo, A.B., Lee, N., and Pelletier S.L. (1991): Design, chemical synthesis, and expression of genes for the three human color vision pigments. *Biochemistry* 30, 11367-11372.

Sakmar, T.P., Franke, R.R., and Khorana, H.G. (1989): Glutamic acid-113 serves as the retinylidene Schiff base counterion in bovine rhodopsin. *Proc. Natl. Acad. Sci. U.S.A.* 86, 8309-8313.

Sakmar, T.P., Franke, R.R., and Khorana, H. G. (1991): The role of the retinylidene Schiff base counterion in rhodopsin in determining wavelength absorbance and Schiff base pK_a. *Proc. Natl. Acad. Sci. U.S.A.* 88, 3079-3083.

Zhukovsky, E.A., and Oprian, D.D. (1989): Effect of carboxylic acid side chains on the absorption maximum of visual pigments. *Science* 246, 928-930.

Cone visual pigments : structure, evolution and function

Toru Yoshizawa

Department of Applied Physics and Chemistry, The University of Electro-Communications, Chofu, Tokyo 182, Japan

Vertebrate retinas have two kinds of visual cells: One is rod responsible for scotopic vision and the other cone responsible for photopic vision. So far the studies on molecular mechanism of vision have exclusively been carried out on scotopic vision. Now the main path of the visual transduction system from absorption of light by a rhodopsin molecule to generation of the receptor potential has been elucidated. On the other hand, the study on photopic vision which is more important for daily life than scotopic vision has been hampered by difficulty of isolation of cone pigments from retinas.

CONE PIGMENTS AND OIL DROPLETS

For many years iodopsin has been believed to be the only cone pigment extractable from chicken retinas. Light microscopic observations, however, displayed that the chicken retina has six kinds of visual cells, i.e., one type of rod, one type of double cone and four types of single cone (Oishi et al., 1990). Though the rod has no oil droplet, the single cones have own colored oil droplets, i.e., red, yellow, clear (fluorescent) and pale-blue. The double cone is composed of principal and accessory members; the former has a green oil droplet and the latter has colorless oil droplet-like organelle. Immuno-histochemical observations revealed that the red cone (a single cone with a red oil droplet) and both members of the double cone were stained with highly specific monoclonal antibodies against iodopsin. Since the other cones were not stained with the antibodies, there would be present other cone pigments different from iodopsin.

Using a new solubilizer (a mixture of 0.75 per cent CHAPS and 1.0 mg/ml phosphatidylcholine) together with a series of column chromatographies (ConA-, CM- and DEAE-Sepharose columns), we had succeeded to isolate rhodopsin and three kinds of cone pigments called chicken blue, chicken green and iodopsin (Okano et al., 1989; 1992). Chicken violet was obtained as a mixture with chicken blue. The ratio of these visual pigments in the extract is estimated to be 49(rhodopsin) : 40(iodopsin) : 5(chicken green) : 5(chicken blue) : 1(chicken violet).

In order to determine the loci of the cone pigments, number of each oil droplet in the retina was counted; the ratio among each of the cones was 26.6(principal, green) : 26.6(accessory) : 14.5(red) : 14.3(yellow) : 12.2(clear) : 5.8(pale-blue). Taking the transmission and number of each oil droplet and the ratio among each cone pigments into consideration on the basis of presence of iodopsin in double and red single cones, it was inferred that chicken green, blue and violet would be present in yellow, clear and pale-blue single cones, respectively.

Since the incident light to a cone in the retina passes through the oil droplet in the inner segment and then reach to the cone pigment in the outer segment, the oil droplet acts like a colored glass filter for cutting off the short wavelength light of the incident light. In order to infer the absorption of the light by a cone pigment in the cone, the absorption spectrum of the cone pigment must be corrected according to the characteristic curve of transmission of the respective oil droplet (Bowmaker & Knowles, 1977).

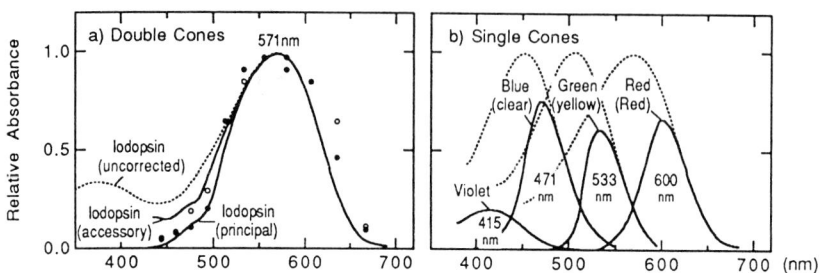

Fig. 1. Chicken photopic sensitivity curves and spectra of iodopsin. o and o are photopic sensitivity curves of ventral and dorsal retinas, respectively, (Wortel et al., 1978). Dotted lines are spectra of iodopsin (a) and other cone pigments (b), and solid lines are their spectra corrected by the respective oil droplets or an organelle in cones denoted by parentheses. (Modified from Yoshizawa & Okano, 1992)

As shown in Fig.1a, the spectra of iodopsin in both principal and accessory members in double cone were not modified near the absorption maximum and in the long wavelength side by the green oil droplet and the oil droplet-like organelle, respectively. The spectrum of iodopsin is so close to chicken photopic sensitivity curves that one may suppose that the double cone would be responsible for light intensity discrimination in photopic vision rather than color vision.

In the cases of red, green and blue single cones, the shapes of the spectra were remarkably modified by the respective oil droplets; not only the absorption maxima were shifted to longer wavelengths, but also the tails of the short wavelength side of the spectra were cut off(Fig. 1b). The spectrum of chicken violet was not corrected because the pale-blue oil droplet has only a little absorbance in the visible range. Thus the overlaps between the two corrected spectra of iodopsin and blue and between the corrected spectrum of chicken green and the spectrum of chicken violet were remarkably reduced. Consequently, a wavelength in the range roughly from 600 nm to 400 nm can be discriminated by a ratio between photosensitivity by only two cones, i.e., iodopsin and chicken green, chicken green and blue, or chicken blue and violet, in the range roughly from 600 nm to 540nm, from 540 nm to 480 nm or from 480 nm to 400nm, respectively. There is no wavelength discrimination in the range above 600 nm or below 400 nm because of absorption by a single cone pigment, i.e., iodopsin and chicken violet, respectively. The chicken wavelength discrimination mechanism over the range from 600 nm to 400 nm by two cone pigments which are selected from the four cone pigments regularly positioned on the wavelength scale, would be better in wavelength resolution than human one on the basis of the three cone pigments whose absorption maxima are irregularly positioned in the wavelength scale and also the overlap of the three cone pigments in the wide range from 540 nm to 400 nm.

STRUCTURES OF CONE PIGMENTS AND MOLECULAR EVOLUTION

Recently we have succeeded to determine the amino acid sequences of four types of cone pigments (Kuwata et al., 1990; Okano et al., 1992) and two types of nocturnal gecko visual pigments, P521 and P467 (Kojima et al., 1992) by cDNA analyses. Since 1972, Crescitelli(1991) has been reporting that the former is similar in physical and chemical properties to iodopsin and the latter to rhodopsin, though

the gecko has pure rod retinas morphologically and electro-physiologically.
According to our amino acid sequence analyses, P521 is definitely similar to iodopsin (82.6 per cent identity) but p467 is closer to chicken green (82.3 per cent identity) than rhodopsins (71.0-75.4 per cent identities). These findings that the gecko rods hold the respective cone-like pigments would add support to the "transmutation theory" (Walls, 1934) that cones of diurnal lizard had morphologically transmuted to rods in forming a nocturnal habit.

Based upon the amino acid identities of various visual pigments which have been analyzed so far, we have constructed a phylogenetic tree of visual pigments by a neighbor-joining method (Okano et al. 1992) (Fig. 2). An ancestral visual pigment first diverged to vertebrate and invertebrate visual pigments. Then the ancestral vertebrate visual pigment had evolved into four groups L, S, M1 and M2 (long, short, middle 1 and middle 2 wavelength groups). Each group is composed of visual pigments showing more than 70 per cent identity, except for group M1 composed of only chicken blue. Thus chicken blue is a considerably different protein from other visual pigments including human blue (49.4 per cent identity). Human blue and chicken violet are classified to group S. Group L is composed of iodopsin, gecko P521, human red, human green and others. Unlike human green, chicken green belongs to group M2 together with gecko P467 and vertebrate rhodopsins.

A recent calculation of the isoelectric points of visual pigments clearly demonstrated that all the cone pigments including the gecko visual pigments were basic proteins, while all the rhodopsins examined were acidic proteins, except for a rhodopsin of lamprey (one of the lowest vertebrate) which lay just between rhodopsins and cone pigments (Okano et al., 1992). Thus the ancestral vertebrate visual pigment, presumably a basic protein, first diverged into group S and the other group as the ancestral types of cone pigments, followed by successive divergences of groups M1 and M2. Then the ancestral type of group M2 would be diverged to vertebrate rhodopsin over an ancestor type of rhodopsins close to lamprey rhodopsin by replacement of some of the basic amino acid residues into acidic ones. Thus, one may imagine that color vision would have appeared earlier than scotopic vision.

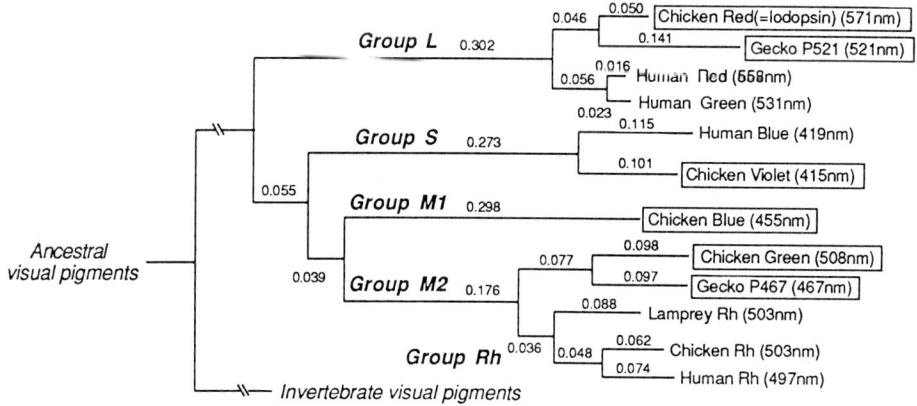

Fig. 2. A phylogenetic tree of visual pigments. (Modified from Okano et al. 1992)

BLEACHING PROCESS OF IODOPSIN AND PHYSIOLOGICAL FUNCTION OF CONE

For understanding the difference in physiological response between rod and cone, the photo-bleaching processes of rhodopsin and iodopsin were analyzed by using pico- and nano-second laser photolyses. As shown in Fig. 3, both the visual pigments have similar intermediates in their bleaching processes, except for lack of meta-intermediate III of iodopsin and short life times of bathoiodopsin and the later intermediates compared with those of the corresponding intermediates of rhodopsin (Shichida & Yoshizawa, 1992).

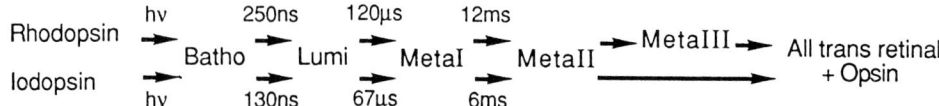

Fig. 3. Photo-bleaching processes of chicken rhodopsin and iodopsin. Time constants of decay of intermediates are shown above and below arrows.

An attention should be paid to meta-intermediates II because of an enzymatic activity catalyzing the GDP-GTP exchange reaction on transducin. Since iodopsin can activate bovine transducin (Fukada et al., 1989) or frog cGMP-phosphodiesterase (Fukada & Yoshizawa, 1981) in the light, one may imagine the activity of metaiodopsin II. If this is the case, the faster formation and decay of metaiodopsin II than metarhodopsin II would be the initial indications of the rapid generation and decay of the cone receptor potential more than the rod receptor potential. Since the activation of transducin is the first step of the amplification in visual transduction system in the outer segment, the short life time of metaiodopsin II would be a sign of less sensitivity of cone than that of rod. In any event the discussion mentioned above is based on the assumption that cones have the similar transduction process to the rods, which still remains to be solved as a future study.

REFERENCES

Bowmaker, J.K., & Knowles, A. (1977): The visual pigments and oil droplets of the chicken retina. Vision Res. 17, 755-764.

Crescitelli F. (1991): The natural history of visual pigments: 1990. Prog.in Retinal Res. 11, 1-32.

Fukada, Y., Okano, T., Artamonov, I.D., & Yoshizawa, T. (1989): Chicken red-sensitive cone visual pigment retains a binding domain for transducin. FEBS Lett. 246, 69-72..

Fukada, Y. & Yoshizawa, T. (1982): Activation of phosphodiesterase by chicken iodopsin. FEBS Lett. 149, 117-122.

Kojima D., Okano, T., Fukada Y., Shichida, Y., Yoshizawa, T. & Ebrey, T.G.(1992): Cone visual pigments are present in gecko rod cells. Proc. Natl. Acad. Sci. USA (in press).

Kuwata, O., Imamoto, Y., Okano, T., Kokame, K., Kojima, D., Matsumoto, H., Morodome, A., Fukada, Y., Shichida, Y., Yasuda, K., Shimura, Y., & Yoshizawa, T. (1990): The primary structure of iodopsin, a chicken red-sensitive cone pigment. FEBS Lett. 272, 128-132.

Oishi, T., Kawata, A., Hayashi, T., Fukada, Y., Shichida, Y., & Yoshizawa, T. (1990): Immunohistochemical localization of iodopsin in the retinas of the chicken and Japanese quail. Cell Tissue Res. 261, 397-401.

Okano, T., Fukada, Y., Artamonov, I.D. & Yoshizawa, T. (1989): Purification of cone visual pigments from chicken retina. Biochemistry 28, 8848-8856.

Okano, T., Kojima, D., Fukada, Y., Shichida, Y. & Yoshizawa, T. (1992) Primary structures of chicken cone visual pigments. Proc. Natl. Acad. Sci. USA (in press).

Shichida, Y. & Yoshizawa, T. (1992): Visual pigments in photoreceptor cells of color vision. Chemistry and Biology (in Japanese) 30, 351-359.

Walls, G.L. (1934): The reptilian retina I. A new concept of visual-cell evolution. Am. J. Ophthalmol. 17, 892-915.

Wortel, J.F., Rugenbrink, H. & Nuboer, J.F.W. (1987): The photopic spectral sensitivity of the dorsal and ventral retinae of the chicken. J. Comp. Physiol. A. 160, 151-154.

Yoshizawa, T. & Okano, T. (1992): Color reception in visual cells: the molecular mechanism and evolution. In <u>Evolution</u> of <u>Vision</u> <u>and</u> <u>Brain</u> (in Japanese), ed A. Mikami. Tokyo: Asakura.

Dissecting the squid phototransduction membrane

Alan Carne, Shaun Conway, Jeffrey N. Keen, Liu Shu-Hua,
Richard A. McGregor, Phillip D. Monk, J. Shaun Lott, John D.D. Pottinger,
Nicholas J.P. Ryba, Alison Sinclair and John B.C. Findlay

Department of Biochemistry and Molecular Biology, University of Leeds, Leeds LS2 9JT, UK

Introduction

While both vertebrate and invertebrate visual transduction systems perform essentially the same function, they contain significant structural and functional differences. Squid retinal photoreceptors are composed of highly ordered, closely packed arrays of microvilli which appear to be riveted together. (Saibil and Hewat, 1987). The phototransduction process involves a light-regulated phosphoinositidase C-based signalling system, linked to membrane depolarisation (Saibil, 1990).

The ability to obtain relatively large amounts of retinal material renders the squid system particularly attractive for direct study of the molecular components of visual signal transduction. Such a study complements recent analyses of *Drosophila* visual mutants (Bloomquist et al., 1988; Montell and Rubin, 1988; Montell and Rubin, 1989) which have provided data on the functional role of several protein components in fly photoreceptors.

Our approach involves combined strategies of protein chemistry and DNA cloning procedures to elucidate the organisation and function of the squid visual transduction system at the molecular level.

Fractionation of Principal Protein Components

The squid retinal layer can be readily isolated as a microvillar membrane fraction by physical detachment from the eye cup in neutral calcium-free buffer, followed by homogenisation and sucrose density centrifugation (Saibil and Hewat, 1987). Treatment of the microvilli preparation with mild neutral detergents such as octylglucoside or sucrose monolaurate (SML), has enabled separation of detergent-soluble protein components and a detergent-insoluble cytoskeletal fraction.

Rhodopsin and the Gα, Gβ and Gγ protein subunits were purified by conventional DEAE-ion exchange and gel filtration chromatography from SML-solubilised squid retinal microvilli, followed in the case of the Gβ and Gγ subunits by electroelution from SDS-PAGE (Pottinger et al., 1991). Protein sequence analysis confirmed the identification of these proteins and allowed the production of deoxyribooligonucleotide probes suitable for cDNA cloning strategies. Squid libraries were screened with these oligonucleotides and the sequences of full length cDNA clones obtained. (Ryba et al., 1991; Hall et al., 1991; Lott et al., in preparation; Ryba et al., submitted).

Rhodopsin

Analysis of the deduced protein sequence indicates that squid rhodopsin has high identity to that of octopus rather than the *Drosophila* pigments. It has only a single phosphorylatable residue in the C-terminal region but can exhibit light-dependent phosphorylation by a soluble kinase. The protein possesses a C-terminal extension comprising a multiple pentapeptide repeat not seen in other than octopus rhodopsin (Ovchinnikov *et al.*, 1988). This feature appears to cluster rhodopsin molecules in the membrane, as visualised by electron microscopy, and may contribute to the unusual immobility of rhodopsin in the bilayer evident in ESR studies. Removal of this domain by proteolysis generates a species, which while still functionally active, now no longer displays the characteristic membrane projections (Saibil *et al.*, submitted). This domain can be expressed in *E. coli* while the intact rhodopsin can be produced in COS 1 or insect larval cell expression systems, as monitored by Western blotting and detected by polyclonal antiserum to squid rhodopsin (Conway *et al.*, personal communication).

G-Protein

The Gα subunit of the squid heterotrimeric G-protein belongs to the Gq class identified with activation of phosphoinositidase C (Strathman and Simon, 1990). It can be expressed in yeast but does not complement deletion of the endogenous mating factor linked-G-protein (Ryba *et al.*, submitted). The Gβ subunit is highly conserved, pointing yet again to the importance of its structure and function (Ryba *et al.*, 1990). Gγ exhibits low homology with other Gγ subunits and in addition contains an unusual N-terminal extension which comprises a region of multiple lysine and glutamic acid residues. (Lott *et al.*, in preparation). This extension is not present in the *Drosophila* Gγ cloned recently and therefore appears, like the C-terminal repeat of rhodopsin, to be a feature unique to cephalopods (Ray and Ganguly, 1992). The primary structure of this region has been confirmed by protein sequencing. Its function has not yet been established but an involvement with calcium and/or interaction with other components in the system are obvious candidates. The cloning from the tissue of a nucleoside diphosphate kinase often thought to be associated with G-protein coupled systems is well advanced (Monk *et al.*, personal communication).

Structural Organisation

Attempts to solubilise the SML-detergent insoluble cytoskeletal protein component, using a variety of pH, ionic strength and detergent conditions, indicate a very tight association of the constituent proteins. Very low pH or SDS-PAGE regimes have provided the only means of disaggregating these cytoskeletal components and therefore analysis has been dependent on this approach. The general strategy employed is to initially fractionate the cytoskeletal proteins by 1-D mini-gel SDS-PAGE (Fig. 1) and electroblot the resolved protein species on to new-generation polyvinylidene-difluoride (PVDF) membranes for direct protein sequencing. In several cases, the N-termini of the intact protein species have not provided any sequence data and appear to be blocked. Subsequent experiments have involved excision of the protein band(s) from the 1st-dimension PAGE followed by *in situ* digestion of the protein in the gel slice using either cyanogen bromide, iodosobenzoic acid, endoproteinase Glu-C or trypsin. The protein fragments thereby generated have been resolved by application of the treated gel slices to a second dimension of SDS-PAGE, followed by electroblotting onto PVDF and sequence analysis. This experimental approach has generated internal partial sequence data for several proteins, allowing the identification of actin and of a protein with strong homology to the protein identified from analysis of the *trp* mutant of *Drosophila* (Montell and Rubin, 1988). The *trp* protein (*trp* : Transient Receptor Potential) in *Drosophila* photoreceptors appears to be required for perpetuation of the signal potential generated by ion flux, and has been postulated to function as a calcium channel (Minke and Selinger, 1992). The squid "*trp*-protein" (90 kDa) has an apparent molecular weight lower than its fly counterpart (140 kDa) and is present in large amounts. It is also unusual in being associated with the cytoskeletal fraction. It can be solubilised under very acidic conditions, suggesting that its appearance with the cytoskeletal components may be due to functional interaction rather than non-specific precipitation. This echoes the observation of immobility made for rhodopsin and emphasises the highly interactive nature of this membrane. It points to the possibility of a complex mechanical as well as biochemical aspect to signal transduction.

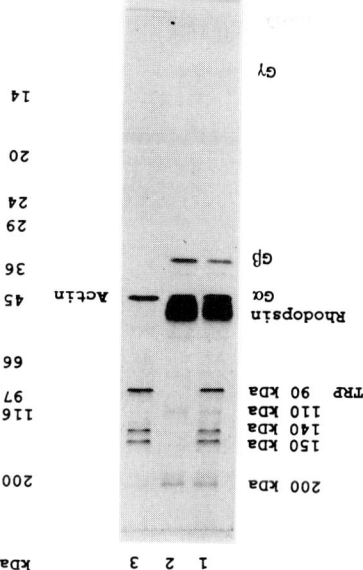

Fig. 1 Squid photoreceptor microvilli preparation separated by SDS-PAGE (5-17.5% acrylamide) and stained with Coomassie blue. Lane 1, whole photoreceptor microvilli preparation; lane 2, sucrose monolaurate detergent-soluble microvilli fraction; lane 3, sucrose monolaurate detergent-insoluble (cytoskeletal) microvilli fraction. TRP, Transient Response Potential protein; Gα, β, γ, heterotrimeric G-protein subunits; 140 kDa, *NinaC*-like, calmodulin-binding protein; 'kDa' column, position of protein molecular weight standards.

Calmodulin-binding Proteins

The 140 kDa cytoskeletal protein exhibits the ability to bind calmodulin which has prompted experiments designed to obtain information on protein interactive domains. Recent experiments involving ^{125}I-labelled-calmodulin and ^{125}I-actin overlays of Western blots of both microvilli proteins and *in situ* digests of the 140 kDa protein, combined with experiments demonstrating phosphorylation of the 140 kDa protein have identified associated functional domains. This is consistent with immunochemical and ultrastructural analyses of squid microvilli which indicate that this 140 kDa component has myosin I-like characteristics, suggesting an actin-myosin structure in the cytoskeleton (Saibil *et al.*, personal communication). Protein sequence analyses of these domains reveal some similarity to the *Drosophila ninaC* visual mutant protein (Montell and Rubin, 1988), which has been proposed recently to be a member of the "unconventional" myosin class (Cheney and Mooseker, 1992). Experiments designed to investigate possible interactions of this 140 kDa protein with other components in the vicinity of the microvillar membrane are underway to elucidate further the interaction within the cytoskeleton. A 130 kDa species obtained in some microvilli preparations behaves in an identical way to the 140 kDa component, suggesting that they are very closely related proteins or that the former is a proteolytic product of the latter.

Work is well advanced on the analysis of other polypeptides from squid microvilli with the aim of identifying and cloning the genes for these proteins. Thus far, the 150 kDa component does not behave like the 140 and 130 kDa species and sequence data show no homology to any other polypeptide so far sequenced. There is some evidence for the phosphoinositidase in this molecular weight region. Sequence data from the 200 kDa region of the gel has also not revealed any homology with known proteins.

Conclusion

Dissection of the squid photoreceptor membrane by protein fractionation, sequence analysis and molecular cloning has led to the identification and characterisation of many of the principal components in the transduction pathway. The membrane also has a universally tight structural organisation and some of the proteins involved in this aspect are being characterised. The interaction between the structural and transduction elements of the membrane raises the possiblity of mechanical as well as chemical coupling processes. Still to be positively identified in the system and cloned are the phosphoinositidase C and arrestin, while proteins at 200 kDa and 110 kDa look to be new proteins of so far undetermined function.

Acknowledgements

We thank the Marine Biological Association, Plymouth, U.K., for help in obtaining live squid, and the Science and Engineering Research Council for financial support.

References

Bloomquist, B.T., Shortridge, R.D., Schneuwly, S., Perdew, M., Montell, C., Steller, H., Rubin, G. and Pak, W.L. (1988): Isolation of a putative phospholipase C gene of *Drosophila, norpA*, and its role in phototransduction. *Cell* 54: 723-733.
Cheney, R.E. and Mooseker, M.S. (1992): Unconventional myosins. *Current Opinion in Cell Biol* 4: 27-35.
Hall, M.D., Hoon, M.A., Ryba, N.J.P., Pottinger, J.D.D., Keen, J.N., Saibil, H.R. and Findlay J.B.C. (1991): Molecular cloning and primary structure of squid (*Loligo forbesi*) rhodopsin, a phospholipase C-directed G-protein-linked receptor. *Biochem. J.* 274: 35-40.
Minke, B. and Selinger, Z. (1992): Intracellular messengers in invertebrate photoreceptors studied in mutant flies. In *Neuromethods* 20: Intracellular messengers, eds A. Boulton, G. Baker and C. Taylor, pp. 517-563. The Humana Press Inc.: New York.
Montell, C. and Rubin, G.M. (1988): The *Drosophila ninaC* locus encodes two photoreceptor cell specific proteins with domains homologous to protein kinases and the myosin heavy chain head. *Cell* 52: 757-772.
Montell, C. and Rubin, G.M. (1989): Molecular characterization of the *Drosophila trp* locus: A putative integral membrane protein required for phototransduction. *Neuron* 2: 1313-1323.
Ovchinnikov, Yu.A., Abdulaev, N.G., Zolotarev, A.S., Artamonov, I.D., Bespalov, I.A., Dergachev, A.E. and Tsuda, M. (1988): Octopus rhodopsin. Amino acid sequence deduced from cDNA. *FEBS Lett* 232: 69-72.
Pottinger, J.D.D., Ryba, N.J.P., Keen, J.N. and Findlay, J.B.C. (1991): The identification and purification of the heterotrimeric GTP-binding protein from squid (*Loligo forbesi*) photoreceptors. *Biochem. J.* 279: 323-326.
Ray, K. and Ganguly, R. (1992) The *Drosophila* G protein γ subunit gene (D-Gγ1) produces three developmentally regulated transcripts and is predominantly expressed in the central nervous system. *J. Biol. Chem.* 267: 6086-6092.
Ryba, N.J.P., Pottinger, J.D.D., Keen, J.N. and Findlay J.B.C. (1991): Sequence of the β-subunit of the phosphatidylinositol-specific phospholipase C-directed GTP-binding protein from squid (*Loligo forbesi*) photoreceptors. *Biochem. J.* 273: 225-228.
Saibil, H. (1990): Cell and molecular biology of photoreceptors. In *Seminars in The Neurosciences*, eds J. Ashmore and H. Saibil, 2: 15-23. Saunders Scientific Publications: Philadelphia.
Saibil, H. and Hewat, E. (1987): Ordered transmembrane and extracellular structure in squid photoreceptor microvilli. *The Journal of Cell Biol.* 105: 19-28.
Strathmann, M. and Simon, M.I. (1990) G protein diversity : A distinct class of α subunits is present in vertebrates and invertebrates. *Proc. Natl. Acad. Sci.* 87: 9113-9117.

The opsin sequence of the mantid *Sphodromantis* sp.

Paul Towner[1] and Wolfgang Gärtner[2]

[1] Department of Biochemistry, University of Bath, BA2 7AY, UK. [2] Max-Planck Institut für Strahlenchemie, D-4330 Mülheim/Ruhr, Germany

The optimal function of visual pigments as extremely sensitive photodetectors is based on the picosecond photoisomerization of their 11-*cis* retinal chromophore which proceeds with a high quantum yield. In spite of the large amount of information on structural and functional features for opsin function, there is still considerable lack of information on protein domains which are essential for the optimization of photochemistry, and on protein stretches employed in signal transduction *e.g.* G-protein binding and activation.

Since only ca. 20 opsin sequences are known, 13 from vertebrates which are all extremely conserved to each other, and only 7 from the evertebrates, a comparative study which may answer at least some of these questions is impeded [1]. The broad variability in visual properties of insects make them prime candidates for the study of the still unknown underlying principles of visual pigment optimization. We have identified the opsin sequence from *Sphodromantis sp.* (MANTODEAE) [2] which was chosen for the following reasons:

i) Spectral sensitivity investigations on a close relative of *Sphodromantis*, the praying mantis *Tenodera australasiae* [3] have revealed the presence of only one visual pigment with λ_{max} around 500 nm. This observation offers a clear advantage over the situation with *Drosophila* where four visual pigment genes have been characterized and one pigment in the R8 ommatidial cells still remains to be identified.

ii) Both insect species with visual pigment sequences so far known (*Drosophila* [4,5] and *Calliphora* [6]) belong to DIPTERA, an (evolutionary) most recently developed insect order. DIPTERA contain the 3-hydroxy derivative of retinal as visual chromophore [7]. 3-OH retinal was acquired by only the more recently evolved insect species with very few exceptions in the ancient groups. MANTODEAE, as members of the evolutionary older groups were identified as retinal animals [8]. Thus, a retinal containing pigment from an ancient species promises greater similarity to a common ancestor of visual pigments. It furthermore allows comparative investigations to the pigments of a 3-OH retinal species like *Drosophila* which have adapted its opsin to the new chromophore.

The sequence of Mant1 was obtained by nested PCR utilizing highly redundant oligonucleotides as primers. The primer sequences were derived from conserved amino acid clusters. The open reading frame encodes an opsin with 376 amino acids (MW: 41.8 kD).

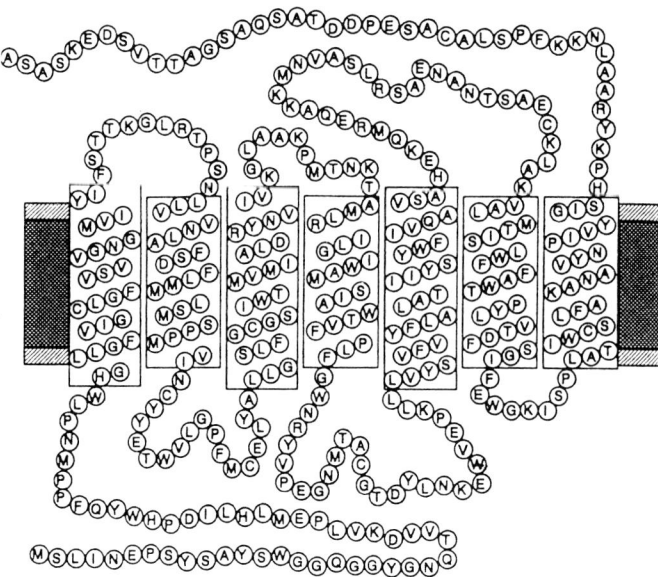

Fig. 1 Structural model of the visual pigment of *Sphodromantis sp.*. Sequence comparison of Mant1 is made to Rh1 from *Drosophila* [4,5].

N-terminus and extracellular loops:
Remarkable changes between Mant1 and Rh1 are found in the N-terminal sequence. A stretch of alanines in Rh1 (A5, A7-A9) is replaced in Mant1 by Glycins (G15,G16,G18,G19,G21). Two Asn residues are available at positions 5 and 22, the latter one being in a good glycosylation signal sequence. The N-terminal portion close to the entrance into the first membrane spanning domain is better preserved (changes to Rh1 in brackets with small letters, semi-cons. changes in capital letters):

N Q(g) T(S) V V D K V L(t) P E(D) M L(a) H L I D(s) P H(y) W Y(n) Q F P P(a) M N(d) P L

All three extracellular loops (2-3, 4-5, 6-7) exhibit high homology to Rh1. Both Cys residues which form a C-C bridge for structural stability and correct folding are at identical positions (125, 202). In the section of loop 4-5 protruding from helix IV, all residues with charged, polar or aromatic substituents are found at identical positions in Mant1 and Rh1:

W-190, R-192, P-195, E-196. Some rearrangement is found in loop 4-5, preferentially close to helix V:

Mant1: D Y L N K E W V E P K - L L
Rh1:   D Y L E R D W - N P R S Y L

Similar changes were detected in loop 6-7 where acidic and basic residues are mutually exchanged (one additional change in an adjacent position):

Mant1: F E W G K I ... (G/K positional change, E/K exchange)
Rh1:   F K F E G L ...

Cytosolic loops and C-terminus:
loop 1-2: This loop is highly conserved with only three exchanges (two Ser/Ala exchanges and one Gly/Ser exchange). Also the second cytosolic loop (3-4) exhibits strong similarity between Mant1 and Rh1: Leu/Met exchange and Ala/Gly together with a semiconserved exchange: K/R. The stretch of this loop, entering helix IV, however, strongly deviates from Rh1, as does the entire helix IV.

A high degree of conservation is also identified for the third loop (5-6), apparently from the same constraint as suggested for loop 3-4 by providing G-protein binding. Its highly charged character in Rh1 is preserved also in Mantids. We find four glutamic acid residues and 5 + 2 Lys/Arg in Mant1 vs. 6 acidic - 5 + 1 - Glu/Asp and 7 + 2 Lys/Arg residues in *Drosophila*, yielding a net positive charge of 3 in either case. Similar in both sequences is the clustering of the charged residues indicating possible ionic interactions to bound G-protein. In comparison to bovine rhodopsin, one finds an additional Ala and a stretch of 10 inserted residues in both insect sequences. This decapeptide in Rh1 was suggested to be involved in G-protein activation in insects due to its similarity to a sequence from the a-subunit of (bovine) transducin [9]. In Mantids, however, we find remarkable divergence in this domain which makes a G-protein activation as proposed for *Drosophila* questionable. Either, a different G-protein sequence or a different G-protein activation mechanism must be assumed.

C-terminus: We find strong deviation between the Mantid and the *Drosophila* sequence for the C-terminal end. The triple charged residues K E K present in Rh1 and bovine rhodopsin, believed to be important for G-protein binding, are absent. The palmitoylated membrane-anchored vicinal cysteines which create a fourth loop for G-protein binding in bovine and *Drosophila* opsin are also absent [10]. We can identify a group of Ser and Thr residues which may serve as target for phosphorylation and transduction termination. From the comparison of the extramembraneous stretches of Mantid and *Drosophila* opsins we conclude that there must be differences in the G-protein interaction.

Membrane sections:
The Lys residue for chromophore binding is clearly identified in helix VII (Lys-321), preceded by Phe-Ala as a conserved motif. Most exchanges from Rh1 to Mant1 within the

seven α helices maintain the hydrophobic character. The highest degree of accordance is found for helices III, VI and VII. Helix III shows three full plus four semicons. changes of out 26 amino acids (the arrangement of amino acids in helices may slightly vary due to algorithm used for hydrophobicity). Helix VI exhibits six full and two semicons. exchanges out of 24 amaino acids. Seven full exchanges in 24 amino acids are found in helix VII.

Strong variation is found in helix I, II and IV. Though speculative, these changes may be considered as adaptation to the differences in chromophore structures, retinal and its 3-OH derivative. In Mant1, a Val (68) replaces a Cys (66, helix I) of Rh1, a Met (101) is placed at the position of a Gly (98, shifted by one position) in helix II, and three amino acids with hetero atoms able to form hydrogen bonds (Ser-136, Cys-142 and Ser-145 in Rh1) are replaced by aliphatic residues (Gly-138, Val-144 and Ala-147) in helix III. The exchange of larger to smaller residues together with an increase in polarity may account for the substitution of retinal (Mantid) by 3-OH retinal (*Drosophila*) due to changed noncovalent interaction of the chromophore and hydrogen bonding.

The charged amino acids within the hydrophobic core are similar in Mant1 and Rh1. An Asp (98) and a Pro (107) could be identified at nearly identical positions in helix II. As in Rh1 a Glu is absent from helix III at the equivalent position of 113 (bovine rhodopsin) which was proposed to serve as counterion for the Schiff base and to support irreversible bleaching [11]. Helix III exhibits an ion pair D-R (149-150) which is found identical in Rh1 and also in bovine rhodopsin (E-R). In helix VI an Asp (298) is introduced at the place of an Asn within an albeit completely scrambled stretch T D F S (Mant1) vs. I N C M (Rh1). Finally, identifying aromatic Trp residues which noncovalently interact with the chromophore we find within the hydrophobic core nine Trp in Mant1 compared to ten in Rh1.

Though still speculative, the few distinct changes between Mant1 and Rh1 in polarity and space filling within the membrane core may well indicate an adaptation of the protein binding site to the altered chromophore structure and properties. Thus, in order to expand the basis for a comparative investigation, other insect species with interesting features in chromophore/protein interaction are being studied.

References:
1: Hargrave, P.A. and McDowell, J.H. (1992) FASEB J. *6* 2323.
2: Towner, P. and Gärtner, W. (1992) submitted.
3: Rossel, S. (1979) J. Comp. Physiol. *131* 95.
4: O'Tousa, J.E. *et al.* (1985) Cell *40* 839.
5: Zuker, C.S. *et al.* (1985) Cell *40* 851.
6: Huber, A. *et al.* (1990) J. Biol. Chem. *265* 17906.
7: Vogt, K. and Kirschfeld K. (1984) Naturwiss. *71* 211.
8. Vogt, K. (1987) Photobiochem. Photobiophys. Suppl. 273
9: Baehr, W. and Applebury, M.L. (1986) Trends Neurosci. *9* 198.
10: Ovchinnikov, Y.A. *et al.* (1988) FEBS Lett. *230* 1.
11: Sakmar, T.P. *et al.* (1989) Proc. Natl. Acad. Sci. USA *86* 8309.

Two cone types expressing different visual pigments in rodents

Pál Röhlich[1], Agoston Szél[1], Victor I. Govardovskii[2] and Theo van Veen[3]

[1] Laboratory I of Electron Microscopy and Second Department of Anatomy, Semmelweis University of Medicine, H-1450 Budapest, P.O. Box 95, Hungary; [2] Sechenov Institute of Evolutionary Physiology, Russian Academy of Sciences, 194223 St. Petersburg, Russia; [3] Department of Zoology, University of Göteborg, S-40031 Göteborg, P.O. Box 25059, Sweden

Rodents have been believed to possess heavily rod-dominated retinas. The cones were generally assumed to represent a single spectral class because the few spectral sensitivity measurements on several rodent species pointed to a cone monochromacy in the green range of the spectrum.

Monoclonal antibodies that label different cone visual pigments were found very useful in recognizing color-specific cones in various mammalian species by immunocytochemistry (Szél & Röhlich, 1989). This approach was applied in the present study to analyze photoreceptor cells in the rat (CFY albino rat), the mouse (*Mus musculus, Mus spicilegus* and a series of laboratory strains: BALB/c, NZB, C57BL, CBA, A, NMRI, AKR, C3H) and the gerbil (Mongolian gerbil: *Meriones unguiculatus*), species commonly used in retinal research. Retinas were fixed in paraformaldehyde or glutaraldehyde and either used as whole-mounts or embedded in epoxy resin for semithin sectioning. Immunocytochemistry was performed using two cone-specific monoclonal antibodies (COS-1 and OS-2)) and an anti-opsin polyclonal antibody (AO), followed by the ABC or the PAP techniques for detecting the antibodies.

All three rodent species were found to exhibit cone photoreceptor cells, although in various cone/rod ratios. The Mongolian gerbil showed the highest density of cones (about 13 per cent), while the rat was rather poor in cones (about 1 per cent). The most interesting finding was that, in addition to cones labelled by COS-1, there were cones which were selectively recognized by the other monoclonal antibody, OS-2. This clearly shows that all three rodent species exhibit two types of cones each containing a different visual pigment (Figs. 1, 2, and 3). Comparisons of sectional profiles of identical cone outer segments (Figs. 1G and 1H, 2A and 2B) on adjacent semithin sections prove that there is no cross-reactivity between the two cone types. Earlier studies on the color-specificity of these antibodies (Szél *et al.*, 1986, 1988) allow us to assume that also in rodents, COS-1 labels the middlewave-sensitive cones and OS-2 stains the shortwave-sensitive cones. Consequently, the three rodent species have two spectral classes of cones, one containing a middlewave- (green) sensitive **and the other one expressing a shortwave-sensitive cone visual pigment.** The frequency of the shortwave-sensitive cones as expressed in the percentage of all cones was 3-5 per cent in the gerbil and 7 per cent in the rat. In both the gerbil and the rat, photoreceptor cell types were rather uniformly distributed throughout the retina with only minor regional differences.

In contrast to the rat and the gerbil, the **mouse** was found to be unique in two respects. One was the unexpected **topographical separation of the two cone types to the superior and inferior halves of the retina.** More precisely, the superior part of the retina (M-field) largely resembled in its photoreceptor composition the retina of other mammalian species, with an overwhelming majority of middlewave-sensitive cones and with a low percentage of shortwave-sensitive ones (Figs. 3A and 3B). However, the inferior field of the retina (S-field) was quite peculiar in that all

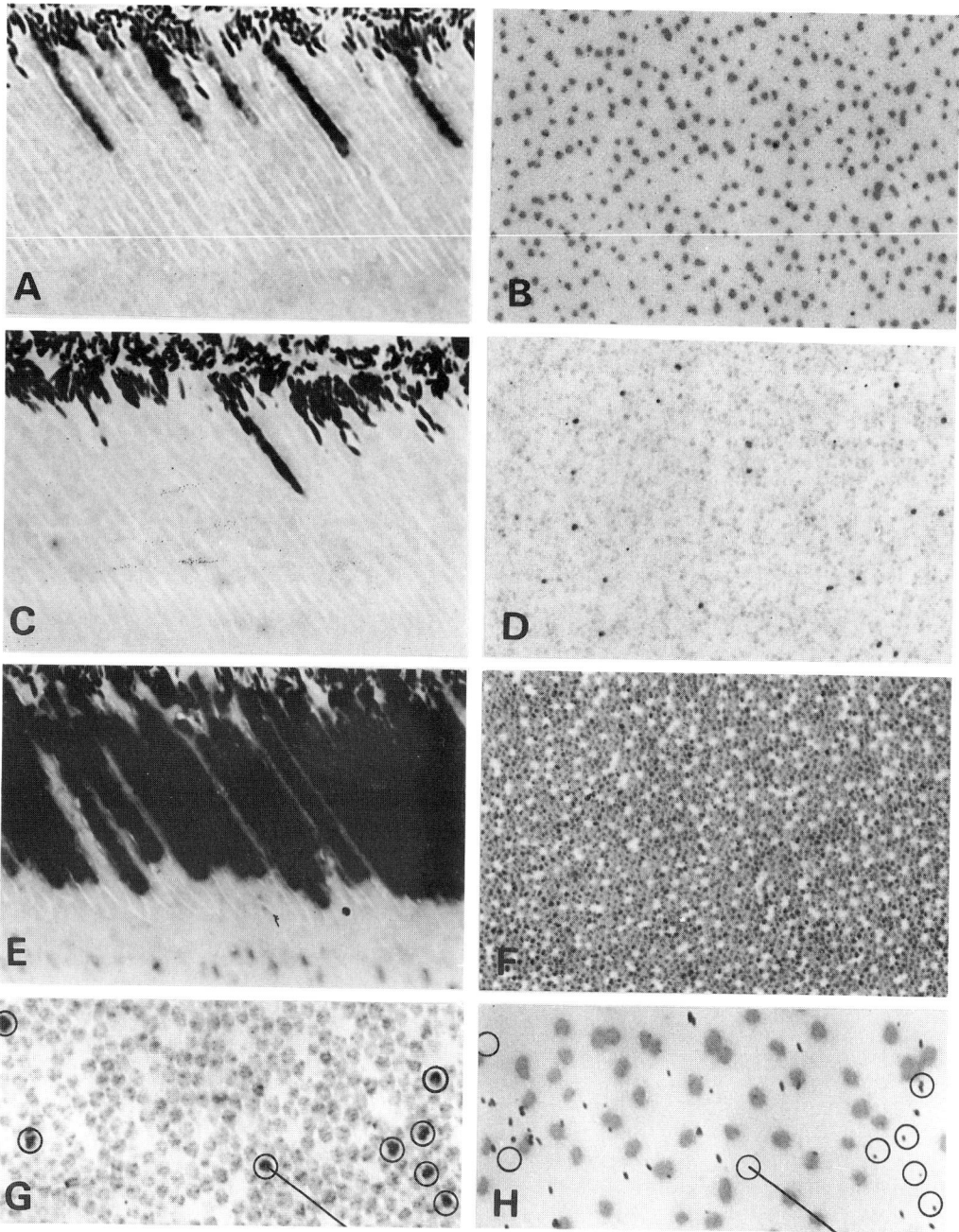

Fig. 1. **Mongolian gerbil.** Figures A, C, and E are micrographs from radial sections (Mag. x1300), figures B, D, and F represent tangential sections of the retina at the outer segment level (Mag. x620). A and B were reacted with COS-1, C and D with OS-2, and E and F with anti-opsin AO. Figures G and H represent adjacent 0.5 um thick tangential sections; G was exposed to OS-2 and H to COS-1 (Mag. x1300). The encircled outer segments on the two figures belong to identical shortwave-sensitive cones; note that the latter are not reactive with COS-1.

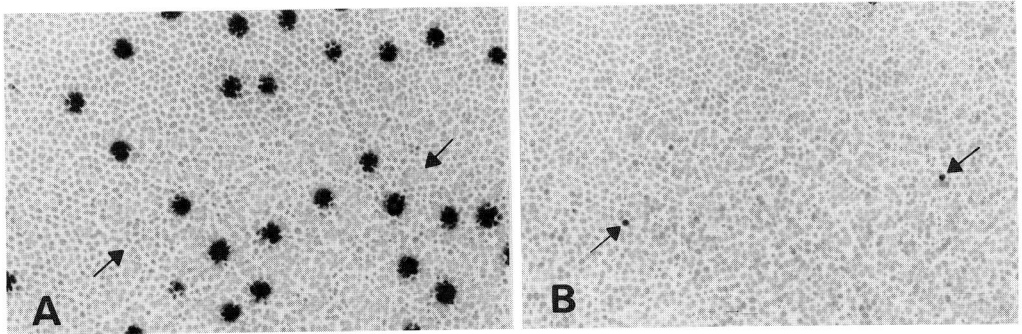

Fig. 2. **Rat.** Adjacent 0.5 um thick sections from the same retinal area. Figure A shows immunocytochemical reaction with COS-1, while B demonstrates OS-2 reactivity (Mag. x900). The two OS-2 positive cones on Fig. 2B are marked with arrows on Fig. 2A.

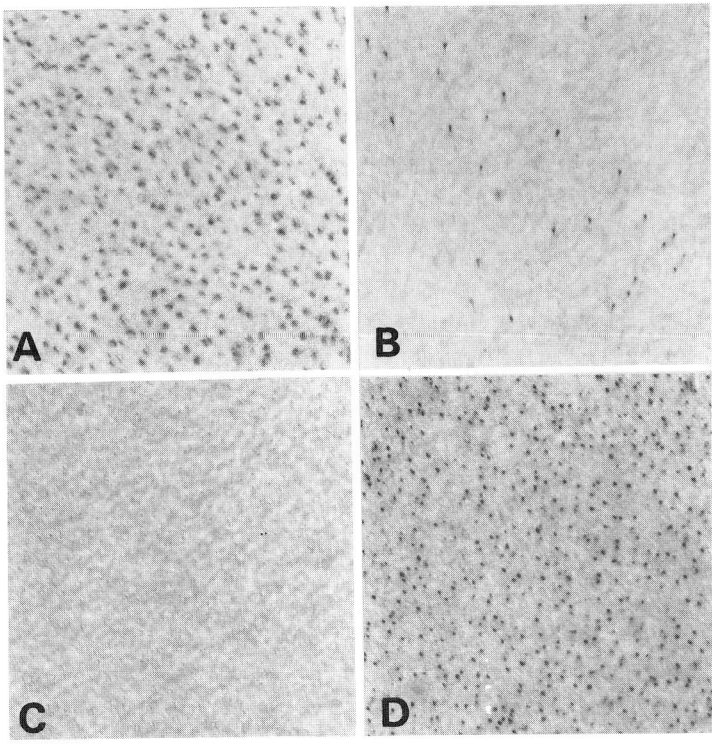

Fig. 3. **Mouse.** Retinal whole-mounts from the M-field (A and B) and S-field (C and D) reacted with COS-1 and OS-2, resp. A and C represent OS-2, while B and C show COS-1 immunoreaction. Note the complete lack of COS-1 reactivity and the high density of OS-2 positive cones in the S-field.

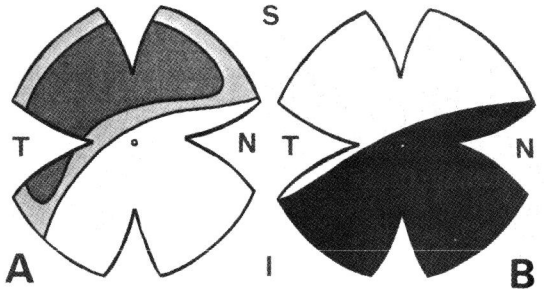

Fig.4. Distribution pattern of COS-1 (Fig. A) and OS-2 positive cones (Fig. B) in the mouse retina. Darker shades represent higher densities. S: superior, I: inferior, N: nasal, T: temporal. Note the obliquely running transition zone where the M- and S-fields overlap.

cones were OS-2 positive, with COS-1 positive cones completely missing (Figs. 3C and 3D). We assume therefore that this field has a photopic sensitivity in the short range of the spectrum. The other characteristic of the mouse retina was that **the density of OS-2 positive (shortwave-sensitive) cones in the S-field (about $18000/mm^2$) greatly surpassed that of the COS-1 positive (middlewave-sensitive) cones** (8000 to $12000/mm^2$) in the M-field. This is an unusually high density not found in other mammalian species. The borderline between the M- and S-fields was an obliquely running transition zone, 50-100 um in width, sloping from the supero-nasal to the infero-temporal direction and passing above the optic disc (Figs. 4A and 4B). The above distribution pattern of cones was consistently present in all mice, irrespective of age, sex, pigmentation and laboratory strain and can be regarded as a special characteristic of the mouse. The presence of an area making roughly the half of the whole retina and containing purely shortwave cones as photopic elements is unique in mammals; its significance is still unclear.

Our results on the presence of shortwave-sensitive cones was surprising since recent data on the gerbil and the rat (Jacobs & Neitz, 1989; Neitz and Jacobs, 1986) indicated a green monochromacy in these animals. Therefore we have undertaken colour-substitution experiments at the ERG level on the gerbil and found evidence for a cooperative activity of middlewave and shortwave-sensitive cones (Govardovskii *et al.,* 1992). Quite recently Jacobs *et al.* (1991) described a retinal photoreception maximally sensitive to ultraviolet light in the rodents. It is reasonable to assume that the shortwave-sensitive cones labelled by OS-2 are identical with the ultraviolet receptors.

The present paper is a summary; a detailed description of the cones of the individual rodent species can be found elsewhere (gerbil: Govardovskii et al, 1992, rat: Szél and Röhlich, 1992, mouse: Szél et al., submitted for publication).

REFERENCES

Govardovskii, V.I., Röhlich, P., Szél, Á. & Khokhlova, T.V. (1992): Cones in the retina of the Mongolian gerbil, *Meriones unguiculatus*: an immunocytochemical and electrophysiological study. *Vision Res.* 32, 19-27

Jacobs, G.H. & Neitz, J. (1989): Cone monochromacy and a reversed Purkinje shift in the gerbil. *Experientia* 45, 317-319

Jacobs, G.H., Neitz, J. & Deegan, J.F. (1991): Retinal receptors in rodents maximally sensitive to ultraviolet light. *Nature* 353, 655-656

Neitz, J. & Jacobs, G.H. (1986): Reexamination of spectral mechanisms in the rat (*Rattus norvegicus*). *J. Comp. Psychol.* 100, 21-29

Szél, Á., Diamantstein, T. & Röhlich, P. (1988): Identification of the blue-sensitive cones in the mammalian retina by anti-visual pigment antibody. *J. Comp. Neurol.* 273, 593-602

Szél, Á. & Röhlich, P. (1989): Colour vision and immunologically identifiable photoreceptor subtypes. In *Neurobiology of Sensory Systems*, ed. R.N. Singh & N.J. Strausfeld, pp. 275-293. New York: Plenum Press

Szél, Á. & Röhlich, P. (1992): Two cone types of rat retina detected by anti-visual pigment antibodies. *Exp. Eye Res.* in press

Szél, Á., Takács, L., Monostori, É., Diamantstein, T., Vigh-Teichmann, I. & Röhlich, P. (1986): Monoclonal antibodies recognizing cone visual pigments. *Exp. Eye Res.* 43, 871-883

III. Mutagenesis studies of bacteriorhodopsin

III. Etudes par mutagenèse de la bactériorhodopsine

Halobacterial vector development and the opportunities for gene expression and analysis

Mike Dyall-Smith, Melissa Holmes, Masahiro Kamekura[1] and Ford Doolittle[2]

Department of Microbiology, University of Melbourne, 3052, Australia; [1] Noda Institute for Scientific Research, 399 Noda, Noda-Shi, Chiba-Ken 278, Japan; and [2] Canadian Institute for Advanced Research and Department of Biochemistry, Dalhousie University, Halifax, B3H 4H7, Canada

SUMMARY

In the last few years a number of plasmid vectors have been developed for use in halobacteria. Selectable marker genes conferring resistance to mevinolin and novobiocin have been incorporated into various replicons (both phage and plasmid) to form stably maintained plasmids with useful cloning sites. This has opened up modern genetics in archaebacteria and also provided the technology to synthesise large quantities of halophilic proteins. Recent examples will be reviewed, including expression of the *hly* gene, coding for halolysin (a halophilic serine protease). Transposons have also been constructed and which should be useful for studying genes that cannot be directly selected.

INTRODUCTION

Genetic analysis of halophilic archaebacteria has been severely hampered by the lack of good genetic tools, such as cloning vectors, transposons etc. Whereas antibiotic resistance genes were readily available for constructing selectable *E.coli* vectors, this has not been possible with halobacteria. Not only are they unaffected by the majority of eubacterial antibiotics, but even if an appropriate eubacterial resistance gene were available, the gene product would be unlikely to function in environment of the halobacterial cytoplasm (about 5M in KCl). Cloning chromosomal genes from drug resistant mutants has overcome this obstacle.

The two genes now commonly used are those for HMG CoA reductase (mevinolin resistance), and DNA gyrase (novobiocin resistance); developed by Lam and Doolittle (1989) and Holmes and Dyall-Smith (1990), respectively. Both genes have been fully characterised and extraneous upstream and downstream regions trimmed. In parallel, the plasmids originally used in constructing vectors, have been sequenced, and redundant regions removed, thus making the most recent versions of these plasmids much smaller and with a larger selection of unique cloning sites (Holmes et al., 1991; Lam and Doolittle, 1989; Lam, personal communication). Selectable vectors with other halobacterial replicons (Blaseio and Pfeifer, 1990) and with a strong promoter incorporated (Nieuwlandt and Daniels,1990) have been constructed in other laboratories. It should be noted that since the resistance markers were originally derived from *Haloferax* chromosomal loci, their introduction on plasmids into *Haloferax* cells may result in recombination with the chromosome, so converting the chromosomal gene (ie. *hmg* or *gyrB*) to the resistant genotype.

As an example of the use of novobiocin resistance plasmids, I will describe the expression of the gene for the serine protease, halolysin. In the second example, the construction of transposons containing the mev^r marker is described.

RESULTS AND DISCUSSION

Expression of halolysin in *Hf volcanii*.

Kamekura and Seno (1990) described an extracellular serine protease, termed halolysin, produced by an extremely halophilic archaebacterium (strain 172P1). The enzyme has been purified and is thermophilic (opt. temp. 75-80°C), and very halophilic, being inactivated at salt concentrations below 3M NaCl. From N-terminal sequence information, the gene (*hly*) was cloned and its nucleotide sequence determined (Kamekura et al., 1992). The full coding sequence would produce a protein of 530 amino acids, but the first 119 amino acids of the precursor are normally cleaved to give a final protease of MW 41,963. The first 40 amino acids of the precursor contains a probable signal sequence, which would allow secretion of the protease. This is the first serine protease sequenced from the archaebacteria.

Since the enzyme was unlikely to function if expressed in cells with a low internal salt concentration, the *hly* gene was engineered into a derivative of pMDS9 (containing the pHK2 replicon and the novobiocin resistance determinant) for expression in *Haloferax volcanii* . The plasmid was first passaged in *E.coli* JM110 (a *dam* strain) to overcome the methylation-dependent restriction in *Hf. volcanii* and then transformed into *Hf volcanii* WFD11 cells (method of Cline et al., 1989). Several transformant colonies displayed clearing zones (on milk agar plates) typical of protease production. Wild type *Hf volcanii* does not produce any clearing. Protease producing transformants were shown to contain the plasmid construct, whereas non-producing transformants contained no plasmid and were presumed to have arisen by homologous recombination via the *gyrBA* genes of the resistance marker.

The amount of enzyme produced by *Haloferax* cells was determined to be 65.2ug/ml (in the supernatant of a 2L culture), which was about 30 fold higher than the amount secreted by the original protease producing strain, 172P1. The expression of the protease gene on a multicopy plasmid in a more convenient host strain has allowed a much higher yield of enzyme, but it may be possible to greatly increase this by using a more powerful (perhaps controllable) promoter than the natural one. Another potential problem is plasmid loss by recombination of the resistance marker with the chromosomal homologue (see the next example).

Transposon mutagenesis of *Ha. hispanica*.

Transposons are mobile genetic elements consisting of IS (insertion sequence) elements and other (often, antibiotic resistance) genes. For example, in Tn10, a tetracycline resistance gene is flanked by 1.3kb insertion sequences (IS10 in this case) which can mobilise the entire compound element, allowing it to move to distant, unrelated sites. Derivatives of eubacterial transposons have have been useful for identifying and rapidly isolating genes; they act as insertional mutagens, and since they carry a selectable marker the resulting mutants are transformed to a resistant phenotype and can be isolated on selective media. Disrupted genes can then be readily cloned, either from genomic libraries using the transposon sequences as a specific probe, or more recently by PCR.

Insertion sequences have been known for some time to cause the very high frequency of mutations (of the order of 10^{-3}) observed in *Halobacterium* sp. Many of these ISH elements have been cloned and sequenced and the availability of drug resistance markers now make it possible to construct synthetic transposons consisting of a marker flanked by ISH elements. Currently, the most popular host strain for genetic work is *Haloferax volcanii* and ISH elements were chosen with this in mind. We selected ones that had published sequences available, shared no homology to ISH elements of *Hf volcanii*, appeared to have little insertion specificity, were moderate to highly mobile, and not too large. The best candidates were ISH2, 26 and 28. ISH2 is particularly mobile and has the novel characteristic of being the smallest known IS element (521bp).

To facilitate the construction, ISH elements with convenient terminal restriction sites were generated using PCR, then cloned as direct repeats in pUC19. The mevinolin resistance marker was inserted between the two ISHs, and the constructs introduced into *Hf. volcanii* cells using PEG. Southern blot analysis of the transformants revealed that all of 20 colonies tested had resulted from recombination of the mevinolin marker (HMGCoA reductase gene) with the chromosomal homologue. Linearizing the plasmid, (hopefully reducing the probability of a double crossover event), produced 10 fold fewer transformants but again these were also shown to have recombined at the *hmg* locus.

Cline and Doolittle (1992) recently showed that pWL102 could transform *Ha hispanica*, a genetically stable halobacterial strain. In these experiments, the mevinolin resistance gene never recombined with the chromosome of this organism, probably because of low homology. Encouraged by these properties, we are currently exploring the use of this strain to reduce the background level of homologous recombination. Our preliminary results indicate that transposition does occur.

Future directions and applications

We are at the beginning of true genetics in archaebacteria. Cloning, expressing and mutational analyses of genes is now possible and will become progressively easier as vectors and host strains are further developed and improved. Some of the most urgent requirements are: a controllable promoter, an indicator gene (equivalent to *E.coli lacZ*), a *recA* host, and selectable marker genes that share little homology to their chromosomal homologues.

REFERENCES

Blaseio, U., and Pfeifer, F. (1990): Transformation of Halobacterium halobium: development of vectors and investigation of gas vesicle synthesis. Proc. Natl. Acad. Sci. U.S.A. 87: 6772-6776.

Cline, S. W., Lam, W. L., Charlebois, R. L., Schalkwyk, L. C., and Doolittle, W. F. (1989): Transformation methods for halophilic archaebacteria. Can. J. Microbiol. 35: 148-52.

Cline, S., and Doolittle, W.F. (1992): Transformation of members of the genus Haloarcula with shuttle vectors based on Halobacterium halobium and Haloferax volcanii plasmid replicons. J. Bacteriol. 174: 1076-1080.

Holmes, M. L., and Dyall-Smith, M. L. (1990): A plasmid vector with a selectable marker for halophilic archaebacteria. J. Bacteriol. 172. 756-61.

Holmes, M. L., Nuttall, S. D., and Dyall-Smith, M. L. (1991): Construction and use of halobacterial shuttle vectors and further studies on Haloferax DNA gyrase. J. Bacteriol. 173: 3807-13.

Kamekura, M., and Seno, Y. (1990): A halophilic extracellular protease from a halophilic archaebacterium strain 172 P1. Biochem. Cell. Biol. 68: 352-9.

Kamekura, M., Seno, Y., Holmes, M.L., and Dyall-Smith, M.L. (1992): Molecular cloning and sequencing of the gene for a halophilic alkaline serine protease (Halolysin) from an unidentified halophilic archaea strain (172P1) and expression of the gene in Haloferax volcanii. J. Bacteriol. 174: 736-742.

Krebs, M. P., Hauss, T., Heyn, M. P., RajBhandary, U. L., and Khorana, H. G. (1991): Expression of the bacterioopsin gene in Halobacterium halobium using a multicopy plasmid. Proc. Natl. Acad. Sci. U. S. A. 88: 859-63.

Lam, W., and Doolittle, W.F. (1989): Shuttle vectors for the archaebacterium Halobacterium volcanii. Proc. Natl. Acad. Sci. U.S.A. 86: 5478-5482.

Nieuwlandt, D. T., and Daniels, C. J. (1990): An expression vector for the archaebacterium Haloferax volcanii. J. Bacteriol. 172: 7104-10.

Expression in *H. halobium* of bacteriorhodopsin mutants obtained by site-directed mutagenesis

Man Chang, Xiaole Fan and Richard Needleman

Department of Biochemistry, Wayne State University School of Medicine, 540 E. Canfield Avenue, Detroit, MI 48201, USA

Techniques for the genetic engineering of proteins are widely available, but prediction of the structural and functional consequences of residue replacements is still a considerable problem. Much information can be gained from the study of model systems, where structure and function can be related in unambiguous ways.

The study of bacteriorhodopsin variants synthesized in *E. coli* has played a prominent role in elucidating its function. Instead of obtaining mutant bacteriorhodopsin by reconstitution of *E. coli* synthesized proteins, we, in collaboration with Janos Lanyí's laboratory, are using mutants obtained by site-directed mutagenesis in *H. halobium* to investigate the mechanism of proton transport by bacteriorhodopsin.

We have described the first system for the plasmid-mediated synthesis of bacteriorhodopsin in *H. halobium* (Ni, et al., 1990). In this system mutant bacteriorhodopsins are synthesized in their natural host and not in *E. coli*. This allows the rapid and facile production of large quantities of mutant proteins (25mgs/l of culture) and in addition produces bacteriorhodopsins in the more stable trimeric form. Most importantly, the proteins produced differ from those synthesized in *E. coli* in ways suggesting that the instability problems and anomalous behavior of the *E. coli* bacteriorhodopsins are due to aberrant protein folding. We describe here some of the biological features of this expression system which may be of interest to workers wishing to synthesize mutant bacteriorhodopsins in *H. halobium*.

The *bop* expression plasmids that we use contain a replication origin for *E. coli*, the *halobium bop* gene, the *H. volcanii* plasmid pHV2, and a gene coding for resistance to the anti-cholesterol drug mevinolin (Mev^r). The plasmids were originally developed by Charlebois, et al., for transformation of *H. volcanii* (Charlebois, et al., 1987). The pHV2 plasmid, which supplies a replication origin for *H. halobium*, is not essential for effective transformation and it is possible to use strictly integrating vectors which carry *bop* on an *E. coli* phagemid, and *Mev^r* (unpublished). While integration is more effective in plasmids carrying larger pieces of *H. halobium* DNA, plasmids having as little as 1.2kb of *H. halobium* DNA can be used. We currently use phagemid variants of these plasmids lacking pHV2 sequences to ensure integration, and an *Haloferax volcanii* gene that confers novobiocin resistance (*Nov^r*, isolated by Holmes and Dyall-Smith, 1990). In contrast to the plasmids described by Krebs and others (Krebs, et al., 1991; Blaseio and Pfeifer, 1990) these vectors have the advantage that they integrate and are therefore relatively stable in the absence of selection for the resistance marker; in addition they produce levels of bacteriorhodopsin comparable to wild type, unlike the vector described by Krebs et al., 1991. For plasmids having both the pHV2 replicon and *bop* DNA, the most prevalent pathway is homologous recombination at the chromosomal *bop* gene leading to a single integrated plasmid, although autonomously replicated plasmids are also obtained. The high integration frequency and the maintenance of autonomously replicating forms of the plasmid which fail to integrate may be due to the action of restriction systems which stimulate recombination by creating free DNA ends. Regardless of the mechanism, in 60 independent integrants studied by Southern transfer analysis, all were integrated at the *bop* gene; no plasmids integrated at other possible sites (i.e., at the *H. halobium* gene equivalent to the *H. volcanii Mev^r*) were recovered (unpublished). This site specific integration can be utilized for gene manipulations including gene disruptions and deletions. For example, we have recently engineered a halorhodopsin overproducing strain by fusing the *bop* promoter to the gene encoding halorhodopsin (*hop*). Since no ISH insertions into *hop* have been characterized, the vector used for isolating halorhodopsin mutants had the desired mutation in the proximal portion of *hop* and a nonsense mutation before lysine216. Integration can lead to the production of mutant halorhodopsin expressed from the *bop* promoter and the same time introduces a nonsense mutation into the resident *hop* gene. Yields of halorhodopsin comparable to those of BR have been obtained in this manner.

Although studies of mutant bacteriorhodopsins produced in *H. halobium* are just beginning to be published, it is already clear from the first few studies that they are significantly different from those produced in *E. coli*. Alterations in the photocycle seen in the *E. coli* mutants must now be confirmed for their naturally synthesized variants in our expression system; the properties of the few mutants synthesized in both systems show important differences in their behavior at different pH and in their photocycles. Strong evidence for such differences in the case of the D212N protein is given in a recent paper of ours (Needleman et al, 1991). In particular, the fact that bacteriorhodopsin synthesized in *H. halobium* strains transformed with *bop* plasmids is highly ordered and has lattice constants indistinguishable from those of native bacteriorhodopsin probably accounts for the instability seen in the *E. coli* synthesized 'form' which is monomeric. *Halobium* D212N is stable, but attempts to produce monomers by detergent lead to rapid bleaching of the protein, even in the dark. In addition, the *halobium* protein is 16nm red shifted from wild type as compared to a 9-13nm blue shift in the *E. coli* protein, and, unlike the *E. coli* protein, is stable in the light, shows a normal light adaptation, and a pH dependent ability to pump protons. Similarly, our D115N bacteriorhodopsin also appears to be very different from the *E. coli* expressed protein. In this mutant the absorption maximum of the monomeric form shifts by 15nm to the blue during the photocycle relative to the wild type protein, suggesting that residues as far away as 10 Ångstroms from the Schiff base can play a role in the critical M1→M2 switch in the photocycle (Váró, et al., 1992) Previous conclusions obtained for the 'E. coli' mutant bacteriorhodopsins must therefore be reevaluated.

All strains, plasmids, and mutants are available to interested investigators.

REFERENCES

Blaseio, U., and Pfeifer, F. (1990): Transformation of *Halobacterium halobium*: Development of vectors and investigation of gas vesicle synthesis. *Proc. Natl. Acad. Sci. USA* 87:6772-6776.

Charlebois, R. L., Lam, W. L., Cline, S. W., and Doolittle, W. F. (1987): Characterization of pHV2 from *Halobacterium volcanii* and its use in demonstrating transformation of an archaebacterium. *Proc. Natl. Acad. Sci USA* 84:8530-8534.

Holmes, M. L., and Dyall-Smith, M. L. (1990): A plasmid vector with a selectable marker for halophilic Archaebacteria. *J. Bact.* 172:756-761.

Krebs, M. P., Hauss, T., Heyn, M. P., RajBhandary, U. L., and Khorana, H. G. (1991): Expression of the bacterioopsin gene in *Halobacterium halobium* using a multicopy plasmid. *Proc. Natl. Acad. Sci. USA* 88:859-863

Needleman, R., Chang, M., Ni, B., Váró, G., Fornés, J., White, S. H., and Lanyí, J. L. (1991): Properties of Asp^{212}→Asn bacteriorhodopsin suggest that Asp^{212} and Asp^{85} both participate in a counterion and proton acceptor complex near the Schiff base. *Jour. Biol. Chem.* 266:11478-11484

Ni, B., Chang, M., Duschl, Lanyí, J. L., and Needleman, R. (1990): An efficient system for the synthesis of bacteriorhodopsin in *Halobacterium halobium Gene* 90:167-172.

Váró, G., Zimányi, L., Chang, M., Ni, B., Needleman, R., and Lanyí, J. L. (1992): A residue substitution near the β-ionone ring of the retinal affects the M substates of bacteriorhodopsin. *Biophys. J* 61:820-826.

Bacteriorhodopsin precursor is partially processed at the N-terminal end in hetereologous *in vitro* and *in vivo* expression systems

Volker Hildebrandt, Ulrich Bauer, Norbert A. Dencher*,
H. Georg Büldt and Paul Wrede

Department of Physics/Biophysics, Freie Universität Berlin, Arnimallee 14, W-1000 Berlin 33, Germany.
* *Hahn-Meitner-Institut, BENSC-N1, Glienickerstrasse 100, W-1000 Berlin 39, Germany*

INTRODUCTION

The bacterio-opsin-precursor (pre-BO) has 262 amino acids and is processed at the amino terminus (-13 amino acids) and at the carboxyl terminus (-1 amino acid), leading to mature BO with 248 amino acids. The amino-terminal presequence deviates from those of other known eubacterial and metazoic presequences. It is unusually short, contains negatively charged amino acids, and a short stretch of hydrophobic amino acids (Dellweg & Sumper, 1980). A comparison of six retinal binding proteins from halobacteria showed, that these pigments share sequence similarities in the N-terminal region (Gropp et al., 1992). The differences between the prequences of retinal proteins and the other known presequences have led to the suggestion that two different translocation mechanisms may exist in halobacteria. There is currently no homologous *in vitro* transcription-translation system available for the expression of bacteriorhodopsin (BR) (Gropp & Oesterhelt, 1989). We therefore translated BO-mRNA derived from different plasmids in a wheat germ system and compared the processing pattern with data obtained from fission yeast. Integration of *in vitro* expressed bacterio-opsin was tested with dog pancreas microsomes. Both pre-BO and mature BO do integrate into the eukaryotic membrane cotranslationally (Bauer et al., submitted elsewhere). Bacteriorhodopsin integrates cotranslationally in the homologous system also (Gropp et al., 1992). *In vivo* experiments of heterologously expressed pre-BO and mature BO demonstrated already (Hildebrandt et al., 1991) that both forms of BO are integrated into the plasma membrane of *Schizosaccharomyces pombe*. Addition of retinal leads to functional active pigments (Hildebrandt et al., submitted elsewhere). But the color of pre-BR is more intensive than the color of mature BR in *S. pombe*, supporting the data of Gropp et al. (1992), that the presequence may have a biological function in halobacteria as well as in fission yeast.

MATERIALS AND METHODS

The plasmid pBSBOp contains the complete bacterio-opsin-gene (bop) inserted into the multiple cloning site of pBSM13. pBSBOm is the same construction, but without the presequence-encoding region (Hildebrandt et al., 1989, 1991). These genes were transcribed with T7-RNA-polymerase. For *in vitro* translation, a commercial wheat germ system from Promega (Serva, Heidelberg, FRG) was used according to the company's instructions. Protein synthesis was monitored by autoradiography and immuno-detection with the antisera directed against the C-terminal and N-terminal end of BR (Wrede et al., 1985).

RESULTS

In vitro translation of BO-mRNA leads to a partially processed protein

Purple membrane (PM) isolated from halobacteria shows on a SDS-PAGE a band corresponding to mature BO and two additional proteins of larger molecular weight (pre-BO and truncated pre-BO) (Fig. 1, lane 1 and 6). According to Wölfer et al. (1988), the band with the highest molecular weight protein is the precursor-BO. The middle band represents a truncated pre-BO and is thought to be a processing intermediate. These three proteins shall be compared with the *in vitro* translation products of the two plasmids pBSBOp and pBSBOm. Transcription of both plasmids with T7-RNA-polymerase yields highly enriched homogeneous mRNA. The expressed product of pBSBOm corresponding to BR (26.5 kDA, lane 4) migrates very close to the lower band (26.3 kDA, lane 5) of the purple membrane, while the product of pBSBOp (Fig. 1, lane 3, overloaded) has a higher molecular weight and is partially processed.

Fig. 1: **Correlation of the *in vitro* translational products** of the plasmids pBSBOp (lane 3) and the pBSBOm (lane 4) with the three protein bands of purple membrane (lane 1, 2, 5 and 6). The upper band is pre-BO, the middle band is a truncated pre-BO and the lower band is the mature BO. Lane 1, 2, 5, 6 are immuno-stained bands of a Western blot (lane 1 and 6: 100 ng; lane 2 and 5: 5 ng purple membrane), the lanes 3 and 4 are from the autoradiogram of samples containig [^{35}]S methionine of the same Western blot.

The autoradiogram (lane 3 and 4) has several smaller protein bands, which may be fragments of BO caused by an incomplete translation. When capped mRNA was used, all of the smaller proteins were absent (data not shown), therefore these bands may also come from new translation initiations at internal AUG (encoding for methionine) of the mRNA.

Identification of partially processed and mature BO in fission yeast

Yeast expression vectors containing the same gene-cassettes (BOp and BOm) were transformed into *S. pombe*. Western blot analysis of the proteins from transformed cells showed a pattern comparable with that obtained in the *in vitro* experiments.

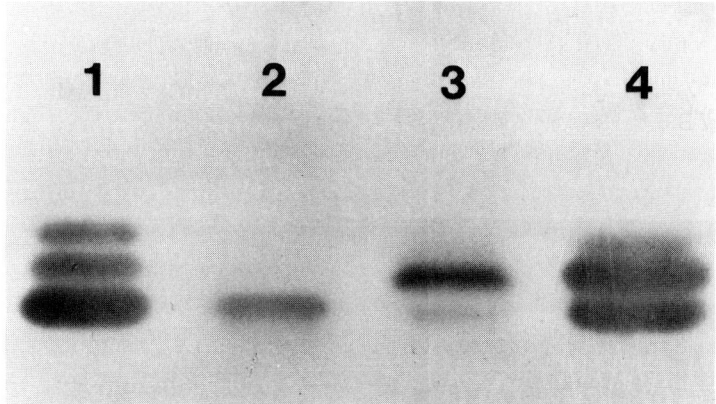

Fig. 2: **Western blot with antisera directed against the C-terminal end.** Lane 1 and Lane 4: 100 ng PM from *H. salinarium* ET1001. Lane 2: 10 μg crude membrane from *S. pombe*-cells expressing the mature protein. Lane 3: 10 μg crude membrane from *S. pombe*-cells expressing the precursor. Yeast cells were grown in 2 % glucose and harvested at early exponential growth phase.

With antisera directed against the N-terminal end of BR the precursors could be identified in *S. pombe* and purple membrane (Fig. 3). The lower band seems to be the mature protein, because it shows no cross-reaction with these antisera and the higher bands are containing at least some amino acids of the presequence. The content and molecular weight of precursors in the samples depend on the growth phase of the hosts (Wölfer et al., 1988, Hildebrandt et al., 1991), as well as the association of the 7S RNA with bacterio-opsin (Gropp et al., 1992).

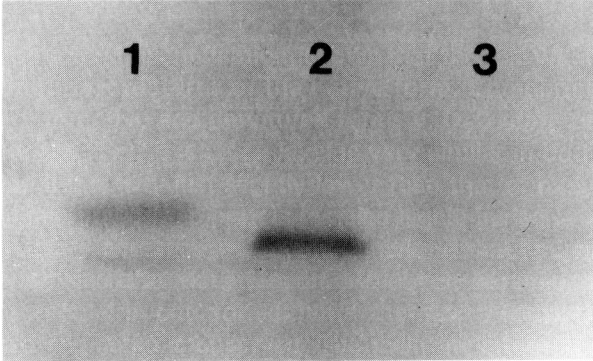

Fig. 3: **Western blot with antisera directed against the N-terminal end.** Lane 1: 200 ng PM. Lane 2: 20 μg crude membrane from *S. pombe*-cells expressing the precursor. Lane 3: 20 μg crude membrane from *S. pombe*-cells expressing the mature protein. This antisera reacts only with the first 13 amino acids of the presequence.

DISCUSSION

In different heterologous systems with lower expression levels we found partially processed and mature BO comparable with BO from *H. salinarium* . Partially processed forms of pre-BO have been analysed by Miercke et al. (1989) intensively. According to their data from purple membrane isolated of *H. salinarium* strains JW-3 and R1, pre-BO is processed in several steps. Highly purified purple membrane is running in four different bands in PAGE with a ratio of 63, 21, 12 and 4 % determined by quantative scanning of the gel. Since the expression of BO in the *in vitro* system is lower, it is possible that only the main bands (63 and 21 %) can be detected in the autoradiogram. These two braod bands are also visuable in the crude membrane fraction of *S. pombe* (Hildebrandt et al., 1989). High level expression of pre-BO in *S. pombe* leads to three and four different BO-bands in the immuno-staining of washed plasma membranes (data not shown). Interestingly a similar processing of the first amino acids takes place immediately in the different biological systems. The "consensus sequence" in pre-BR identified by Gropp et al. (1992), seems to be cleaved off according to the sequencing data from the truncated precursors by Miercke et al. (1989).

REFERENCES

Bauer, U., Hildebrandt, V., Dencher, N. A. & Wrede, P. (submitted elsewhere): *In Vitro* Synthesis of Bacterio-opsin: Integration into Microsomal Membranes.

Dellweg , H. G. & Sumper, M. (1980): Identification of a Bacterio-opsin Species with a N-terminal extended Amino Acid Sequence. FEBS Lett., **116**, 303-306.

Gropp, R. & Oesterhelt, D. (1989): In vitro translation of halobacterial mRNA. FEBS Lett., **259**, 5-9.

Gropp, R., Gropp, F. & Betlach, M. C. (1992): Association of halobacterial 7S RNA to the polysome correlates with the expression of the membrane protein bacterioopsin. Proc. Natl. Acad. Sci. USA, **89**, 1204-1208.

Hildebrandt, V., Polakowski, F. & Büldt, G. (1991): Purple fission yeast: overexpression and processing of the pigment Bacteriorhodopsin in *Schizosaccharomyces pombe*. Photochem. Photobiol., **54**, 1009-1016.

Hildebrandt, V., Ramezani-Rad, M., Swida, U., Wrede, P., Grzesiek, S., Primke, M. & Büldt, G. (1989): Genetic transfer of the pigment bacteriorhodopsin into the eukaryote *Schizosaccharomyces pombe*. FEBS Lett., **243**, 137-140.

Hildebrandt, V., Fendler, K., Heberle, J., Hoffmann, A., Bamberg, E. & Büldt, G. (submitted elsewhere): Bacteriorhodopsin expressed in *Schizosaccharomyces pombe* efficiently pumps protons over the plasma membrane.

Miercke, L. J. W., Ross, P. E., Stroud, R. M. & Dratz, E. A. (1989): Purification of Bacteriorhodopsin and Characterization of Mature and Partially Processed Forms. J. Biol. Chem., **264**, 7531-7535.

Wölfer, U., Dencher, N. A., Büldt, G. & Wrede, P. (1988): Bacteriorhodopsin precursor is processed in two steps. Eur. J. Biochem., **174**, 51-57.

Wrede, P., Büldt, G., Engelhard, M., & Hess, B. (1985): Synthesis of the signal peptide of bacteriorhodopsin. In: PEPTIDES: Structures and Functions, eds. Deber, C. M., Hruby, V. J. & Kopple, K. D. pp. 879-882. Pierce Chem. Co..

Comparative studies on bacteriorhodopsin-type pumps: basic physicochemical data for proton pumps

Yasuo Mukohata, Kunio Ihara, Yukiya Miyashita, Tomohiro Amemiya, Tomoyasu Taguchi, Minoru Tateno and Yasuo Sugiyama

Department of Biology, Faculty of Science, Nagoya University, Nagoya 464-01, Japan

In *Halobacterium halobium (salinarium)*, are present two retinal-protein ion-pumps, bacteriorhodopsin (bR; Oesterhelt and Stoeckenius, 1971) and halorhodopsin (hR; Mukohata *et al.* 1980), and two retinal-protein photosensors, sensory rhodopsin (sR; Tsuda *et al.* 1982; Bogomolni and Spudich, 1982) and phoborhodopsin (pR; Takahashi *et al.* 1985). In *Halobacterium* sp. aus-1 and aus-2 (Mukohata *et al.* 1988) are present archaerhodopsin(-1) (aR-1; Sugiyama *et al.* 1989) and archae-rhodopsin-2 (aR-2; Uegaki *et al.* 1991), respectively. They are light-driven proton pumps. The discovery of these new bR- family members from other strains of halobacteria initiated new lines of study of bacterial rhodopsins.

One line is the alignment of the primary structures of bacterial rhodopsins of similar functions to find out the conserved amino acid residues, under the preposition that the amino acids conserved in nature are essential for the structure and the function of bacterial rhodopsins. Almost all the essential amino acid residues which have been assigned by various means are actually found among the conserved amino acids (Mukohata *et al.* 1991a,b). The alignment of various types of bacterial rhodopsins will lead to find out the amino acids essential for their specific functions. Along with this line of study some primary structures of bacterial rhodopsins in strains of Halobacteriaceae have been analyzed (e.g., Lanyi *et al.* 1990).

The other line of study is the comparison of physicochemical properties and functions of those bacterial rhodopsins to find differences, which are then reduced to the change(s) in amino acid residue(s) and in the seven-helices architecture caused by such substitution of amino acid(s) (Mukohata *et al.* 1991b). In this case, the assignment can be tested by substituting the amino acid(s) by gene technology.

Our aR-1 and aR-2 have led us into these two lines of study. In this paper, we describe some basic data on the physicochemical properties of three proton pumps, bR, aR-1 and aR-2. The contents of retinal isomers in these pumps in the light- and the dark-adapted states and the absorption spectrum of bR with 100% of 13-*cis* retinal are also reported.

MATERIALS AND METHODS

Preparation of proton pump rhodopsins
Halobacterium halobium R_1M_1 (for bR), *Halobacterium* sp. aus-1 (for aR-1) and aus-2 (for aR-2) were cultured as described previously (Mukohata *et al.* 1991b). BR (purple membrane) was isolated by an ordinary procedure, stored in 50 mM NaCl in a refrigerator and used within one week after preparation. The claret membranes containing aR-1 (noted as aR in the previous papers) and aR-2 were isolated by the same method for bR/purple membrane. Protein was determined by the Lowry method. In the following experiments, unless otherwise noted, all three proton pumps were handled by identical procedures so that bR can be referred to aR-1 and aR-2.

Spectral parameters of aR-1 and aR-2
In order to obtain the absorption maxima (λmax) and the molar extinction coefficients (ϵ) for aR-1 and aR-2 in the claret membranes which contain isoprenoid pigments together with archaerhodopsins, the claret membranes were bleached in 1 M NH_2OH (pH 7.0) under yellow light illumination from 750W slide projector for 6 hrs at room temperature. The bleached membranes were washed twice with 50 mM NaCl. The difference spectra (near absorption maxima) were then recorded with stepwise additions of all-*trans* retinal to the bleached membranes in the dark or in the light. The absorbance increases were plotted against the retinal used for titration and the ϵ for aR-1 and aR-2 were calculated on the basis of M^{-1} of retinal.

Proton pump activities of aR-1 and aR-2

By procedures similar to those used for spectral measurements, cell envelope vesicles of *Halobacterium* sp. aus-1 and aus-2 were bleached. The proton pump activities of aR-1 and aR-2 were measured in the presence of 1 mM triphenylmethyl phosphonium (to diminish the influence of possible halorhodopsin-type pumps) as the initial rate of light-induced pH decrease.

Titration of the blue forms of proton pumps

The blue forms of aR-1, aR-2 and bR were prepared by passage of the corresponding membrane through a cation-exchange column (AG 50WX2, Bio-Rad). The blue membranes were washed with deionized water and then titrated spectrophotometrically with a $LaCl_3$, $CaCl_2$ or NaCl solution.

Determination of retinal isomers

A bR sample (10 µM in 10 mM phosphate buffer pH 7.2, 50 mM NaCl and 5 mM $MgCl_2$) in a jacketed cuvette was placed in a spectrophotometer (Shimadzu UV-300/SAPCOM-1) and illuminated through an optical guide by a 150W halogen lamp furnished with a yellow glass filter (Toshiba Y-42) for 1 min to produce 100% light-adapted bR. At designed time intervals after illumination, absorption spectra and the computed difference spectra were recorded, and soon after each scan of spectrophotometry 100 µl each of bR was sampled. This sampled bR was injected in 250 µl of an extraction mixture [a 1:1 mixture of ethanol and *n*-hexane (Scherrer *et al.* 1989)] at 0°C in a 1.5 ml Eppendorf tube, shaken by a flash mixer for 30 sec, centrifuged at 15,000 rpm for 1 min and kept in a dry block at 0°C. The retinal isomers in the upper layer (20 µl) were applied on HPLC (Beckman 112/420, Ultrasphere ODS 5 x 250 mm). Retinal isomers were eluted with *n*-hexane containing 4% diethyl ether at 0.2 ml/min, detected photometrically at 365 nm (Jasco UVIDEC-100-IV) and recorded (Shimadzu, C-R4A). The amounts of individual retinal isomers were computed from the area belonging to each peak. The temperature of the cuvette was kept constant at 25°C by circulating water through the jacket of the cuvette. All these procedures were carried out in the dark except for occasional illumination of dim red light for manipulation.

RESULTS AND DISCUSSION

Spectral parameters of aR-1 and aR-2

In Table I the λmax and the ε values of three proton pumps in the light-adapted and the dark-adapted states are listed. Although aR-1 is embedded in the claret membrane with relatively large amounts of isoprenoid pigments, a value of about 555 nm for the λmax of aR-1 is also obtained with the aR-1 which is expressed in *E. coli* and isolated and reconstituted with all-*trans* retinal (Sugiyama and Mukohata, to be published). The values of the dark-adapted pumps depend on the ratios of 13-*cis*:all-*trans* retinal in the rhodopsins and the ratios obtained for three proton pumps are actually different from one another. These smaller λmax values of the dark-adapted proton pumps than those of the light-adapted ones suggest that the λmax's for the aR-1 and aR-2 with 100% 13-*cis* retinal as the chromophore are also smaller than those with 100% all-*trans* retinal as shown for bR below.

Retinal isomers in the light-adapted and the dark-adapted rhodopsin pumps

The yield of extraction of total retinal from bR with the mixture of ethanol and *n*-hexane was 21-27 %. This value was close to that of Scherrer *et al.* (1989) who initiated the use of this semi-hydrophilic extraction medium. The efficiency of retinal extraction seemed to be sufficient to estimate the ratio of retinal isomers. The retention times of HPLC for the peaks of all-*trans* and 13-*cis* retinal were 16 min and 19 min, respectively. A 98% portion of total extracted retinal were recovered in all-*trans* and 13-*cis* form. The rest of the isomers (2% of total peak areas) could not be identified.

At the light-adapted states of all three proton pumps, the chromophore was found to be only in the all-*trans* configuration.

Table I. Spectral parameters and the ratios of retinal isomers of three proton pumps

	Light-adapted		*Dark-adapted*		
	λmax (nm)	ε ($M^{-1}cm^{-1}$)	λmax (nm)	ε ($M^{-1}cm^{-1}$)	all-*trans*:13-*cis*
aR-1	565	7.5×10^4	555	6.8×10^4	52 : 48
aR-2	560	7.8×10^4	555	7.6×10^4	75 : 25
bR	568	6.3×10^4 [a]	558	5.3×10^4 [b]	35 : 65

[a] from Oesterhelt and Hess (1973) [b] from Mukohata *et al* (1981).

At the fully dark-adapted states, a portion of the chromophores were in the 13-*cis* configuration. The ratios of all-trans:13-*cis* retinal are shown in Table I. The result for bR supports the result (34:66) of Scherrer *et al.* (1989) but not that (49:51) of Mowery *et al.* (1979). AR-1 and aR-2 showed different ratios.

The difference in the 13-*cis* retinal contents in three proton pumps would be ascribed to the difference in their amino acid sequences especially in the retinal pocket (Henderson, 1990). The only difference is found in Met-145 in bR and aR-1, which is substituted by Phe in aR-2. This substitution, however, can be one but not all, of the causes for the low 13-*cis* content in the dark-adapted aR-2, because aR-1, which has the identical sequence of the retinal pocket with bR, still contains a lower amount of 13-*cis* than bR does.

Absorption spectrum of the bR with 13-*cis* retinal

In the course of dark adaptation, the absorption spectrum of the light-adapted bR with almost 100% all-*trans* retinal as the chromophore (all-*trans* bR) is lowered and shifts toward shorter wavelengths with an isosbestic point at about 531 nm. The absorbance values at various wavelengths are plotted against the contents of 13-*cis* retinal which were determined just after each scan of spectral recording in the dark-adapting bR (Fig. 1A). As assumed from the one isosbestic point, the dark-adapting bR contains either one of the two predominant isomers, all-*trans* and 13-*cis*. This is shown in Fig. 1A by the linear relationship between the absorbance and the content of one of the two isomers.

The lines drown in reference to fixed wavelengths are then extraporated to 100% of 13-*cis*. This gives the absorbance of bR with 100% 13-*cis* as the chromophore (13-*cis* bR). These values are plotted against wavelengths (Fig. 1B) to obtain the absorption spectrum of 13-*cis* bR. The λmax and the ε of 13-*cis* bR are found to be 554 nm and 5.2×10^4 $M^{-1}cm^{-1}$. The λmax of all-*trans* bR was obtained similarly as 569 nm (and ε = 6.3×10^4 $M^{-1}cm^{-1}$ from Oesterhelt and Hess, 1973).

Titration of the blue forms of proton pumps with cations

The pH titration profiles of bR was nearly a single sigmoid, whereas those of aR-1 and aR-2 showed at least two transitions (Mukohata *et al.* 1991b). The blue forms of aR-1, aR-2 and bR are commonly produced by lowering pH and/or removing cations. These blue forms could be titrated back to the purple forms with cations. The titration curves obtained with La^{3+} were not so different from one another, whereas the curves obtained with Ca^{2+} (Fig. 2) and Na^+ were different. The blue-to-purple transitions of aR-1 and aR-2 require much higher concentrations of cations than that of bR does. The difference in the amino acid sequences of the loops and the margins exposed externally is to be discussed in future.

Proton pumping activities of aR-1, aR-2 and bR proton pumps

Relative rates of proton pumping of aR-1, aR-2 and bR in the NH_2OH-bleached/retinal-reconstituted cell envelope vesicles at pH 7.0 are 1.5 : 1.8 : 1, respectively. Since the amino acid residues of proton channel (and retinal pocket) in three proton pumps are almost perfectly conserved (Mukohata *et al.* 1991a,b), this difference may be ascribed to the efficiency of reconstitution and/or the degree of deterioration of vesicle membrane. Interaction of bacterioruberin with aR-1 and aR-2 (but not with bR) was shown by induced circular dichroism (Mukohata *et al.* 1991b). The structural differences between the claret membranes and the purple membrane due to bacterioruberin may cause such difference in proton pumping activities.

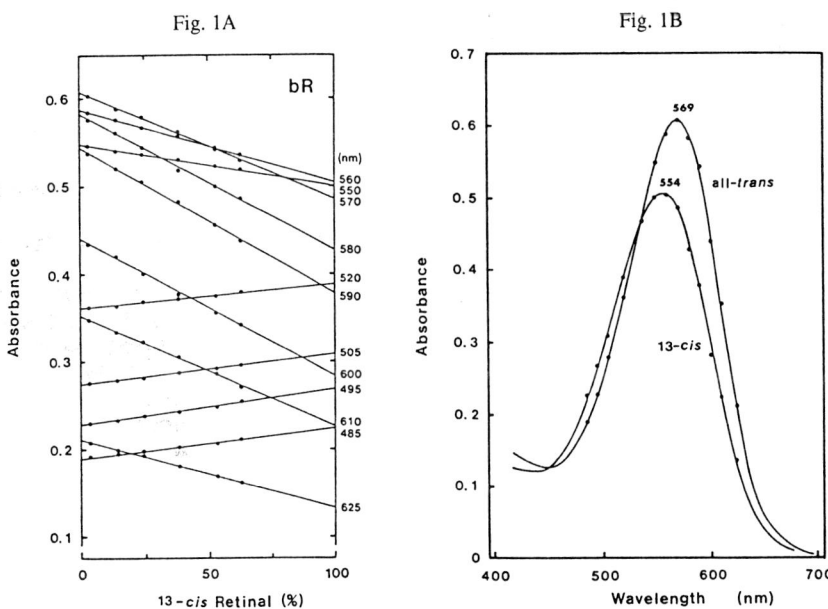

Fig. 1A

Fig. 1B

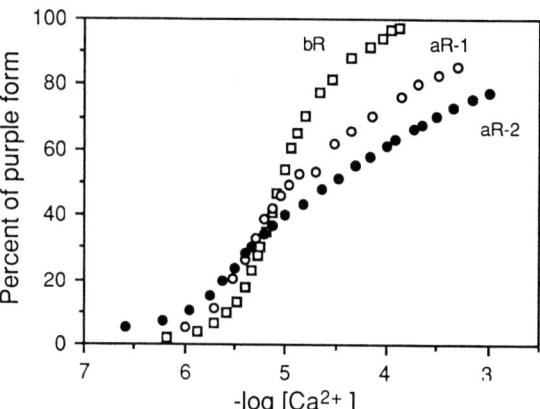

Fig. 2. The titration curves of the blue forms of proton pumps with Ca^{2+}.

REFERENCES

Bogomolni, R. A. and Spudich, J. L. (1982): Identification of third rhodopsin-like pigment in phototactic *Halobacterium halobium*. *Proc. Natl. Acad. Sci. U.S.A.* **79**, 6250-6254.

Henderson, R., Baldwin, J. M., Ceska, T. A., Zemlin, F., Beckmann, E. and Downing, K. H. (1990): Model for the structure of bacteriorhodopsin based on high-resolution electron cryo-microscopy. *J. Mol. Biol.* **213**, 899-929.

Lanyi, J. K., Duschl, A., Hatfield, G. W., May, K. and Oesterhelt, D. (1990): The primary structure of a halorhodopsin from *Natronobacterium pharaonis* structural functional and evolutionary implications for bacterialrhodopsins and halorhodopsins. *J. Biol. Chem.* **265**, 1253-1260.

Mowery, P. C., Lozier, R. H., Chae, Q., Tseng, Y. W., Taylor, M. and Stoeckenius, W. (1979): Effect of acid pH on the absorption spectra and photoreactions of bacteriorhodopsin. *Biochemistry* **18**, 4100-4107.

Mukohata, Y., Matsuno-Yagi, A. and Kaji, Y. (1980): Light-induced proton uptake and ATP synthesis by bacteriorhodopsin-depleted *Halobacterium*. In "*Saline Environment*" eds. Morishita, H. and Masui, M., pp..31-37. Tokyo: Business Center for Academic Society.

Mukohata, Y., Sugiyama, Y., Kaji, Y., Usukura, J. and Yamada, E. (1981): The white membrane of crystalline bacterioopsin in *Halobacterium halobium* strain R$_1$mW and its conversion into purple membrane by exogenous retinal. *Photochem. Photobiol.* **33**, 593-600.

Mukohata, Y., Sugiyama, Y., Ihara, K. and Yoshida, M. (1988): An Australian halobacterium contains a novel proton pump retinal protein archaerhodopsin. *Biochem. Biophys. Res. Commun.* **151**, 1339-1345.

Mukohata, Y., Sugiyama, Y. and Uegaki, K. (1991a): Archaerhodopsins: nobel bacterial proton pumps. In "*Light in Biology and Medicine*" ed. Douglas, R.H., Moan, J. and Ronto, G., pp. 557-566. New York: Plenum.

Mukohata, Y., Ihara, K., Uegaki, K., Miyashita, Y. and Sugiyama, Y. (1991b): Australian *Halobacteria* and their retinal-protein ion pumps. *Photochem. and Photobiol.* **54**, 1039-1045.

Oesterhelt, D. and Stoeckenius, W. (1971): Rhodopsin-like protein from the purple membrane of *Halobacterium halobium*. *Nature* **233**, 149-152.

Oesterhelt D. and Hess, B. (1973): Reversible photolysis of the purple complex in the purple membrane of *Halobacterium halobium*. *Eur. J. Biochem.* **37**, 316-326.

Scherrer, P., Mathew, M. K., Sperling, W. and Stoeckenius, W. (1989): Retinal isomer ratio in dark-adapted purple membrane and bacteriorhodopsin monomers. *Biochemistry* **28**, 829-834.

Sugiyama, Y., Maeda, M., Futai, M. and Mukohata, Y. (1989): Isolation of a gene that encodes a new retinal protein archaerhodopsin from *Halobacterium* sp aus-1. *J. Biol. Chem.* **264**, 20859-20862.

Takahashi, T., Tomioka, H., Kamo, N. and Kobatake, Y. (1985): A photosystem other than PS370 also mediates the negative phototaxis of *Halobacterium halobium*. *FEMS Microbiol. Lett.* **28**, 161-164.

Tsuda, M., Hasemoto, N., Kondo, M., Kamo, N., Kobatake, Y. and Terayama, Y. (1982): Two photocycles in *Halobacterium halobium* that lacks bacteriorhodopsin. *Biochem. Biophys. Res. Commun.* **108**, 970-976.

Uegaki, K., Sugiyama, Y. and Mukohata, Y. (1991): Archaerhodopsin-2, from *Halobacterium* sp. aus-2 further reveals essential residues for light-driven proton pumps. *Arch. Biochem. Biophys.* **286**, 107-110.

Properties and the primary structures of new bacterial rhodopsins

Jun Otomo, Hiroaki Tomioka, Yasuko Urabe and Hiroyuki Sasabe

Frontier Research Program, Riken Institute, Wako, Saitama 351-01, Japan

Summary

Several new halobacterial strains were isolated from crude solar salts commercially produced in Mexico and Australia. The presence of bacteriorhodopsin (BR)- and halorhodopsin (HR)-like pigment in their total membrane fraction was examined by flash spectroscopy and light-induced ion pumping activity. Two of these strains contained both BR- and HR-like pigments; the others contained only BR-like pigment. In addition, the presence of sensory rhodopsin (SR)- and phoborhodopsin (PR)-like pigments in these membrane was detected by flash spectroscopy. The kinetics of the cyclic photoreaction of the SR-like pigment was more than 15 times slower than that of *H. halobium*. A PR-like pigment existed in these strains, and the kinetics of the cyclic photoreaction of the PR-like pigment was similar to that found in *H. halobium*. The primary structures of helices A to G of all BR-like pigments identified in newly isolated halobacteria have been determined from the nucleotide sequences of genes coding these BR-like pigments. Using PCR method, the genes of BR-like pigments were directly amplified from the total genomic DNA of seven newly isolated halobacteria. The sequences of BR-like pigments in all strains were different from that of *H. halobium* and both *Halobacterium* sp. aus-1 and aus-2. However, functionally important residues were conserved in all strains. The sequence of HR-like pigment in one of these strains was also determined using PCR method. The sequence is different from that of both *H. halobium* and *Natronobacterium pharaonis,* and some residues proposed for the anion binding site are similar to those of *Natronobacterium pharaonis*.

Introduction

Since the first discovery of retinal containing protein, the light-driven proton pump bacteriorhodopsin (BR), in *Halobacterium halobium*, halorhodopsin (HR) with a light-driven chloride pump and two pigments with sensory functions, sensory rhodopsin (SR) and phoborhodopsin (PR) (or sensory rhodopsin-II) have also been found in the same halobacterial species (Oesterhelt & Stoeckenius, 1971; Matsuno-Yagi & Mukohata, 1977; Bogomolni & Spudich, 1982; Tomioka et al., 1986). Recently these bacterial rhodopsins were also found in other halobacterial species, such as archaerhodopsin in halobacteria from Australia, halorhodopsin- and phoborhodopsin-like pigments in haloalkaliphilic bacteria, Natronobacterium pharaonis (Mukohata et al., 1988; Bivin & Stoeckenius, 1986; Duschl et al., 1990; Mukohata et al., 1991). The amino acid sequences of these bacterial rhodopsins have been determined (Dunn et al., 1981; Blank & Oesterhelt, 1987; Blank et al., 1989; Sugiyama et al., 1989; Lanyi et al., 1990; Uegaki et al., 1991). The mechanism of the light-driven proton pump for BR has been extensively studied by site-directed mutagenesis (Khorana, 1988). The mutational approaches of HR, SR and PR, however, have so far not been reported, because of difficulties in expression and purification of the protein. The important residues for chloride ion transport and sensory functions have hence not yet been identified. In this context, amino acid comparisons between bacterial rhodopsins isolated from the other species could be effective for identification of functionally important residues.

Results and Discussion

Seven different single colonies from crude solar salts were isolated, and their features are summarized in Table 1. Purple membrane was isolated only from strain damp by washing in water. From strains mac and mex, a red colored patch was isolated instead of purple membrane.

Table 1. Summary of features and presence of BR-, HR-, SR- and PR-like pigments of seven halobacterial strains newly isolated from crude solar salts.

Name	Collected country	Color	Shape	Motility	Purple membrane	BR	HR	SR	PR
damp	Australia	Purple	Rods	+	+	+++	(+)	+	(+)
mac	Australia	Red	Rods	+	-	++	++	+	ND
mex	Mexico	Red	Rods	+	-	++	++	+	ND
port	Australia	Red	Rods	±	-	++	ND	ND	+
shark	Australia	Orange	Rods	+	-	+	ND	+	++
mex2	Mexico	Orange	Spheres	+	-	+	ND	+	++
shark2	Australia	Orange	Rods	+	-	+	ND	+	++

Salts from Guerrero Negro in Mexico, Dampier, Macleod, Port Hedland and Shark Bay in Australia were collected. New halobacterial strains were named after the place at which the salt was produced. The cell motility was monitored using the cell, which was selected by the chemotactic ring in semisolid agar plate. The presence of four pigments was estimated from flash-induced absorption change measurements and ion pump activity measurements. The presence and relative amounts of these four pigments are indicated by the number of plus sign.

Cell envelope vesicles of these new halobacterial strains and *H. halobium* R1 were prepared, and flash-induced absorption changes were measured. The flash-induced difference spectra of these strains have shown that there are BR-, SR- and PR-like pigments in these envelope vesicles. In addition, the light-induced ion pump activity of these cell envelope vesicles were also investigated. The data of these pumping activities suggested the presence of BR- and HR-like pigments in these new halobacterial strains (Otomo et al., 1992). The presence of four retinal pigments in these halobacterial strains were suggested in Table 1.

Using polymerase chain reaction (PCR), genes coding helices A to G of BR-like pigments have been succeeded in amplifying from all new halobacterial strains (Otomo et al., submitted). The nucleotide sequences of these genes have been determined, and these amino acid sequences have been deduced. The sequences of BR-like pigments in strains mac, mex, port, shark, shark2, and mex2 were different from that of *H. halobium* and both *Halobacterium* sp. aus-1 and aus-2. However, functionally important residues, such as Asp85, Asp96, Asp212 and Arg82 were conserved in all strains. Figure 1 shows the conserved amino acids in all BRs in the helical segments. Most of the conserved amino acid residues are faced forward the retinal, suggesting that these amino acids are important for proton pumping and interaction with the retinal.

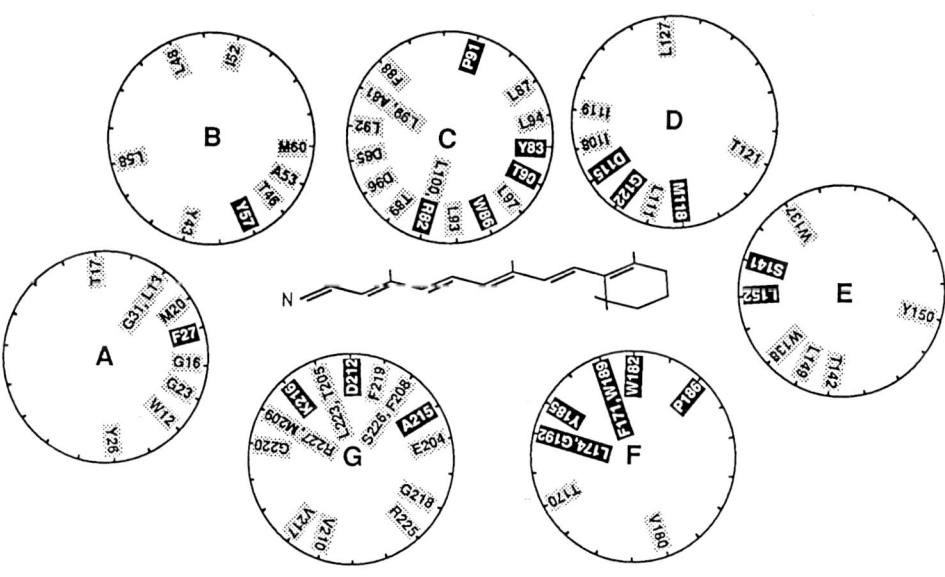

Fig. 1 A helical wheel projection model of BR. The amino acids in the helical segments proposed by Henderson et al. (1990) and all-trans retinal are shown. All conserved residues in halobium BR and five BR-like pigments (including aR-1 and aR-2) are indicated. The black colored parts indicate conserved residues in halobium BR, five BR-like pigments (including aR-1 and aR-2), two HRs and SR. The positions from 221 to 228 are compared among halobium BR, aR-1 and aR-2.

In addition, amino acid sequences of HR-like pigments in strains mac and mex have been determined. The amino acid sequence of mex HR is different from that of both *H. halobium* and *Natronobacterium pharaonis*. The homology of the amino acid sequences of three HRs (halobium HR, pharaonis HR and mex HR) are shown in figure 2.

Fig. 2. Comparisons of the amino acid sequences of three HRs. Percentages of homology at amino acid level are shown.

mex (m)			
pharaonis (p)	67		
haloblum (h)	65	62	
HR (%)	m	p	h

References

Bivin, D. B. & Stoeckenius, W. (1986): Photoactive retinal pigments in haloalkaliphilic bacteria. *J. Ccn. Microbiol.*, **132**, 2167-2177.
Blanck, A. & Oesterhelt, D. (1987): The halo-opsin gene. II. Sequence, primary structure of halorhodopsin and comparison with bacteriorhodopsin. *EMBO J.* **6**, 265-273.
Blanck, A., Oesterhelt, D., Ferrando, E., Schegk, E. S. & Lottspeich, F. (1989): Primary structure of sensory rhodopsin I, a prokaryotic photoreceptor. *EMBO J.*, **8**, 3963-3971.
Bogomolni, R. A. & Spudich, J. L. (1982): Identification of a third rhodopsin-like pigment in phototactic *Halobacterium halobium*. *Proc. Natl. Acad. Sci. U.S.A.*, **79**, 6250-6254.
Dunn, R., McCoy, J., Simesk, J. M., Majumdar, A., Chang, S. H., RajBhandary, U. L. & Khorana, H. G. (1981): The bacterioopsin gene. *Proc. Natl. Acad. Sci. U.S.A.*, **78**, 6744-6748.
Duschl, A., Lanyi, J. K. & Zimanyi, L. (1990): Properties and photochemistry of a halorhodopsin from the Haloalkalophile, *Natronobacterium pharaonis*. *J. Biol. Chem.*, **265**, 1261-1267.
Henderson, R., Baldwin, J. M., Ceska, T. A., Zemlin, F., Beckmann, E. & Downing, K. H. (1990): Model for the structure of bacteriorhodopsin based on high-resolution electron cryo-microscopy. *J. Mol. Biol.*, **213**, 899-929.
Khorana, H. G. (1988): Bacteriorhodopsin, a membrane protein that uses light to translocate protons. *J. Biol. Chem.*, **263**, 7439-7442.
Lanyi, J. K., Duschl, A., Hatfield, G. W., May, K. & Oesterhelt, D. (1990): The primary structure of a halorhodopsin from Natronobacterium pharaonis. *J. Biol. Chem.*, **265**, 1253-1260.
Matsuno-Yagi, A. & Mukohata, Y. (1977): Two possible roles of bacteriorhodopsin; a comparative study of strains of *Halobacterium halobium* differing in pigmentation. *Biochem. Biophys. Res. Commun.*, **78**, 237-243.
Mukohata, Y., Sugiyama, Y., Ihara, K. & Yoshida, M. (1988): An Australian Halobacterium contains a novel proton pump retinal protein: archaerhodopsin. *Biochem. Biophys. Res. Commun.*, **151**, 1339-1345.
Mukohata, Y., Ihara, K., Uegaki, K., Miyasaka, Y. & Sugiyama, Y. (1991): Australian *Halobacteria* and their retinal-protein ion pumps. *Photochem. Photobiol.*, **54**, 1039-1045.
Oesterhelt, D. & Stoeckenius, W. (1971): Rhodopsin-like protein from the purple membrane of *Halobacterium halobium*. *Nature New Biol.*, **233**, 149-152.
Otomo, J., Tomioka, H. & Sasabe, H. (1992): Bacterial rhodopsins of newly isolated halobacteria. *J. Gen. Microbiol.*, **138**, 1027-1037.
Sugiyama, Y., Maeda, M., Futai, M. & Mukohata, Y. (1989). Isolation of a gene that encodes a new retinal protein, achaerhodopsin, from Halobacterium sp. aus-1. *J. Biol. Chem.*, **264**, 20859-20862.
Tomioka, H., Takahashi, T., Kamo, N. & Kobatake, Y. (1986). Flash-spectrophotometric identification of a fourth rhodopsin-like pigment in *Halobacterium halobium*. *Biochem. Biophys. Res. Commun.*, **139**, 389-395.
Uegaki, K., Sugiyama, Y. & Mukohata, Y. (1991). Archaerhodopsin-2, from Halobacterium sp. aus-2 further reveals essential residues for light-driven proton pumps. *Arch. Biochem. Biophys.*, **286**, 107-110.

IV. Photocycle

IV. Cycle photochimique

Arg82Ala mutant of bacteriorhodopsin expressed in *H. halobium* : drastic decrease in the rate of proton release and effect on dark adaptation

Sergei P. Balashov, Rajni Govindjee, Masahiro Kono, Eugene Lukashov, Thomas G. Ebrey, Yan Feng*, Rosalie K. Crouch* and Donald R. Menick*

*Department of Physiology and Biophysics, University of Illinois at Urbana-Champaign, Urbana, IL 61801, USA and * Medical University of South Carolina, Charleston, SC 29425, USA*

In the R82A mutant of bacteriorhodopsin expressed in H.halobium the rate of proton release is delayed by two orders of magnitude, so the uptake of protons precedes the release. This suggests that Arg82 is associated with proton release in bR. The pH dependence of the rate of dark adaptation is drastically changed in the R82A mutant and correlates with the pK of the purple-to-blue transition. Apparently dark adaptation is catalyzed by the same group (Asp85) as the purple-to-blue transition, and transient protonation of Asp85 is necessary for thermal isomerization (dark adaptation). There is no net change in the protonation state of Tyr residues upon dark adaptation of R82A or WT. The R82A mutant shows some features [decreased stability at high pH, effect of buffer (CHES) on the N to M back reaction] which indicate that R82 is important not only for maintaining the functional state of bR at low pH (by shifting the pK of Asp85 from 7.5 to 2.8) but also at higher pH (9-11).

The studies of mutants of BR produced by site directed mutagenesis have revealed several key amino acids which are involved in proton transfer and maintaining the functional state of the pigment (for reviews see Mathies et al., 1991; Rothschild et al., 1992; Ebrey, 1992). These studies, together with a moderate resolution structure for bR (Henderson et al., 1990) have led to the conclusion that the Schiff base interacts with Asp85, Asp212 and Arg82 and probably Tyr57, Tyr185 and a divalent cation. It was concluded that Asp85 is a part of a complex counterion and acts as the primary acceptor of protons in the process of light-induced proton transfer. Moreover it is protonated in the purple-to-blue transition. Both the change in color of the pigment and proton transfer were found to be affected by Arg82 (Stern and Khorana, 1989). Substitution of a positively charged Arg82 by neutral alanine or glutamine results in a drastic shift of the pK of the purple-to-blue transition from pH 3 to pH 6.5-7 so that at neutral pH about half of the pigment is in the blue form, which does not form M and is inactive in proton transfer (Stern and Khorana, 1989; Subramaniam et al., 1990; Drachev et al. 1992). At pH 7.5 the R82Q mutant pigment incorporated in phospholipid vesicles pumps protons with a diminished efficiency compared to the wild type (Stern and Khorana, 1989; Miercke et al., 1991). However, the transient proton release and uptake from R82Q pigment was almost abolished; a small transient uptake signal was observed in R82A, the decay of which correlated with absorption changes at 650 nm (Otto et al., 1990). Dark adaptation was found to be much faster in R82A (Dunach et al., 1990), and the 13-cis-isomer was found, in light-adapted state of R82Q (Lin et al., 1991a; Thorgeirsson et al., 1991). The rate of M formation in R82Q mutant was found to be close to that in controls (Drachev et al., 1992) or faster (Lin et al., 1991b). These studies were done on mutants which were produced by expression of the synthetic gene of bR in E.coli. The pigment was obtained by subsequent reconstitution of the protein with retinal and lipids in vesicles. This pigment (e-bR) has slightly altered features as compared to bR: the absorption maximum is shifted to shorter wavelengths, the pigment is less stable and has narrower pH range of proton pumping, some of the rates of the photocycle reactions are altered.

Recently the bR gene has been expressed into *H.halobium* (Ni et al., 1990; Needleman et al., 1991), so that mutant pigments could be studied in their native lipid environment in purple membrane. In this paper the R82A mutant purple membrane isolated from *H.halobium* was examined with special interest in transient proton release and uptake kinetics, light and dark adaptation processes, and photochemical features as compared to the WT and e-bR.

Blue to purple transition in the mutant. Upon increasing the pH in the dark from pH 4.3 to 8.9 the absorption maximum of R82A mutant membranes shifts from 585 nm to 554 nm. The pK of the transition between blue and purple forms is approx. 7.5 in 0.15 M KCl and 15% glycerol which is close to the values reported previously for e-bR mutant. The formation of the long wavelength absorbing species at low pH (blue membrane) is apparently caused by the protonation of Asp85. The pK of the transition is around 2.8 in the WT (in 150 mM KCl). The shift of the pK to 7.5 in the case of R82A indicates that Arg82 causes a shift in the pK of Asp 85 by 4.7 units.

Light and dark adaptation. Difference absorption spectra upon dark adaptation at pH 7.8-10.0 in the mutant were similar to those in the WT. In both cases there were no significant changes at 240 nm, suggesting no change in the protonation state of a tyrosine residue during dark–adaptation.

pH dependence of the rate of dark adaptation in R82A. We have found that the rate of dark adaptation of the R82A mutant is very pH dependent at neutral and alkaline pH. At pH 7.8 and 20°C the half time of dark adaptation was 7 min (in the WT it was 122 min). The log of the rate constant of dark-adaptation decreases almost linearly with increase in pH from 7.8 to 10.5 (10 times per pH unit). At pH 8.8 the half time is 75 min, at pH 9.6 it is 5.6 hours, and at pH 10.0, 16 hours. This dependence on pH is very different from that of the WT, which has almost linear pH dependency of the log of the rate in the pH range 3-6. Between pH range 6 and 8.5 the rate is almost unchanged, while at pH > 8.5 a decrease in the rate is observed (Ohno et al, 1977). Thus the shift in pK of purple-to-blue transition (and apparently the pK of Asp85) from pH 2.8 to neutral pH correlates with the analogous shift of pH dependency of the rate of dark adaptation. This suggests that the protonated state of Asp85 catalyses dark isomerization in bR. That the rate of dark adaptation greatly increases at the pH range close to the pK of Asp85 can be explained by the simple suggestion that thermal isomerization is much more likely in the absence of a hydrogen bond between Asp85 and the Schiff base (which happens when Asp85 is protonated).

The isomer ratio in the LA and DA and absorption spectra of all-trans and 13 cis-isomeric states were determined from the flash induced absorbance changes at 650 nm (formation of the bathoproduct of 13-cis-bR) and 410 nm (M formation) at pH 8.8. The fast light-induced increase in absorbance at 650 nm was absent in the LA states both for the WT and the R82A mutant. This indicates that there is no 13-cis-bR in the LA state of the mutant. In the DA state, similar to the WT, the amount of M was 50±2% of that in the LA state. These data indicate that there is 50±2% trans-bR in the DA state and close to 100% (>98%) in the LA state. The absorption spectra of LA and DA states were obtained from the isomer ratios. Both are 4-6 nm shifted to the blue as compared to the absorption maxima in the WT: 565 nm (versus 569 nm) for the LA state, and 544 nm (versus 550) for 13-cis-pigment.

The decay time of the bathoproduct K^c. In the DA state the R82A mutant showed a large light-induced increase in absorbance at 650 nm due to the formation of the bathoproduct of 13-cis-bR, K^c. The lifetime of K^c is more than an order of magnitude shorter in the mutant (2.1 ms vs 30 ms at pH 10.0, 17 mM CHES) as compared to the WT. In the WT the formation of K^c is biphasic: the fast (unresolved) phase is accompanied by a slower additional increase in absorbance at 650 nm (half rise time is approx 2 ms). Similar but faster kinetics were observed at 650 nm by Otto et al. (1990) and were attributed to L or early O formation. The present data indicate that these absorbance changes at 650 nm and kinetically related proton changes described in Otto et al. (1990) probably are associated with the 13-cis photocycle (K^c) rather than the trans-cycle.

Photochemical conversions of the R82A mutant. *Rate and yield of M formation.* The rise of M is very fast in R82A. 70% of M is formed with a rise time of 1.0 µs and 30% with a rise time 25 µs. In the WT M rise time is 100 µs at neutral pH and decreases to 6 µs at pH 10. In the mutant the rate of M formation did not change significantly in the pH range between 7 and 10.5. The *yield* of M is pH dependent in R82A, correlating with the amount of the purple form of the pigment. At pH 4.3 there is almost no purple form, and there is no M. At pH 8.8 the blue form is absent. The yield of M at this pH is equal to that of the WT. The equality of the yields under these conditions indicates that the photochemical yield of M in the purple form of R82A is the same as in the wild type (0.64), assuming the extinction coefficient of the M's is the same.

M decay. N and O intermediates. Buffer effect on N to M back reaction. Between pH 7 and 8, M decay is monoexponential with a life time of 5±0.5 ms. At pH 10, in the absence of buffer, M decay has an additional slow component (> 100 ms, about 40%) which corresponds to the analogous component of 570 nm recovery, which is caused by the decay of N. In the presence of the pH buffer CHES (17 mM) this very slow phase in M decay disappears. This is different from the wild type which has the same slow kinetic component both in the presence and absence of buffer. The buffer affects the slow component of M decay only in R82A mutant. The slow component of M decay at high pH coincides with the decay of N and apparently is due to the back reaction from N to M. The lack of slow component in the R82A mutant indicates that the rate of back reaction from N to M in the presence of buffer is much less than the forward one. We did not observe an O intermediate (defined as a long wavelength species arising after M is formed) in the R82A mutant, probably because the rate of O to bR transition is very fast.

Proton release and uptake in the suspensions of R82A membranes. Figure 1 shows the flash-induced transient absorption changes of pyranine in the suspensions of WT and R82A purple membrane. The fast (less than 1 ms) decrease in absorbance of the dye observed in the WT corresponds

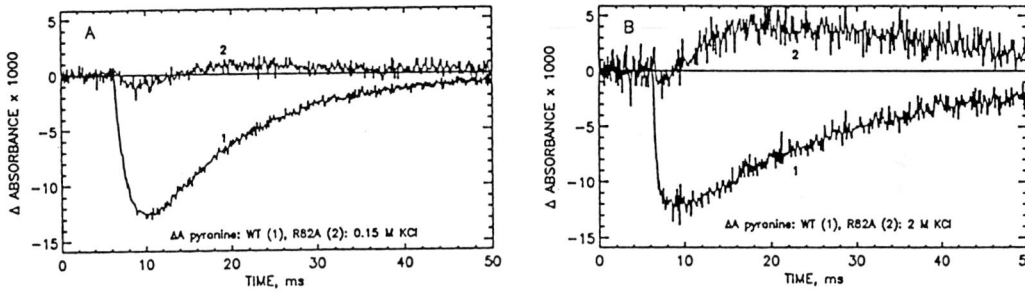

Fig. 1. Flash-induced transient absorption changes of pyranine in the suspensions of WT (1) and R82A (2) purple membrane in 150 mM KCl, pH 7.2 (A) and 2M KCl, pH 7.3 (B). The dye signals are normalized to the same amount of M formed. The decrease in absorbance corresponds to proton release.

to the release of proton from the purple membrane. Subsequent restoration of absorbance (rise time 12 ms) corresponds to the uptake of proton by the purple membrane. The transient proton release and uptake signals were highly inhibited in R82A at low salt concentration and pH 7.2. The maximum amplitude of the signal was only about 8% of the WT (when normalized to same amount of M). At pH 7.2 about 50% of the pigment is in blue form, and the rate of dark adaptation is fast so the amount of the active purple form in the sample is probably about only 30-40% of the WT. However the signals were normalized to the amount of M formed, to account for the actual amount of the active form in the sample. Upon raising the salt concentration to 2M KCl the signal from R82A increases several times and is 35-40% that of the WT (Fig. 1B). The signal from R82A has a positive (uptake) phase (rise time 6 ms) followed by a slow relaxation phase (H+ release) with the decay time of about 30 ms. These data indicate that the proton release process is highly constrained in the R82A mutant. However, experiments with liposomes showed that the mutant pumps protons with an efficiency at least 40% of that of the WT. This indicates that proton release is not completely inhibited in the mutant but highly delayed. Since the transient uptake signal is also small, one can speculate that proton release and uptake take place with almost the same rate in the mutant (in 150 mM salt). In 2M salt the positive (6 ms) and negative (30 ms) phases of the signal are seen, so one can suggest that the release of protons from the membrane is delayed as compared to the uptake.

Thus the lack of a positive charge in the vicinity of the Schiff base in the R82A mutant results in 100 times increase in the rate of M formation and Schiff base deprotonation (1 µs versus 100 µs) and approx. 100 times delay in proton release from the membrane. This suggests the possibility that R82 is on the release pathway of proton and probably directly catalyses proton release. One may speculate that since the positive charge on R82 changes the pK of D85 by 4.7 units, protonation of D85 during M formation also can cause considerable shift of pK of the guanidinium group (or water molecule associated with it, see Rothschild et al., 1992) to low pH and as a result cause a release of a proton from

R82A (or a water molecule). In the absence of R82 the release of a proton may occur from D85 (and may be some other intermediate group) after reprotonation of the Schiff base and reformation of the initial pigment.

Acknowledgements We wish to thank Dr. R. Needleman for providing the shuttle vector containing the *bop* gene, pMC-1, and for the generous help and assistance in making the mutant.

REFERENCES

Drachev, L.A., Kaulen, A.D., Khorana, H.G., Mogi, T., Postanogova, N.V., Skulachev, V.P., and Stern, L.J. (1992): The role of arginines 82 and 227 in the bacteriorhodopsin proton pump. *Photochem. Photobiol.* 55: 741-744.

Dunach, M., Marti, T., Khorana, H.G., and Rothschild, K.J. (1990): UV-visible spectroscopy of bacteriorhodopsin mutants: Substitution of Arg-82, Asp-85, Tyr-185, and Asp-212 results in abnormal light-dark adaptation. *Proc. Natl. Acad. Sci. USA* 87: 9873-9877.

Ebrey, T.G. (1992): Light energy transduction in bacteriorhodopsin. In *Thermodynamics of cell surface receptors*, ed M. Jackson. in press, CRC Press.

Henderson, R., Baldwin, J.M., Ceska, T.M., Zemlin, F., Beckmann, E., and Downing, K.H. (1990): Model for the structure of bacteriorhodopsin based on high resolution electron cryomicroscopy. *J. Mol. Biol.* 213: 899-929.

Lin, S.W., Fodor, S.P.A., Miercke, L.J.W., Shand, R.F., Betlach, M.C., Stroud, R.M., and Mathies, R.A.. (1991a): Resonance Raman spectra of bacteriorhodopsin mutants with substitutions at Asp-85, Asp-96 and Arg-82. *Photochem. Photobiol.* 53: 341-346.

Lin, G.L., El-Sayed, M.A., Marti, T., Stern, L.J., Mogi, T., and Khorana, H.G. (1991b): Effects of individual genetic substitutions of arginine residues on the deprotonation and reprotonation kinetics of the Schiff base during the bacteriorhodopsin photocycle. *Biophys. J.* 60: 172-178.

Mathies, R.A., Lin, S.W., Ames, J.B., and Pollard, W.T. (1991): From femtoseconds to biology: mechanism of bacteriorhodopsin's light-driven proton pump. *Ann. Rev. Biophys. Chem.* 20: 491-518.

Marti, T., Rosselet, S.J., Otto, H., Heyn, M.P., and Khorana, H.G. (1991): The retinylidene Schiff base counterion in bacteriorhodopsin. *J. Biol. Chem.* 266: 18674-18683.

Miercke, L.J.W., Betlach, M.C., Mitra, A.K. Shand, R.F. Fong, S.K., and Stroud, R.M. (1991): Wild type and mutant bacteriorhodopsins D85N, D96N, and R82Q: purification to homogeneity, pH dependence of pumping, and electron diffraction. *Biochemistry* 30: 3088-3098.

Ni, B., Chang, M., Duschl, A., Lanyi, J.K., and Needleman, R. (1990): An efficient system for the synthesis of bacteriorhodopsin in Halobacterium halobium. *Gene* 90: 169-172.

Needleman, R., Chang, M., Ni, B., Varo, G., Fornes, J., White, S., and Lanyi, J. (1991): Properties of Asp^{212} --> Asn bacteriorhodopsin suggest that Asp^{212} and Asp^{85} both participate in a counterion and proton acceptor complex near the Schiff base. *J.Biol. Chem.* 266: 11478-11484.

Ohno, K., Takeuchi, Y., and Yoshida, M. (1977): Effect of light-adaptation on the photoreaction of bacteriorhodopsin from Halobacterium halobium. *Biochim. Biophys. Acta* 462: 575-582.

Otto, H., Marti,T., Holz, M., Mogi, T., Stern, L.J., Engel, F., Khorana, H.G., and Heyn, M.P. (1990): Substitution of amino acids Asp-85, Asp-212, and Arg-82 in bacteriorhodopsin affects the proton release phase of the pump and pK of the Schiff base. *Proc. Natl. Acad. Sci. USA* 87: 1018-1022.

Rothschild, K.J., He, Y.-W., Sonar, S., Marti, M., and Khorana, H.G. (1992): Vibrational spectroscopy of bacteriorhodopsin mutants. Evidence that Thr-46 and Thr-89 form part of a transient network of hydrogen bonds. *J. Biol. Chem.* 267: 1615-1622.

Stern, L. and Khorana, H.G. (1989): Structure-Function Studies on Bacteriorhodopsin. X. Individual substitutions of arginine residues by glutamine affect chromophore formation, photocycle, and proton translocation. *J. Biol. Chem.* 264: 14202-14208.

Subramaniam, S., Marti, T., and Khorana, H.G. (1990): Protonation state of Asp(Glu)-85 regulates the purple to blue transition in bacteriorhodopsin mutants Arg82->Ala and Asp85-> Glu: The blue form is inactive in proton translocation. *Proc. Natl. Acad. Sci. USA* 87: 1013-1017.

Thorgeirsson, T.E., Milder, S.J., Miercke, L.J.W., Betlach, M.C., Shand, R.F., Stroud, R.M., and Kliger, D.S. (1991): Effects of Asp-96->Asn, Asp-85->Asn, and Arg82->Gln single-site substitutions on the photocycle of bacteriorhodopsin. *Biochemistry* 30: 9133-9142.

Tyr57Asn mutant of bacteriorhodopsin: M formation and spectral transformations at high pH

Rajni Govindjee, Eugene Lukashev, Masahiro Kono, Sergei P. Balashov, Thomas G. Ebrey, Jörg Soppa*, Jörg Tittor* and Dieter Oesterhelt*

*Department of Physiology and Biophysics, University of Illinois, Urbana-Champaign, IL, USA and *Max-Planck Institut für Biochemie, Martinsried, Germany*

Here we present spectroscopic and photochemical features of Y57N mutant of bR from *Halobacterium sp.* GRB. At pH below 7 a long-lived L intermediate is observed but no M is formed. The bathoproduct of 13-cis bR, K^c, dominates in the difference spectrum taken 1 ms after the flash. K^c decays within 30 ms. In a longer time scale (500 ms) slow decaying absorbance changes at 570 nm are seen which we assign to the photocycle of all-trans bR, and in particular to the transformation of the L intermediate back to initial bR. At pH > 8 some M is formed. However, the maximal amount of M is quite small (20% of that of the WT). Addition of glycerol or Triton X-100 (0.1%) results in the formation of some M at neutral and low pH. Alkaline titration shows that in Y57N, unlike the WT, there is no red shift of the chromophore absorption band and no tyrosine deprotonation (with pK around 9). This suggests that Y57 is one of the tyrosine residues deprotonating in bR at high pH.

The role of tyrosines in the BR photocycle is unclear. Mogi et al. (1987) saw no substantial difference in proton pumping ability upon substitution of tyrosine residues by phenylalanine. In contrast to Y57F, no M intermediate nor proton pumping activity was observed in Y57N (Soppa et al. 1989). Moreover, light/dark adaptation was altered. There was a higher ratio of 13-cis to all-trans isomer in the dark compared to the WT and 25% of 13-cis isomer was observed in the LA sample of Y57N. Braiman et al. (1988) also observed altered light/dark adaptation in Y57F.

Herzfeld et al. (1990) using NMR technique did not find any tyrosinates in dark adapted bR below pH 12. Also Ames et al. (1990), in the UV resonance Raman spectra, did not observe any tyrosinates up to pH 11. In contrast we have observed tyrosinate formation in bR with apparent pKs of about 9.0, 10.3 and 11.3 (in 167 mM KCl) by monitoring the absorbance changes in 230-350 nm region (Balashov et al., 1991). At pH 10.3, the difference spectrum is identical to that of free tyrosinate in solution. The source of this discrepancy is not clear. The pK of a small red shift of the chromophore absorption band seen at alkaline pH (Maeda et al., 1986; Balashov et al.,1991) coincided with the low pK tyrosine and the appearance of the fast rising component in the formation of the M intermediate (Kono et al., 1992).

In this study we wish to address the contributions of Tyr 57 to the alkaline behavior of BR by analyzing the spectral properties and photochemical reactions of the Y57N mutant.

Photochemical reactions in Y57N

Photoreactions at neutral pH. At pH 6.8 the light adapted form of Y57N has an absorption maximum around 560 nm. The half-time of dark adaptation is 40 min at 21°C. The Y57N mutant does not form M at neutral pH. Low temperature spectral measurements showed the formation of the early photoproducts K (at -160°C under 510 nm illumination) and L (at -65°C under 610 nm illumination). At room temperature at pH 6.5, the difference spectrum of light-induced absorbance changes has a maximum at 620 nm and a minimum at 560 nm (Fig. 1). The absorbance increase at 620 nm is apparently due to the bathoproduct of 13-cis bR, K^c, which completely decays within 30-40 ms (Fig. 2A). The Y57N mutant does not show a significant flash induced absorbance increase around 410 nm (Fig. 2B, curves 1, 2). In a longer time scale (500 ms) slowly decaying absorption changes (decrease in absorbance at 570 nm and increase at 490 and 390 nm) are observed, which are presumably associated with the all-trans cycle. Absorption changes at 570 nm (Fig. 2C, curve 1) and 490 nm recover with $t_{1/2}$~150 ms. Most likely these changes are caused by the formation of L and its decay to bR bypassing M, similar to that observed in bR at low temperatures (Litvin et al., 1975).

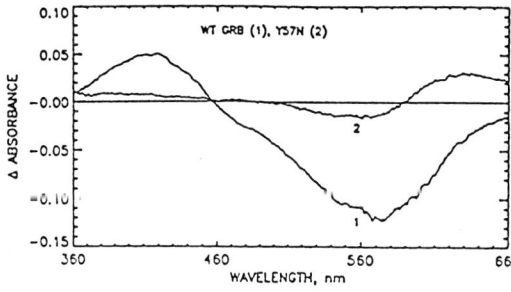

Fig. 1. Flash-induced difference absorption spectra measured 1 ms after the flash (532 nm). pH 7, 20°C.

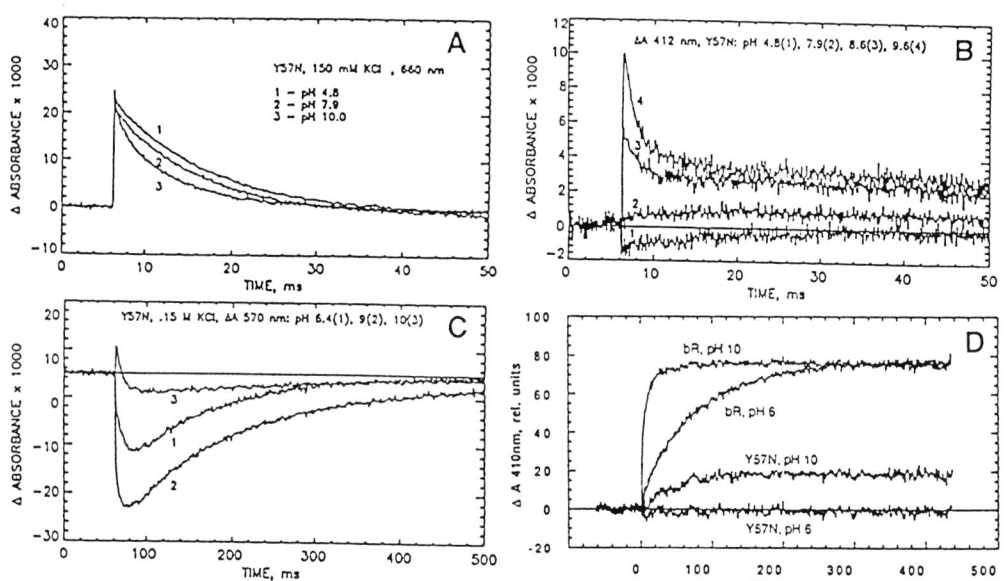

Fig. 2. Flash-induced absorption changes in Y57N at 412, 660 and 570 nm. $\lambda_{actinic}$ = 532 nm, 20°C, 150 mM KCl, pH as indicated.

pH dependence of the yield of M formation. At pH>8 a small amount of M intermediate is formed (Fig. 2B and D). The maximal amount of M is observed at pH 9.5 (150 mM KCl), though it is still less than 20% of the amount seen for WT (Fig. 2D). The formation of M is accompanied by ca. 100 µs photoelectric signal, the decay of which correlates with proton release in bR (Liu, 1990). The rise time of M in Y57N at pH 10 (~ 60 µs) is close to that in bR at pH 6 (85 µs) and much longer than in bR at pH 10 (6 µs). M decay is clearly biphasic (Fig. 2B), which suggests the formation of the N intermediate. It can be seen also in the slow decaying phase at 570 nm at pH 9 (Fig. 2C, curve 2).

Effect of glycerol and Triton X-100 on M formation. We have found that the ability to form M at neutral pH can be partially restored in the mutant by the addition of glycerol or Triton X-100. The rise time of M in glycerol and Triton X-100 (about 50-80 μs) is similar to that for WT. Y57F mutant studied by Mogi et al. (1987) and Braiman et al. (1988) and Y57N have several similar features: altered light adaptation and decreased yield of M intermediate. The reported difference in the proton pumping efficiencies, however, is very surprising. Y57N did not show any pumping (in cells), while Y57F (in vesicles) was almost as efficient as e-bR (Mogi et al., 1987). The effect of Triton X-100 on M formation provides a partial explanation for the different results. Apparently Y57N in the presence of Triton is closer to the monomeric Y57F in vesicles. However both in glycerol and in Triton X-100 the amount of M formed in Y57N is much less than in the WT.

Alkaline titration of the absorption spectrum

WT: Upon increasing the pH of a light adapted suspension of WT GRB purple membrane the main absorption band shifts a couple of nms to longer wavelengths. This red shift can be seen as an increase in absorbance around 615 nm in the difference spectra (pHi-pH7). At the same time there is a small increase in absorbance around 450-460 nm (Fig. 3A). In the near ultraviolet region there is an increase in absorbance around 240 nm, a decrease around 275 nm and an increase at 288 and 297 nm (Fig. 3C). These difference spectra suggest the deprotonation of tyrosine residues at high pH. Fig.4 shows that approximately 3 tyrosine residues deprotonate in WT GRB with pKs approximately 9.6, 10.5 and 11.9 (calculated from the absorption changes at 240 nm). These values are similar to those reported for the bR from *Halobacterium halobium*, strain S9 (Balashov et al. 1991).

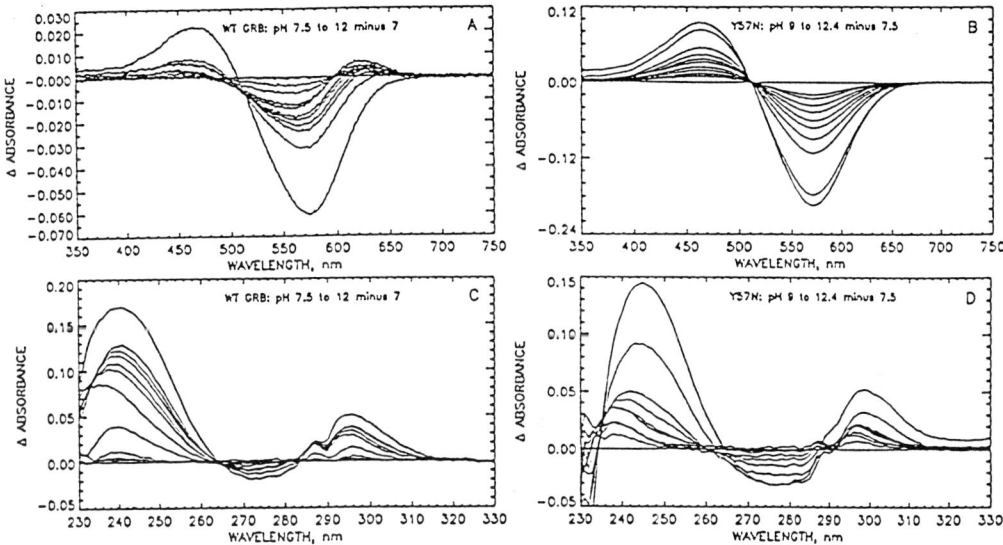

Fig. 3. Difference absorption spectra of light-adapted purple membrane from Halobacterium sp.GRB (A, C) and Y57N mutant membranes (B, D) in 167 mM KCl at pHi -pH7. Samples were light-adapted at pH 7 and then pH was changed in the dark. pHi inside to outside: WT, 7.2. 8.3, 9.0, 10.0, 10.5, 11.0, 11.3, 11.5, 11.6, 11.8, 12.2 and Y57N, 9.0, 9.5, 9.8, 10.2, 10.6, 11., 11.3, 11.6, 12.1, 12.3.

Y57N: No red shift of the main absorption band (i.e. no increase in the difference spectra at 615 nm) is observed when the pH of the light-adapted sample is changed from 7 to ~12 (Fig. 3B). The decrease in the main absorption band is accompanied by a monotonic increase in absorbance around 460-470 nm with an isosbestic point around 510 nm.This is caused by the formation of P480 species, the amount of which at a given pH is 3 times larger in Y57N. In the ultraviolet region, the difference spectra (pHi-pH7) show a maximum around 238-240 nm and a maximum around 296 nm with a minor shoulder at 288 nm (Fig. 2D). In Y57N, only 2 tyrosines deprotonate per BR with pK's 10.6 and 12.1 (Fig. 4). The low pK tyrosine seen in the WT is missing in the Y57N mutant. Thus, it is tempting to conclude

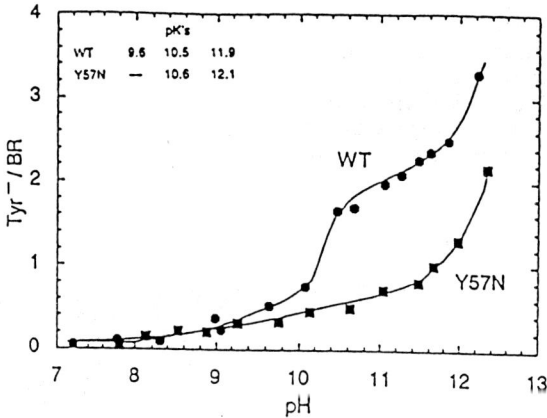

Fig.4. Tyrosines per bR in WT GRB and Y57N as a function of pH. Data taken from Fig. 3C and D (calculated from the absorption changes at 240 nm).

that this is Y57, and that the deprotonation of Y57 is responsible for the red-shifted absorption seen upon the formation of alkaline BR.

REFERENCES

Ames, J. B., Bolton, S. R., Netto, M. M., and Mathies, R. A. (1990): Ultraviolet resonance Raman spectroscopy of bacteriorhodopsin: Evidence against tyrosinate in the photocycle. *J. Am. Chem. Soc. 112*: 9007-9009.

Balashov, S. P., Govindjee, R., and Ebrey, T. G. (1991): Red shift of the purple membrane absorption band and the deprotonation of tyrosine residues at high pH: Origin of the parallel photocycles of trans-bacteriorhodopsin. *Biophys. J. 60*: 475-490.

Braiman, M. S., Mogi, T., Stern, L. J., Hackett, N. R., Chao, B. H., Khorana, H. G., and Rothschild, K. J. (1988): Vibrational spectroscopy of bacteriorhodopsin mutants: I. Tyrosine-185 protonates and deprotonates during the photocycle. *Proteins: Structure, Function, and Genetics 3*: 219-229.

Herzfeld, J., Gupta, S. K. D., Farrar, M. R., Harbison, G. S., McDermott, A. E., Pelletier, S. L., Raleigh, D. P., Smith, S. O., Winkel, C., Lugtenburg, J., and Griffin, R. G. (1990): Solid-state 13C NMR study of tyrosine protonation in dark-adapted bacteriorhodopsin. *Biochemistry 29*: 5567-5574.

Kono, M., Misra, S., and Ebrey, T.G. (1992): Light-induced currents in bacteriorhodopsin: pH and salt effects. *Biophys. J. 61*: 529a.

Liu, S.Y. (1990): Light-induced currents from oriented purple membrane. I. Correlation of the microsecond component (B2) with the L-M photocycle transition. *Biophys. J. 57*:943-950.

Litvin, F.F., Balashov, S.P., and Sineshchekov, V.A. (1975): The investigation of the primary photochemical conversions of bacteriorhodopsin in purple membranes and cells of Halobacterium halobium by the low temperature spectrophotometry method. *Bioorganic Chemistry (USSR) 1*: 1767-1777.

Maeda, A., Ogura, T., and Kitagawa, T. (1986): Resonance Raman study on proton-dissociated state of bacteriorhodopsin: Stabilization of L-like intermediate having the all-trans chromophore. *Biochemistry 25*:2798-2803.

Mogi, T., Stern, L. J., Hackett, N. R., and Khorana, H. G. (1987): Bacteriorhodopsin mutants containing single tyrosine to phenylalanine substitutions are all active in proton translocation. *Proc. Natl. Acad. Sci. USA 84*: 5595-5599.

Soppa, J., Otomo, J., Straub, J., Tittor, J., Meessen, S., and Oesterhelt, D. (1989): Bacteriorhodopsin mutants of Halobacterium sp. GRB. II. Characterization of mutants. *J.Biol.Chem. 264*: 13049-13056.

Conformers and multiple primary photoproducts of bacteriorhodopsin and N intermediate at low temperature

Sergei P. Balashov, Eleonora S. Imasheva, Nina V. Karneyeva, Felix F. Litvin and Thomas G. Ebrey*

Biology Faculty, M.V. Lomonosov Moscow State University, 119899, Moscow, Russia and *Department of Physiology and Biophysics, University of Illinois at Urbana-Champaign, Urbana, IL 61801, USA

At 90K the bathoproduct of all-trans-bR, K^t, consists of at least three spectrally different states, which are formed from different precursors - conformers of all-trans-bR. These conformers are stable at 90K, but are in equilibrium at 210K. The bathoproduct of 13-cis-bR, K^c, is also heterogenous. Upon illumination at 77K at 510 nm the photocycle intermediate N transforms into a bathoproduct K^N, which is actually a mixture of at least two species, K^N_I and K^N_{II}. K^N can be photoreversed with illumination at >650 nm. With high light intensity irradiation at 500-650 nm the intermediate N is transformed into a product, designated as P^N, with an absorption maximum around 610 nm. In contrast to K^N, P^N is not photoreversible. Only a fraction of N can be converted into P^N, which indicates that there are two fractions of N, or two N intermediates in the photocycle, N_1 and N_2. At temperatures higher than -100°C K^N transforms into an intermediate with an absorption maximum at 565 nm, which is an L intermediate of N, L^N. No significant amount of M like intermediate was formed upon illumination of N at -60°C.

The functioning of bacteriorhodopsin involves conformational changes of both the chromophore and the protein moiety. The internal mobility of protein groups apparently results in conformational heterogeneity of the protein. The different conformeric states are in fast equilibrium at room temperature, however, they can be distinguished at low temperatures (77-90K) (Balashov et al., 1986, 1988, 1991; Ormos et al., 1987). Upon illumination, the various conformers are transformed into a mixture of different bathoproducts. Illumination of this mixture with far red light enabled us to separate spectrally distinguishable bathoproducts and their parent pigments (Balashov et al., 1988; 1991). The approach based on the identification of the different photoactive states of the pigment (even though spectrally identical) from the back photoconversion of their primary photoproducts under selective excitation was applied to the L and M intermediates (Litvin and Balashov, 1977) and more recently to the N intermediate (Balashov et al., 1990). In addition to the main intermediates of the photochemical cycle of bR, bR ↔ K → L → M → N → O → BR (Lozier et al., 1975; Litvin et al., 1975; Mathies et al., 1991) several other intermediates were found (Balashov et al., 1981; 1988) which are formed upon the excitation of L and M intermediates [Fig. 1; see also Hurley at al., (1978); Kalisky et al., (1978)]. In this paper we describe multiple photoproducts and conformeric states of bR and the N intermediate.

Multiple bathoproducts and conformers of BR

At 90 K the photoproduct of the primary light reaction of all- trans-bacteriorhodopsin (trans-bR), the bathoproduct K^t, produced by 500 nm illumination of a water-glycerol suspension of light-adapted purple membranes, is not homogeneous. It consists of at least three spectrally different components (conformers), K^t_I, K^t_{II} and K^t_{III}, having absorption maxima at ca 595, 605 and 615 nm, respectively. The difference between absorption maxima of the bathoforms is quite large, about 10 nm (the maxima in the difference absorption spectra are shown in Fig. 2). The three Ks were observed at different pHs (5, 6.7 and 10.5).

The bathoproducts are formed from different precursors, conformers of trans-bR (Balashov et al., 1991). These conformers are stable at 90 K, but are in equilibrium with each other at 210 K. We suggest that these conformers of bR and K result from different conformations of the chromophore binding site, probably the Schiff base counterion. The existence of several conformers of trans-bR may result in parallel photocycles of bR, at least at low temperature. 13-cis-bacteriorhodopsin also yields at least three different bathoproducts, K^c_I, K^c_{II} and K^c_{III} (the difference spectra of these bathoproducts are shown in Fig. 2, curves 4,5 and 6). Multiplicity of the primary photoproducts is also observed upon excitation of M (see the scheme in Fig.1) and N at 77-90K.

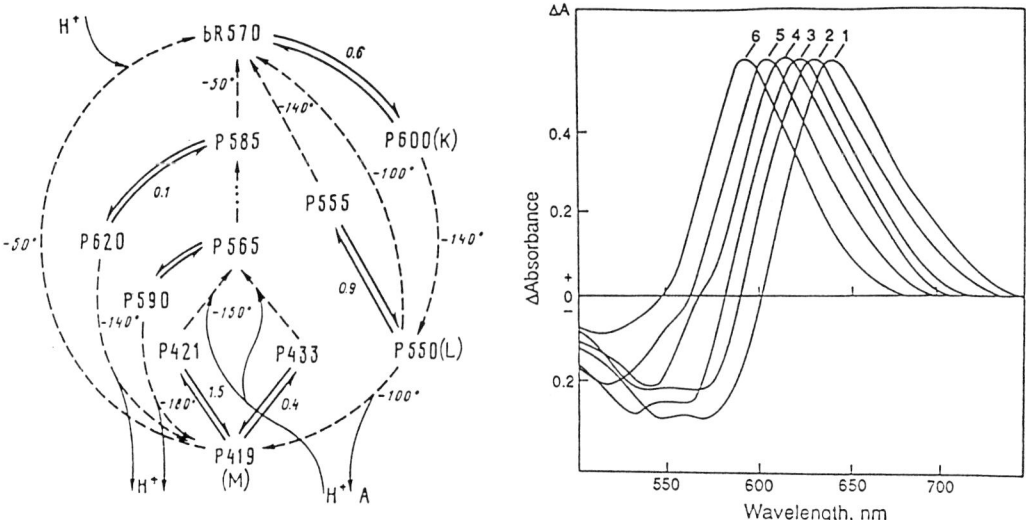

Fig. 1 (Left). Low temperature photoconversions of bR and the L and M intermediates. The numbers indicate the quantum yield ratios of the forward and back photoreactions.

Fig. 2 (Right). Difference absorption spectra (normalized and shown with opposite sign) corresponding to the photoconversion of the bathoproducts of all-trans-bR at 90K, K^t_I (1) ; K^t_{II} (2) and K^t_{III} (3) and 13-cis-bR K^c_I (4), K^c_{II} (5) and K^c_{III} (6). The spectra were obtained by the following procedure. The light adapted membranes were illuminated at 90K with 510 nm light to produce bathoproduct K^t. Then the sample was illuminated at wavelengths >720 nm and the kinetics of back conversion of K^t to bR was measured at 640 nm, which revealed three kinetic and spectral components corresponding to the transformation of K^t_I, K^t_{II} and K^t_{III}. The fractions of absorption changes associated with these bathoproducts were: 34%, 51% and 15%, respectively. In a similar way, using illumination at > 680 nm, the bathoforms K^c_I, K^c_{II} and K^c_{III} were revealed in the dark adapted sample after illumination with 510 nm at 90K and at > 720 nm in order to remove the bathoproducts of all-trans-bR (Tokunaga et al., 1976). The fractions of absorption changes at 620 nm for K^c_I, K^c_{II} and K^c_{III} were 40%:46%:14%.

Photoconversion of N intermediate at low temperature. Conformers of N

At high pH and high salt concentration the lifetime of the N intermediate increases, and its excitation becomes possible under physiological conditions. Kouyama et al. (1988) suggested that the N intermediate is photoactive and upon illumination can form an L-type and slow decaying M-type intermediate. In contrast to this Drachev et al. (1987) did not find any M type intermediate upon excitation of N. In order to investigate the photoconversion of N we accumulated and trapped it at low temperatures in aqueous suspensions of purple membrane, pH 10.2, 0.2 M KCl, (Balashov et al., 1990). Here we describe the intermediates and states revealed upon excitation of N.

Primary photoreaction N → K^N At 77K the N intermediate has an absorption maximum at 570 nm. Upon illumination of N with 510 nm light at 77K it transforms into a bathoproduct K^N (absorption maximum at 605 nm). K^N is spectrally similar to the bathoproduct of 13-cis-bR but differs from the bathoproduct of L. The formation of K^N is photoreversible. Illumination at wavelengths >640 nm completely converts K^N back to N (the difference spectrum of K^N minus N is shown in Fig. 3A, curve 1). Both forward and back photoreactions, N → K^N, have high quantum efficiencies at 77K (>0.2). Illumination of K^N at longer wavelengths (> 680 nm) converts only part of K^N, which indicates that K^N, similar to the bathoproducts of 13-cis- and all-trans-bR consists of at least two states, K_I^N and K_{II}^N, having slightly different absorption spectra.

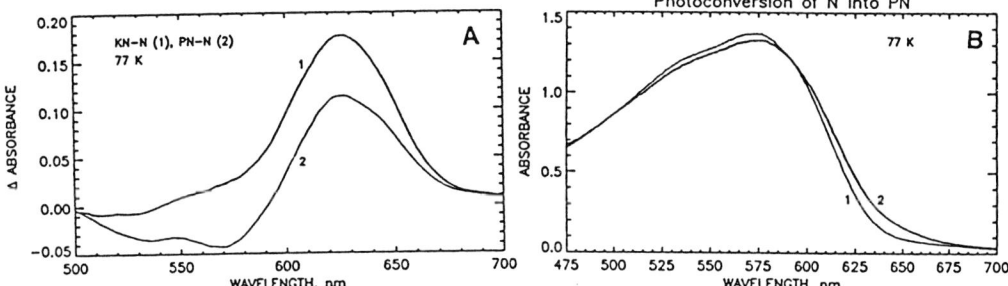

Fig. 3. A: Difference absorption spectra showing the photoconversions of N at 77K: N → K^N (curve 1) obtained after irradiation of N for 3 min with 510 nm (2 W/m^2); N → P^N (curve 2) obtained as curve 2 minus curve 1 in panel B. B: Absorption spectra of the samples containing 70% N and 30% trans-bR at 77K before (curve 1) and after (curve 2) 30 min high intensity irradiation at 520-700 nm (50 W/m^2), which results in transformation of 60% of N into P^N. Sample 2 was also illuminated 1 min at > 650 nm for conversion of the bathoproducts K^t and K^N back to the initial states.

N → P^N, photoreaction with a low quantum yield. Prolonged irradiation of N at 77K with high intensity yellow light (30 min, 50 W/m^2) N is transformed into a red shifted photoproduct, P^N. The position of the absorption maximum is around 600-620 nm. In contrast to the bathoform K^N, P^N is not photoactive. It does not convert back to N upon irradiation with red light (even after prolonged illumination). The quantum efficiency of N → P^N photoreaction is about 10^{-3}. Upon heating to -60°C P^N converts mainly back to N. The chromophore is in 13-cis-configuration in N (Mathies et al., 1988). The photoreaction N → P^N apparently is similar to the photoreaction of 13-cis-bR → P582 which takes place at 77K and results in the formation of a photoproduct (with a quantum yield of 6·10^{-4}) with an absorption spectrum very similar to all-trans-bR, but shifted 4 nm to longer wavelengths (Balashov et al., 1988). One may suggest that the transformations of N into P^N also involve isomerization and that P^N has a trans-configuration of the chromophore and thus similar to the intermediate O, which has an all-trans configuration of the chromophore. In fact both P^N and O have red shifted absorption bands around

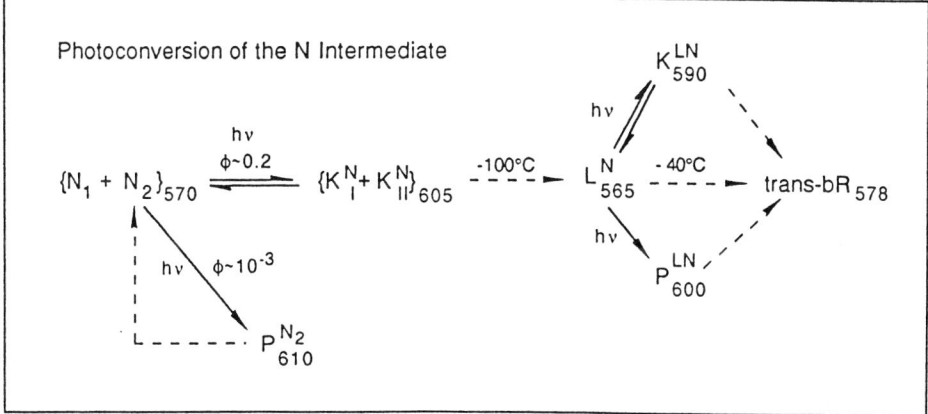

Fig. 4. Scheme of reactions and intermediates in the photoconversion of N. Photoreactions are designated by solid arrows and dark reactions by dashed arrows.

600-620 nm. The O intermediate is also apparently nonphotoactive. Thus it is reasonable to suggest that P^N is in fact O, or an intermediate similar to O.

Evidence for two N intermediates. Only a fraction of N can be converted into P_N (about 50%) upon prolonged irradiation at 77K. (The amount of N in the sample was estimated from the amplitude of the photoreversible absorption changes associated with the N ↔ K^N photoconversion). This indicates that there are two populations of N, or two Ns in the photocycle, N_1 and N_2. It is likely that N_1 is in equilibrium with M, and N_2 with O (since it is photoconvertable into the O-like intermediate P^N):

$$M \rightleftharpoons N_1 \rightleftharpoons N_2 \rightleftharpoons O$$

We suggest that N_2 differs from N_1 in the conformation of the protein part of the pigment, so that only N_2 can accommodate the trans-configuration of the chromophore. The finding of the two N forms at low temperatures is in agreement with the suggestion of Mathies et al. (1991) on the 2 N states in the photocycle of bR.

L^N, L form of N. At -100C K^N converts in the dark into a product with an absorption maximum at 540 nm. It may be called the L form of N, or L^N. L^N is stable at -60°C. Upon illumination it transforms into a red shifted bathoproduct which may be called K^{LN}. It is also stable at -60°C. Under illumination at 650 nm K^{LN} converts back to L^N. Illumination of N at -60°C results in the formation of a mixture of L^N and K^{LN}. We did not observe the formation of a significant amount of M from N at -60°C. Under high intensity irradiation L^N undergoes a phototransformation $L^N \rightarrow P^{LN}$, similar to the reaction N → P^N.

REFERENCES

Balashov, S.P., Imasheva, E.S., Litvin, F.F., and Lozier, R.H. (1990): The N intermediate of bacteriorhodopsin at low temperatures. *FEBS Lett. 271:* 93-96.

Balashov, S.P., Karneyeva, N.V., Litvin, F.F., and Ebrey, T.G. (1991): Bathoproducts and conformers of all-trans and 13-cis-bacteriorhodopsin at 90 K. *Photochem. Photobiol. 54:* 949-953.

Balashov, S.P., Karneyeva, N.V., Imasheva, E.S., and Litvin, F.F. (1986): Two forms of trans-bacteriorhodopsin at 77K. *Biophysics 31:* 1070-1073 (Engl. Transl., Pergamon Journals Ltd.).

Balashov, S.P., and Litvin, F.F. (1981): Photochemical conversions of bacteriorhodopsin. *Biophysics 26:* 566-581 (Engl.Transl., Pergamon Press Ltd).

Balashov, S.P., Litvin, F.F., and Sineshchekov, V.A. (1988): Photochemical processes of light energy transformation in bacteriorhodopsin. *Physicochemical Biology Reviews. Vol 8*, ed V.P. Skulachev, pp. 1-61. Switzerland, Harwood Academic Publ.

Drachev, L.A., Kaulen, A.D., Skulachev, V.P., and Zorina, V.V. (1987): The mechanism of H^+ transfer by bacteriorhodopsin. The properties and the function of intermediate P. *FEBS Lett. 226:* 139-144.

Hurley, J.B., Becher, B., and Ebrey, T.G. (1978): More evidence that light isomerises the chromophore of purple membrane protein. *Nature 272:* 87-88.

Kalisky, O., Lachish, U., and Ottolenghi, M. (1978) Time resolution of a back photoreaction in bacteriorhodopsin. *Photochem. Photobiol. 28:* 261-263.

Kouyama, T., Kouyama, A., Ikegami, A., Mathew, M.K., and Stoeckenius, W. (1988): Bacteriorhodopsin photoreaction: identification of a long-lived intermediate N (P, R350) at high pH and its M-like photoproduct. *Biochemistry 27:* 5855-5863.

Litvin, F.F., Balashov, S.P., and Sineshchekov, V.A. (1975): The investigation of the primary photochemical conversions of bacteriorhodopsin in purple membranes and cells of Halobacterium halobium by the low temperature spectrophotometry method. *Bioorganic Chemistry (USSR) 1:* 1767-1777.

Litvin, F.F., and Balashov, S.P. (1977): New intermediates in the photoconversion of bacteriorhodopsin. *Biophysics 22:* 1111-1114.

Lozier, R.H., Bogomolni, R.A., and Stoeckenius, W. (1975): Bacteriorhodopsin: a light-driven proton pump in Halobacterium halobium. - *Biophys. J. 15:* 955-962.

Ormos, P., Braunstein, D., Hong, M.K., Lin, S.-L., and Vittitow, J.(1987): Hole burning in bacteriorhodopsin. In *Biophysical Studies of Retinal Proteins*, eds T.G.Ebrey, H.Frauenfelder, B.Honig and K. Nakanishi, pp.238-246. Urbana, USA, University of IL.

Mathies, R.A., Lin, S.W., Ames, J.B., and Pollard, W.T. (1991): From femtoseconds to biology: Mechanism of bacteriorhodopsin's light-driven proton pump. *Ann. Rev. Biophys. Chem. 20:* 491-518.

Tokunaga, F., Iwasa, T., and Yoshizawa, T. (1976): Photochemical reaction of bacteriorhodopsin. *FEBS Lett. 72:* 33-38.

Biological significance of the *trans–cis* isomerization of retinal in proton transfer processes of bacteriorhodopsin

Hidemi Okazaki[1], Chia-Wun Chang[1], Toshihiro Akaike[1], Osamu Oshida[2] and Tamio Yasukawa[2]

[1] *Kanagawa Academy of Science and Technology Takatsu-Ku, Kawasaki, Kanagawa 213, Japan.* [2] *Division of Chemical and Biological Science and Technology, Faculty of Technology, Tokyo University of Agriculture and Technology, Nakamachi, Koganei, Tokyo 184, Japan*

The determination of 3-dimentional structure of bacteriorhodopsin bR in its ground state (Henderson et al., 1990) enable us to discuss the mechanism of proton transfer in more details. The structures of intermediate states in photocycle, however, have not been known. In a preceeding paper (Oshida et al., 1991), we reported some preliminary results on the structure of initial K and L states obtained by molecular dynamics simulation. On the other hand, in their extensive review on dynamical aspects of bR photocycle, Mathies et al. (1991) reported structural changes of bR calculated by an energy minimization procedure. Generally speaking, the determination of 3 dimentional structure of complex system by energy minimization is difficult because of multiple minima. Since the conformation of retinal and its specific interactions with neighbouring amino acid residues play critical roles in proton transfer process, we have attempted to discuss in details the conformation of retinal and the biological significance of 13-trans to cis transformation in the initial stage of photocycle by computer simulation.

Conformation of 13-cis retinal Schiff base in bR and its pKa

In the initial state of photocycle, ground state bR_{568} is converted into J, and then into K on a 3 ps time scale and into L_{550} with the life time of $\sim 2\mu s$ (Petrich et al., 1987). Low temperature resonance Raman spectra suggested that K intermediate is a highly distorted 13-cis species (Braiman and Mathies, 1982). During this short time, the large ionon ring of retinal moiety, as well as the backbone of Lys-216, cannot move appreciably. Hence, to begin with, we generated an isolated all-trans retinal–Lys Schiff base model molecule whose ionon ring and backbone atoms $-NH-C\alpha-CO-$ of N-acetyl,acetoamide of Lys are fixed just at the spacial points of corresponding groups of bR_{568} (all-trans). Then, an isolated 13-cis retinal–Lys Schiff base was formed initially using artifically weakened force constants for bond stretching, bending, and rotation (10% as low as normal values) and temperature was raised to 800K. By employing simulated annealing technique (Davis, 1987), temperature was reduced gradually to 300K. Force constants are also gradually strengthened to normal values in the course of perturbative motions of the conjugate chain part of retinal and side chain of Lys under the constraint that ionon ring and backbone atoms of Lys are fixed to the original sites. In the course of thermal fluctuations of model chain, 100 samples having different conformations were sampled and energy-minimized. By this procedure, 5 different well populated energy-minimized structures were obtained. Then, the retinal–Lys Schiff base moiety in bR_{568} was replaced by the lowest energy 13-cis moiety of which force constants and nonbonded atom interaction potentials are again weakened. This artificial bR containing 13-cis retinal underwent again annealing from 800K to 300K accompanied with gradual strengthening of force constatns to normal values. In the course of this simulated annealing processes, residues of opsin beyond 10Å from any atoms of retinal–Lys Schiff base were tentatively fixed at original sites. After this system reached a steady state, it was energy-minimized. Computations of loop parts of bR outside the membrane are not known and are tentatively determined by consulting substructure forms of homologous chains obtained from PDB.

One of basic problems in discussing the mechanisms of photo-induced proton transfer is the biological significance of trans–cis isomerization of retinal. In the present work, we have discussed the effects of molecular environments on pKa of Schiff base of trans and cis form retinal moieties by using a semi-empirical quantum calculation program MNDO-PM3.

The environmental effects on pKa of Schiff base have been examined by calculating the heat of formation Hf (electronic energy + core-core repulsion) for assemblies of model compounds of bR segments. Namely, each amino acid residue around retinal is replaced by its N-acetyl, N-methyl acetamide derivative with the same coordinates for respective corresponding atoms in energy-optimized bR. In table 1 are shown 1) Hf of isolated retinal-Lysine derivative (NH^+) and its deprotonated counterparts (-N-) in trans and cis form, 2) those of retinal-Lys$^+$ + Asp$^-$85, 3) those of retinal-Lys$^+$ + Asp$^-$212, 4) those of retinal-Lys$^+$ + Asp$^-$85 + Asp$^-$212, and 5) those of retinal-Lys$^+$ + Asp$^-$85 + Asp$^-$212 + Arg$^+$82. Free energy of deprotonation process may approximately be equated to the difference ΔHf (deprotonation) ($= Hf(-N-) + Hf(H^+_{solv}) - Hf(NH^+)$). Since exact estimate of $Hf(H^+_{solv})$ is problematic, we discuss only the difference of pKa between trans and cis retinal ΔpKa;

$$\Delta pKa = 0.362 * \{\Delta Hf(cis) - \Delta Hf(trans)\}/RT$$

The results given in Table 1 indicate that 1) there is scarce difference in Ka between isolated trans and cis-retinal-Lys$^+$ model systems, and 2) ΔpKa is sensitive to molecular environments and, in all cases, the cis form has larger Ka value than the trans form. The problem in these calculations is that the conformations of retinal and amino acid residues employed in PM3 calculation are obtained from 13-cis bR which was structure-optimized by molecular mechanics.

Electrostatic potential field in 13-cis species

The proton transfer process may be sensitive to the electrostatic field in bR since proton is charged. Hence, we calculated the electrostatic field in bR. For this calculation, the ESP charge distribution (Besler et al., 1990) in isolated 13-trans and 13-cis retinal–Lys Schiff base model compounds were calculated by MOPAC PM3. As for atomic charges of amino acid residues, standard values of DISCOVER were used. Electrostatic potential inside and around bR was calculated by a finite difference approximation (Gilson et al., 1985). with $\epsilon_{int} = 8$ for interior of bR, and $\epsilon_{ext} = 80$ for exterior of the protein with grid size of 1.0Å . The ionic strength of aqueous solution outside bR was equated to 0.100M.

In biological systems, bR is embedded in biomembrane. It is impractical to simulate the complex system composed of bR and biomembrane at atomic level. So, membrane is represented by a continuum of 40Å in thickness with $\epsilon = 8$. Fig. 1 shows the electrostatic potential surface of 13-cis species, where 1) -NH- group of Schiff base and 2,3) middle points of two oxygen atoms of carboxylate group of side chains of Asp85 and Asp212 are located on the plane. This figure clearly shows that potential around Asp85 and Asp212 are appreciably negative. Hence, a proton on the Schiff base may be transferred easily to either of these groups so far as electrostatic force is concerned. FTIR spectroscopy indicates that Asp85 is protonated in the L → M step (Ames and Mathies, 1990). This step is a relative slow reaction (~ 50μs) and a large scale reorganization may accompany around chromophore during this step. This reorganization was followed by perturbative simulations of thermal fluctuation accompanied with gradual elimination of proton at Schiff base coupled with gradual addition of proton to carboxylate ion of Asp85 or Asp212, employing molecular dynamics program DISCOVER of Biosym Inc.

Table 1. Heat of formation for model systems calculated by MNDO-PM3 (kcal/mol).

System[a]	all-trans			13-cis			$\Delta Hf(cis-trans)$[c]
	N:	NH	$\Delta(N:-NH)$[b]	N:	NH	$\Delta(N:-NH)$[b]	
1)	33.99	175.20	−141.21	33.19	174.50	−141.31	−0.10
2)	−188.60	−112.40	−76.20	−189.70	−110.00	−79.70	−3.50
3)	−196.40	−136.10	−60.30	−188.20	−114.70	−73.50	−13.20
4)	−330.50	−332.60	2.10	−359.10	−333.10	−26.00	−28.10
5)	−368.63	−325.33	−43.30	−315.86	−261.71	−54.15	−10.85

a: refer to text,   b: $\Delta(N:-NH) = Hf(N)-Hf(NH+)$,
c: $\Delta Hf(cis-trans) = \Delta(N:-NH)cis - \Delta(N:-NH)trans$.

The electrostatic potential field around chromophore of protonated Asp85 species and protonated Asp 212 species were also calculated respectively. In both species, protonated carboxylic group have slightly positive potential of ~5kT. Total electrostatic energies calculated by $E_s = \frac{1}{2}\sum q_i \phi_i$ were, 2930 (13-cis Schiff base species), 2815 (protonated Asp85), and 2942 Kcal/mol (protonated Asp212), respectively. These figures indicate that proton transfer from Schiff base to Asp85 is much easier than to Asp212. In the present study, electrostatic potential was calculated by the finite difference solution method with grid size of 1.0Å and, therefore, more elaborate studies are needed for more detailed discusssions.

Free energy changes in proton transfer processes

Chemical processes are determined by free energy, rather than enthalpy. Hence, we have attempted to estimate Helmholtz free energies by using the relative free energy module of DISCOVER. Free energy changes of proton transfer from Schiff base to infinite separation were calculated to be 6.1±2.5 for 13-trans species and 1.9±2.1 Kcal/mol for 13-cis species by gradual reduction of electric charge and van der Waals potential of proton of Schiff base. This is not free energy changes of actual proton transfer process since it does not include the solvation free energy of proton $\Delta F(H)_{solv}$. Then, free energy of protonation of Asp^-85 and Asp^-212 were calculated. In this process, free energy of desolvation of proton is not included and it can be equated to $-\Delta F(H)_{solv}$. Hence, total free energy changes of proton transfer from Schiff base to Asp^- in respeceive systems can be calculated without estimating $\Delta F(H)_{solv}$. Final results are summarized in Table 2. These results indicated that only the proton transfer from 13-cis species to Asp^-85 is allowed from the standpoint of free energy change. Details of results and discussion will be presented elsewhere in a paper under preparation.

Table 2. Free energy changes of deprotonation from Schiff base and protonation of Asp85 and Asp212 (kcal/mol).

Species	ΔF (deprotonation)	ΔF (protonation)		ΔF (overall transfer)
13-trans	6.1±2.5	Asp85	0.6±0.4	6.7±2.9
		Asp212	5.9±1.2	12.0±3.7
13-cis	1.9±1.2	Asp85	-3.8±0.3	-1.9±1.5
		Asp212	4.6±1.1	6.5±2.3

REFERENCES

Ames, J.B., and Mathies, R.A. (1990): The role of back-reactions and proton uptake during the N→O tansition in bactcriorhodopsin's photocycle: a kinctic rcsonancc Raman study. Biochemistry 29: 7181-7190

Besler, B.H., Mertz, K.M.J., and Kollman, P.A. (1990): Atomic charges derived from semiempirical methods. J.Compt.Chem. 11: 431-439

Braiman, M., and Mathies, R.A. (1982): Resonance Raman spectra of bacteriorhodopsin's primary photoproduct: evidence for a distorted 13-cis retinal chromophore. Pro.Natl.Acad.Sci. USA 79: 403-407

Davis, L. (Ed.) (1987): Genetic algorithm and simulated annealing. California: Morgan Kaufmann pub.

Gilson, M., Rashin, A., Fine, R., Honig, B. (1985): On the calculation of electrostatic interactions in proteins. J.Mol.Biol. 183: 503-516

Henderson, R., Baldwin, J.M., Cesca, T.A., Zemlin, f., Beckmann, E., and Downing, K.H. (1990): Model for the structure of bacteriorhodopsin based on high-resolution electron cryomicroscopy. J.Mol.Biol. 213: 899-929

Mathies, R.A., Lin, S.W., Ames, J.B., and Pollard, W.T. (1991): From femtosedonds to biology: Mechanism of bacteriorhodopsin's light-driven proton pump. Ann.Rev.Biophys.Biophys.Rev. 20: 491-518

Oshida, O., Kataoka, R., Yasukawa, T., Okazaki, H., and Akaike, T. (1991): Mechanism of proton transfer in bacteriorhodopsin. I. The initial proton transfer process bR→K→L. Rep.Prog.Polym.Phys.Jpn. 34: 553-556

Petrich, J.W., Breton, J., Martin, J.L., Antonetti, A. (1987): Femtosecond absorption spectroscopy of light-adapted and dark-adapted bacteriorhodopsin. Chem.Phys.Lett. 137: 369-375

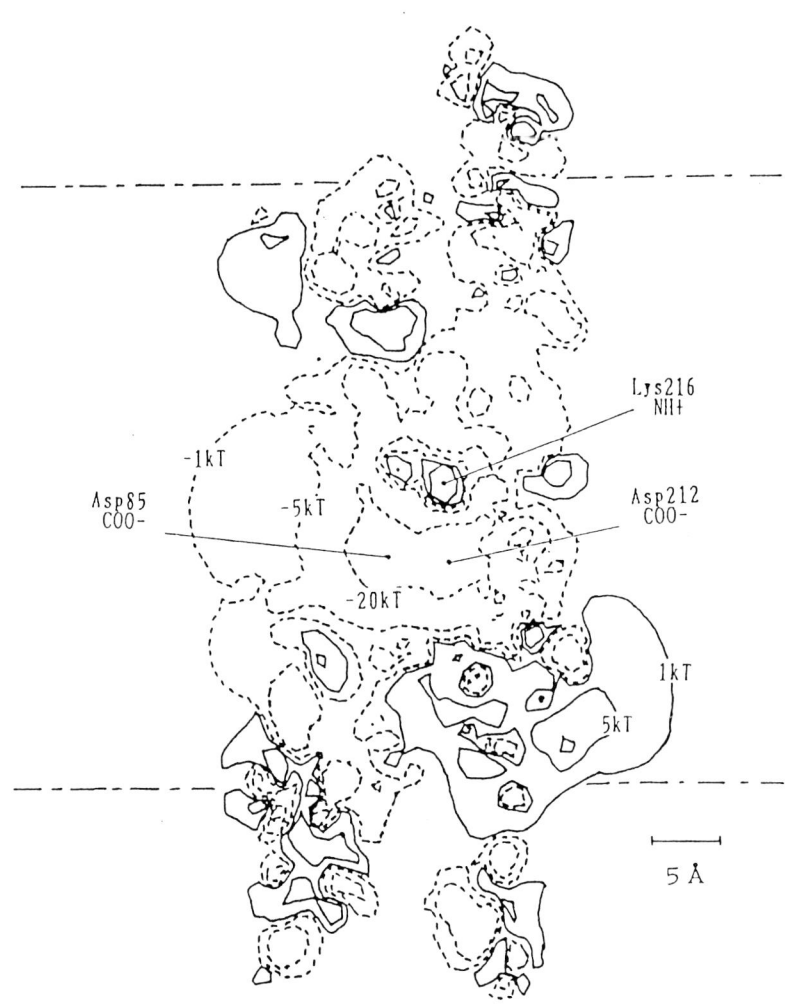

Fig. 1; Electrostatic field in 13-cis bR calculated by finite difference approximation with grid size 1.0Å.

The azide-effect in D96→N/G mutated bacteriorhodopsins monitored by timeresolved FTIR-difference-spectroscopy

Johannes Le Coutre and Klaus Gerwert

Max-Planck-Institut für Ernährungsphysiologie, Rheinlanddamm 201, D-4600 Dortmund, Germany

The slowed down M-decay in D96→N/G mutated Bacteriorhodopsin can be accelerated by azide to almost wild-type kinetics [1]. Since this effect also occurs in proton transfer events in mutated photosynthetic reaction centers [2] it's investigation seems to be of general interest in enzymatic proton transfer reactions. The molecular mechanism of this effect is investigated by FTIR-spectroscopy. Absorbance changes due to azide reactions are presented and their kinetics will be compared to the mutant and wild-type photocycle.

Protonation changes of internal aspartic acids were shown in FTIR experiments [3]. Using mutated bacteriorhodopsin, D85 and D96 were identified as catalytic proton-binding-sites in the proton-release- and -uptake-pathway and their transient protonantion changes were connected to the photocycle [4, 5]. D96→N and D96→G still pump protons although their M-decay compared to wild-type bR is slowed down to about three orders of magnitude. Acceleration of M-decay in D96→N and in D96→G can be achieved by addition of azide or other anions of weak acids. The efficiency of the anions mainly depends on their size and their pKs [1]. Figure 1 compares wildtype-, mutant- and the restored kinetics in the visible spectral range at 410 nm which represent rise and decay of the M-intermediate and at 570 nm which represent depletion and recovery of the BR-ground-state. Experiments are performed in hydrated films required for FTIR-spectroscopy.
Model-IR-measurements with azide and hydrogen azide were performed to identify the asymmetric stretching vibrations. As these vibrations depend on the environment, the experiments were made in H_2O and different concentrations of DMSO and isopropanol (fig. 2).

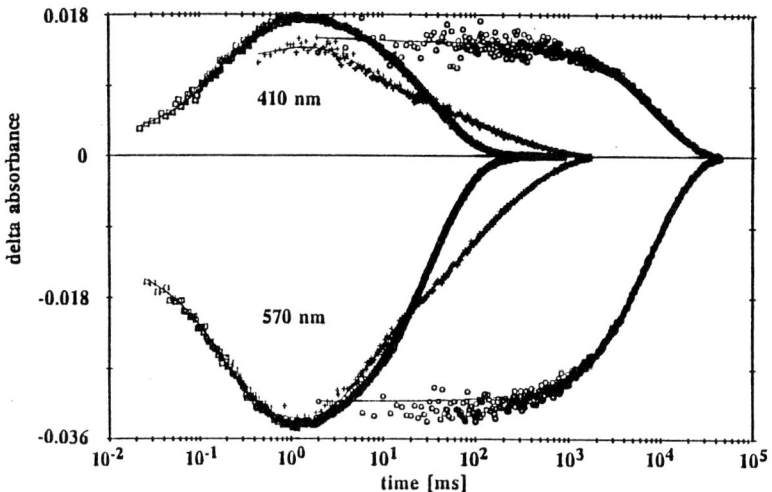

Fig.1 M-decay (410 nm) and recovery of the BR-ground-state (570nm) measured in hydrated films. □: wild-type-bR 10°C, pH7; o: D96→N 10°C, pH7.5; +: D96→N + azide 10°C, pH7.

Isopropanol with a low dielectric constant ($\epsilon = 18.3$) and DMSO ($\epsilon = 46.7$) as an aprotic solvens should be able to mimic an protein-like environment. Increase of the concentrations of either Isopropanol or DMSO shifted the vibrations towards lower frequencies.

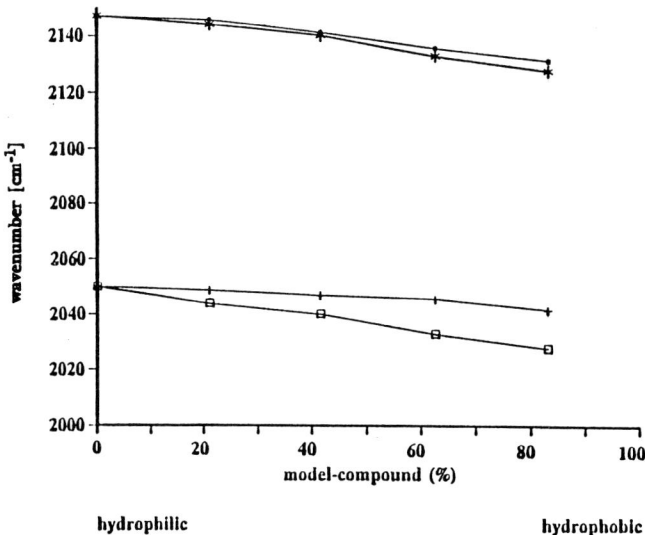

Fig.2 Band-shifting of the (hydrogen-)azide vibrations using various concentrations of the model-compounds isopropanol and DMSO as solvens. •: N_3H in isopropanol; *: N_3H in DMSO; +: N_3^- in isopropanol; □: N_3^- in DMSO.

Static measurements with mutant-bR and azide were made on the one hand to observe the azide peaks at high spectral resolution, on the other hand to connect them to photocycle intermediates. In order to improve the signal/noise ratio up to 40000 scans were averaged. The spectra at pH 7 showed a small difference-band between 2052 cm^{-1} and 2030 cm^{-1} (fig.3).

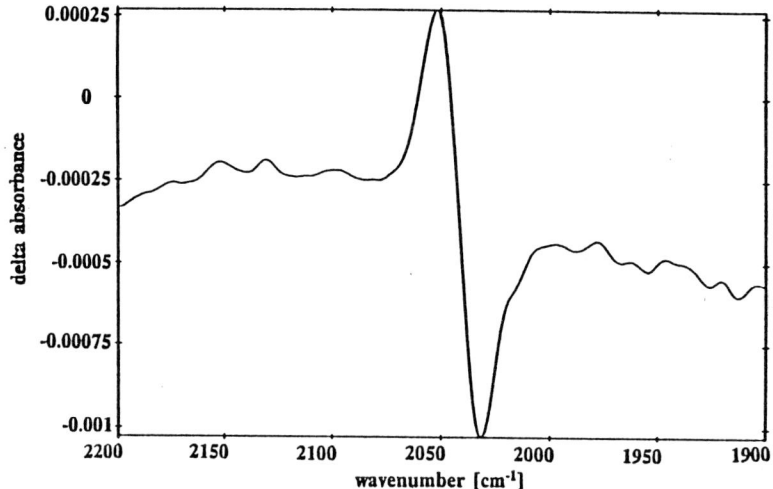

Fig.3 Difference-band in the azide-region showing an environmental shift of the azide anion. Conditions for the static measurement were D96→N + azide (1:50), pH 6.8, -11.5 °C illuminated with a halogen lamp using an OG-515 filter. scans were averaged 39600 times.

Timeresolved FTIR difference spectroscopy shows kinetic properties of the azide-effect and enables us to connect the mode of action of azide to the photocycle. Although D96→N and D96→G behave similar in flash-photolysis-experiments, time-resolved FTIR-investigations revealed differences in their photocycle-kinetics. D96→N doesn't accumulate the N-intermediate in contrast to wild-type-bR and D96→G . This behaviour reflects the extremely slowed down M-decay in D96→N, which results in an slow N-rise compared to the N-decay. In the presence of azide the N-intermediate accumulates as it does in wild-type-bR.

The measurements show participation of azide during the M-decay (fig. 4). Considering the measurements with model compounds this environmental change reflects a movement of the azide-anion from an hydrophobic surrounding to a hydrophilic one during the M-decay.

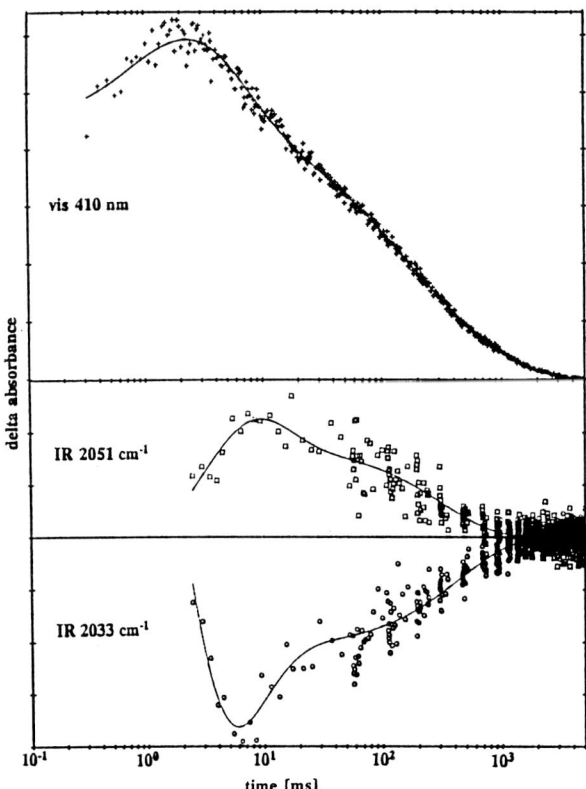

Fig.4 kinetics of the environmental shift of the azide anion compared to the M-decay. Conditions were D96→N + azide (1:100), pH7.5, 11°C.

Two general modes of action are conceivable: i. azide undergoes a protonation change during the photocycle and works as internal proton-donor like D96 does. ii. Azide facilitates reprotonation of the schiff-base via a structural change of the protein-backbone. Since we observe azide in the BR-ground-state, a pure imitation of D96 can be excluded. As we don't observe a corresponding hydrogen-azide-band, further experiments will have to be performed in order to distinguish protonation changes from environmental shifts of the azide-anion.

References:
[1] Tittor J., Soell C., Oesterhelt D., Butt H.-J. and Bamberg E. (1989), *The EMBO Journal* 8: 3477-3482
[2] Takahashi E. and Wraight C.A. (1991), *FEBS LETTERS* 283: 140-144
[3] Engelhard M., Gerwert K., Hess B., Kreutz W. and Siebert F. (1985), *Biochemistry* 24: 400-407
[4] Gerwert K., Hess B., Soppa J. and Oesterhelt D. (1989), *Proc. Natl. Acad. Sci. USA* 86: 4943-4947
[5] Gerwert K., Souvignier G. and Hess B. (1990), *Proc. Natl. Acad. Sci. USA* 87: 9774-9778

Fourier transform infrared studies on light energy transfer process of bacteriorhodopsin

Akio Maeda

Department of Biophysics, Faculty of Science, Kyoto University, Kitashirakawa-Oiwake-cho, Sakyo-ku, Kyoto 606-01, Japan

Light-adapted form of bacteriorhodopsin (BR) transports protons across the purple membrane upon light absorption. The conversion process of light energy absorbed by the chromophore is reflected in a series of intermediates of the photocycle of BR. Light-induced trans-cis isomerization of the chromophore results in the movement of the protonated Schiff base, by successively forming bathochromic intermediates, J, K and KL in the pico- and nano-second ranges. Over the microsecond time range, the slightly blue-shifted intermediate, L, appears. Its formation is accompanied by pronounced perturbation in the aspartic acid residues (Braiman et al., 1988), which are embedded in the membrane and necessary for proton pumping (Mogi et al., 1988). The subsequent deprotonation of the Schiff base to form M is important to enforce proton release to the exterior of the membrane. A structural analysis of L is therefore required for the understanding of the mechanism in the deprotonation of the Schiff base during the process from L to M. The Schiff base is reprotonated by the proton transferred from Asp-96 (Otto et al., 1989), and N forms. The deprotonated Asp-96 then takes the proton from the opposite side and returns to BR via O.

The Fourier transform infrared (FTIR) spectroscopic method has provided useful knowledges through the assignments of the vibrational bands of both the chromophore and the functional residues of the protein (Kitagawa and Maeda, 1989). In the present paper, I will review our recent studies on structure changes of the Schiff base (Maeda et al., 1991; Pfefferle et al. 1991), aspartic acids (Maeda et al., 1992b) and internal water molecules (Maeda et al,. 1992a) in the K to N process.

THE SCHIFF BASE (Maeda et al., 1991)

Changes in the interaction of the protonated Schiff base with the protein residue are expected to play a central role in light energy conversion. The strongly polar Schiff base of the chromophore is the site for electrostatic interaction with the surrounding protein moiety. The H-bonding strength of the Schiff base with the protein will be reflected in the frequency of the bands containing the

N-H bending vibration mode. Two N-H in-plane bending vibrations of **L** and **N** are located at the higher frequency side than those of **K**, respectively. Also, the N-^{2}H in-plane bending vibration of **L** is much higher in frequency than that of **K**. Thus, H-bonding between the protonated Schiff base and the protein residues becomes stronger upon conversion from **K** to **L**. In this respect **N** is similar to **L**.

The intensities of the bands containing the C_{15}-H in-plane bending vibration of **L** are smaller than those of **K** and **N**. Stronger intensity may arise from polarization of the C_{15}-H bond by a repulsive interaction between C_{12}-H and C_{15}-H of the 13-cis protonated Schiff base. Steric repulsion of **L** will be relieved by a distortion of the C_{14}-C_{15} single bond. The distortion dictated by the strong interaction with the protein residue may lead to the deprotonation of the Schiff base (Fahmy et al., 1989).

ASPARTIC ACIDS (Pfefferle et al., 1991; Maeda et al., 1992b)

In the functionally important two aspartic acid residues, Asp-85 is unprotonated in **BR** and accepts the proton of the Schiff base. Asp-96 is protonated in **BR** and reprotonates the Schiff base (Braiman et al. 1988). FTIR spectrum of **L** shows a strong perturbation of Asp-96 along with Asp-115. However, the perturbation was interpreted as either the partial deprotonation or environment changes. ^{13}C-Aspartic acid-labeled bR and a mutant of Asp-96 do not exhibit the shift of the C=O stretching vibration of the deprotonated aspartic acid. Mutation studies further show that the O-H stretching vibration of Asp-96 shifts to the lower frequency side upon **L** formation. The C==O stretching vibration of Asp-96 at 1748 cm^{-1} of **L** does not shift upon ^{2}H$_2$O substitution, in contrast to the corresponding band of **BR** at 1741 cm^{-1}. Model studies on acetic acid in organic solvents indicate that such an insensitivity is due to H-bonding of the O-H of carboxylic acid. These results show that Asp-96 of **L** is protonated and its O-H is strong H-bonding donor.

The perturbation of Asp-96 completely extinguishes upon **L** to **M** conversion. In the C=O stretching vibrational region of **M**, only the positive band at 1762 cm^{-1} due to the protonation of Asp-85 is observed. An intense signal of Asp-96 at 1742 cm^{-1} appears upon conversion to **N**. The deprotonation of Asp-96 of **N** is confirmed by the ^{13}C-shift of the symmetric C=O stretching vibration in the unprotonated carboxylic acid region. This also confirms the positive band at 1755 cm^{-1} to the protonation of Asp-85. The positive band at 1737 cm^{-1} of **N** is attributed to the protonation of Asp-212. The intensity of the 1737 cm^{-1} band is much lower in **M**, indicating that the protonation of Asp-212 occurs mainly in **N**.

WATER STRUCTURE (Maeda et al., 1992a; 1992b)

Vibrational modes of the chromophore of **L** and **N** are similar and Asp-96 undergoes the perturbation only when the Schiff base is protonated. These suggest that the Schiff base points to the side of Asp-96 of **L** and **N**. The distance between the Schiff base and Asp-96 is, however, too long for direct interaction (Henderson et al., 1990). The involvement of a few water molecules in the strong H-bonding of **L** was shown by H$_2^{18}$O shifts of the O-H stretching vibration. In contrast, the H-bonding of the water becomes more weak upon **M** formation. Water structure change associated with L formation was unaffected by the mutation of Asp-96. In conjunction with the previous results on the water in organic solvents and in the protein (Sakabe et al., 1985), these results suggest that unbound water molecules are present closely to the Schiff base in a narrow channel surrounded by a hydrophobic domain.

FUNCTIONAL SITE IN THE PHOTOCYCLE

From these results, a following scheme (Fig. 1) for the proton pumping process of bacteriorhodopsin can be envisaged. (1) **BR**. The N-H bond of the Schiff base points to the side where both Asp-85 and -212 are located. (2) **L**. Upon light absorption, the chromophore isomerizes to the 13-cis form, and the N-H bond points roughly to the opposite side. The positive charge of the Schiff base moves away from the Asp-85 and -212 region and approaches closer to Asp-96. The chromophore twists in the C_{12}-H to C_{15}-H region by forming strong H bonding of the Schiff base with Asp-96 through intervening water molecules. These non-chromophore residues are orienting so as to give twists in the chromophore by strong H bonding. Some N-H bond becomes free and polarized in the site around the chromophore. (3) **M**. Such a twisting leads to the deprotonation of the Schiff base. It then results in the loss of interaction with the protein and relieves the twists. Asp-85 is ready to accept the proton from the Schiff base by losing the electrical interaction with the Schiff base. Except for the protonation of Asp-85, no appreciable

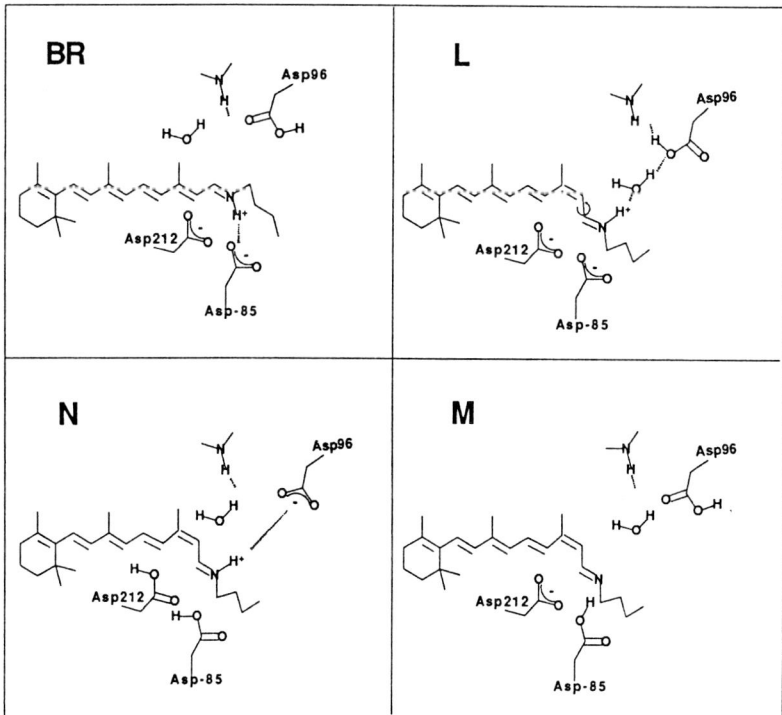

Fig. 1.   Functional site in **BR, L, M** and **N**.

changes were seen in the protein residues and the peptide bonds in view of infrared spectrum. (4) **N**. The Schiff base has restored a high affinity to the proton and accepts the proton from Asp-96. The N-H bond interacts strongly with the deprotonated Asp-96. The interaction does not cause any twisting, probably owing to the relaxation of the protein moiety. Asp-85 is still protonated by increasing the H bonding strength of its C=O. Asp-212 is also protonated. Thus, the protonation state of Asp-96, -85 and -212 are opposite to those in **BR**. Shifts of a few amide I bands may indicate the stronger H bonding of these C=O and the appearance of an intense amide II band may be due to polarization of some peptide N-H. An electrically perturbed state is a characteristic feature of **N** and may be necessary for the reprotonation. We are currently studying the process for the reprotonation of the Schiff base by use of mutant proteins.

References

Braiman, M. S., Mogi, T., Marti, T., Stern, L. J., Khorana, H. G., and Rothschild, K. J. (1988): Vibrational spectroscopy of bacteriorhodopsin mutants: light-driven proton transport involves protonation changes of aspartic acid residues 85, 96, and 212. Biochemistry 27, 8516-8520.

Fahmy, K., Siebert, F., Grossjean, M. F., and Tavan, P. (1989): Photoisomerization in bacteriorhodopsin studied by FTIR, linear dichroism and photoselection experiments combined with quantum chemical theoretical analysis. J. Mol. Struct. 214, 257-288.

Henderson, R., Baldwin, J. M., Ceska, T. A., Zemlin, F., Beckman, E., and Downing, K. H. (1990): Model for the structure of bacteriorhodopsin based on high-resolution electron cryo-microscopy. J. Mol. Biol. 213, 899-929.

Kitagawa, T., and Maeda, A. (1989): Vibrational spectra of rhodopsin and bacteriorhodopsin. Photochem. Photobiol. 50, 883-894.

Maeda. A., Sasaki, J., Pfefferle, J. -M., Shichida, Y., and Yoshizawa, T. (1991): Fourier transform infrared spectral studies on the Schiff base mode of all-trans bacteriorhodopsin and its photointermediates, **K** and **L**. Photochem. Photobiol. 54, 911-921.

Maeda, A., Sasaki, J., Shichida, Y., and Yoshizawa, T. (1992a): Water structural changes in the bacteriorhodopsin photocycle; analysis by Fourier transform infrared spectroscopy. Biochemistry 31, 462-467.

Maeda, A., Sasaki, J., Shichida, Y., Yoshizawa, T. Chang, M., Ni, B., Needleman, R., and Lanyi, J. K. (1992b): Structure of aspartic acid-96 in the L and **N** intermediates of bacteriorhodopsin; analysis by Fourier transform infrared spectroscopy. Biochemistry in press.

Mogi, T., Stern, L., Marti, T., Chao, B. H., and Khorana, H. G. (1988): Aspartic acid substitutions affect proton translocation by bacteriorhodopsin. Proc. Natl. Acad. Sci. USA 85, 4148-4152.

Otto, H., Marti, T., Holz, M., Mogi, T., Lindau, M., Khorana, H. G., and Heyn, M. P. (1989): Aspartic acid-96 is the internal proton donor in the reprotonation of the Schiff base of bacteriorhodopsin. Proc. Natl. Acad. Sci. USA 86, 9228-9232.

Pfefferle, J. -M., Maeda, A., Sasaki, J., and Yoshizawa, T. (1991): Fourier transform infrared study of the **N** intermediate of bacteriorhodopsin. Biochemistry 30, 6548-6566.

Sakabe, N., Sakabe, K., and Sasaki, K. (1985) X-ray studies of water structure in 2 Zn insulin crystals. Proc. Int. Biomol. Struct. Interactions, Suppl. J. Biosci. 8, 45-55.

Infrared spectroscopy of the reactions of the M form of bacteriorhodopsin: conclusions about the mechanism of proton pumping

Pál Ormos, Kelvin Chu* and Judith Mourant*

*Institute of Biophysics, Biological Research Center of the Hungarian Academy of Sciences, Temesvari krt. 62, H-6701 Szeged, Hungary and *Department of Physics, University of Illinois, 1110 West Green Street, Urbana, IL 61801, USA*

Introduction

The generally accepted reaction scheme:

bR - K - L - M - N - O - bR

is regarded as a good approximation of the Bacteriorhodopsin photocycle. The description of the molecular changes during the transitions between the intermediates is a key to the understanding of how the proton pump works. Of particular interest in the photocycle are the steps involving the M intermediate. M is the only form where the Schiff-base - through which the retinal chromophore is connected to the protein - is deprotonated. During the pumping sequence this proton is released to the extracellular side of the membrane and reprotonation takes place from the opposite side. A protein conformational change is thought to play a key role in changing the connectivity of the Schiff-base to ensure that deprotonation and reprotonation occur to different groups (Fodor **et al.**, 1989). In infrared spectroscopic experiments performed in the temperature region 240K - 260K we observed that the events following the formation of M completely change characteristics in this temperature interval (Ormos, 1991): At 240K during the decay of M the Schiff-base reprotonates from Asp-85, the primary proton acceptor, therefore pumping does not take place. At 260K, however, the spectral changes indicate a Schiff-base reprotonation through Asp-96 (Braiman **et al.**, 1988), the "normal" sequence of events during pumping (Butt **et al.**, 1990). The transition in the behavior is coupled to a well defined protein conformational change seen in the infrared spectrum that is represented by the exchange of two Amide I vibrations at 1660 cm^{-1} and 1670 cm^{-1}. We assigned this transition to the protein switch. A protein switch with this property is clearly crucial for the function and therefore its characterization would bring us closer to understand the proton pumping by bacteriorhodopsin as well as the general principles of the action of transport proteins.

It is not obvious that the protein relaxation has to be synchronized to the known steps at all. Or if it takes place during the lifetime of M for example, it may result in extra M states. There have been numerous reports on multiple M forms, the significance of their existence, however, is not clear. It is not decided whether they are in parallel paths or in sequence after each other. In the latter case the sequentially arranged M forms offer an elegant picture for the action: the transition between the two sequential M intermediates would be the protein switch (Váró & Lányi, 1991).

Experiments and Results

We applied infrared difference spectroscopy to clarify the coupling of the protein switch to the known steps in the photocycle. Our goal was to clearly separate the infrared spectra of the L, M and N intermediates in a wide range of temperatures and to see whether the protein change observed in (Ormos, 1991) can be unambiguously correlated with any transition between them. We used the photosensitivity of M for this purpose: It is known that if M is excited by light the molecule returns to the bR ground state very fast (Kalsisky et al., 1977). This photoreaction offers an advantageous way to determine the spectrum of M: 1. it is independent of any kinetic model and 2. the mixture of M and any other form can be easily separated. The method and the experiments are described in detail in (Ormos et al., 1992) here we point out the main features. Infrared difference spectroscopy was performed on fully hydrated films of bR between 215K and 265K: First the sample was illuminated by green light, this resulted in a photoproduct which is a mixture of M and some other form. Subsequent blue illumination only drives the M form away from the mixture and therefore the difference due to the blue illumination yields the spectrum of pure M. Once the spectrum of M is known that of the other component can be determined. This procedure was applied at all studied temperatures and the spectra of L, M and N were obtained.

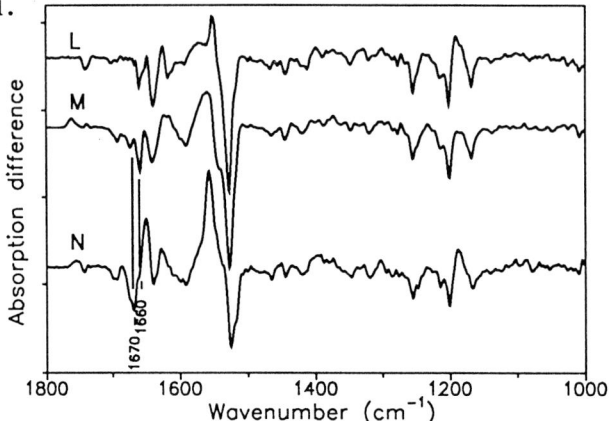

Fig.1. The difference spectra L-bR, M-bR and N-bR separated from the spectra of the mixture of intermediates produced by illumination at different temperatures.

Surprisingly, the spectra obtained proved to be temperature independent, including that of M, independently whether it was in equilibrium with L or N. The result's important conclusion is that the spectra show that the protein switch characterized in (Ormos, 1991) is between the M and N states as indicated by the exchange of the marked Amide I bands. Let us investigate the contribution of the forms at different temperatures. With the component spectra of L, M and N known the relative contribution of the particular forms at any temperature can be determined. In the characterization of the mixture we use the result that the spectra are temperature independent. We assume that the concentration of the form is proportional to the amplitude of the negative band at 1528 cm^{-1} (the ethylenic band of the missing bR) in the difference spectrum. The results are shown in Fig.2.

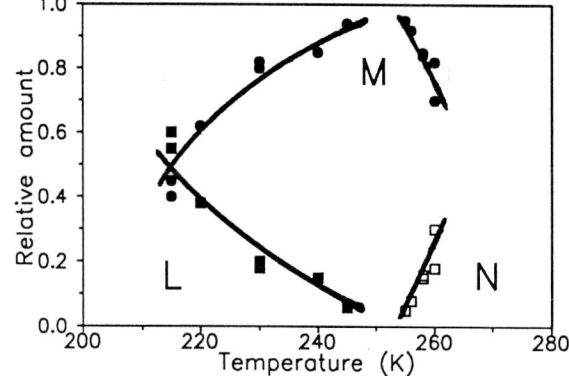

Fig.2. Relative contribution of the L, M and N intermediates in the difference spectra obtained after green illumination at different temperatures

The values represent average concentration ratios for the time of the scan (an average over 1 minute following 20 s illumination). The strong temperature dependence of the relative N concentration is remarkable: the amount of N dramatically increases above 255K.

Discussion

Earlier, when we first observed the temperature dependent branching after the formation of M into the non-pumping and pumping paths (Ormos, 1991) we noticed the large temperature dependence of the change of kinetics around 250K: the transition of the M decay from the non-pumping to the pumping route takes place within about 20K. The transition temperature coincides with the freezing of bound water and therefore it obviously represents a melting of the system. Differential Scanning Calorimetry experiments (our unpublished observation) have shown that indeed a phase transition takes place around 250K. These results also stress that for the pumping it is crucial that the protein is in a flexible state and this is when N can form. The observations support the picture of the protein switch with the following sequence of events: 1.: The Schiff-base loses its proton to the primary proton acceptor, 2.: The flexible protein undergoes a relaxation changing the accessibility of the Schiff-base, and 3.: The Schiff-base reprotonates from a proton donor other than the original acceptor. The results tell how these pumping events are coupled to the known steps. We obtained identical M spectra at every temperature both below and above the transition temperature of 250K; it is the decay of M that changes its characteristics. The conformational change that correlates with the transition from the non pumping to the pumping state is during the M to N transition. According to the logic of the pumping sequence outlined above steps 2 and 3 occur simultaneously during the M to N step. Fig.3. schematically represents the possible decay paths of M. At low temperatures where the protein is rigid - frozen - the non-pumping M - bR backreaction takes place. Around 250K where the protein "melts" the protein relaxation becomes fast and the pumping side of the branch becomes dominant. The separation of the protein relaxation and Schiff-base reprotonation in the figure is done on a didactic basis only, in our experiments they occur simultaneously.

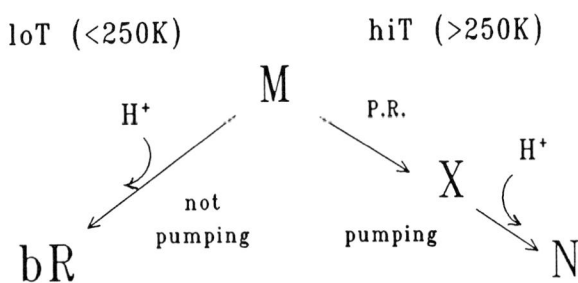

Fig.3. The two possible (pumping and not pumping) pathways for the decay of the M intermediate

The protein relaxation is the rate limiting step: as soon as it takes place reprotonation occurs. At higher temperatures, however, the two steps may separate due to different temperature dependence of the rates. The experiment should be performed at room temperature to check this possibility.

References

Braiman, M. S., Mogi, T., Marti, T., Stern, L. J., Khorana, H. G and Rothschild, K. J. (1988): Vibrational spectroscopy of bacteriorhodopsin mutants. II. Light-driven proton transport involves protonation changes of aspartic residues 85, 96, and 212. Biochemistry 27 8516-8520

Butt H-J., Fendler K., Bamberg E., Tittor J. and Oesterhelt D. (1989): Aspartic acids 96 and 85 play a central role in the function of bacteriorhodopsin as a proton pump. EMBO J 8 396-400

Fodor, S. P. A., Ames, J. B., Gebhard, R., van den Berg, E. M. M., Stoeckenius, W., Lugtenburg, J. and Mathies, R. A. (1988): Chromophore Structure in Bacteriorhodopsins N intermediate: Implications for the Proton-Pumping Mechanism. Biochemistry 27 7097-7101

Kalisky, O., Lachish, U. and Ottolenghi M. (1977): Time resolution of a back photoreaction in Bacteriorhodopsin. Photochem. Photobiol. 28 261-263

Ormos, P. (1991): Infrared spectroscopic demonstration of a conformational change in bacteriorhodopsin involved in proton pumping Proc. Natl. Acad. Sci. USA 88 473-477

Ormos, P., Chu, K. and Mourant, J. (1992): Infrared study of the L, M and N intermediates of Bacteriorhodopsin using the photoreaction of M. Biochemistry in press

Váró, Gy. and Lányi, J. (1991): Thermodynamics and Energy Coupling in the Bacteriorhodopsin Photocycle. Biochemistry 30 5016-5022

Optical transient studies on the photochemical cycle of bacteriorhodopsin (0.5 µs – 500 ms, pH 4 – 10)

W. Eisfeld and M. Stockburger

Max-Planck Institut für biophysikalische Chemie, Abteilung Spektroskopie, Am Fassberg, W-3400 Göttingen, Germany

The work presented here completes our time–resolved resonance Raman (RR) studies on Bacteriorhodopsin (bR) presented in this volume by Pusch et al. (herein called *paper I*). The photocycle of bR was studied in aqueous suspensions of purple membranes by conventional optical transient spectroscopy (flash photolysis). For photolysis of the parent chromophore, BR_{570}, a frequency-doubled Nd:YAG laser system (532 nm, 10 ns pulse width) was used. Optical transients were measured at 20°C in the time domain 0.5 µs – 500 ms for seven different wavelengths (410 – 700 nm) in the range pH 4 – 10. The data were analysed by fitting the transients with a sum of exponentials.

In the whole pH range the decay of the intermediate M_{412} can be described by the superposition of a slow (M^s, τ^s) and a fast (M^f, τ^f) exponential component (*Fig. 1*). Whereas in the range pH 4 – 7.5 τ^f and τ^s lie between 2 and 10 ms, τ^s drastically increases for pH>7.5. Of great importance for the analysis of the photocycle is the pH–dependence of the amplitudes M^s_{rel} and M^f_{rel} in *Fig. 1b* which can be described by an apparent pK_a of $\sim$7.5 (see *paper I*).

The rise and decay of the intermediate O_{640}, probed at 700 nm, is displayed in *Fig. 2a*. The time constants of the two events differ in the entire pH range by at least a factor of 3 and thus can be well distinguished. The maximum of the fractional concentration of O which is achieved during the photocycle is displayed in *Fig. 2b*. This quantity is related to the fraction of bR which is photolyzed in a single laser pulse. Using the rise and decay times one can conclude that for pH<6 about 20% of the photolyzed bR go through the O intermediate. The amplitude of O_{max} decreases with increasing pH. This behavior can be described by a titration–like function with an apparent pK_a of $\sim$7.2.

In *Fig. 3* the time constants for the reconstitution of BR_{570}, monitored at 640 nm in the range pH 7 – 10, are displayed. For comparison the pH dependence of the slow decay of M (τ^s in *Fig. 1a*) is indicated by the dotted line. For pH>8, τ^s refers to the direct transition from M^s to BR_{570}. It is evident from *Fig. 3* that in this range the measured time constants are significantly greater than τ^s and therefore cannot be ascribed to the M^s-to-BR transition. Since for $\lambda > 580$ nm the extinction coefficient of BR_{570} exceeds that of N_{560}, it was proposed by Kouyama et al. (1988) that the time constants measured for $\lambda > 580$ nm and pH>8 refer to the reconstitution of BR_{570} from the long–lived N_{560} intermediate. Here we follow their interpretation which is in accord with our time-resolved RR experiments.

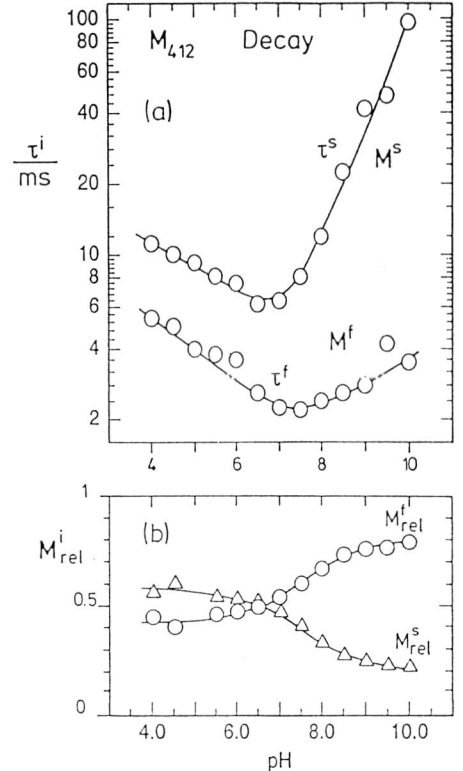
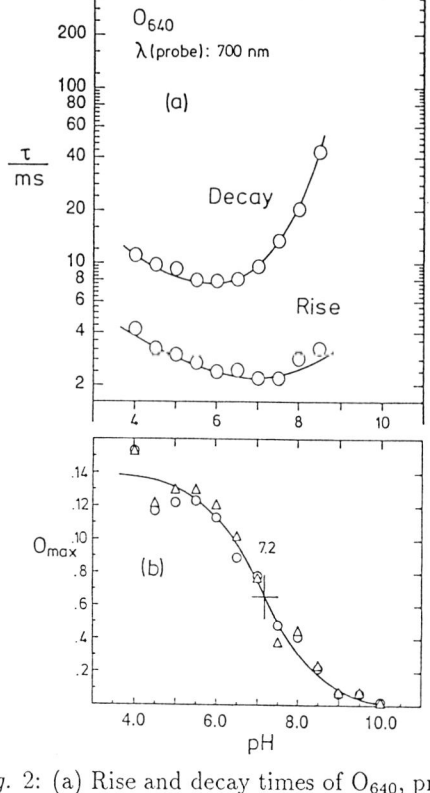

Fig. 1: (a) The two decay constants of M, τ^s and τ^f, deduced from optical transients at 412 nm. (b) Relative amplitudes: $M^f_{rel}=M^f/(M^f+M^s)$, $M^s_{rel}=1-M^f_{rel}$.

Fig. 2: (a) Rise and decay times of O_{640}, probed at 700 nm. (b) O_{max} is the maximum of the fractional concentration obtained in a single laser pulse.

Fig. 4 shows results obtained for probe wavelengths in the range 580 – 600 nm. For pH<6 the transients can be fitted by a single time constant τ^s, whereas for pH>6 a second time constant τ^f of significant amplitude can be identified. From the data for the decay of M^s and O in Figs. 1 and 2 we conclude that τ^s (pH<6) reflects the two reconstitution steps M^s-BR and O-BR. From the decay of M^s (Fig. 1) and from Fig. 3 we further conclude that τ^s (pH>8) refers to reconstitution steps M^s-BR and N-BR.

The time constant τ^f in Fig. 4 we exclusively assign to transitions from M^f to N. This is based on the fact that a direct reconstitution of BR from M^f could not be identified. The amplitude of τ^f (A^f_{rel} in Fig. 4b) reflects the formation of N. It is concluded from Fig. 4b that N is not populated below pH 6 and increases steadily for pH>6. This is in agreement with our finding from RR experiments that the formation of N is correlated with apparent pK_a values of 5.6, 6.2 and 7.5 (see paper I).

It can be concluded from our data that the intermediate O_{640} is a direct product of M^f. Thus for pH<6, O_{640} is the only product which follows on M^f and the rise time of O matches the decay time of M^f.

The assignment of O_{640} to a direct product of N (as frequently proposed in the literature) is in contradiction with our results. Thus, for pH<5, where O has its maximum population (Fig. 2b), the intermediate N is not populated at all. In addition, the maximum concentrations of O and N as a function of pH change in the opposite direction (see *Fig. 2b* and *Fig. 4b*).

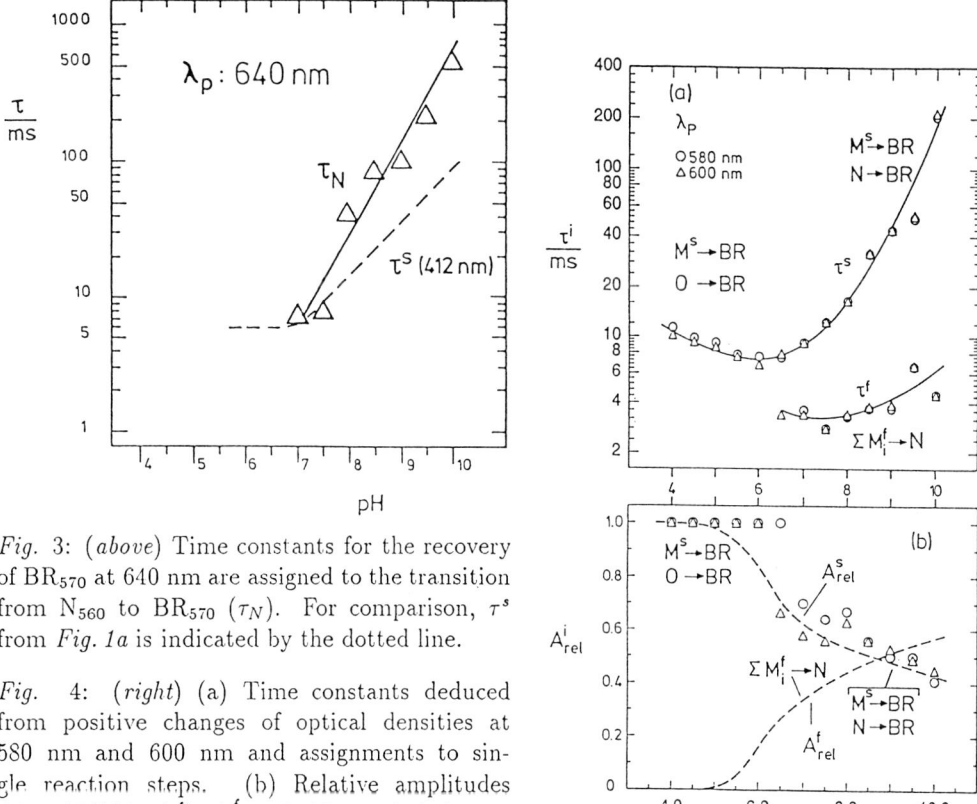

Fig. 3: (above) Time constants for the recovery of BR_{570} at 640 nm are assigned to the transition from N_{560} to BR_{570} (τ_N). For comparison, τ^s from *Fig. 1a* is indicated by the dotted line.

Fig. 4: (right) (a) Time constants deduced from positive changes of optical densities at 580 nm and 600 nm and assignments to single reaction steps. (b) Relative amplitudes $A^s_{rel}=A^s/(A^s+A^f)$, $A^f_{rel}=1-A^s_{rel}$, and their correlation with single reaction steps.

On the basis of our time-resolved RR and optical transient studies the reaction scheme I (*Fig. 3, paper I*) was proposed. Here we discuss the proton-translocating steps wich occur in the different reaction pathways of *scheme I* and which might be important for the biological function. An inspection of *scheme I* reveals the two principally different pathways displayed in *scheme II* (*Fig. 5*).

Let us first consider the reaction in *scheme IIa* with an N intermediate in which two types of proton transfer reactions can be distinguished: Firstly, the internal motion of a proton from the Schiff base to the counterion (Asp85) and its reversal. This is essentially concluded from the fact that the formation of N is independent of the external pH and therefore can only occur from an internal proton donor. Secondly, the translocation of a proton across the membrane in phase with the photocycle. Evidence for this function is provided by the observation that the transition from N to BR, i.e. the re-isomerization of retinal, is significantly delayed with increasing pH. From the data in *Fig. 3* it can be concluded that the N-to-BR transition is determined by the diffusion of a proton from the external medium to the membrane. We therefore conclude that the re-isomerization step is controlled by the protonation of an "internal reactive site" in the vicinity of the chromophore. It is proposed that this site is deprotonated at an earlier stage synchroneously with the deprotonation of the Schiff Base in the L-to-M transition. In this way the vectorial motion of a proton from the cytoplaymic to the extracellular side of the membrane in phase with the photocycle is described. This proton has also a catalytic function.

Scheme IIb describes the reaction pathways in which no N intermediates occur. The rate-determining step again is the cis/trans re-isomerization as induced by the reprotonation of the reactive site. For pH<7.2 the internal group A_zH acts as the proton donor so that the transition from M^f to O via the intermediate M^* is fast. For pH>7.2 A_zH is deprotonated and the reprotonation of the reactive site occurs from the outside. Then the re-isomerization of M is also diffusion-controlled (M^f is replaced by M^s and O is no longer accumulated).

In the picture of *Scheme II* the protons which move through the reactive site are the ones wich are pumped vectorially across the membrane.

More details of our work are described by Pusch et al. (1992).

SCHEME II

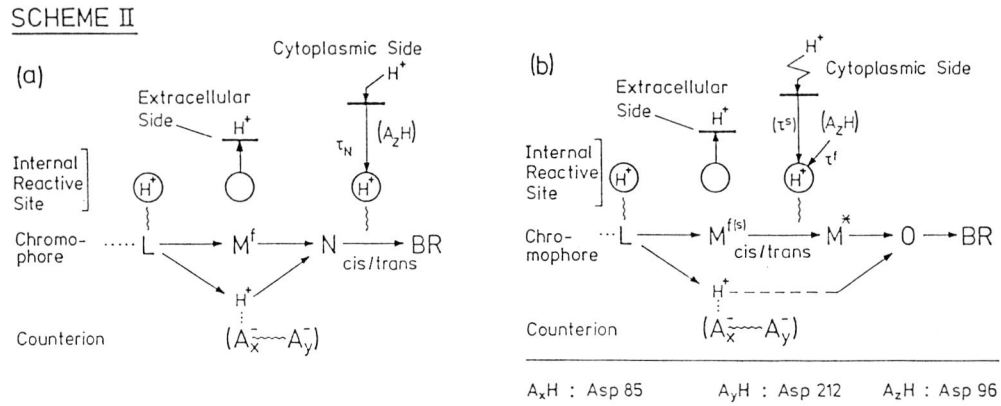

Fig. 5: Reaction Scheme II.

REFERENCES

Pusch, C., Diller, R., Eisfeld, W., and Stockburger, M. (1992): The Light-Induced Proton-Pump of Bacteriorhodopsin Studied by Resonance Raman and Optical Transient Spectroscopy. To be published in *Biochemistry*.

Kouyama, T., Nasuda-Kouyama, A., Ikegami, A., Mathew, M. K., and Stoeckenius, W. (1988): Bacteriorhodopsin Photoreaction: Identification of a Long-Lived Intermediate N (P, R_{350}) at High pH and Its M-like photoproduct. *Biochemistry 27*, 5855 – 5863.

The light-induced proton-pump of bacteriorhodopsin studied by resonance Raman and optical transient spectroscopy

C. Pusch, R. Diller, W. Eisfeld, R. Lohrmann and M. Stockburger

Max-Planck Institut für biophysikalische Chemie, Abteilung Spektroskopie, Am Fassberg, W-3400 Göttingen, Germany

Excited by light, the retinylidene Schiff base chromophore (BR_{570}) of Bacteriorhodopsin (bR) runs through a series of intermediate states (K_{590}, L_{550}, M_{412}, N_{560}, O_{640}) and, under physiological conditions, is reconstituted within a few milliseconds. This cyclic reaction (photocycle) controls bR's unique function as a proton pump. A detailed analysis of the various reaction steps of the photocycle is therefore a prerequisite for an understanding of the proton pump mechanism. It appears that up to the stage of L the photocycle can be described by the linear sequence

$$BR \underset{(\sim 5ps)}{\overset{h\nu}{\rightsquigarrow}} K_{590} \underset{(1.2 \mu s)}{\overset{\tau_1}{\longrightarrow}} L_{550}$$

where the reaction step from BR to K is dominated by photoisomerization of the retinal chain around the 13-14 double bond. The subsequent step from K to L consists of the relaxation from a distorted conformation to a more planar structure. The time constants of 5 ps and 1.2 μs refer to room temperature and neutral pH.

In the reaction step from L to M a proton is released from the Schiff base group and during the transition from M to N the Schiff base is reprotonated. It was found that the decay of L occurs with at least two different time constants (Diller and Stockburger, 1988). The biphasic decay of M with a fast (M^f, τ^f) and a slow (M^s, τ^s) decay component, respectively, is well known. It can be expected that proton translocation steps inside the protein strongly depend on the protonation state of internal side groups and consequently on the external pH. In addition, the protons which are vectorially pumped from the external phase through the membrane might also influence the reaction of the chromophore. Systematic studies of the photocycle as a function of the external pH therefore should help to disentangle the complex kinetic behavior of the chromophore.

We have carried out time-resolved resonance Raman (RR) as well as optical transient experiments with purple membrane suspensions over the range pH 4 – 10 and in the time domain 10 μs – 1 s. First, a few typical pH dependent phenomena shall be described, and then a complete reaction scheme of the photocycle will be presented. Results of our optical transient experiments are described by Eisfeld and Stockburger in this volume (herein called *paper II*). More details of the whole work are on the way to be published (Pusch et al., 1992).

Fig. 1 shows RR spectra obtained in a pump–probe experiment with a delay time δ between pump and probe event of 0.8 ms (for a description of the experimental techniques see Diller and Stockburger, 1988). The spectra, in which the contribution of BR_{570} has been subtracted, display the characteristic C=C stretching bands of the intermediates L, M and N. At pH 5 the N-intermediate is not populated and L and M have nearly the same concentration. At pH 7.55 the major part of L is converted into N and the minor part to a new intermediate called $\tilde{M}$. The transition from L to N can be associated with the deprotonation of an internal group (A_yH) with an apparent pK_a value of 6.2, and the transition from L to $\tilde{M}$ refers to a pK_a of ~ 5.6 (Pusch et al., 1992).

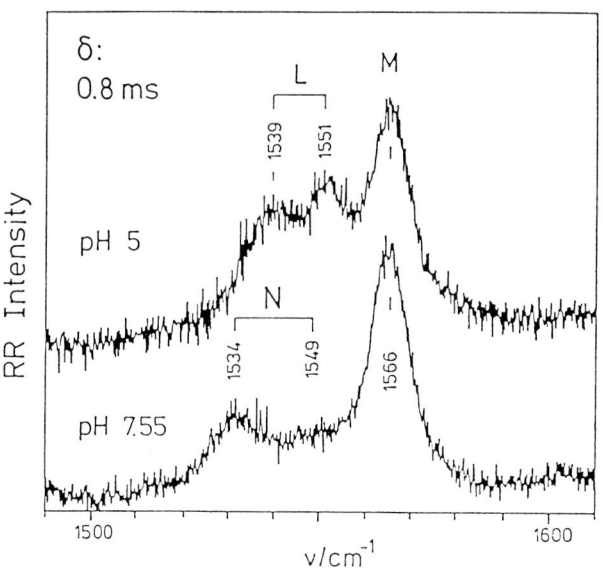

Fig. 1: RR spectra with the C=C stretching bands of L, M and N obtained in a pump–probe flow experiment (rotating cell) from an aqueous purple membrane suspension (22°C, OD = 1 at 570 nm). Pump beam: 633 nm, 80 mW, 160 μm in diameter. Probe beam: 476 nm, 3 mW, 80 μm in diameter. Flow velocity: 1.26 ms^{-1}. The contribution of BR is subtracted.

The following conclusions are important: Firstly, at pH 5, M is formed as a "fast" product of L ($\tau_2 \sim 80\mu s$ at 20°C) for only one half of the bR molecules (bR(α)). For the second half of bR (bR(β)) the reaction is significantly retarded at the stage of L. Secondly, the existence of the internal group A_yH in its deprotonated form is a necessary condition for the continuation of the reaction at the stage of L. Whereas in bR(α) the group A_yH is already deprotonated at pH 5 so that M can readily be formed, this group is still protonated in bR(β). Only when the pH is increased so that A_yH becomes deprotonated, the reaction of L (to N or to $\tilde{M}$) also procedes in bR(β). This behavior suggests that under thermal equilibrium conditions bR exists in at least two different subspecies bR(α) and bR(β) with significantly different pK_a values for the internal group A_yH (<4 in bR(α) and >5.5 in bR(β)). A further subdivision of bR(β) is given by the pK_a values 5.6 and 6.2 and the related transitions of L to $\tilde{M}$ or to N.

The decay of M as displayed in *Fig. 1* of *paper II* can be described by at least two different time constants in the whole pH range. The pertinent amplitudes $M_{rel}^f = M^f/(M^f+M^s)$ and $M_{rel}^s = 1 - M_{rel}^f$ display a characteristic pH–dependence. The titration–like shape of this function suggests that the increase of M^f with respect to M^s is essentially determined by the dissociation of an internal group with a pK_a of ~ 7.5.

The temporal evolution of L, M and N for delay times in the range 0.3 – 7 ms and pH 7.6, as inferred from time-resolved RR measurements, is displayed in *Fig. 2*. At the lowest value of $\delta = 0.3$ ms, the fast decaying L component ($\tau_2 \approx 80\mu s$) is nearly completely converted to M. The residual L component decays with a time constant of $\sim 360\mu s$, and it appears that the fast rise of N (N_{fr}) is directly correlated with the "slow decay" of L. A closer inspection of the rise of N_{fr}, however, reveals a very short-lived M intermediate (M^{vf}, vf:"very fast", see *scheme I*). In order to fit the observed time-dependence of N a second component with a "slow rise" (N_{sr}) is needed. This component can be correlated with the decay of M^f. Our analysis thus reveals at least two main pathways for the formation of N. This is different from earlier proposals in which the formation of N from M^f was considered as the only way (Kouyama et al., 1988; Ames and Mathies, 1990).

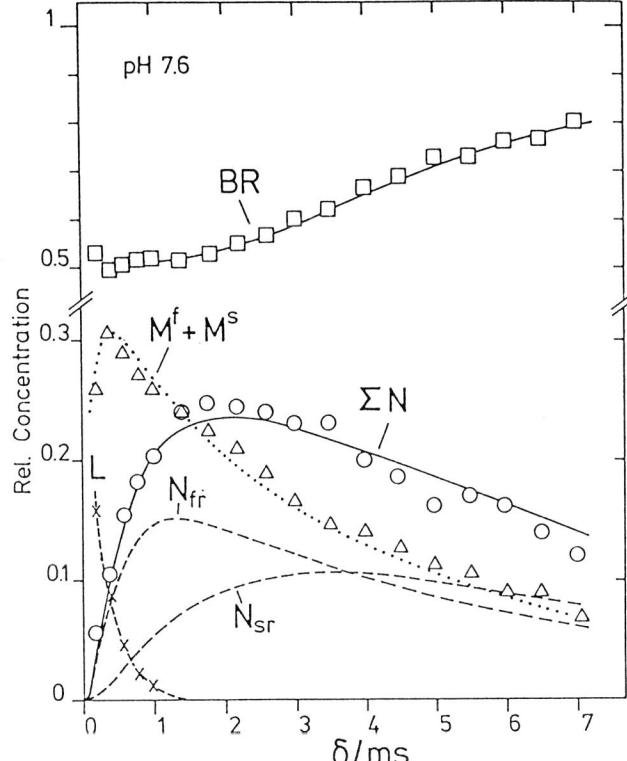

Fig. 2: Relative concentrations of BR, L, M and N from a time-resolved RR pump-probe experiment. Experimental conditions as in Fig. 1. Relative concentrations were inferred from the integrated intensities of the characteristic C=C stretching bands with RR cross sections relative to BR of 1 for L and M and 1.5 for N at 476 nm. The dashed lines for N were calculated accordingly to scheme I ($\tau_2 = 80\mu s$, $\tau^f = 2\,ms$ for N_{sr}; 360 μs and 170 μs for N_{fr}). The dotted line was obtained from optical transient experiments.

Reaction scheme I (Fig. 3) is based on the subdivision of bR into bR(α) and bR(β) according to significantly different pK_a values of the internal group A_yH (see above).
Let us begin with bR(β) which can be further subdivided. In bR(β,1) the pK_a of A_yH is ~6.2. Below this value the reaction is retarded at the stage of L. Above this value the "fast rise" of N (N_{fr} in Fig. 3) is initiated. In bR(β,2) the pK_a of A_yH is ~5.6. Below this value the reaction is also retarded at the stage of L. Above this value $\tilde{M}^f$ is formed which then contributes to the "slow rise" of N (one part of N_{sr} in Fig. 2).
In bR(α) the group A_yH is always deprotonated so that the reaction is not retarded at the stage of L. A further subdivision is based on the observation that in the range pH 3 – 5 in which N is not formed the decay of M^f can be directly correlated with the rise of the intermediate O. This reaction, however, only occurs for a fraction of bR(α) molecules, i.e. for bR(α,2) (see paper II). In the residual part, i.e. in bR(α,1), M^s must be directly converted to BR since no intermediate product of M^s could be identified in this pH range.
When the pH is raised above 7 dramatic changes occur. Thus the disappearance of O can be associated with a pK_a of ~7.2 (paper II, Fig. 2b). Accordingly M^f changes to M^s in bR(α,2) (see scheme II in paper II). On the other hand, in bR(α,1), M^f increases at the expense of M^s according to a pK_a of ~7.5 (Fig. 1b, paper II) which leads to another "slow rise" component of N.

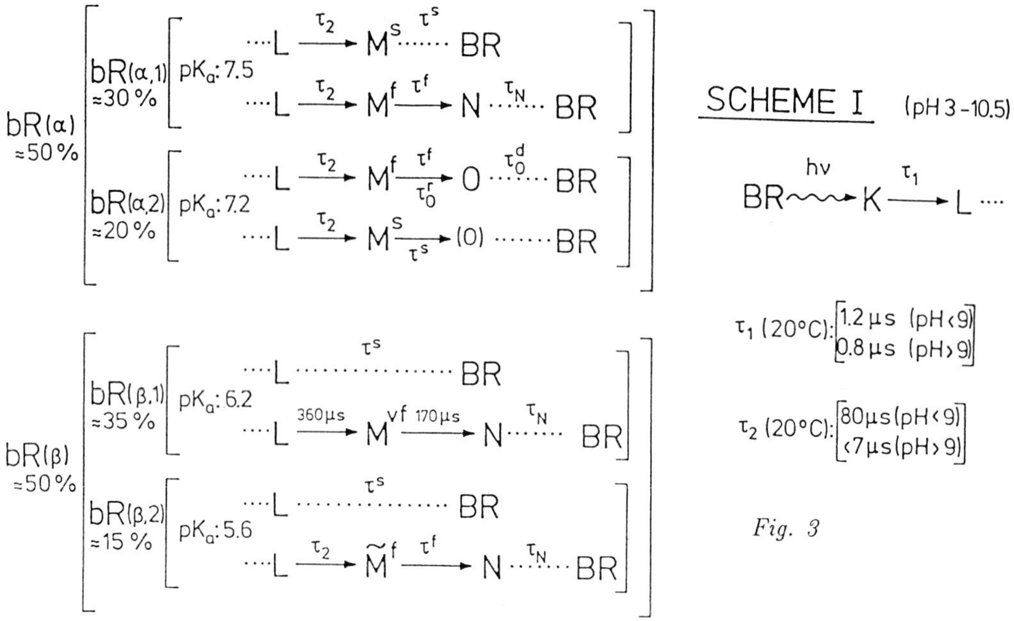

$$\text{bR}(\alpha) \approx 50\% \begin{bmatrix} \text{bR}(\alpha,1) \Big| \begin{matrix} \text{pK}_a: 7.5 \\ \approx 30\% \end{matrix} \begin{bmatrix} \cdots L \xrightarrow{\tau_2} M^s \xrightarrow{\tau^s} \cdots \text{BR} \\ \cdots L \xrightarrow{\tau_2} M^f \xrightarrow{\tau^f} N \xrightarrow{\tau_N} \cdots \text{BR} \end{bmatrix} \\ \text{bR}(\alpha,2) \Big| \begin{matrix} \text{pK}_a: 7.2 \\ \approx 20\% \end{matrix} \begin{bmatrix} \cdots L \xrightarrow{\tau_2} M^f \xrightarrow{\tau^f_0} O \xrightarrow{\tau^d_0} \cdots \text{BR} \\ \cdots L \xrightarrow{\tau_2} M^s \xrightarrow{\tau^s} (O) \cdots \text{BR} \end{bmatrix} \end{bmatrix}$$

$$\text{bR}(\beta) \approx 50\% \begin{bmatrix} \text{bR}(\beta,1) \Big| \begin{matrix} \text{pK}_a: 6.2 \\ \approx 35\% \end{matrix} \begin{bmatrix} \cdots L \xrightarrow{\tau^s} \cdots \text{BR} \\ \cdots L \xrightarrow{360\mu s} M \xrightarrow{vf\ 170\mu s} N \xrightarrow{\tau_N} \cdots \text{BR} \end{bmatrix} \\ \text{bR}(\beta,2) \Big| \begin{matrix} \text{pK}_a: 5.6 \\ \approx 15\% \end{matrix} \begin{bmatrix} \cdots L \xrightarrow{\tau^s} \cdots \text{BR} \\ \cdots L \xrightarrow{\tau_2} \widetilde{M}^f \xrightarrow{\tau^f} N \xrightarrow{\tau_N} \cdots \text{BR} \end{bmatrix} \end{bmatrix}$$

SCHEME I   (pH 3-10.5)

$$\text{BR} \xrightarrow{h\nu} K \xrightarrow{\tau_1} L \cdots$$

$\tau_1 (20°C): \begin{bmatrix} 1.2\ \mu s\ (\text{pH}<9) \\ 0.8\ \mu s\ (\text{pH}>9) \end{bmatrix}$

$\tau_2 (20°C): \begin{bmatrix} 80\ \mu s\ (\text{pH}<9) \\ <7\ \mu s\ (\text{pH}>9) \end{bmatrix}$

Fig. 3

In summary we conclude that bR consists of various fairly stable subspecies which undergo different cyclic reactions. Such species are characterized by different substructures of dissociable internal groups which lead to different pK_a values and reactivities. By analogy with kinetic measurements on site-specific mutants (Needleman et al., 1991), we assign the group A_yH to the carboxylic side chain of the residue Asp212. The assignment of the groups with $pK_a \sim 7.5$ in $bR(\alpha,1)$ and $pK_a \sim 7.2$ in $bR(\alpha,2)$ requires further experimental evidences. It is important to note that the different substructures and protonation states of the reactive internal groups could not be identified via the RR spectra of the chromophore. This means that the structural diversity of such groups does not markedly influence the structure of the chromophore although it strongly influences its kinetic behavior.

REFERENCES

Pusch, C., Diller, R., Eisfeld, W., and Stockburger, M. (1992): The Light-Induced Proton-Pump of Bacteriorhodopsin Studied by Resonance Raman and Optical Transient Spectroscopy. To be published in *Biochemistry*.

Diller, R., and Stockburger, M. (1988): Kinetic Resonance Raman Studies Reveal Different Conformational States of Bacteriorhodopsin. *Biochemistry 27*, 7641 – 7651.

Kouyama, T., Nasuda-Kouyama, A., Ikegami, A., Mathew, M. K., and Stoeckenius, W. (1988): Bacteriorhodopsin Photoreaction: Identification of a Long-Lived Intermediate N (P, R_{350}) at High pH and Its M-like photoproduct. *Biochemistry 27*, 5855 – 5863.

Ames, J. B., and Mathies, R. A. (1990): The Role of Back-Reactions and Proton Uptake during the N→O Transition in Bacteriorhodopsin's Photocycle: A Kinetic Resonance Raman Study. *Biochemistry 29*, 7181 – 7190.

Needleman, R., Chang, M., Ni, B., Váró, G., Fornés, J., White, S. H., and Lanyi, J. K. (1991): Properties of $Asp^{212} \rightarrow Asn$ Bacteriorhodopsin Suggest that Asp^{212} and Asp^{85} Both Participate in a Counterion and Proton Acceptor Complex near the Schiff Base. *J. Biol. Chem. 266(18)*, 11478 – 11484.

Evidences for structural changes at the chromophoric site of bacteriorhodopsin during the K-to-L transition

R. Lohrmann and M. Stockburger

Max-Planck Institut für biophysikalische Chemie, Abteilung Spektroskopie, Am Fassberg, W-3400 Göttingen, Germany

The light-induced proton pump of Bacteriorhodopsin is controlled by the photochemical cycle of its retinylidene Schiff base chromophore. It is well established that the reaction step from the parent state BR_{570} to the intermediate state K_{590}, which occurs within ~5 ps at room temperature, involves *trans-cis* photoisomerization around the 13,14 double bond of the retinal chain. We believe that the subsequent dark reaction from K_{590} to L_{550} (1.2 μs at 293 K) is also of functional importance. To substantiate this idea we have studied resonance Raman (RR) spectra of the late K_{590} as a direct precursor of L_{550} using the spinning-cell technique in combination with the 514.5-nm line of a continuous wave Ar^+-laser serving both for photolysis and RR-excitation. Important conclusions on structural changes of the chromophore and the protein environment could be inferred from the vibrational spectra.

Various difference procedures were developed to obtain the 'pure' K_{590} spectra from the composite spectra of the photolyzed sample which for principle reasons (photoreversal) always involves a large contribution (~50%) of parent BR_{570}. If appropriate pump conditions are used and the dwell time, Δt, of a sample element in the laser beam is ~2 μs, a nearly 2:1:1 mixture of BR, K and L is established in the beam (Lohrmann et al., 1991). The spectrally separated C=N stretching bands of BR and K, which lie between 1610 and 1640 cm^{-1} give a well-defined substraction criterion to obtain the pure spectra of the intermediates K and L.

In Fig. 1 overview spectra of BR, K and L are shown in H_2O and D_2O, respectively. Characteristic features in the spectra of K are several hydrogen-out-of-plane (HOOP) bands of medium intensity in the range of 950-990 cm^{-1} (The fairly strong band at 1008 cm^{-1} refers to a bending mode of the CH_3 groups of retinal). In the case of BR only a single HOOP band is found at 960 cm^{-1} which was assigned to a 11H,12H HOOP mode. In general HOOP modes are not expected to be RR-active in planar polyenes. Therefore it can be concluded from the K-spectra that the chromophore is in a distorted conformation. The fact that in the spectra of L no resonance-enhanced HOOP modes of the retinal chain are observed, implies, that during the K-to-L transition (1.2 μs, 293 K) the conformational distortion of the chain is released.

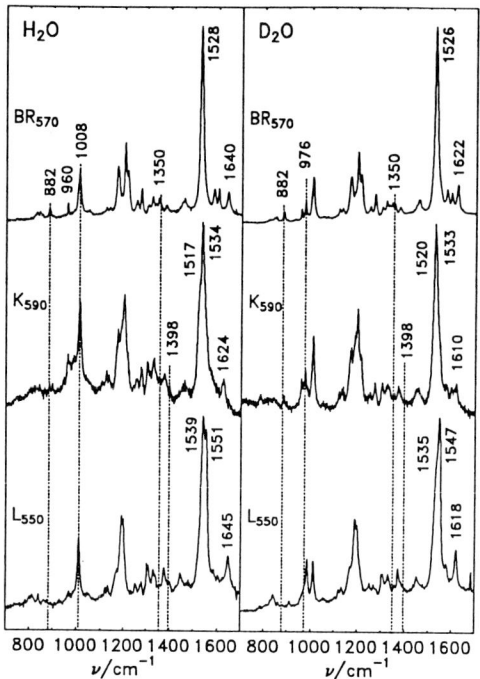 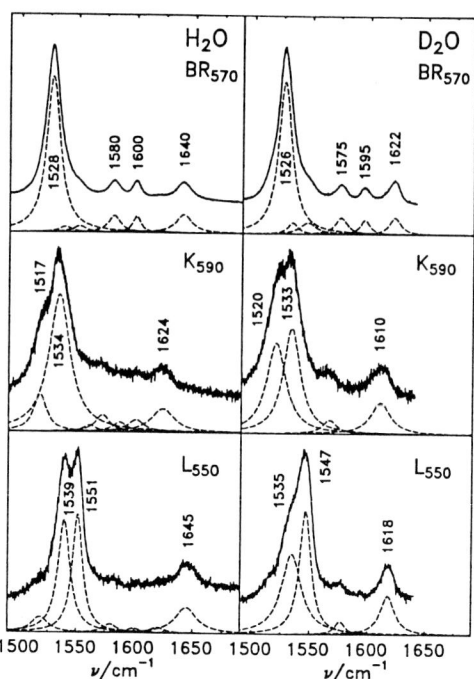

Fig. 1: 'Overview spectra' of BR, K and L in H_2O and in D_2O with an excitation wavelength $\lambda_{ex} = 514.5$ nm.

Fig. 2: Region of the C=C and C=N stretches of BR, K and L in H_2O and in D_2O ($\lambda_{ex} = 514.5$ nm)

In D_2O suspension the proton at the Schiff base nitrogen is exchanged for a deuteron whereas all other hydrogen atoms of the chromophore remain unaffected. The observed spectral isotope effects therefore must be correlated with the vibrational modes of the Schiff base group. In the case of BR the bands at 1640 cm^{-1} and 1350 cm^{-1} in H_2O are shifted in D_2O to 1622 cm^{-1} and 976 cm^{-1}, respectively.

The band at 1640 cm^{-1} was assigned to a normal mode in which the C=N stretching motion is coupled to a N–H in-plane bending motion (here called [C=NH$^+$] mode) whereas the bands at 1350 (976) cm^{-1} mainly reflect localized N–H (N–D) in-plane bending modes. It is generally accepted that the down-shift from ν[C=NH$^+$] = 1640 cm^{-1} to ν[C=ND$^+$] = 1622 cm^{-1} is due to the decoupling of the C=N stretch from the low-frequency N–D bend in the [C=ND$^+$] normal mode. This implies that the [C=ND$^+$] mode essentially reflects a 'pure' C=N stretching motion.

Analogous isotope effects were observed for the Schiff base bands of K and L. Thus in K the [C=NH$^+$]/[C=ND$^+$] modes lie at 1622/1610 cm^{-1} and in L they are found at 1645/1618 cm^{-1}, respectively. Whereas one observes significant frequency changes of such modes this is not the case for the N–H and N–D bending modes which for all species with protonated Schiff bases (BR, K, L, N) were found at ~1350 cm^{-1} and ~976 cm^{-1}, respectively.

The strongest bands in the spectra which refer to 'C=C stretching modes' of the retinal chain reflect important structural changes of the chromophore during the photocycle. We therefore have studied these bands together with the adjacent [C=NH$^+$]/[C=ND$^+$] bands on an enlarged scale.

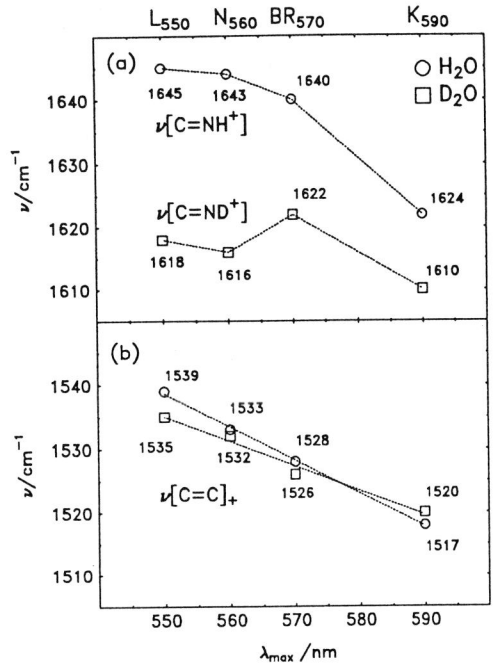

Fig. 3: Vibrational frequencies of C=C and C=N stretching modes in correlation with λ_{max} of the optical absorption bands
(a) $\nu[C=NH^+]$ and $\nu[C=ND^+]$
(b) $\nu[C=C]_+$

The results are depicted in Fig. 2. In the case of BR a single intense band is found at 1528 cm^{-1} which is assigned to a normal mode in which several C=C stretches of the retinal chain move in phase ($[C=C]_+$ mode). There is only a minor deuteration shift to 1526 cm^{-1} in D$_2$O. In the spectra of K and L this band is split into a doublet. On the basis of normal mode calculations the low-frequency component can be assigned to $[C=C]_+$ and the high-frequency component to a mode in which the C=C stretches of the retinal chain are not in phase ($[C=C]_-$).

It has been frequently reported that an empirical inverse linear correlation exists between the frequency of the strong C=C stretch of the retinal Schiff base and the λ_{max} of the optical absorption band. We found that this correlation also holds for the bR-chromophores BR, K, L and N if it is specified for the low-frequency component $\nu[C=C]_+$ and λ_{max} (Fig. 3b). It is well established from quantum chemical calculations that this behavior can be simulated by a model in which the distance between the positively charged Schiff base and a single negative counterion is varied. Thus the λ_{max} increases and the $\nu[C=C]_+$ decreases with increasing distance between the charged groups. The one-counterion model predicts that the inverse linear correlation also holds for the λ_{max} and $\nu[C=NH^+]$ values of the chromophores. However, as can be seen in Fig. 3a the data don't fit into such a simple scheme. In the following an extended model will be proposed which allows to explain the spectral changes and consequently also to follow the temporal evolution of the chromophore in the binding pocket during the reaction sequence BR, K, L.

It is well established that in the parent state, BR$_{570}$, the positively charged chromophore is stabilized by a negative counterion. This has been identified as the carboxylate side chain of the residue Asp 85. It can be assumed that during the all-*trans* to 13-*cis* photoisomerization of the retinal chain the distance between the positive and the negative charge is increased. For the transition from BR to K the one-counterion model would thus predict an up-shift in λ_{max}

and a down-shift in $\nu[C=C]_+$ as well as in $\nu[C=ND^+]$ (this mode has to be considered since, as outlined above, it reflects a 'pure' C=N stretch). This is observed indeed. It means that the one-counterion model explains the spectral changes in the BR-to-K transition correctly. In the transition from K to L $\nu[C=ND^+]$ shifts from 1610 cm^{-1} to 1618 cm^{-1}. If this shift is explained by the one-counterion model it would mean that the Schiff base and the counterion would again approach each other. The observed up-shift for K to L (8 cm^{-1}) is lower than the down-shift for BR to K (12 cm^{-1}). If λ_{max} and $\nu[C=C]_+$ of L would be determined alone by the distance of the counterion from the Schiff base one would expect $\lambda_{max} > 570$ nm and $\nu[C=C]_+ < 1528$ cm^{-1}. However, one observes for L a significant blue-shift of λ_{max} to 550 nm and a concomitant up-shift of $\nu[C=C]_+$ to 1539 cm^{-1}. Such big shifts must be due to an additional interaction of the chromophore and its environment in the binding pocket. In an earlier paper (Lohrmann et al., 1991) it was proposed that such shifts are caused by a positively charged group which is approached by the β-ionone ring of the retinal during the conformational relaxation in the K-to-L transition. This conclusion is based on spectroscopic studies of model compounds with the respective charge distribution (Bassov et al., 1987).

The existence of a positive charge in the binding pocket which strongly influences the π-electrons of the retinal chain in the L state is corroborated by the spectral data of the intermediate N_{560} which are also given in Fig. 3. It is well known from RR-spectroscopic evidences that the conformational structure of L_{550} and of N_{560}, which appears at a later stage of the photocycle, are closely related. This is in particular the case for the Schiff base group. (frequencies and isotope effects of its characteristic vibrations nearly coincide, cf. Fig. 3). The most significant difference between L and N lies in the down-shift of $\nu[C=C]_+$ by ~ 6 cm^{-1} (Fig. 3) in N. This can be explained only by an interaction which is not localized in the vicinity of the Schiff base. In our model this spectral shift would imply that in the state of N_{560} the β-ionone ring is removed from the positively charged group, or the positively charged group has disappeared.

In conclusion it is proposed on the basis of RR-spectroscopic data that during the transition from K_{590} to L_{550} the chromophore undergoes considerable conformational changes which introduces new electrostatic interactions with its environment in the binding pocket. Such interaction may be important for controlling the biological function.

REFERENCES

Baasov, T., Friedemann, N. and Sheves, M. (1987): Factors affecting the C=N stretching in protonated Retinal Schiff base: A model study for Bacteriorhodopsin and visual pigments. *Biochemistry*, 26, 3210-3217

Lohrmann, R., Grieger, I. and Stockburger, M. (1991): Resonance Raman Studies on the Intermediate K_{590} in the Photocycle of Bacteriorhodopsin. *J. Phys. Chem.*, 95, 1993-2001

Detection of conformational changes during the photocycle of bacteriorhodopsin by laser-induced optoacoustic spectroscopy (LIOAS)

Mathias Rohr, Peter Schulenberg, Wolfgang Gärtner and Silvia E. Braslavsky

Max-Planck Institut für Strahlenchemie, Stiftstrasse 34-36, D-4330 Mülheim, Ruhr, Germany

The light driven photocycle of bacteriorhodopsin (BR) starts with a sub-ps all-*trans* $\xrightarrow{h\nu}$ 13-*cis* photoisomerization of the retinal chromophore, followed by a series of conformational changes of chromophore and protein, which are accompanied by de- and reprotonation of the Schiff base and charged amino acids [1]. Spectroscopic methods have revealed detailed information on the dynamic processes of chromophore and protein. The investigation of the primary photoreaction, however, is complicated by the rapid formation of the first photoproducts, J and K, which, due to their spectral similarity to BR, may undergo photoreactions within the duration of the exciting flash. Sub-ps excitation and acoustic detection of the heat released by the intermediates (Laser-induced optoacoustic spectroscopy, LIOAS) circumvents the above mentioned drawback [2]. If performed in temperature dependent manner, LIOAS can yield volume changes associated with a photochemical reaction. Furthermore, spectrally silent protein conformation changes which are not monitored through a change of the spectroscopic properties and may escape from absorption or vibrational spectroscopic detection, are readily identified by thermal detection methods.

The photocycle of BR was investigated in the ns → µs-time scale by LIOAS with fs- and ns-flash excitation. Whereas fs-excitation exclusively initiates the forward reaction BR $\xrightarrow{h\nu}$ K, high photon density and laser flash durations of several ns establish a photoequilibrium between BR and K (BR $\xrightleftharpoons{h\nu}$ K) and thus monitor the combined volume changes of forward and backward reaction. Light adapted BR samples ($A_{570} < 0.3$) in native form (purple membrane patches) and CHAPS solubilized preparations (final concentration: 20 mM) were studied. The laser systems used for fs- (500 fs) and ns- (8 ns) excitation were described in detail previously [2]. Both systems were operating at 585 nm with a maximal effective pulse frequency of 52 Hz. The heat integration time was adjusted to 670 ns by means of a pinhole. The samples for temperature dependent measurements were thermostatted between 0-20 ± 0.1 °C. $CoCl_2$ was used as calorimetric reference.

In LIOAS, the heat emitted by a sample after absorption of a laser pulse generates an optoacoustic signal which is based on the thermally induced volume change ΔV_{th}. This volume change gives rise to a pressure wave detectable with a piezoelectric transducer [3] and is proportional to the term $\beta/C_p\rho$ (eq.1, β: cubic expansion coefficient, C_p: molar heat capacity at constant pressure, ρ: density). The optoacoustic signal amplitude H is proportional through an instrumental constant k to the heat dissipated to the solvent within the acoustic transit time of the sound wave which in aqueous solutions is in the ns to μs range (eq. 2, for detailed description see [2]).

$$\Delta V_{th} \sim \alpha \ \frac{\beta}{C_p\rho} \ E_l(1-10^{-A}) \quad ; \quad H = k \ \alpha \ \frac{\beta}{C_p\rho} \ E_l(1-10^{-A}) \qquad (1)$$

α: fraction of absorbed energy $[E_l(1-10^{-A})]$ dissipated into the medium as "prompt heat", E_l: laser fluence, A: absorbance of the sample). In addition to ΔV_{th}, a volume change originating from a molecular process like a conformational change can also contribute to the optoacoustic signal ΔV_r, and therefore eq.1 has to be extended. The strong temperature dependence of the dominating β-value of H_2O between 0 °C and room temperature allows to extract ΔV_r as slope from a plot of the ratio of the energy normalized signals for sample, H_n^S, and reference, H_n^R, vs $C_p\rho/\beta$ (eq.2):

$$\frac{H_n^S}{H_n^R} = \frac{H^S/[(1-10^{-A^S})E_l]}{H^R/[(1-10^{-A^R})E_l]} = \alpha + \frac{\Delta V_r}{(1-10^{-A^S})E_l} \ \frac{C_p\rho}{\beta} \ . \qquad (2)$$

A linear correlation between the fluence-normalized optoacoustic signal H/E_l and β for BR and the reference compound $CoCl_2$ confirmed the assumption that the thermoelastic properties of the dilute BR solutions are correctly described using the β-value for water. Contrary to these relations, plots of the fluence-normalized sample and reference signal H_n^S/H_n^R vs. $C_p\rho/\beta$ (cf. eq.2) were not linear for various BR concentrations (A_{585} = 0.09 - 0.32). The extrapolation to low values of $C_p\rho/\beta$ yielded a decreasing α-value for increasing concentrations implying a dependence of the stored energy (1-α) on the concentration of BR. This observation was further investigated with CHAPS treated BR-solutions which disintegrates interactions between purple membrane patches and solubilizes BR in a monomeric form. Accordingly, CHAPS containing solutions showed no concentration effect (Fig.1). This result permits to ascribe the concentration dependence of H_n^S/H_n^R to an interaction between purple membrane patches or retinal chromophores within the trimeric BR. Consequently, plots of H_n^S/H_n^R vs. A have to be extrapolated to zero concentration in order to extract a value free from any intermolecular interaction. These extrapolated H_n^S/H_n^R-values as a function of $C_p\rho/\beta$ now yield a linear correlation with a slope different from zero (Fig.2).

The negative slope indicates a molecular contraction of $\Delta V_r = 3.783 \cdot 10^{-6}$ cm^3/J upon BR excitation. Since the signals were linear for the absorbed fluence E_a at the excitation wavelength (585 nm), a volume change of $\Delta V_r = 1.28 \cdot 10^{-30}$ m^3/molecule was calculated which - assuming an isotropic volume change - gives a value for a one dimensional motion of $\sqrt[3]{\Delta V_r} = 1.09$ Å/molecule.

Fs-pulses exclusively excite the all-*trans* to 13-*cis* BR $\xrightarrow{h\nu}$ K reaction. Like in the ns-experiments, a concentration-dependent, nonlinear relation of H_n^S/H_n^R vs. $C_p\rho/\beta$ was also observed. After extrapolation to zero concentration, a linear correlation results (Fig.2). Also for fs-excitation a contraction of $\Delta V_r/E_a = 1.453 \cdot 10^{-6}$ cm^3/J was determined, corresponding to $\Delta V_r = 0.49 \cdot 10^{-30}$ m^3/molecule and a value of $\sqrt[3]{\Delta V_r} = 0.79$Å/molecule for a one dimensional motion.

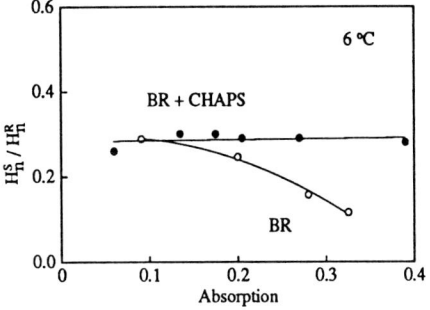 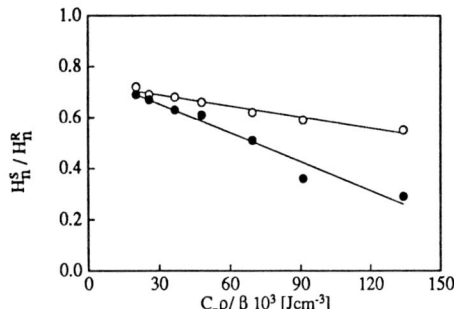

Fig.1: Ratio of normalised BR and reference signals H_n^S/H_N^R vs. absorption, with (●) and without (○) CHAPS, ns-excitation

Fig.2: Plot of the extrapolated values of H_n^S/H_n^R vs. $C_p\rho/\beta$ for ns- and fs-excitation, $\lambda_{exc} = 585$ nm, $\tau_a' = 670$ ns

It may be assumed that the reverse process (reformation of BR from K) compensates the volume change of the forward reaction by simply reforming the BR ground state. However, since both isomerizations (all-*trans* $\xrightarrow{h\nu}$ 13-*cis*) and (13-*cis* $\xrightarrow{h\nu}$ all-*trans*) proceed through an orthogonal state and thus require additional space, there is no difference in sign for ΔV_r in forward or backward photoreaction of BR. Thus, for the ns-experiments (BR $\xrightleftharpoons{h\nu}$ K) a larger overall movement, as observed, should result. Apparently, protein environment and chromophore conformation in the BR ground state adopt a "preactivated" conformation ensuring the high quantum yield and the absolute steric reliability of the resulting isomer in the photoproduct. Since the chromophore induces geometrical changes of closely adjacent amino acids during the formation of the J-, K- and KL-intermediates, a re-isomerization (backwards reaction) finds a new structured binding site and again induces changes in the environment. Therefore, the definition of the observed volume changes as "activation" or "reaction" volumes has to yield different values for the forward and backward reaction due to the differently activated BR and K-states.

A remarkable contribution of protein movements to the optoacoustic signal within the heat integration time (670 ns) can be ruled out since a reorganization of the protein environment as a main source for ΔV_r on a ns time scale should yield an additional term of equal size for the ns- and fs-experiments. The values of ns- and fs-LIOAS experiments, however, differ by a factor of ca. 2.5. Contributions of protein conformation changes to the thermal signal which now become the dominating effect were detected from measurements in the μs time range. These processes were identified by the beam deflection technique. During the time range observed (1 to 200 μs), the appearance and disappearance of the L-intermediate is well resolved by two life times of 1.2-1.5 μs and 80-90 μs, respectively (Fig.3). Most interestingly, an additional time of

20-25 µs with an amplitude of ca. 20 % of the decay process is detected. Such event gives a clear indication for a protein conformational change during the lifetime of the L-intermediate which - being *spectrally silent* - has not been identified before. This process concurs with protein conformation changes recently proposed in a C,T-model of the BR-photocycle [4].

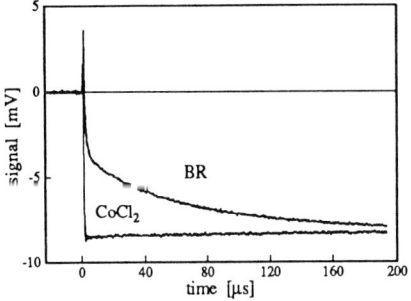
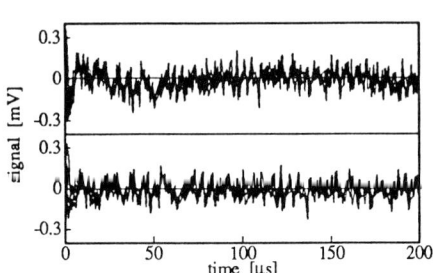

Fig.3a: Beam deflection signals of BR samples and CoCl$_2$ reference, $\lambda_{exc} = 550$ nm, $A_{570} = 1.0$, ns-excitation.

Fig.3b: Fit residuals of the BR signal by two (upper trace) and three (lower trace) exponentials.

The bilobic shape of the visible CD-spectrum of trimeric BR has been attributed to exciton coupling of the chromophores, in particular, since this spectral feature disappears during monomer formation. Also the different chromophore isomer composition in monomeric or trimeric BR forms, and a recent quantitative reexamination of the chromophore composition in dark-adapted BR of 2:1 for 13-*cis* : all-*trans* retinal [5] were assumed to indicate a mutual influence of the three retinals in the BR trimer. In this study, however, we arrive at similar results from either monomerized BR (CHAPS treatment) or from extrapolating the BR concentration to zero (without "disintegrating" the BR-trimers in this theoretical approach). We therefore have to ascribe the observed concentration dependent optoacoustic signals to interactions between purple membrane (stacks) and not to retinals within BR trimers.

References

[1] Mathies, R.A.; Steven, W.L.; Ames, J.B.; Pollard, W.T. *Annu. Rev. Biophys. Biophys. Chem.*, **1991**, 20, 491–518.

[2] Rohr, M.; Gärtner, W.; Schweitzer, G.; Holzwarth, A.R.; Braslavsky, S.E. *J. Phys. Chem. (in press)*, **1992**.

[3] Patel, C.K.N.; Tam, A.C. *Rev. M. Phys.*, **1981**, 53, 517–550.

[4] Fodor, S.P.A.; Pollard, W.T.; Gebhard, R.; van den Berg, E.M.M.; Lughtenburg, J.; Mathies, R.A. *Proc. Natl. Acad. Sci.*, **1988**, 85, 2156–2160.

[5] Scherrer, P.; Mathew, M.K.; Sperling , W.; Stoeckenius, W. *Biochemistry*, **1989**, 28, 829–834.

A new approach to analyse kinetic data of bacteriorhodopsin, factor analysis and decomposition

Benedikt Hessling, Georg Souvignier and Klaus Gerwert

Max-Planck Institut für Ernährungsphysiologie, Rheinlanddamm 201, 4600 Dortmund 1, Germany

A key element in the understanding of the detailed Bacteriorhodopsin's (bR) proton pump mechanism is the assignment of intramolecular reactions to specific photocycle intermediates. Due to the biphasic M-rise and -decay an unidirectional photocycle has to be excluded. This implies that the number of apparent rate constants being observed in photocycle kinetics is smaller than the number of intrinsic rate constants needed to give a reasonable description of the photocycle (Souvignier & Gerwert, 1992). As a consequence modelling of the photocycle reaction is an underdetermined problem and absorbance spectra of intermediates turn out to be model-dependent. To overcome this dilemma we used a different approach. By factor analysis and decomposition we calculated from the measured intermediate mixtures the pure intermediate spectra without any assumption on specific photocycle models. Thereby the minimal number of pure intermediates is yielded, that are kinetically and spectrally distinct. Restrictions in the calculations are given for well-known specific bands of the respective pure intermediate spectra. The idea of factor analysis and its mathematical background is explained and a first application to bR absorbance spectra in the visible and infrared is given.

The absorbance changes A_{ik} during a photocycle of l involved intermediates can be described by the following n times m equations:

$$A_{ik} = \sum_{j=1}^{l} \epsilon_{ij} c_{jk} \; ; \; i=1,n \; k=1,m$$

n   number of measured wavelengths
m   number of measured spectra
A_{ik} absorbance at time t_k and wavelength ν_i
ϵ_{ij} difference of extinction coefficient between intermediate j and BR_{570} at wavelength ν_i
c_{jk} concentration of the intermediat j at the time t_k

or in matrix form: [A] = [E] [C]

where Q_j is the j-th collumn of matrix [Q]. These collumns, called eigenvectors, form an orthogonal set which is normalized to an orthonormal set. Since $[Q]^T = [Q]^{-1}$

$$[Q]^{-1} [Z] [Q] = [Q]^{-1} [A]^T [A] [Q] = [Q]^T [A]^T [A] [Q]$$

$$= [U]^T [U] \quad ; \quad [U] = [A] [Q]$$

Multiplying [U] by $[Q]^{-1}$ we find

$$[A] [Q] [Q]^{-1} = [U] [Q]^{-1} = [U] [Q]^T$$

and we identify $[Q]^T$ with [C] and [U] with [E]. Therefore we conclude that the transpose of matrix [Q], which diagonalizes the covariance matrix [Z], represents the matrix [C], its rows form a set of orthonormal eigenvectors. Multiplying the calculated matrizes [E] and [C] we can reproduce the data matrix via a short circuit. Up to this stage of factor analysis we describe exactly the original data by a set of mutually orthonormal eigenvectors. Due to experimental error, however, the number of eigenvectors found does not depend on the number of involved intermediates, but on the dimension of matrix [A] (without error the number of eigenvalues unequal zero would represent directly the number of distinct intermediates). Each successive eigenvector accounts for the greatest possible variance (a formal definition is given in Malinowski, 1980) in the data. The most important eigenvector associated with the greatest eigenvalue is oriented in a factor space, constituted by all eigenvectors, so as to account for the maximum possible variance. It defines the best one-factor model. Since the least important eigenvectors associated with the smallest eigenvalues in practice regenerate the experimental error we recalculate the data matrix using only the most important eigenvector Q_1 to gain E_1 and C_1. We continue adding eigenvectors associated with the largest eigenvalues, sequentially, until we satisfactorily reproduce the data (see Fig. 2):

$$[E_1 \ E_2 \ ... \ E_n] \begin{bmatrix} C_1 \\ C_2 \\ \vdots \\ C_n \end{bmatrix} = [E^*][C^*] = [A^*] \approx [A]$$

The minimum number of eigenvectors required represents the number of factors involved. Though, after dropping negligible eigenvectors, the eigenvectors left still have no physical meaning. In order to gain physically significant factors an iterative procedure is used. First a relatively good estimation for the pure intermediate spectra is found as a linear combination of the calculated E_i

$$E_i' = a_i E_1 + b_i E_2 + ...$$

making use of earlier measurements and data from literature. During iteration those coefficients a_i, b_i, .. are varied that i) well-known characteristic bands of intermediates are present, ii) visible intermediate spectra correspond to spectra out of literature and iii) a positive concentration of intermediates is found.

Here the i-th row of matrix [A] represents the time dependence of absorbance changes at wavelength ν_i and the k-th collumn can be read as the absorbance spectrum at the time t_k. Though we know [A] within experimental error the elements of [E] and [C] are unknown. Starting from the original data [A] a matrix [Z] called covariance matrix is calculated. With the help of standard mathematical techniques [Z] is decomposed into a set of "abstract" factors [E] and [C], which, when multiplied, reproduce the original data (see Fig. 1). These factors [E] and [C] have a priori no physical meaning, but they show the minimal number of intermediates that describe the absorbance changes without any specific assumption on the photocycle. In order to give a physical meaning they are transformed into "real" factors, that reproduce our data within experimental error.

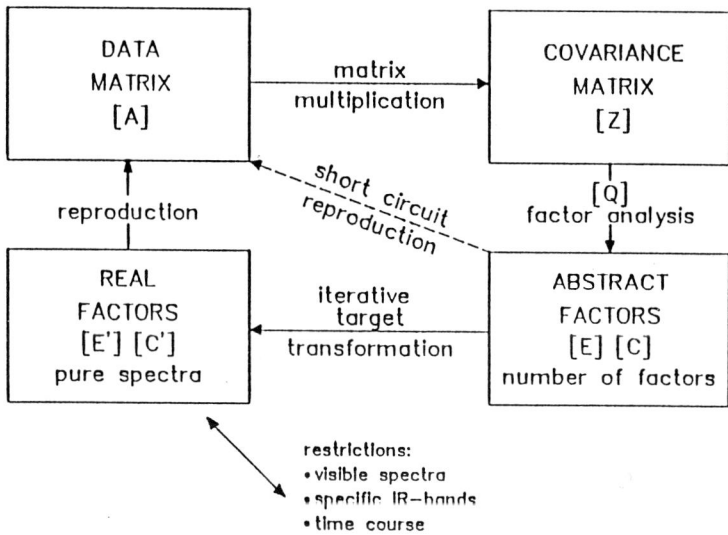

Fig. 1. Key steps in factor analysis.

[Z] is constructed by premultiplying the data matrix [A] by its transpose

$$[Z] = [A]^T [A]$$

With the help of numerical methods a matrix [Q] can be found that diagonalizes [Z] such that

$$[Q]^{-1} [Z] [Q] = [\lambda_j \delta_{jk}] = [\lambda]$$

Here δ_{jk} is the Kronecker delta and λ_j is an eigenvalue of the set of equations

$$[Z] Q_j = \lambda_j Q_j$$

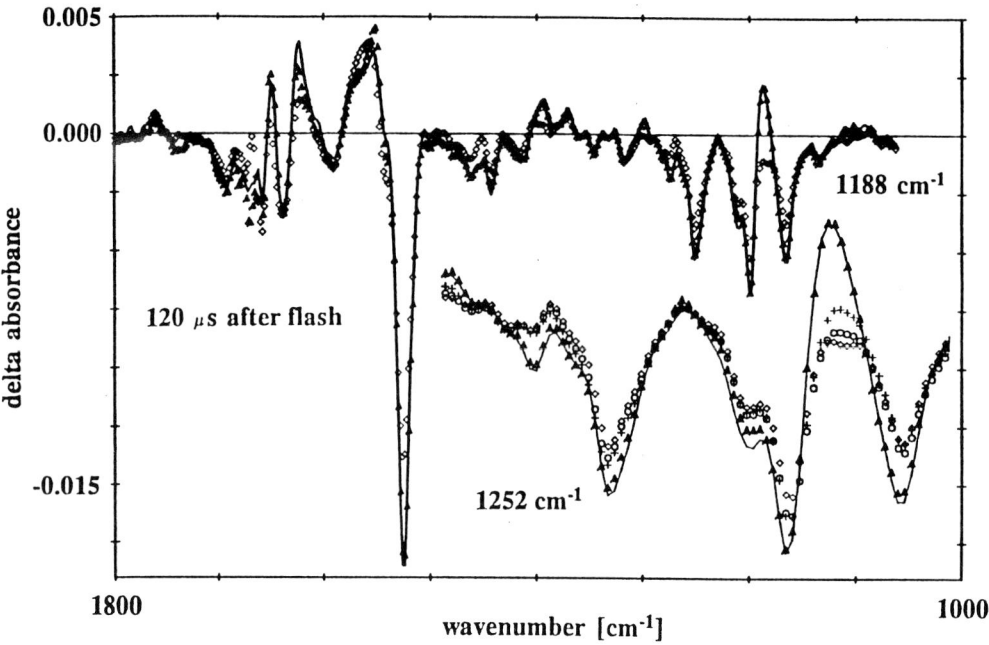

Fig. 2. Spectrum (1800-1060 cm^{-1}, 1300-1150 cm^{-1} enlarged) of a L-BR-measurement as part of data matrix [A] and its reproduction by one (+), two (o), three (◊) and four (▲) "abstract" factors.

The procedure described was applicated to the absorbance changes of bR from 50 µs to 700 ms after flash excitation, containing 380 wavenumbers in the infrared (1800-1060 cm^{-1}) measured with stroboscope-FTIR and 12 wavelengths in the visible range (410-680 nm). Two different measurements at pH 7 and pH 6 are regarded in one calculation. As shown in Fig. 2 four factors are required to reproduce absorbance changes during the L-BR reaction pathway via M, N and O. Three factors do not satisfactorily describe the difference spectrum measured 120 µs after flash excitation due to significant deviations around 1560 cm^{-1}, 1503 cm^{-1} and 1190 cm^{-1}. Adding of a 5-th factor (data not shown) only regenerates experimental error. To decide about the significance of an additional factor a careful comparison between original data and recalculated data is required. We conclude that 4 intermediats are required to sufficiently describe the data. Importantly this was found without supposing any specific photocycle and even without the assumption of first-order reactions. In the next step it should be possible to calculate the pure spectra of the involved intermediates.

References:
Malinowski, E. R. (1980), Factor Analysis in Chemistry, John Wiley & Sons Ltd.
Souvignier, G., Gerwert, K. (1992), Biophys.J. (in press)

Relationship of M-intermediates in bacteriorhodopsin photocycle

Lel A. Drachev, Andrey D. Kaulen and Andrey Yu. Komrakov

A.N. Belozersky Institute of Physico-Chemical Biology, Moscow State University, Moscow, 119899, Russia

The photocycle of the wild-type bacteriorhodopsin and D96N mutant was investigated with a flash-photolysis technique. M formation at 400 nm and L decay at 520 nm was well fitted with two components having time constants $\tau_1= 65$ and $\tau_2= 250$ µS for the wild-type bR and with three components having $\tau_1= 55$ µS, $\tau_2= 220$ µS and $\tau_3= 1$ mS for D96N bR. Proton release measured with pyranine in the absence of buffer was identical for both protein types and was approximated using one component having $\tau= 1$ mS. The temperature dependence for components of the M formation of D96N bR and for the rate of pyranine protonation revealed a coincidence of the proton appearance in the bulk medium with the third M component of the D96N bR. In the presence of 4 mM MES the proton appearance in the bulk matched all the components of M formation of the D96N bR and wild-type bR. A 1-mS-component was found to be present in the photocycle of the wild-type bR as a lag-phase in the decay of the photoresponse at 400 and 520 nm, and as a component in the photoresponse at 335 nm in addition to the fast components associated with the bR→K and K→L transitions. Under the conditions of 2 mM Lu^{3+} ions or 80% glycerol, this component appeared in the M formation of the wild-type bR (Fig.1, curves 1,2). Fig. 1 shows the photoresponse at 400 nm of the wild-type bR under 40% sucrose, 2 mM $LuCl_3$, pH 7.0 conditions (curve 5) and after addition of 10 mM EDTA (curve 6). One can see that the component at 40-mS time scale appeares at the expense of the first two components. The 1-mS-component differs from the 55-µS and 220 µS components in D96N bR. Its maximum is at 404 nm compared with the maximum at 412 nm for 55 and 220 µS components. Under D_2O conditions the first two components were slowed down by a factor of 5, while the third one - by a factor of 3, as well as the rate of deiterium release in the bulk medium compared with the proton release was slowed down by a factor of 3. Under 60% sucrose and 80% glycerol conditions the first two components were found to be faster than the same components in a water solution, while the third component was slowed down.

The decay time of M of the D96N bR was decreased from 200 mS

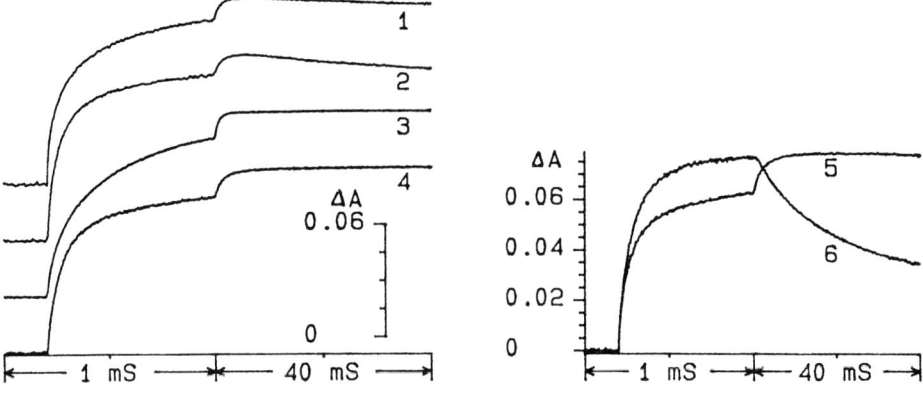

Fig.1. Photoresponses at 400 nm for the wild-type bR (1, 2, 5, 6) and D96N bR (3, 4) under the conditions of: 80% glycerol (1, 4); 2 mM Lu^{3+}; 40% sucrose, 2 mM Lu^{3+} before (5) and after addition of 10 mM EDTA (6).

at pH 5.0 to 1.5 mS in the presence of 7.5 mM azide. It was disclosed that for different kinetic models, suitable for the 3-exponential M-formation, the best approximation was obtained when azide was suggested to protonate the Schiff base of all M forms simultaneously and with the similar time constants. This allowes us to explain the dramatic decrease in the amplitude of the photosignal at 400 nm. When azide concentration was 1M the rate of the M-decay was found to be 20 µS. After rapid reprotonation of M, a red-shifted intermediate with a differential maximum at 630 nm was formed (Fig.3A, curve 1). Under D_2O conditions the rates of the first two components were increased from 55 and 230 µS to 250 µS and 1.6 mS, whereas the rate of azide action did not change significantly. That made it possible to obtain O-like intermediate rise signals with identical rate constants as for M intermediate. The first two components of the SB deprotonation (L-decay) were observed to be insensitive to the presence of azide, while the third component disappeared in the L-decay and appeared in the formation of the second batho-form, with a differential maximum at 638 nm, directly from the second O-like intermediate with a 3-4 mS time constant under H_2O conditions, and with a 7-8 mS one under D_2O conditions it transformed into the N-like form (Scheme 3). This corresponds well to the M→N transition of the wild-type bR. These data indicate that the D96N bR with the protonated Schiff base (SB) (by azide) exhibits very similar transitions when M-forms are replaced by O-like intermediates in comparison with transformations of M intermediates in the wild type bR. We interprete these observations in terms of protein conformational changes. Such changes may be initiated at the early stages of the photocycle (trans-cis isomerisation, for example). The first conformational transition with the time constant 1 mS in the case of the

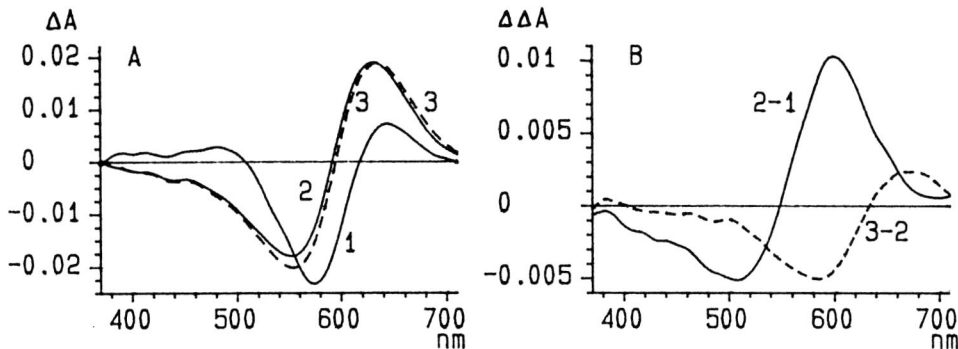

Fig.2 A: Differential spectra measured for the D96N bR under the conditions of 1M azide, pH 5.0. mS after flash: 1- 0.05; 2- 1.1; 3- 2.5; B: Differencies between spectra presented in A.

wild-type bR, reflects the SB switch from the outward proton channel, and the second one with the time constant 4-5 mS may reflect the SB approach to the internal proton donor when the later is in the account. The proton release into the bulk medium in the absence of buffers matched the first conformational transition. This may be explained by two alternatives. 1) The protein conformational transition causes changes in the membrane-water interface followed by breakings in the surface proton-conducting chain and enhancing the probability for the proton transfer into the bulk; 2) The probability for the proton release into the bulk increases with the SB removing from the outward proton channel with a time constant 1 mS. Second conformational transition at 4-5 mS is likely to facilitate SB reprotonation and N formation. It is associated with the appearance of the positive phase of the light scattering change signal in pm suspension.

According to our findings concerning the two phases in the L-decay independent on azide concentration we present here a convenient model for M rise in the wild type (1) and D96N bR (2)

$$L_1 \xrightarrow{60\mu S} M_1 \xrightarrow{1mS} \searrow \atop M' \quad (1) \qquad L_1 \xleftrightarrow{55\mu S} M_1 \xrightarrow{1mS} \searrow \atop M' \quad (2)$$

$$L_2 \xrightarrow{250\mu S} M_2 \xrightarrow{1mS} \nearrow \qquad L_2 \xleftrightarrow{220\mu S} M_2 \xrightarrow{1mS} \nearrow$$

The azide action mode described above is presented at scheme (3) on the basis of scheme (2):

$$\begin{array}{c} L_1 \xleftrightarrow{55\mu S} M_1 \xrightarrow{1mS} \searrow \\ L_2 \xleftrightarrow{220\mu S} M_2 \xrightarrow{1mS} \nearrow \end{array} \xrightarrow{M'} \xrightarrow{az,20\mu S} O638 \xrightarrow{4mS} N \quad (3)$$

with $az,20\mu S \to O630 \xrightarrow{1mS}$ branches above and below.

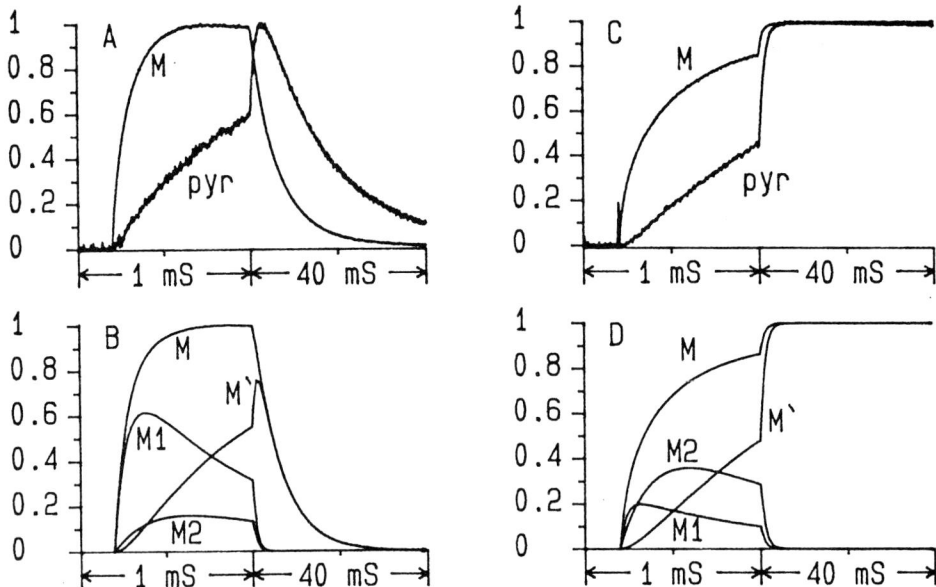

Fig.3. M-formation measured at 400 nm and proton release measured with pyranine in the absence of buffer for the wild-type bR (A) and the D96N bR (C). Grafical representation for schemes (1) and (2) (B and D respectively).

Illustrations for schemes (1) and (2) are presented in Fig.3. Our scheme may be regarded as a combination of Butt et el, 1989, a, b and Varo and Lanyi, 1991 schemes. The discrepancy of our and Varo and Lanyi schemes concerning the indication of the component in M-formation ts attributed to the irreversible M⟶M' step. The component they pointed out as a conformational switch is likely to be one of the two parallel M-formation components. This is in line with the Butt et al. scheme. We have suggested that the conditions of glycerol or Lu^{3+} ions cause a conformational distortion of the wild-type bR and the back-reaction M⟶L(K) was induced, so the 1-mS component appears in the photoresponse at 400 nm as a "real" component, while in the photoresponse of the undistorted bR 1-mS component it is presented as a lag-phase at 400 nm.

REFERENCES

Butt, H-J., Fendler, K., Der, A. and Bamberg, E. (1989a), Biophys.J., 56, 851-859.

Butt, H-J., Fendler, K., Der, A., Bamberg, E., Tittor, J. and Oesterhelt, D. (1989b), EMBO J., 8, 1657-1663.

Tittor, J., Soell, C., Oesterhelt, D., Butt, H-J., and Bamberg, E. (1989), EMBO J., 8, 3477-3482.

Varo, G., and Lanyi, J., K. (1991), Biochemistry, 30, 5008-5022.

M-type intermediate formation during 13-*cis*-bacteriorhodopsin photocycle and light-dark adaptation

Andrey D. Kaulen, Lel A. Drachev and Swetlana V. Dracheva*

A.N. Belozersky Institute of Physico-Chemical Biology, Moscow State University, Moscow 119899, and
*Institute of Theoretical and Experimental Biology, Puschino, Moscow Reg. 142292, Russia

Earlier we have demonstrated that dark-adapted 13-cis-retinal containing bR (13-cis-bR) is able to form the deprotonated shortwave M intermediate [Drachev et al.,1988; Kaulen et al.,1990]. The efficiency of M intermediate formation in purple membranes during the 13-cis-bR photocycle is approximately 35% of the efficiency of M intermediate formation during the trans-bR photocycle, and in asolectine liposomal preparation it reaches 75%. In the monomeric bR solubilized with 2% Triton X-100 the efficiencies of M intermediate formation in 13-cis- and all-trans-bR photocycles are practically equal. The M intermediate formation during the 13-cis-bR photocycle is a pH-dependent process. The pK of this process in a solubilized preparation <5; in a liposomal preparation pK=7.6, and in purple membranes 8.5<pK<9.5 depending on the ionic strength. We have demonstrated that the 13-cis-bR photocycle which includes formation of the M intermediate is accompanied by transmembrane proton transfer. We have drawn an analogy between the pH-dependent transition of 13-cis-bR from the active form, which is able to produce the M inter mediate, to the inactive form, and the pH-dependent transition of the neutral trans-bR form into the blue acidic form, which is unable to produce the M intermediate. It is easy to suppose the existence of an Y group in the outward proton pathway, and that the protonation of this group leads to the inhibition of proton transport during the trans-bR photocycle. A similar supposition can be done for 13-cis-bR: the proton-transporting activity of 13-cis-bR is due to the protonation of some Y' group in the outward proton pathway.

The simplest suggestion is, that both Y and Y' groups receives a proton directly from the Schiff base. Now we can speak about the existence of an Y group with confidence (Asp-85, probably, plays the role of this group), but some questions arise concerning the Y' group. The first question is - if Y and Y'groups are one and the same group with different pK in the two forms of bR? The second question is - is the Y'group really exists, or the pH-dependence of the 13-cis-bR photocycle simply reflects pK of an exciting chromophore?

The first supposition is rejected by the fact that both trans- and 13-cis-bR are converted to the blue acidic form at low pH. The solution of the second question has lead us to the conclusion that the Y^1 group really exists. Our arguments are based on the fact that the group with similar pK influences the dark processes of the chromophore isomerization.

Firstly, we have discovered the existence of the pH-dependence of the of trans- and 13-cis-bR ratio in dark-adapted purple membranes. At pH range from 4 to 8 this ratio is practically constant and the content of trans-bR is 42-45%. Lowering of pH leads to the inhibition of M intermediate formation by trans-bR simultaneously with bR transition to the blue acidic form, and to an increas in the trans- bR content up to 60-65%. All these transitions are synchronously shifted to a low pH region by increasing the ionic strength. The further increase in pH higher than 8 leads to the appearance of 13-cis-bR ability to form the M intermediate; simultaneously in dark- adapted bR the content of trans-bR decreases to 31-33%. Synchronous lowering of pK of these transitions by increasing the ionic strength is also observed. So, the protonation of the Y and Y^1 groups leads to the increase in trans-bR in dark-adapted bR. The level of protonation of these groups has influence on the equilibrium between trans- and 13-cis-bR. It is possible to distinguish between three main equilibriums: (1) at the blue acidic form where these groups are protonated, (2) at the neutral purple form, where the Y group is deprotonated and the Y^1 group is protonated and at last (3) the equilibrium at a high pH region, when these groups are deprotonated.

Data on the rate of dark adaptation obtained by Ohno et.al.(1977) should be noted. According to their results the rate of dark adaptation is practically constant at pH range from 6 to 8 and significantly increased at low pH and decreased at high pH. Analyzing these data Warshel and Ottolenghi came to the conclusion that at least two groups participate in the dark isomerization of the chromophore. The pK of one group is equal to the pK of the protein transition to the blue form; and the pK of the other group is similar to the pK of transition of 13-cis-bR from inactive to active form. Summarizing these data, we can conclude that both the Y and Y^1 groups influence on the position of equilibrium between 13-cis- and trans-bR and participate in the catalisys of the chromophore group isomerization. Probably, these groups are localized not far from the Schiff base. The protonation of these groups leads to an increase in delocalization of the positive charge in the retinal polyen chain, to a decrease in double bonds order and as a result to facilitation of isomerization around these bonds. Such facilitation of isomerisation around double bonds C=N and C13=C14 under the increasing in the distance between anion and the protonated Schiff base was described by Shives and Balashov(1974).

If the Y group differs from group Y^1, and these groups are localized not far from the chromophore, one can expect that deprotonation of the Y^1 group resulting in deprotonation of the Schiff base during the 13- cis-bR photocycle must lead to some changes in thetrans-bR photocycle. Indeed, apparently the deprotonation of Y^1 leads to a significant increase in the rate of the Schiff base deprotonation during the trans-bR photocycle, but does not influence on the efficiency of this process.

It is well known, that the increase in pH leads to the acceleration of the M intermediate formation during the trans-bR photocycle. It is supposed, that this transition of the trans-bR neutral purple form to the alkaline form is the result of Tyr residue deprotonation (Balashov et.al.,(1991)). We have discovered that the kinetic of the M intermediate formation under the intermediate pH values is well approximated by a sum of two kinetics - the kinetics of M formation during the photocycle of the purple bR form, and the kinetics of M formation in the alkaline form. It permitted us to determine the contribution of the alkaline form under the different pH values, and, correspondingly, the pK of transition of the neutral to the alkaline form. It was found that the pK of the transition of trans-bR to the alkaline form and the pK of 13-cis-bR transition to the active proton-translocating form are similar. The values of these pK are identically changed with the change in the ionic strength. However, despite some vagueness in determining the pK of these processes it is possible to conclude that the pK of trans-bR transition to the alkaline form is shifted for 0.2- 0.3 pH unites to a higher pH region in comparison with the pK of 13-cis-bR transition to the active form. Nevertheless, we consider that both transitions are due to the protonation of one and the same group. The pK of this group in the two bR forms can be different as a result of various surrounding (f.e., various interaction with chromophore groups with different isomeric composition). The results obtained by the analysis of Triton X-100 solubilized bR prove this point of view. Already at rather low pH values (<5) in solubilized bR the rate of the M intermediate formation similar to the rate of the M intermediate formation of the alkaline form of bR in purple membranes. Characteristics of slow stages of the photocycle are similar in solubilized and in alkaline forms of bR. In solubilized bR already at low pH the 13-cis-bR photocycle is accompanied by deprotonated M intermediate formation. It is interesting to note, that the rate of dark adaptation in solubilized and alkaline forms of bR are slowed down by increasing pH in all pH range.

Thus, it should be mentioned, that the data obtained are in favour of existence of at least two groups in the outward proton pathway, which are necessary for its function. The deprotonated state of one of these groups (group Y, apparently Asp-85) is necessary for proton transport and M intermediate formation during trans- (and, probably, 13-cis-) bR photocycles. Deprotonation of the other group is accompanied by acceleration of the M intermediate formation during trans-bR photocycles and by appearance of the Schiff base deprotonation during the 13-cis-bR photocycle. According to the data of Balashov et.al.,(1991) the Tyr residue must be this group. However, taking into account our results on the solubilized bR, we consider that some carboxil-containing residue is a better candidate on this role. The pK of this residue is significantly increased by intraprotein interaction, if the protein is included into purple membrane rigid structure.

We have carried out a comparative analysis of trans- and 13-cis-bR photocycles. In dark-adapted and light-adapted (containing 15% of 13-cis-bR) preparations the photochemical conversions can be represented like a superposition of two processes - 13-cis and trans-bR photocycles. As a whole, these photocycles are similar, but have some differences in kinetic parameteres.

In these photocycles M intermediates have practically identical spectra, but the M intermediate formation in the 13-cis-bR photocycle is 10-fold slower, than in the trans-bR photocycle. The M intermediate formation in both photocycles leads to the proton release from the protein. On the next stage of the trans-bR photocycle the equilibrium between M and N(P) intermediates is established, and in the case of the 13-cis-bR photocycle the equilibrium between M and bathochromic O-like intermediates is established. Both these processes are practically independent of pH. The O intermediate in trans-bR photocycle is not observed. In the trans-bR photocycle the equilibrium is shifted to the M intermediate, and in 13-cis-bR the equilibrium is shifted to the O-like intermediate. Apparently, both these processes are due to the intra-protein Schiff base protonation with Asp-96 participation. The character of the changes in β-absorption band has shown that the isomeric state of the chromophore does not change during the M-->N transition in the trans-bR photocycle, but in 13-cis-bR the intra-protein protonation is accompanied by changing the isomeric state of the chromophore. The O- like intermediate reterns to the 13-cis-bR ground state, and, correspondently, N returns to the trans-bR. Both these processes are accompanied by proton uptake by the protein molecule. Both processes become ten-fold slower at increasing pH for 1 unit. It is interesting to note, that O intermediate decay is 100-fold faster than N intermediate decay. Partially this difference is due to the difference between the M<->O and M<->N equilibrium, but partially - to the difference between, the structures of the proton wires in two bR forms. The previous data have shown that the proton uptake is carried out by the group with low pK. The difference in the rates of the last photocycle stage is due to the difference of the pK of this group in the cases of trans- and 13-cis-bR.

REFERENCES

Balashov, S.P.,Govindjee, R., Ebrey, T.G. (1991): Red shift of the purple membranes absorption band and the deprotonation of tyrosine residues at high pH. Biophys.J. 60.pp.475-490.

Drachev, L.A., Kaulen, A.D., Skulachev, V.P., Zorina, V.V. (1988): Electrogenic photocycle of the 13-cis-retinal-containing bacteriorhodopsin with an M intermediate involved. FEBS lett., 1., pp.1-4.

Kaulen, A.D., Drachev, L.A., Zorina, V.V. (1990): Proton transport and M-type intermediate formation by 13-cis-bacteriorhodopsin. Biochim. et biophys.acta. 1018., pp.103-113.

Ohno, K., Takeuchi, Y., Yoshida, M. (1977): On the photocycle and light adaptation of dark-adapted bacteriorhodopsin. Biochim. et biophys. acta., 462., pp.575-582.

Sheves, M., Baasov, T. (1984): Factors affecting the rate of thermal isomerization of 13-cis-bacteriorhodopsin to all-trans. J. Amer. Chem. Soc. 106., pp.6840-6841.

M-intermediate in the 13-*cis*-cycle of bacteriorhodopsin analogs

Lyubov V. Khitrina, Lel A. Drachev, Sergey V. Eremin*, Andrey D. Kaulen and Andrey A. Khodonov*

*A.N. Belozersky Institute of Physico-Chemical Biology, Moscow State University, Moscow, 119899 and *Institute of Fine Chemical Technology, Moscow, 117571, Russia*

Photocycles of bacteriorhodopsin, fluorophenylbacteriorhodopsin and 11,12-didehydrobacteriorhodopsin with 13-*cis*-chromophore at pH 6 have no intermediates similar to the M-form (Dencher et al., 1976; Stoeckenius et al., 1979; Drachev et al., 1987, 1988a; Danshina et al., 1989). Lately Drachev et. al. (1988b, 1990) have shown that on alkalinization (pK $\geqslant$ 8.5) of 13-*cis*-bacteriorhodopsin its photocycle acquires an M-type intermediate. Bacteriorhodopsin containing pure 13-*cis*-chromophore can be obtained only by regeneration of apo- or white membranes with 13-*cis*-retinal (Dencher et al., 1976; Stoeckenius et al., 1979; Drachev et al., 1990). Since isomerization and regeneration rates for 13-*cis*-bacteriorhodopsin are comparable, the accuracy of the measurements is limited. However for some bacteriorhodopsin analogs (Maeda et al., 1984; Drachev et al., 1987; Danshina et al., 1989) isomerization and regeneration rates differ by several orders. In the present work we carried out direct measurements of 13-*cis*-cycles of such pigments at different pH.

synthesized as described in (Khodonov et al., 1987; Drachev et al., 1987; Danshina et al., 1989). Apomembranes were prepared by light-dependent hydroxylaminolysis. Experiments with analogs were carried out at dim red light. Pigments were regenerated by adding I, II or retinal to apomembranes (Drachev et al., 1987; Danshina et al., 1989). The photocycle was studied by flash-photolysis (Drachev et al., 1987; Danshina et al., 1989). Measurements were performed by single flashes without averaging.

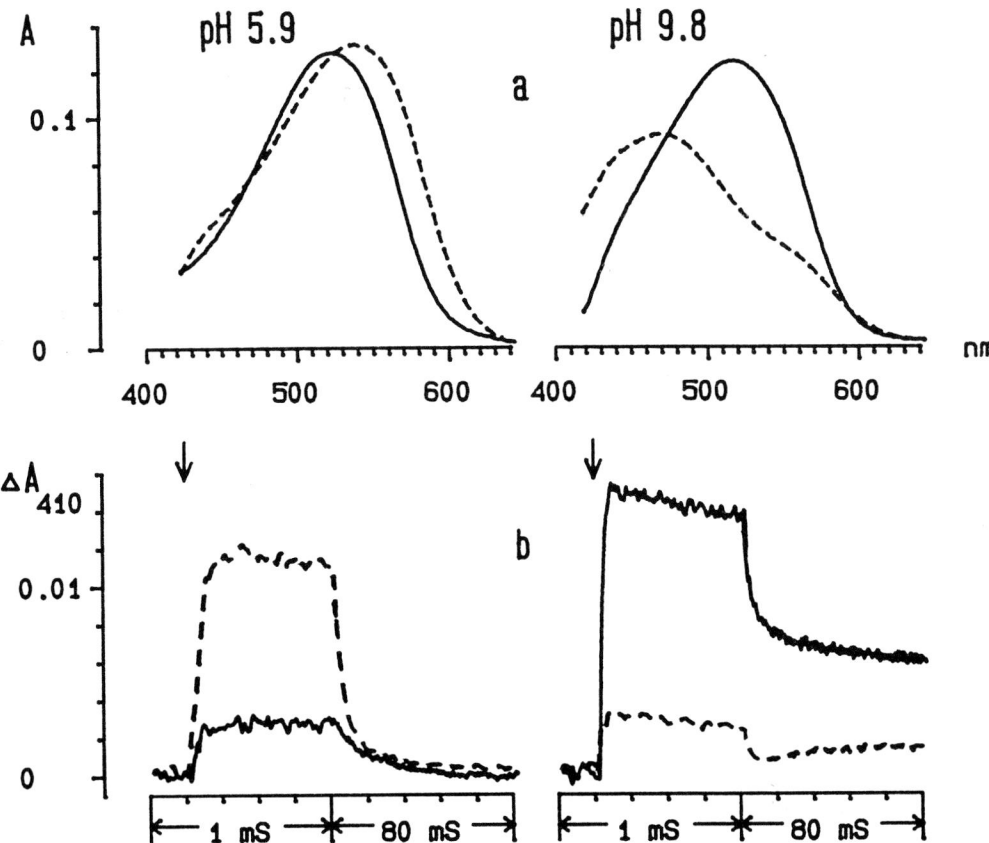

Fig.1 Spectral characteristics of 11,12-didehydrobacteriorhodopsin sheet suspensions at 20°C, pH 5.9 and pH 9.8. (a) Absorption spectra taken against apomembranes. (b) Optical density changes induced by 15 nS laser flashes (λ = 532 nm, the energy in green light, 50 mJ) and monitored at 410 nm. The flash is indicated by an arrow. Solid line: 13-*cis*-isomer. Dashed line: isomerisation product of the initial suspension of 13-*cis*-analog (the suspension in bidistilled water at 1°C was illuminated for 75 min by a 2500 W Xenon lamp with heat filtration). Before measurements both preparations were ten-fold diluted by warm solution finally containing 200 mM potassium phosphate, 200 mM citric acid and 200 mM boric acid. pH was adjusted with KOH before membrane addition and checked in the spectrophotometric cuvet after the measurement.

Fig.1b shows that alkalinization of 13-*cis*-11,12-didehydrobacteriorhodopsin suspension leads to a considerable increase in the 410 nm photoresponse amplitude (M-intermediate). The response at pH 5.9 (Fig. 1b) is caused by an admixture of the *all-trans*-isomer (Danshina et al., 1989).
Our data show that long illumination of 13-*cis*-11,12-didehydrobacteriorhodopsin at pH 6 converts about 85 % of chromophore into the *all-trans*-isomer (Danshina et al., 1989). We used this transformation to prepare the *all-trans*-pigment. The absorption spectrum shows that at pH 9.8 the analog, containing the *all-trans*-chromophore, is considerably denatured (Fig. 1a). The amplitude of its M-intermediate also strongly decreases. It is noteworthy that the purple complex of the light adapted bacteriorhodopsin does not break down at the pH increase up to 9.8 even when the preparation is regenerated from apomembranes with retinal. Due to this decrease in the denaturation pK of the *all-trans* analog, there is actually no doubt about the origin of the M-intermediate arising at alkalinization of 13-*cis*-11,12-didehydrobacteriorhodopsin. It is clear that the M-intermediate is formed due to the 13-*cis*-cycle and not to its dark isomerization. Our data described in (Danshina et al., 1989) and the results given in Fig. 1b show that the M-intermediate amplitudes in the cycles of the *all-trans*-11,12-didehydrobacteriorhodopsin at pH 6 and of 13-*cis*-isomer at pH 9.8 are actually very similar.
The appearance of an M-type intermediate at growing pH can be observed with the 13-*cis*-fluorophenylbacteriorhodopsin as well.
Thus, the data obtained on analogs with retarded dark adaptation completely confirmed that at high pH values the photocycle of 13-*cis*-bacteriorhodopsin acquires an M-intermediate.

REFERENCES

Danshina, S.V., Drachev, A.L., Drachev, L.A., Kaulen, A.D., Mitsner, B.I., Khitrina, L.V. and Khodonov, A.A. (1989): 11,12-Didehydroretinal-containing *all-trans*- and 13-*cis*-analogues of bacteriorhodopsin. *Bioorgan. Khim. (USSR)* 15, 307-312.
Dencher, N.A., Rafferty, C.N. and Sperling W.(1976): 13-*cis* and *trans* bacteriorhodopsin photochemistry and dark equilibrium. *Her. Kernforsch. Julich.* 1374, 1-42.
Drachev, A.L., Zorina, V.V., Mitsner, B.I., Khitrina, L.V. Khodonov, A.A. and Chekulaeva, L.N. (1987): Photocycle and electrogenesis of 13-*cis*- and *all-trans*-aromatic analogs of bacteriorhodopsin. *Biokhimiya (USSR)* 52, 1559-1569.
Drachev, A.L., Drachev, L.A., Kaulen, A.D., Skulachev, V.P. and Khitrina L.V. (1988a): Phases of the 13-*cis*-bacteriorhodopsin photoelectric response. *Biokhimiya (USSR)* 53, 707-713.
Drachev, L.A., Kaulen, A.D., Skulachev, V.P. and Zorina, V.V. (1988b): Electrogenic photocycle of the 13-*cis* retinal-containing bacteriorhodopsin with an M intermediate involved. *FEBS Lett.* 239, 1-4.
Drachev, L.A., Kaulen, A.D. and Zorina, V.V. (1990): Proton transport and M-type intermediate formation by 13-*cis*-bacteriorhodopsin. *Biochim. Biophis. Acta* 1018, 103-113.
Khodonov, A.A., Mitsner, B.I., Zvonkova, E.N. and Evstigneeva,

R.P. (1987):Aromatic analogues 13-*cis*- and *trans*-retinals. *Bioorgan. Khim. (USSR)* 13, 238-251.

Maeda, A., Asato, A.E., Liu, R.S.H. and Yoshizawa, T.(1984): Interaction of aromatic retinal analogues with apopurple membranes of Halobacterium halobium. *Biochemistry* 23, 2507-2513.

Stoeckenius, W., Lozier R.H., and Bogomolni, R.A.(1979): Bacteriorhodopsin and the purple membrane of halobacteria. *Biochim. et Biophys. Acta* 505, 215-278.

Two quantum absorption of ultrashort laser pulses by the bacteriorhodopsin chromophore

I.V. Chizhov[(1)], M. Engelhard[(2)], A.V. Sharkov[(2)] and B. Hess[(3)]

[(1)] General Physics Institute and [(2)] P.N. Lebedev Physical Institute of the Russian Academy of Sciences, Moscow, Russia. [(3)] Max-Plank Institut für Ernährungsphysiologie, Dortmund, Germany

The recently observed irreversible photobleaching (or photodestruction) of the bR chromophore by the intense green laser flashes (Hess et al., 1990; Govindjee et al., 1990; Czege & Reinisch, 1991) has the certainly proved two quantum absorption mechanism. However there were not clear what chromophore structure corresponds to the spectra of photoproducts and what state of the bR chromophore is responsible to the absorption of second quantum. In our experiments we intended to find out the second problem. For this aim the effect of different light intensities and durations of pulses was measured. The experiments were carried with: a - 10 ns, $1-5 \cdot 10^6$ W/cm^2, 532 and 610 nm; b - 30 ps, $0.5 \cdot 10^9$ W/cm^2, 532 nm and c - 200 fs, $6 \cdot 10^{10}$ W/cm^2, 615 nm laser pulses. The frequency of pulse repetition was 10 Hz for 200 fs laser and 0.5-1 Hz for 10 ns and 30 ps ones, the temperature of samples was 20°C.

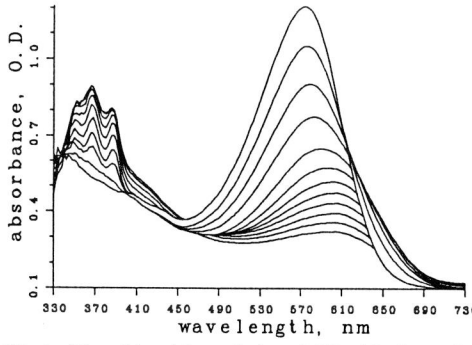

Fig.1. Photobleaching of the visible bR band by nanosecond green laser pulses. Each spectrum under the initial light-adapted bR is drawn after equal doses of light quanta. The dose is counted as D=nEA=6 J/cm^2, where n - number of pulses; E - flux of energy in one pulse; A - relative absorbance at the wavelength of excitation. Average intensity in pulse I=$2 \cdot 10^6$ W/cm^2, initial number of pulses n_0=300.

Whereas the spectra of photoproducts upon the pico- and nanosecond irradiation were practically identical femtosecond pulses produced the photobleached chromphore which significantly differed on absorption spectra. Thus the evaluation of data obtained from such experiments must be based the different schemes of two quantum absorption.

In Fig. 1 the effect of bleaching by nanosecond 532 nm laser pulses is shown. Isobestic point at 400 nm is well established till moving of the visible maximum from 570 to 600 nm and accompanying formation of the UV 340,360,380 nm three-peaked band. Further irradiation results in elimination of 600 nm band and distortion of the fine structure of UV one. Disappearence of isobestic point on this second stage could be probably explained by the growth of scattering of the sample from the aggregation of purple membrane sheets. Analogous behaviour was observed in experiments with picosecond pulses (not shown).

The two quantum nature of irreversible absorbance change might be easy proved by the analisys of efficiency of the irreversible bleaching. In Fig. 2 the relative absorbance change at 570 nm divided on the flux of photons is shown as a function of the light intensity.

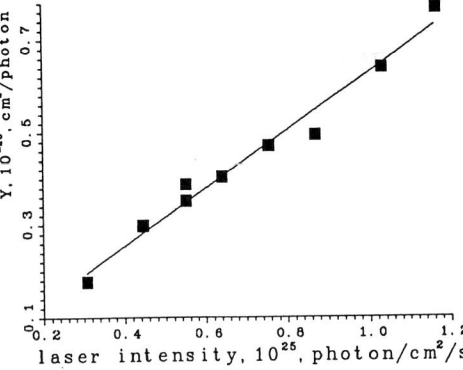

Fig.2. The relative absorbance change at 570 nm normalized on the summary flux of the photons (output of photoproduct Y(I), Angelov et al. (1980)) versus the average photon flux intensity in one laser pulse. The measurements were the following. Optically thin sample (O.D.=0.3) was irradiated by the series of green 10 nanosecond laser flashes with energy in pulse varied from 10 to 50 mJ/cm² to provide the irreversible absorbance change upon 20 %. For the different energies amount flashes varied from 200 (50 mJ/cm²) to 4000 (10 mJ/cm²).

The linear dependence Y(I) shows clearly that the photobleaching has the two quantum character (i.e. the unnormalized absorbance change depends on I^2). The drop of function Y is equal $\phi_{bl} \cdot \sigma_1 \cdot \sigma_2 \cdot \tau_1$ (Angelov et al., 1980) where ϕ_{bl} is quantum yield of bleaching; σ_1 and σ_2 are the cross sections of ground state and electronically excited state of bR which absorbs the second quantum, respectively; τ_1 is the effective (see below) life-time of the excited state. The theory of two quantum absorption provide also the level of the saturation of function Y(I) when the ground and excited states of chromophore are close to the photostationary equilibrium $Y_{sat} = \phi_{bl} \cdot \sigma_1 \cdot \sigma_2 / 2(\sigma_1 + \phi_{bl} \cdot \sigma_2)$. Our experiments with 30 picosecond laser pulses give this value about 10^{-18} cm²/photon at the light intensity about $2 \cdot 10^{27}$ photon/cm²/s (or 10^9 W/cm²). Thus taking cross section of ground state $\sigma_1 = 1.8 \cdot 10^{-16}$ cm² we can estimate the effective life-time of excited state $\tau_1 = 1.7 \cdot 10^{-12}$ s and $\phi_{bl} \cdot \sigma_2 = 2 \cdot 10^{-18}$ cm².

The interpretation of derived values depends on the choice of electronically excited state of the chromophore which absorbs the second quantum. Govindjee et al. (1990) supposed that the second quantum of nanosecond laser pulse (30-90 mJ/cm²) would be absorbed by the bR* state. However we made a crucial experiment on the bleaching by laser pulses with duration about 200 fs. These pulses provide photobleaching effect which differs significantly by the final visible spectra from that one under the pico- and nanosecond laser pulses. In Fig. 3 these spectra are shown combined with ground spectra of bR.

The femtosecond pulses produce also nonfunctioning chromophore state but with the maximum of absorption at 550 nm (the three-peaked UV band is poorly seeing but still recognizable). Thus we can conclude that the second quantum of femtosecond pulse dominantly excites the intermediates of the photocycle with life-time less than 0.2 ps and probably also the primary electronically excited state of the bR chromophore. This excitation provides the change of the chromophore structure that is enable for photocycle and has the visible absorption band at the 550 nm.

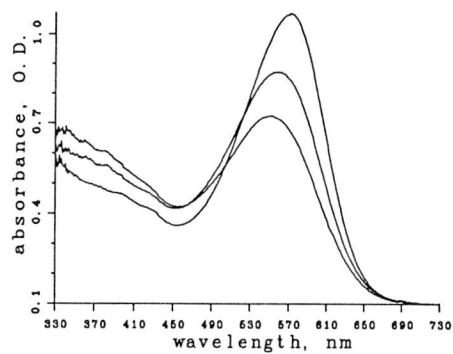

Fig.3. The absorbption spectrum of bR before (upper curve in the 570 nm region) and after (lower ones) irradiation of sample by the series of 200 fs, 615 nm, $6 \cdot 10^{10}$ W/cm^2 laser pulses. First bleached spectrum is measured after 1800 pulses, second one after 9000. (For prevention that observed effect is not caused by the change of the wavelength we made also the experiments with dye 610 nm, 10 ns laser pulses which showed the identical spectra as 530 nm, 10 ns laser)

Thus the second quantum of pico- and nanosecond laser pulses excites mainly the intermediate of the bR photocycle with the time of formation not less than about 0.5 ps and the decay time in the nanosecond range. The most probable candidate is the K610 intermediate of photocycle. Thus the derived above values have the following meaning. The $\tau_1 = 1.7 \cdot 10^{-12}$ s might be the fastest time of electronic relaxation of the K610 intermediate or the time which limits the formation of K from bR*. We suppose that the second explanation is more probable because as it's known in literature the next transformation of K takes time in some nanoseconds (KL intermediate) or even slower. Then the quantum yield of such a conversion would be less than about 10^{-3} if the other ways of relaxation are so fast. The value $\tau_1 = 1.7 \cdot 10^{-12}$ s is very close to the time of formation of K that strongly supports our assumption. Regarding quantum yield of photobleaching we can estimate it as $2 \cdot 10^{-2}$ because in the frame of our scheme the σ_2 is the cross section of the K intermediate and equal at 530 nm about $1 \cdot 10^{-16}$ cm^2.

In the all above described experiments the yield of the nonfunctioning products after absorption of the second quantum is higher than after photoexcitation of the intermediates which are formed after K610. Indeed, the light absorption by the later intermediates changes the relaxation pathways in the photocycle but the degradation of the chromophore was not observed. Because the probability of the irreversible change in the chromophore structure must be higher if to excite the primary excited electronic state therefore we can suggest also that the KL590 intermediate with lifetime about microsecond is the first electronically relaxed state of the bR chromophore in photocycle.

Hess, B., Chizhov I.V. & Engelhard M. (1990): Photobleaching of the bR chromophore by nanosecond or shorter laser pulses. *IV Int. conf. on retinal proteins*, Santa Cruz, CA, Abstracts.

Govindjee, R., Balashov S.P. & Ebrey T.G. (1990): Quantum efficiency of the photochemical cycle of bacteriorhodopsin. *Biophys. J.* 58, 597-608.

Czege, J. & Reinisch L. (1991): Photodestruction of bacteriorhodopsin. *Photochem. Photobiol.* 53, 659-666.

Angelov, D.A., Krukov P.G., Letokhov V.S., Nikogosyan D.N. & Oraevskii A.A. (1980): Selective interaction of ultrashort UV laser pulses with the components of macromolecules. *Quantum electronics (USSR)* 7, 1304-1318.

Structures and Functions of Retinal Proteins. Ed. J.L. Rigaud. Colloque INSERM / John Libbey Eurotext Ltd.
© 1992, Vol. 221, pp. 175-178

Light density controls the mechanism of the photocycle of bacteriorhodopsin

Zsolt Tokaji[1] and Zsolt Dancsházy[1,2]

[1] *Institute of Biophysics, Biological Research Center, Hungarian Academy of Sciences, Szeged, Hungary* and [2] *National Heart, Blood and Lung Institute, NIH, Bethesda MD, USA*

SUMMARY

We have found that by simply varying the photon density (intensity) of the exciting, short (5-10 ns) laser pulses, or the intensity of the continuous, background illumination, the relative weights (yields) of the late intermediates (especially the "fastly" and "slowly" decaying M intermediates: M_f and M_s resp.) in the photocycle of bacteriorhodopsin (BR) change strongly. The rates of all of the intermediates remain unchanged. The same phenomenon has been found in double pulse experiments, when a part of the ground state BR has been depleted by a pre-exciting flash. A new, time dependent anisotropy change has been found in the 1-100 ms time range. Since in all of our experiments we used BR samples incorporated into gel, and we could not detect the anisotropy changes at very weak exciting intensities, they are unlikely to be explainable by rotation of the purple membranes or by simple by the twisting of the retinal in BR. These findings can not be explained by any of the presently used BR photocycle schemes. We suggest that BR may be heterogeneous, but mostly a strong, long distance cooperative interaction between the BR molecules is responsible for the alteration of the photocycle of BR by the actinic light.

INTRODUCTION

We have been puzzled in the past 4 years by the results of one of the most basic experiments which one can and should carry out in photochemistry and photobiology: the exciting light density dependence. In the case of the photocycle of BR, it showed a very surprising new phenomenon: depending on the light intensity, the relative yields of the late intermediates of the photocycle had been strongly altered.

MATERIALS AND METHODS

In Szeged, Hungary, we have developed a unique absorption kinetic measuring system, enabling us to take data with a very high signal/noise ratio up to 500 (Tokaji and Dancsházy, 1991). We were able to follow the rise and the decay of all of the intermediates of the photocycle from 1 µs to 10 sec, using logarithmic time based data collection. The absorbance changes obtained could be quantitatively analyzed from 1/1000 of the saturating light intensity, where the signal was about 0.1 mOD at 410 nm. The continuous side illumination was provided at 514 nm, from an Ar CW laser.

RESULTS AND DISCUSSIONS

We studied the phenomenon at 8 different monitoring wavelengths (275, 297, 335, 412, 545, 570, 610 and 650 nms), at different pH values (pH 6-10) and ionic strength (0-4 M NaCl), with and without continuous background illuminations (0-35 mW/cm2), in the temperature range of -20 - +40 °C. We also applied the most sensitive, double laser pulse excitation technique with a variable delay. We have checked the effect on a variety of carefully purified and biochemically characterized samples of BR prepared by us and also BR samples kindly provided by other laboratories (Drs. T.G. Ebrey, J.K. Lanyi and W. Stoeckenius). Under all experimental

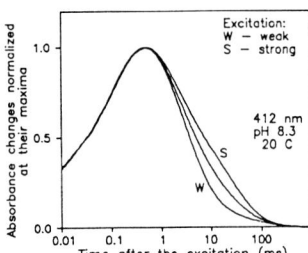 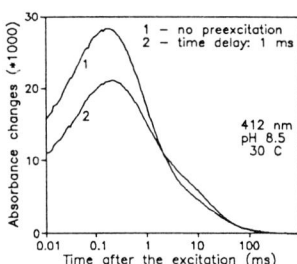 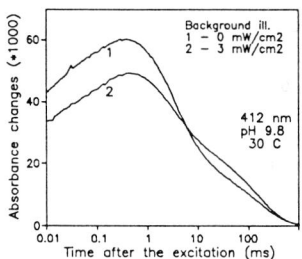

Fig.1. Influence of the different lights on the kinetics of the photocycle. Alteration of the M kinetics a.) by the intensity of the exciting flash, b.) by pre-excitation of the sample, c.) by a weak continuous (green, \ =514 nm)) side illumination. The common tendencies are: the independence of the rising parts (demonstrated only in a.), the increase in the relative weight of M_S. The amplitudes (*1000) in a.) before normalization were 57.7, 29.7 and 4.1 for strong, medium and weak excitations, respectively. Note that the absorbance of the samples were different. The relative weights of Ms are: a.) 47.5 %, 32.9 % and 16 %, b.) 39.4 % and 22.9 %, c.) 47.8 % and 31 %. The lifetimes of M_f and M_S are: a.) 3.3 and 38 ms, b.) 1.2 and 23 ms, c.) 4.8 and 170 ms. The 1.0 actinic light density in a.) corresponds to about 7.5 photons absorbed/BR. The pre-exciting flash (10 ns, λ =505 nm) induced a fraction cycling of ca. 25 % in b.)

conditions mentioned above, the light intensity dependence of the yield of the late intermediates of the BR photocycle has been found clearly very pronounced.

Our data were analyzed with various fitting procedures, including the most sophisticated ones (exponential, "maximum entropy", and SVD fits). Independently, all of these different analyzing techniques yielded similar conclusions: there is a significant change in the ratio of the reaction pathways, relative weights of intermediates of the BR photocycle upon changing the actinic light density.

The kinetics of the decay of the M intermediate which, consist of two main components, a fast (M_f) and a slow (M_S), depend on the actinic light density as it can be seen in Fig.1.a. and has been published recently (Tokaji and Dancsházy, 1991). Similar changes of the M kinetics can be induced, if the sample is pre-excited by another flash at a certain time before the beginning of the kinetic measurement. This influence of the pre-excitation disappears only by the completion of the photocycle (Tokaji and Dancsházy, 1991). By pre-excitation of the sample the yield of M_S can be enhanced, see Fig.1.b., although in the early time range the overall amplitude of the signal is naturally smaller since a part of BR has already been bleached by the pre-exciting flash. This "net overproduction" of M_S can also be induced, if the purple membranes are illuminated by a weak continuous green light, as it can be seen in Fig.1.c. (This result is in contrast with the one published by Bitting et al. (1989)). Note that neither an increase in the actinic light density, nor the presence of a strong pre-exciting flash influences the lifetimes of the components (see later). The continuous side illumination decreases the decay times only if it is strong enough (in our conditions >5mW/cm2). This a different effect and, in the first approximation it is unrelated to the phenomena discussed here. In conclusion any illumination, which increases the number of the cycling molecules increases the relative weight of M_S.

The relative weight of M_f versus the actinic light density is shown in Fig.2.a. at pH 7 both at parallel and perpendicular polarizations of the measuring beam. The inset shows some pieces of the original data traces taken at the parallel polarization of the measuring beam. Figures 2.b. and c. demonstrate the independence of the lifetimes of the kinetic components from the density of the exciting flash. One of the most important informations in Fig.2.a. besides the drop in the relative weight of M_f, is that the relative weights of M_f are different in parallel and perpendicular polarizations, except at low light intensities. This fact indicates corresponding anisotropy changes. In Fig.2.d. the time dependence of the anisotropy changes (characterized by dichroic ratios, D) are shown at different exciting light intensities. (Note that similar anisotropy changes -but measured only at one exciting flash intensity- have been observed recently by G.I. Groma; personal communication) The biggest changes can be observed at medium flash intensity (c.a. 0.1-0.5 photon absorbed/BR). The drop in the anisotropy occurs at about a few milliseconds, i.e. some time after the decay of the fast component of the M intermediate. Before this time, or at low light intensities during the whole photocycle the dichroic ratio is practically constant, in agreement with the previous finding that the transition dipole of the retinal is fixed and the same as in the ground-state (Czégé et al., 1982 and J. Czégé, S. Száraz and A. Dér, personal communication, 1992). The possible solutions for the anisotropy changes are: at strong

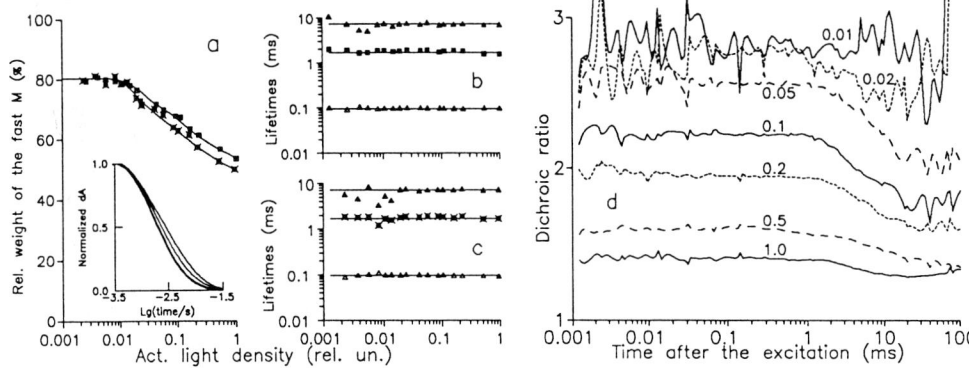

Fig.2. Evaluation of the actinic light density dependence of the M kinetics (412 nm) measured with polarized actinic and measuring beams at pH 7.0, in 1 M NaCl, at 30 °C. a.) The relative weight of M_f at parallel (■) and at perpendicular (✕) polarizations (determined from fits with fixed lifetimes taken from b.) and c.), solid lines). The inset shows some of the original data traces measured at parallel polarization (at weak, medium and strong excitation). The lifetimes of the slowly rising, and the two decaying components of the M kinetics are shown b. at parallel and c. at perpendicular polarizations. d.)Time dependence of the dichroic ratios measured at 412 nm, at pH 9.0, at different actinic light densities. The 1.0 actinic light density (20mW/cm^2, λ =505 nm) corresponds to about 2.5 photons absorbed/BR.

excitation either some tilting of the retinal occurs, or the anisotropy changes are induced by a nonlinear process (i.e. either the light induced heterogeneity of the ground state, or a cooperativity), whose existence was demonstrated recently (Tokaji and Dancsházy, 1991). In this latter case the orientation of the retinal can remain unchanged.

Quantitative, nonlinear models for the phenomenon are under development. These are based on that two parallel M intermediates exist, and that M_s is produced mostly if the excitation is strong, in other words (if the cooperativity occurs during the excitation) by the last photons of the flash. In this case, depending on that either the excitation of a previously non-excited molecule being in the neighborhood of another excited one (cooperativity), or the second excitation of a previously excited molecule (light induced heterogeneity) induces the formation of M_s, the dichroic ratio of M_s relatively decreases, or increases. The reason is that if the excitation is not strictly weak (i.e. D<3), the sample becomes photoselected for the later photons of the flash. This photoselection is induced by that a part of the molecules has already been cycling. Mainly those, whose transition dipole is close to parallel with the polarization of the exciting flash. At cooperativity these molecules can not be Ms, while at the light induced heterogeneity these can only be turned into M_s by the absorption of the later photons of the flash. Consequently, the time dependence of the anisotropy changes shown in Fig.2.d. is an evidence for cooperativity.

The possibility, that M_s could contain a reoriented retinal is controversial, as it should be seen already at a few hundred microseconds by a decreased dichroic ratio, when almost all the cycling molecules are in the M state at room temperature (Váró and Lányi, 1990). Note that unbranched photocycle models could describe the change in the dichroic ratio only by an additional M intermediate (with the tilted retinal in it) accumulating in the millisecond time scale. Such a slowly rising M has never been observed or suggested yet.

The results presented above indicate, that M_f and M_s have different features in the respect of their actinic light density dependence and their anisotropy. The third process which takes place in the millisecond time scale is the decay of the N intermediate. Actinic light density dependence experiments provided that its dependence is similar to that of M_f (data not shown). This relation can more convincingly be seen in double pulse experiments, the results of which are shown in Fig.3. At this wavelength (570 nm) the decay of the N intermediate can also be seen, and its amplitude similarly increases versus the delay time as that of M_f. The "net overproduction" of M_s induced by the pre-excitation can also be pronouncedly seen at this wavelength. These changes disappear only by the completion of the photocycle. Pre-excitation does not alter the lifetimes of the components, as it is shown in Fig.3.b.

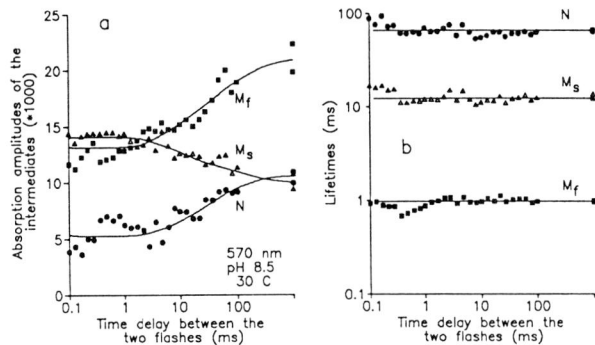

Fig.3. Influence of the pre-excitation on the absorbance changes at 570 nm. The pre-exciting flash (10 ns, $\lambda = 505$ nm) induced a fraction cycling of ca. 25 %. The same relations of M_f and M_s as in a.) has been determined recently in similar experiments with a 412 nm measuring wavelength (Tokaji and Dancsházy, 1991). Note that the amplitude of the third component (N) was negligible at that wavelength.

Figure 3. demonstrates that the second part of the photocycle can not be described by two components at alkaline pH, although under these circumstances the O intermediate does not accumulate, and hence can be omitted from the photocycle schemes. This indicates that those photocycle models which contain only two steps for the BR recovery from M, e.g. in the most frequently used model the M<->N equilibrium, and the N->BR reaction, are unable to account for the phenomena. Note that these models usually contain still another M-like (M_1 or X, Váró and Lányi, 1990, and Lozier et al., 1992) intermediate, but this is the way how these models describe the biphasicity of the L decay, and even a large number of back reactions is unable to account for the kinetics of the photocycle and the absorption spectra of the intermediates at the same time (Lozier et al, 1992).

CONCLUSIONS

We propose as possible explanations for our data that: a.) the BR itself could be heterogeneous (in its conformational states/substates) (Dancsházy et al., 1988), or as it was recently shown by other laboratories, even in its primary protein structure. b.) The BR molecules are in a strong, long distance cooperative interaction with each other in the purple membrane. (Tokaji and Dancsházy, 1991). c.) Local pH and/or electric charge changes related to the M intermediate formation/decay strongly influence the photochemistry of the neighboring BR molecules. Presently we think, all of these effects are involved in the phenomena discussed, and play crucial roles in the photochemistry of BR. Though perhaps cooperativity is the dominant factor.

Acknowledgement - This work was supported mostly by the Hungarian Academy of Sciences. Zs. D. is very grateful to the Fogarty International Center and NHLBI, NIH for the one year fellowship awarded to visit NIH. We are very grateful to Dr. J. Hofrichter for some pilot and to Drs. R.W. Hendler and R.I. Shrager for some more detailed analysis part of our data base by singular value decomposition (SVD) technique. We wish to thank for the help in electronics, instrument and software developments of Drs. J. Gárgyán and G.I. Groma.

REFERENCES

Bitting, H.C., Jang, D-J. and El-sayed, M.A. (1990): On the multiple cycles of bacteriorhodopsin at high pH. Photochem. Photobiol. 51:593-598.

Czégé, J., Dér, A., Zimányi, L. and Keszthelyi, L. (1982): Restriction of motion of protein side chains during the photocycle of bacteriorhodopsin. Proc. Natl. Acad. Sci. USA 79:7273-7277.

Dancsházy, Zs., Govindjee, R. and Ebrey, T.G. (1988): Independent photocycles of the spectrally distinct forms of bacteriorhodopsin. Proc. Natl. Acad. Sci. USA 85:6358-6361.

Lozier, R.H., Xie, A., Hofrichter, J. and Clore, G.M. (1992): Reversible steps in the bacteriorhodopsin photocyle. Proc. Natl. Acad. Sci. USA 89:3610-3614.

Tokaji, Zs. and Dancsházy, Zs. (1991): Light-induced, long-lived perturbation of the photcycle of bacteriorhodopsin. FEBS Lett. 281:170-172.

Váró, Gy. and Lányi, J. (1990): Pathways of the rise and decay of the M photointermediate(s) of bacteriorhodopsin. Biochemistry 29:2241-2250.

Dynamic holograms in bacteriorhodopsin: theory and application for phase-modulated optical beams detecting

Nikolai M. Kozhevnikov

Department of Experimental Physics, St. Petersburg State Technical University, 195251 St. Petersburg, Russia

Abstract. The fundamentals of bacteriorhodopsin (BR) - doped media application in adaptive coherent optical devices are outlined in the report. The unique feature of dynamic holographic couplers (DHC) used in these devices is briefly described in Section 2. The kinetics of holographic processes in BR under saturation effects is discussed in Section 3 which is followed by Section 4 specifically oriented toward DHC-BR used in measuring interferometers and fiber-optical sensors.

1. Introduction

Biological materials based on the photochromic protein bacteriorhodopsin (BR) are now extensively studied as recyclable media for optical information processing. Several important publications, eg Brauchle *et al.* (1991) have already demonstrated the BR applicability for Fraunhofer spatial filtering. Another promising application of these media is connected with phase-modulated (PM) optical beams detecting. It was shown by Hall *et al.* (1980) that dynamic holographic couplers (DHC) used for PM beams mixing could allow the information high-frequency signals demodulation and provide adaptive suppression of inherent low-frequency phase fluctuations of the optical beams propagating through inhomogeneous media or reflecting from unstable objects. The successful experimental investigations confirmed DHC-BR high sensitivity and effective phase-noise cancelling (Barmenkov *et al.*, 1987).

2. DHC as PM signals demodulators

If an inertial photosensitive medium is illuminated with the interference pattern (IP) of two high-frequency PM optical beams then the average steady-state absorption (AG) and phase (PG) volume gratings appear in the medium. The instant IP fast spatial displacements relatively to these gratings stimulate the output beams intensities oscillations (the information signal) which could be registrated with a dual photoreceiver. At the same time the slow IP moving due to the beams low-frequency phase-difference fluctuations is followed by the gratings and gives no contribution to the photoreceiver output.

The complete theoretical analysis of PM beams mixing in a local photosensitive medium has been performed by Kozhevnikov (1991). The most important conclusions followed from this analysis relate to the harmonic optical

PM. 1) The amplitudes of the fundamental and the second harmonics of the output beams intensities oscillations linearly depend on the PG and AG amplitudes. Based on these dependences a very sensitive technique of media optical non-linear photorefractive and photochromic response diagnostics has been suggested by Gehrtz et al. (1987) and experimentally verified by Abdulaev et al. (1988). 2) DHC provide a PM signal demodulation if $\Omega_s \gg \Omega_0 = \tau^{-1}$ where Ω_s and Ω_0 are the PM frequency and the medium cut-off freqency, τ is the gratings relaxation time. The strict condition of a strong harmonic phase noise suppression is $A\Omega_n\tau < 1$ where A and Ω_n are the amplitude and the frequency of the noise. 3) The sensitivity threshold of DHC-BR is estimated by the value 10^{-4} radn per unit optical power and unit frequency bandwidth.

3. The kinetics of DH recording and erasure in BR

BR under weak optical illumination is a typical medium with Kerr photorefractive and photochromic response which is conditioned by the difference between cis- and trans-molecules concentrations. That is why the volume gratings amplitudes in this case linearly depend on the recording beams intensities and the kinetics of these gratings formation is exponential. If the beams intensities increase then the saturation effects restrict the gratings amplitudes and complicate the kinetics of holographic processes. The peculiarities of these processes become clear if we solve the 1-D kinetic equation for cis-molecules concentration as it has been done by Barmenkov & Kozhevnikov (1991). The most important dependences followed from this solution are plotted in Fig.1 where the saturation parameter $R_0 = I_0/I_s$, $I_s = h\nu/\sigma\tau$ is the saturation intensity, σ is the absorption cross-section, I_0 is the average intensity in the IP, m is the IP modulation depth.

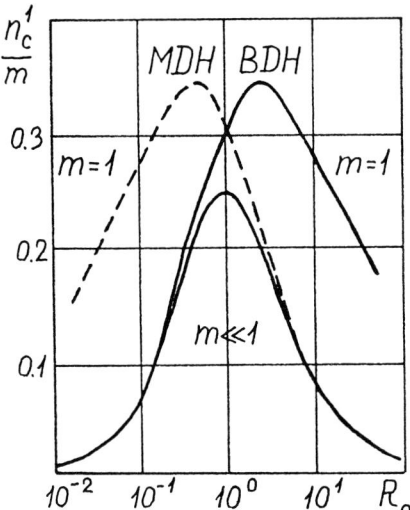
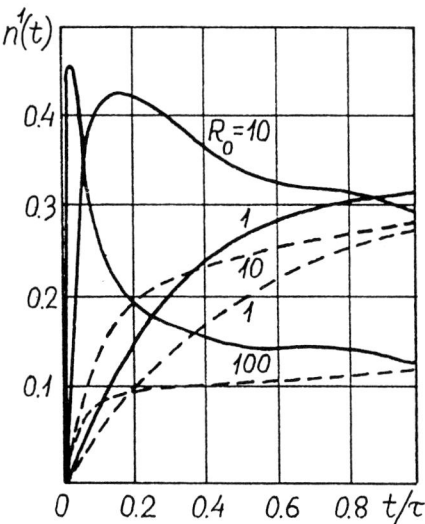

Fig.1. The steady-state (n_c^1) and the temporal $(n^1(t))$ dependences of the fundamental spatial harmonic amplitude of cis-BR concentration which is proportional to the DH amplitude and the output signal modulation depth.

The discovered peculiarities of the holographic processes in BR have been experimentally verified and the optimal conditions of DHC-BR operation have been defined.

4. Adaptive DHC-interferometers for PM optical beams processing

Several adaptive measuring interferometers and fiber-optical sensors which demonstrated the DHC-detectors advantages over conventional coherent optical receivers have been investigated. BR-doped fast-responsive photorefractive media (polymeric films and liquid suspensions) have been used in these devices for DH recording and PM optical beams coupling.

4.1. Multi-channel fiber-optical sensor (FOS) for harmonic PM signals measurement

This FOS has been designed for tiny phase-shifts measurements between several high-frequency sinusoidal PM signals. A possible application of this FOS is connected with the definition of an acoustic wave propagation direction. The FOS setup is shown in Fig.2,a. Multi-mode gradient silica optical fibers (core diameter $50 \mu m$) were used in the signal and the reference arms. The fibers were symmetrically Y-spliced. The signal fibers output ends were polished and fused with the $\lambda/4$ collimating gradans. To stimulate real measurements the signal fibers were partially looped onto piezoceramical cylinders providing optical PM.

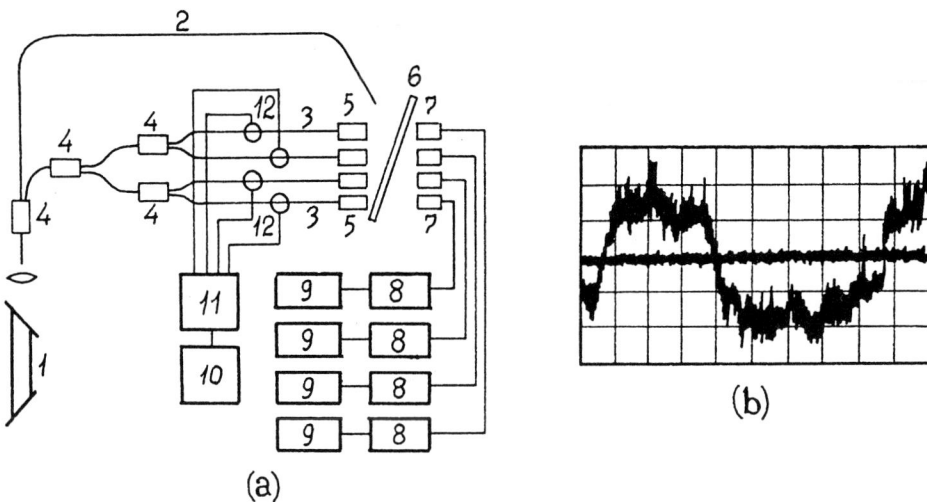

Fig.2. Adaptive FOS experimental setup (a) and a typical output signal (b). (a): 1 - He-Ne laser, 2 - reference arm, 3 - signal arms, 4 - Y-splitters, 5 - $\lambda/4$ gradans, 6 - polymeric BR-doped film, 7 - photoreceivers, 8 - narrow-band amplifiers, 9 - voltmeters, 10 - high-frequency oscillator, 11 - phase-shifter, 12 - piezoceramical cylinders.

The FOS sensitivity threshold was 10^{-3} radn. The relative accuracy of the phase-shift measurements was better than 0.1 per cent for all channels. Adaptive suppression of random phase fluctuations have been provided within the bandwidth from zero up to 10 Hz (films) - 100 Hz (suspensions). Fig.2,b illustrates the low-frequency phase-noise adaptive reduction with DHC-BR (curve 1) comparatively to the conventional technique based on a quasi-static hologram (curve 2).

4.2. Adaptive interferometer for submicron step-displacement measurements

The schematic drawing of the adaptive interferometer designed for a reflecting surface step-displacement measurements is presented in Fig.3,a. A single-mode low-noise He-Ne laser ($\lambda = 0.63 \mu m$, light power 1.5 mW) was used as an optical source. The reference beam was phase-modulated with a piezoelectric ceramic operated with a high-frequency voltage oscillator. The signal beam was reflected from the mirror which could be step-displaced with the voltage impulses operated ceramic. A photodetector output signal at the second harmonic was synchroniously amplified and monitored.

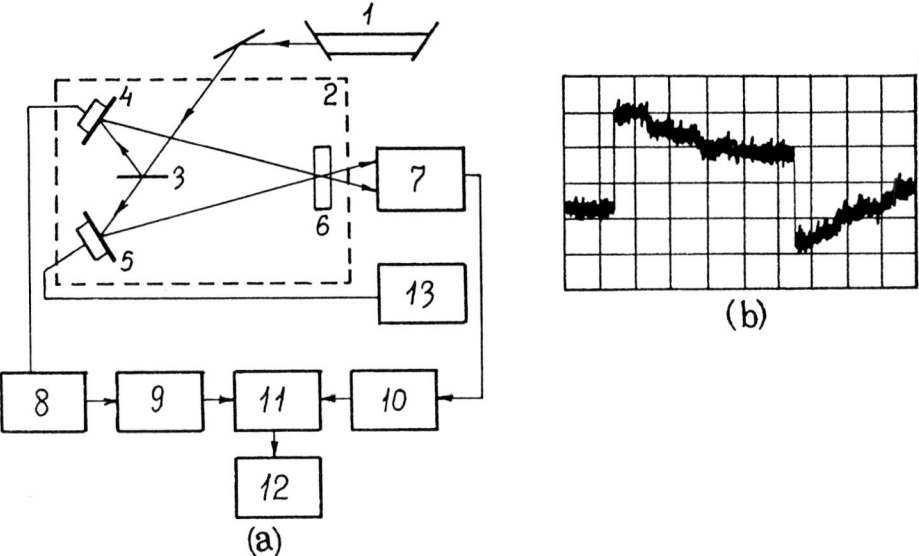

Fig.3. Adaptive interferometer for submicron step-displacement measurements: experimental setup (a) and a typical output signal (b). (a): 1 - He-Ne laser, 2 - vibro-protected table, 3 - beam-splitter, 4,5 - mirrors, 6 - BR-doped film, 7 - dual photoreceiver, 8 - sine-wave generator, 9 - frequency doubler, 10 - narrow-band amplifier, 11 - synchronious detector, 12 - oscillograph, 13 - rectangular pulse generator

A typical oscillogram obtained under calibrated 50 angstrom displacement of the reflecting surface is shown in Fig.3,b. This oscillogram demonstrates the sensitivity threshold of 5 angstrom which is closed to the calculated value under shot-noise limitation.

References

Abdulaev,N.G.,Barmenkov,Yu.O.,Zaitsev,S.Yu.,Zosimov,V.V.,Zubov,V.P.,Kozhevnikov,N.M.,Lipovskaya,M.Yu.,& Lyamshev,L.M.(1988): Photorefractive sensitivity of polymer films containing bacteriorhodopsin. Sov.Phys.Tech.J. 33, 508-510.

Barmenkov,Yu.O.,Zosimov,V.V.,Kozhevnikov,N.M.,Kotov,O.I.,Lyamshev,L.M.&Nikolaev,V.M.(1987): Detection of phase-modulation signal from a fiber-optical interferometer by means of a dynamic hologram in bacteriorhodopsin. Sov.Phys.Acoust. 33, 334-335.

Barmenkov,Yu.O.,& Kozhevnikov,N.M.(1991): Photoresponse saturation at the holographgic recording in bacteriorhodopsin. Sov.Phys.Tech.J. 61, 116-120 (in Russian)

Brauchle,Ch.,Hampp,N.,& Oesterhelt,D.(1991): Optical applications of bacteriorhodopsin and its mutated variants. Advanced Materials. 3, 420-428.

Gehrtz,M.,Pinsl,J.,&Brauchle,Ch.(1987): Sensitive detection of phase and absorptive gratings: Phase-modulated homodyne detected holography. J.Appl.Phys.B. 43, 61-77.

Hall,T.J.,Fiddy,M.A.,& Ner,M.S.(1980): Detector for an optical fiber acoustic sensor using dynamic holographic interferometry. Opt.Lett. 5,485-487.

Kozhevnikov,N.M.(1991): Dynamic holographic microphasometry. Proc.SPIE. 1507.

Engineering of a light-sensitive molecular device

Samuel Scherling and Hans Sigrist

Institute of Biochemistry, University of Berne, Freiestrasse 3, CH-3012 Bern, Switzerland

Arachaebacterial proteins or fragments thereof are structurally simple and exceptionally stable. They provide the structural characteristics required for the engineering of photoresponsive systems. The V-2 fragment of bacteriorhodopsin (bR) - produced by protease V8 cleavage of bR - has been isolated and purified in organic solvents without the use of detergents. The secondary structure of the peptide is preserved in solvents which mimic membrane phases. Following permethylation of intact bR in purple membranes and protease V8 digestion, the V-2 fragment has been acetylated at its N-terminal end. Selective thiocarbamoylation of the remaining free ε-amino function (Lys 216) is attained with all-*trans*-retinylisothiocyanate and 4-N,N-dimethyl-azobenzene-4'-isothiocyanate (DABITC). Azobenzenes are photochromic compounds which reversibly change their color when exposed to light of distinct wavelengths. Fluorescence energy transfer measurements indicate that the N-dimethyl end of covalently peptide-linked *trans*-DABITC - but not the shortened *cis*-form - is in close proximity to the tryptophan residues of helix F (Trp 182, Trp 189).

INTRODUCTION

Biomolecules are like electronic elements and are constructed to process information. Considerable interest has been aroused for the potential use of light transducing proteins to perform optical switching functions. Biomolecules may serve as active components in molecular electronic devices either in their non-modified native form or structurally modulated by chemical derivatisation or protein engineering. One example of a biologically optimized system is bacteriorhodopsin (Henderson et al., 1990; Birge, 1990) the light transducing protein in the purple membrane of *Halobacterium halobium*. A bilayer lipid membrane containing oriented bR produces an electrical signal when illuminated. This photoelectric signal is attributed to internal charge movements associated with light-induced conformational changes of bR and charge movements such as binding and release of protons (Hong and Okajama, 1986; Trissl, 1985). These processes are relevant to the design of molecular optoelectronic devices. Moreover, pioneering studies have indicated that membrane proteins including bR can be incorporated into polymerizable lipid systems. Reconstitution of bR (Pabst et al., 1983; Ahl et al., 1990; Borle et al., 1992) and the rod outer segment rhodopsin (Tyminski et al., 1988) into partially polymerized membranes has been achieved.

In native bR the light-induced all-*trans* to 13-*cis* isomerization of Schiff base-bound retinal occurs in the picosecond time range. Essential molecular movements take place in the two C-terminal helices of bR. Helix F provides the retinal holding device (Trp 182, Tyr 185, Trp 189); the retinal binding site is located in helix G (Lys 216). It is therefore conceivable that the minimal structure required for a photosensitive device may consist of a properly modified C-terminal segment of bR which includes helix F and G.

Optimally, the device should be structurally simple, thermostable, chemically stable and exert fast and reversible photoreactions. Arachaebacterial proteins or fragments thereof are structurally simple and provide exceptionally high inherent stability (Huang et al., 1981). Moreover, bR and its protease V8 fragments demonstrate extraordinary refolding capacity and retain native secondary structures in apolar organic solvents. The V-2 fragment of bR has been characterized as the truncated C-terminal segment (Wüthrich and Sigrist, 1990). It is a most appropriate peptide building block for the construction of a light-sensitive molecular device (Fig. 1).

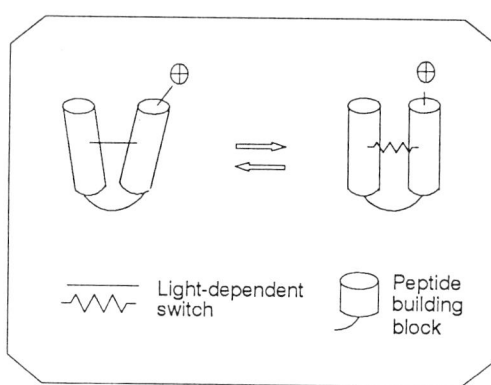

Fig. 1. Excised from the stable bacteriorhodopsin (bR) a peptide is chemically modified to an extent that light-dependent reversible charge dislocation can be realized in a semisynthetic construct, azobenzenes and retinal analogues serving as light-dependent switches. Retinylisothiocyanate is a promising candidate for initiating molecular motion in biodevices by all-*trans* / 13-*cis* isomerization. Analogously azobenzenes reversibly isomerize light-dependently from the extended *trans*-form to the condensed *cis*-form. This property makes bifunctional azobenzene derivatives (e.g. azido-azobenzene-isothiocyanate) attractive reversible switching elements for molecular devices.

RESULTS AND DISCUSSION

In the course of the structural and functional investigation of bR proteolytic peptides have been prepared from this membrane protein and photochromic crosslinking reagents have been synthesized. They are now utilized to construct a light sensitive molecular switch.

A peptide building block excised from bR. Evidence has been presented that transmembrane segments of bR can be isolated and purified in organic solvents without the use of detergents. Of specific interest is the hydrophobic peptide V-2, produced by protease V8 digestion of bR (Sigrist et al., 1988), which is soluble in apolar solvents. The secondary structure of this peptide is preserved in solvents which mimic membrane phases (Wüthrich and Sigrist, 1990). The inherently stable peptide provides the chromophore binding site (Lys 216) and a retinal binding pocket (helix F). Chemical engineering of this peptide has been carried out. Following permethylation of native bR in purple membranes and protease V8 digestion, the V-2 fragment is acetylated at its N-terminal end and isolated. Selective thiocarbamoylation of the remaining free ε-amino function (Lys 216) is achieved with either retinylisothiocyanate or 4-N,N-dimethyl amino-azobenzene-isothiocyanate.

Chromophores for covalent modification. With reference to future applications of structurally stabilized retinal proteins, the goal has been set to chemically stabilize the retinal chromophore in bR. For these purposes all-*trans*-retinylisothiocyanate has been synthesized. The newly described retinal analogue forms a stable thiourea linkage with primary amines. The thiourea bond is irreversible under conditions which guarantee the structural integrity of the polypeptide chain. The reagent may thus prove useful in initiating molecular motion in stabilized bR by all-*trans* / 13-*cis* isomerization.

Azobenzenes are photochromic compounds which reversibly change their color when exposed to light of distinct wavelengths. Spectral changes are due to *cis/trans* isomerization of the azo-bridge substituents. Irradiation of azobenzene in toluene at 365 nm shifts the equilibrium to 91 % of the *cis* isomer. When irradiated with light at 436 nm, however, 86 % of the *trans* form is present (Fischer et al., 1955). The wavelength-dependent isomerization makes azobenzenes photoresponsive modulators (Pieroni and Fissi, 1992). In this study they are used as switching elements. 4-N,N-dimethyl amino-azobenzene-isothiocyanate modification of the engineered peptide building block yields selective Lys 216 thiocarbamoylation. Helicity of the labeled peptide is fully retained. Fluorescence energy transfer measurements indicate that the N-dimethyl end of coupled *trans*-DABITC - but not the shortened *cis*-form - is in close proximity to the tryptophan residues 182 and 189 on helix F (Trp 182, Trp 189) (Fig. 2).

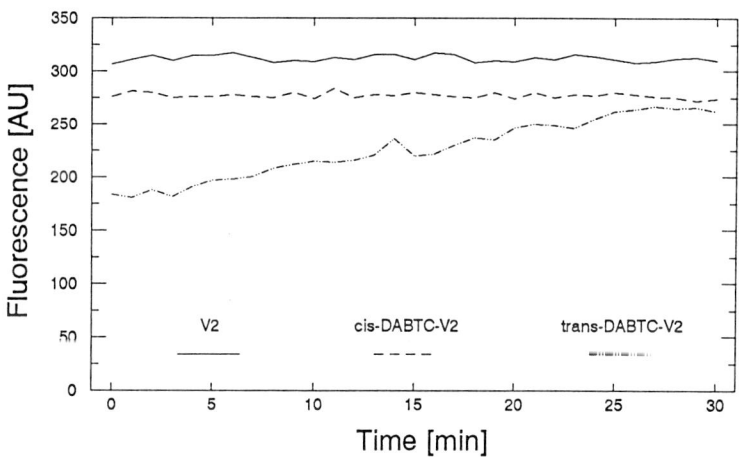

(Fig. 2). Interhelical alignment of either azochromophore isomer has been analyzed by tryptophan fluorescence (λ_{ex} 285 nm, λ_{em} 335 nm). Fluorescence energy transfer measurements were carried out with non-modified V 2, azobenzene-modified V 2 either in the *cis*- (*cis*-DABTC-V2) or *trans*-form (*trans*-DABTC-V2). Initial quenching was observed for *trans*-DABTC-V2 but not for the corresponding *cis* construct. Time dependent dequenching of *trans*-DABTC-V2 suggests that the emitted light (335 nm) induces *trans* -> *cis* isomerization.

Data obtained provide the experimental basis for introducing an azobenzene-derived crosslinker which is able to form a covalent link between the two helices. For site-directed linking of the two helices 4-azido-azobenzene-4'-isothiocyanate (AZBITC) has been synthesized. Both the monofunctional (DABITC) and the heterobifunctional reagent (AZBITC) have similar properties with respect to chemical stability and reactivity. Photoactivation of coupled AZBITC will yield preferred nitrene insertion into the Trp/Tyr cluster, completing therewith the covalent link between helix F and G. Ultimately, selective activation of the azochromophores in aligned switch peptides may induce *cis/trans* isomerization and the concerted movement of peptide-borne charges.

ACKNOWLEDGEMENTS

The authors thank E. Kislig for technical assistance. This work was supported by the Swiss National Science foundation (Grant 31-30847-91).

REFERENCES

Ahl, P.L., Price, R., Smuda, J., Graber, B.P. and Sing, A. (1990): Insertion of bacteriorhodopsin into polymerized diacetylenic phosphatidylcholine bilayers. *Biochim. Biophys. Acta* 1028, 141-153.

Birge, R.R. (1980): Nature of the primary photochemical events in rhodopsin and bacteriorhodopsin. *Biochim. Biophys. Acta* 1016, 293-327.

Borle, F., Michel, H.P. and Sigrist, H. (1992): Phospholipid polymerization and reconstitution of bacteriorhodopsin in photopolymerizable lipid vesicles. *J. Membr. Sci.* in press.

Fischer, E., Frankel, M. and Wolovsky, R. (1955): Wavelength dependence of photoisomerization equilibra in azo compounds. *J. Chem. Phys.* 23, 1367-1369.

Henderson, R., Baldwin, J.M., Ceska, T.A., Zenslin, F., Beckman, E. and Downing, K.H. (1990): Model for the structure of bacteriorhodopsin based on high-resolution electron cryo-microscopy. *J. Mol. Biol.* 213, 899-929.

Hong, F.T. and Okajama (1986): Electrical double layers in pigment-containing biomembranes. In *Electrical double layers in Biology*, ed. M. Blank, pp. 129-147, Plenum, New York.

Huang, K.S., Bayley, H., Liao, M.J., London, E. and Khorana, H.G. (1981): Refolding of an integral membrane protein. Denaturation, renaturation and reconstitution of intact bacteriorhodopsin and two proteolytic fragments. *J. Biol. Chem.* 256, 3802-3809.

Pabst, R., Ringsdorf, H., Koch, H. and Dose, K. (1983): Light-driven proton transport of bacteriorhodopsin incorporated into long-term stable liposomes of a polymerizable sulfolipid. *FEBS Lett.* 154, 5-9.

Pieroni, O. and Fissi, A. (1992): Synthetic photochromic polypeptides: possible models ofr photoregulation in biology. *J. Photochem. Photobiol. B: Biol.* 12, 125-140.

Sigrist, H., Wenger, R.H., Kislig, E. and Wüthrich, M., (1988): Refolding of bacteriorhodopsin. Protease V8 fragmentation and chromophore reconstitution from proteolytic V8 fragments. *Eur. J. Biochem.* 177, 125-133.

Trissl, H.W. (1985): Primary electrogenic processes in bacteriorhodopsin probed by photoelectric measurements with capacitative metal electrodes. *Biochim. Biophys. Acta* 806, 124-135.

Tyminsky, P.N., Latimer, L.H. and O'Brien, D.F. (1988): Reconstitution of rhodopsin and the cGMP cascade in polymerized bilayer membranes. *Biochemistry* 27, 2696-2705.

Wüthrich, M. and Sigrist, H. (1990): Peptide building blocks from bacteriorhodopsin: Isolation and physicochemical characterization of two individual transmembrane segments. *J. Prot. Chem.* 9, 201-207.

V. Bacteriorhodopsin, halorhodopsin : ion translocation, charge movement

V. *Bactériorhodopsine, halorhodopsine : translocation d'ions et mouvements de charges*

The photovoltage of bacteriorhodopsin in X- and Z-type monolayers and Z-type multilayer Langmuir-Blodgett film

Marjo Ikonen, Alexey Sharonov*, Nikolai Tkachenko* and Helge Lemmetyinen

University of Helsinki, Department of Physical Chemistry, Meritullinkatu 1C, 00170 Helsinki, Finland.
*Permanent address : General Physics Institute, Vavilov Street 38, 117942 Moscow, Russia

SUMMARY The photoresponse of bacteriorhodopsin (bR) has been studied in Langmuir-Blodgett films (LB films) both at room humidity and in water saturated conditions. bR was deposited as x- or z-type monolayers and as z-type multilayers between stearic acid LB films. The photoelectric signals of x- and z-type monolayers had opposite polarities and both had two exponential kinetics with time constants of about 10 and 70 microseconds. The amplitude of the slow part of the signal, in the proton pumping direction, was 1.2 mV. The ratio between the amplitudes of the fast negative component, in an opposite direction to that in which protons were pumped, and the slow positive part of the photoresponse signal, was 1:3. The photoresponse signal saturated with a photon density of 0.5 Å^{-2}. The activation energy of bR photocycle was 75±5 kJ/mol in a temperature range between 17 and 35 °C.

INTRODUCTION

The light-driven proton pump of bacteriorhodopsin (bR), the protein of purple membrane (PM) from Halobacterium halobium, has in the past few years received a lot of attention in connection with its possible use in light active bio-organic electrical devices (Birge 1990; Hampp & Bräule, 1990; Miyasaka et al., 1992; Takei et al., 1992). It is also well know that bR is relatively stable molecule in different types of environments. One of the advantages of air-dried oriented PM and bR LB films is that the photocycle of bR works at room humidity and temperature.

In this work it has been shown that it is possible to orientate the dipoles of the bR molecules in LB films. The fact that electrical measurements can be made for bR monolayer gives the possibility of observing the kinetics of protons, which is usually difficult by traditional optical methods.

MATERIALS AND METHODS

The electrodes used were a transparent vacuum deposited ITO on a quartz substrate and an InGa metal alloy drop. To achieve the studied bR-metal sandwich cell system, first 10 and 20 odd or even y-type stearic acid (STA) layers were deposited on ITO. These STA layers were covered by bR:soya-PC films (Ikonen et al., 1992). The bR layers were covered by a DPPC monolayer (Ikonen et. al.,1992) and with several layers of STA to protect the bR from the liquid metal electrode and to be ensure that any holes in bR mono- or multilayers would not cause a shot circuit in the measuring system.

Capacitances and resistances of the samples were estimated by passive electrical measurements in two ways: by applying a low frequency step voltage and measuring the relaxation time, and by applying a sinusoidal signal and measuring the current through the system (Ikonen et al., 1992). Capacitances and resistances of the samples can be estimated also by measuring relaxation times of the photoresponse signals with different loading resistances (Ikonen et al., 1992).

The time-resolved measurements of the photovoltage of a bR sandwich cell were made in a time range from $1*10^{-6}$ s to $10*10^{-3}$ s. The action spectrum was measured at a wavelength range from 440 to 630 nm of mono- and multilayeric bR LB films to be sure that the observed photoresponse signals originated from the bR molecules. The shape of the action spectrum was similar to the absorption spectrum of the pigment measured from bR LB film (Fig. 1).

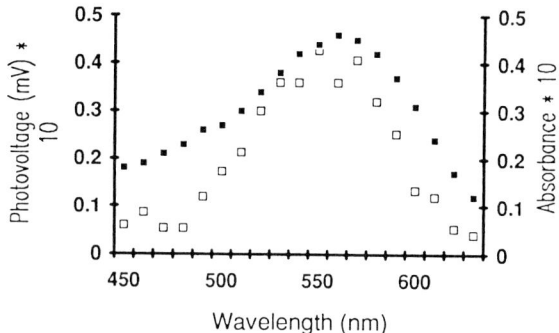

Fig. 1. An action ( ) and a ground state absorption ( ) spectra of bR in the studied LB film.

RESULTS AND DISCUSSION

The photoelectric signals of mono- and multilayeric bR LB films were recorded in room conditions. The amplitude due to five layers of bR was five times as high as that due to a monolayer.

The photovoltage were recorded for both x- and z-type deposited monolayeric bR LB films, Fig.2a. The signals have opposite polarities due to the deposition direction and thus correspond to opposite orientations of the bR molecules in the films. By adding these signals from each other the background signal is obtained, Fig.2b. The actual photovoltage signals were obtained when the background signal was substracted from the initial curves, Fig. 2c.

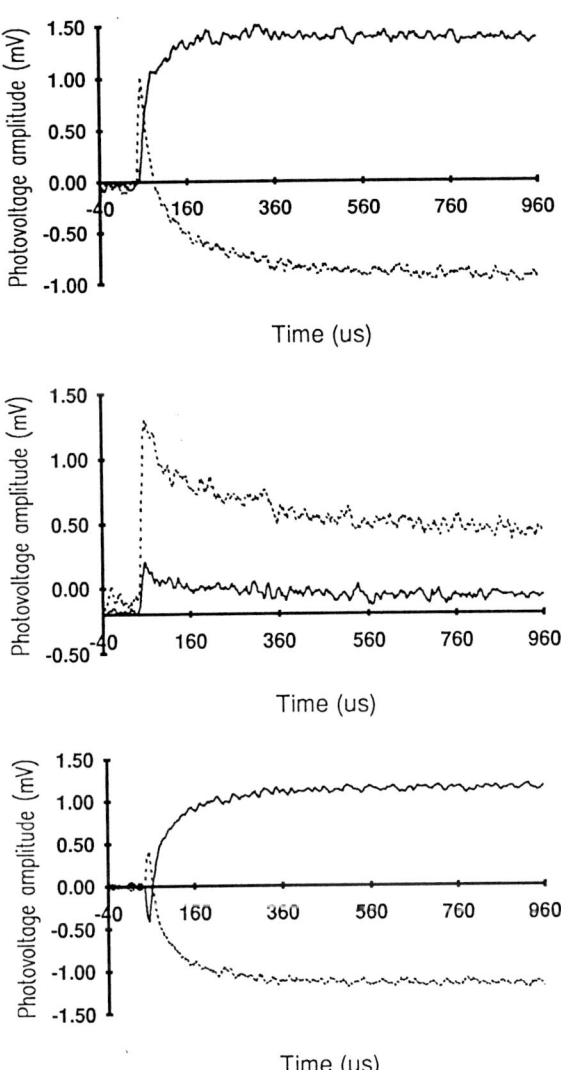

Fig. 2. (a) Measured photovoltage signals of x-(dashed line) and z-type (solid line) bR monolayers, (b) observed background signals obtained by adding the signal x to the signal z (dashed line) and by measuring the signal from a STA LB film (solid line), and (c) the photovoltage signals of bR monolayers obtained by substracting the background signal from the measured signals.

The saturation of the photoresponse signal of bR in LB films can be consider to take place at the excitation energy of about 1.7 mJ/cm^2 taken from the experimental data (Ikonen et al., 1992). This equals to a photon density of about one photo per 2.1 Å^2, which is the absorption cross section.

At a statistical reliable level there are at least two processes in the time range from 1 to 100 10^{-6} s with the time constants of about

191

10 and 70 microseconds corresponding to the L-M transition in the bR photocycle.

The amplitude of 0.4 mV observed for the fast negative component of the photovoltage signal was 20 times less than that calculated theoretically for the studied system (Ikonen et al., 1992). The ratio of the amplitudes of the fast negative component and the slow positive part of the photovoltage signal of bR in LB film was about 0.33 indicating that the induced motion of the proton is about 4.5 Å (Keszthelyi et al., 1982).

The kinetics of the photovoltage signal was measured at a temperature range from 17 to 35 °C and fitted with two exponential approximation. The activation energies and frequency factors were calculated according the Arrhenius equation. A value of 75 ± 5 kJ/mol was obtained for the activation energies of the photocycle both in air-dried mono- and multilayeric bR LB films. The frequency factor has a value of about $7*10^{-20}$ s^{-1}. The obtained values are both somewhat higher than those calculated for bR by others (Varo & Keszthelyi, 1985; Varo, 1981).

REFERENCES

Birge R. (1990): Photolysis and molecular electronic applications of the rhodopsin. *Annu. Rev. Phys. Chem.* 41, 683-733.

Furuno T., Takimoto K., Kouyama T., Ikegami A. & Sasabe H. (1988): Photovoltaic properties of purple membrane Langmuir- Blodgett films. *Thin Solid Films* 160, 145-151.

Hampp N. & Bräule C. (1990): Bacteriorhodopsin and its functional variants: Potential applications in modern optics. *In Photochromism, Molecules and Systems*, Ch. 29. Ed H. Durr and H. Bouas-Laurent, Elsevier.

Ikonen M., Peltonen J., Vuorimaa E. & Lemmetyinen H. (1992): The photocycle of bacteriorhodopsin in Langmuir-Blodgett films. *(Will be published in Thin Solid Films)* 207.

Ikonen M., Sharonov A., Tkachenko N. & Lemmetyinen H. (1992): The photoresponse of bacteriorhodopsin in mono- and multilayer Langmuir-Blodgett films. *(submitted for publication)*

Keszthelyi L., Ormos P. & Varo G. (1982): Fast components of the electric response signal of bacteriorhodopsin protein. *Acta phys. Acad. Sci. Hung.* 53,143-157.

Miyasaka T., Koyama K. & Itoh I. (1992): Qantum conversion and image detection by bacteriorhodopsin-based artificial photoreceptor. *Science* 255, 342-344.

Varo G. (1981): Dried oriented purple membrane samples. *Acta biol. Acad. Sci. Hung.* 32,301-310.

Varo G. & Keszthelyi L. (1985): Photoelectric signals from dried oriented purple membranes of halobacterium halobium. *Biophys. J.* 43,47-51.

Takei H., Lewis A., Chen Z. & Nebezhal I. (1991): Implementing receptive fields with excitatory and inhibitory photoelectrical responses of bacteriorhodopsin films. *App. Opt.* 30, 500-509.

Structures and Functions of Retinal Proteins. Ed. J.L. Rigaud. Colloque INSERM / John Libbey Eurotext Ltd.
© 1992, Vol. 221, pp. 193-196

Charge motions in the D85N and D212N mutants of bacteriorhodopsin

Csilla Gergely and György Váró

Institute of Biophysics, Biological Research Center of the Hungarian Academy of Science, Szeged, H-6701 Hungary

INTRODUCTION

The charge motions in acrylamide gel samples containing oriented bacteriorhodopsin have been intensively studied (Dér et al.,1985; Keszthelyi and Ormos, 1989). The time resolution of the optical and electric signal measurements spans twelve orders of magnitude, from picoseconds (Groma et al.,1988; Simmeth and Rayfield, 1990) to seconds. The first explanations of these signals were based on the unidirectional kinetic model of the bacteriorhodopsin (Keszthelyi and Ormos, 1989), and the constant value of the electric permittivity through the whole membrane. These were oversimplifications of the problem.

The appearance of the new bacteriorhodopsin model with reversible kinetics between the intermediates (Váró and Lanyi, 1991; Mathies et al.,1991), a model more and more accepted, raises the necessity of the re-evaluation of the charge motion measurements. As the electric measurements give a single set of data in function of time, reflecting all the charge motions occurring during the pumping process, the interpretation of such a complex signal in whole at once is difficult or impossible for the native bacteriorhodopsin.

We have searched for such conditions in bacteriorhodopsin function, when the photocycle is truncated and only a few intermediates are present. There are two possibilities: the study of dried bacteriorhodopsin sample at different relative humidities (Váró and Keszthelyi, 1983) or the measurements on different mutants. Here we report the first electric measurements and their interpretation on mutants D85N and D212N made in *Halobacterium halobium* strain L-33. Both aspartic acids in question take part in the quadrupole formed with the Schiff-base and the arginine 82 (Otto et al.,1990; Needleman et al.,1991). Beside the asp 85 has an important role as the proton acceptor from the Schiff-base (Otto et al.,1990; Lanyi et al.,1992). It is known from the literature (Needleman et al.,1991) and from a personal communication of J.K. Lanyi that the photocycle of these mutants, over neutral pH, stops at the L intermediate.

MATERIALS AND METHODS

The gel preparation was made on the basis of the method described by others (Dér et al.,1985). All the optical and electric measurements were made with a similar setup as described elsewhere (Váró and Keszthelyi, 1983). The data acquisition was by a computer controlled transient digitizer (Thurlby DSA 524). The measuring and data analysis programs were written in Quick Basic. The data acquisition consisted of optical measurements at three wavelength and an electric measurement, at three or four increasing time-bases, covering the whole photocycle. These data were converted to a logarithmic time-base with an algorithm which transforms the successive measurements with linearly equidistant points into logarithmically equidistant, 20 points per order of magnitude, by averaging the intermediate data-points. So beside transforming the data to a logarithmic time-scale, a smoothening of the signal was achieved as well. A data set is shown on Fig. 1.

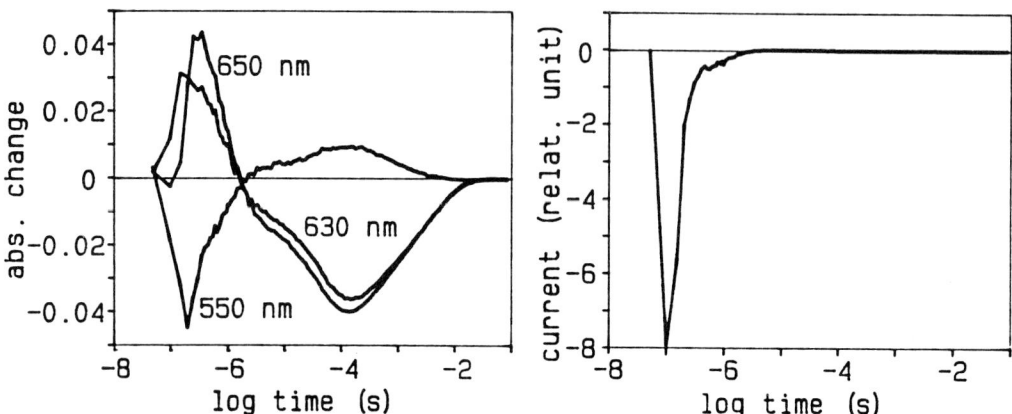

Fig. 1 The optical and electric signals measured on D85N mutant in 100 mM NaCl, 50 mM phosphate buffer at pH 7 and 20 °C.

The optical data were fitted to the equilibrium models as described earlier (Váró and Lanyi, 1990) and by using these kinetic constants to fit the electric signal and its integrated form, the dipole strength of every intermediate was calculated. Here we used the dipole strength as defined by others (Trissl, 1990) and a working hypothesis: the steps size of the photocycle in the mutants are similar than in native bacteriorhodopsin.

RESULTS AND DISCUSSIONS

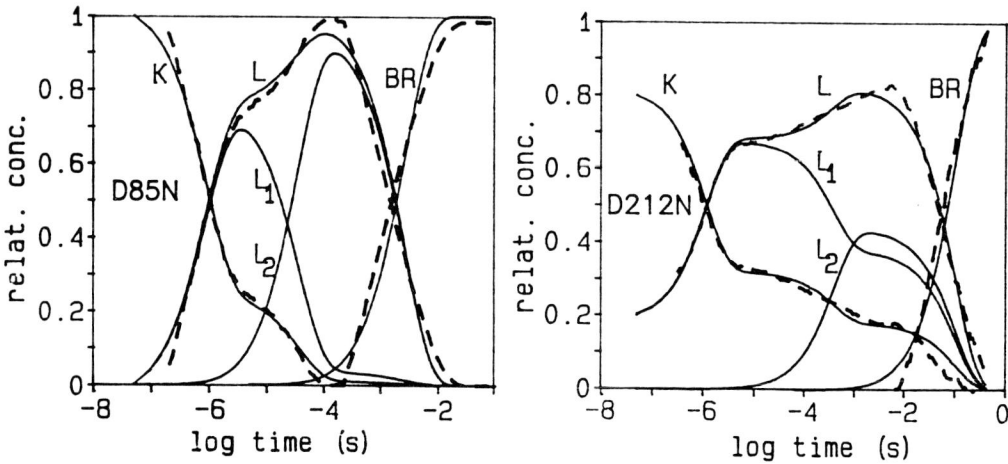

Fig. 2 Concentration change of the intermediates calculated from the measured data (dotted line) and the fit to the model (continuous line) of the D85N and D212N mutants.

For both D85N and D212N mutants the fit of the measured data was made to a truncated photocycle, which contained only **K** and **L** intermediates as described elsewhere (Needleman et al.,1991):

$$K \longleftrightarrow L_1 \longleftrightarrow L_2 \longrightarrow BR$$

This model is valid at higher than neutral pH and the obtained fits are shown on Fig. 2. The calculated reaction rates of the model were used to fit the electric signal and its integrated form (Fig. 3) to calculate the dipole strengths of the intermediates (see Table).

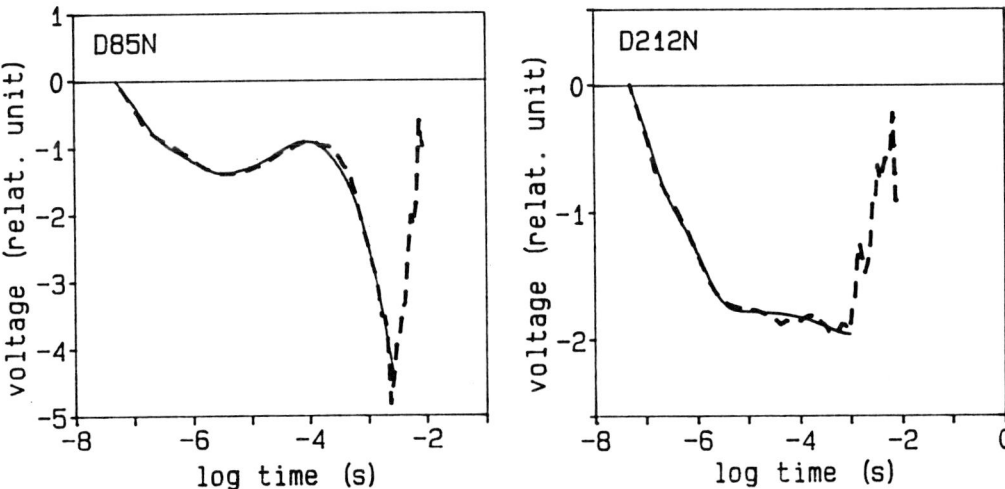

Fig. 3 The integral of the electric signal (dotted line) and its fit to the model (continuous line) of the D85N and D212N mutants.

Table. The calculated dipole strength of the photocycle intermediates of the D85N and D212N mutants.

Intermediates	D85N	D212N
K	-0.029	-0.028
L_1	-0.041	-0.056
L_2	-0.028	-0.054
X	-0.11	

The dipole strengths of the **K** intermediates are the same in both cases, showing that the isomerization of the retinal occur in the same way. The dipoles corresponding to the L_1 have different strength, showing that the relaxation of the isomerization moves the charges of the quadrupole near the Schiff-base. The difference in the dipole strength of the L_2 intermediate, which appears after a conformation change in the mutant, bacteriorhodopsin reveals a different role of the two substituted charges. While the asp 212 is supposedly on the moving part of the protein during the conformation change, the asp 85 does not move.

The very slow charge motions of the photocycle are not resolved, but from the fact that in the mutant D85N the last step does not go back to the zero level and this mutant doesn't pump proton, it is clear that there are charge motions which are not observed in the optical measurements. Maybe these motions are remote from the retinal, close to the surface of the membrane.

It can be seen from this study, the use of the mutants gives a possibility not only to clarify the dipole strength of every intermediate but also to clarify the amino acid which takes part in the charge motion.

The authors are grateful to professor J.K. Lanyi for supplying the mutants and all the necessary informations of optical measurements. This work was supported by the grant OTKA T5073.

REFERENCES

Dér, A., P. Hargittai and J. Simon. 1985. Time-resolved photoelectric and absorption signals from oriented purple membranes immobilized in gel. *J. Biochem. Biophys. Methods.* 10(5-6): 295-300.

Groma, G.I., F. Ráski, G. Szabo and G. Váró. 1988. Picosecond and nanosecond components in bacteriorhodopsin light-induced electric response signal. *Biophys. J.* 54: 77-80.

Keszthelyi, L. and P. Ormos. 1989. Protein electric response signals from dielectrically polarized systems. *J. Membr. Biol.* 109: 193-200.

Lanyi, J.K., J. Tittor, G. Váró, G. Krippahl and D. Oesterhelt. 1992. Influence of the size and protonation state of acidic residue 85 on the absorpion spectrum and photoreaction of the bacteriorhodopsin chromophore. *Biochim. Biophys. Acta* 1099: 102-110.

Mathies, R.A., S.W. Lin, J.B. Ames and W.T. Pollard. 1991. From femtoseconds to biology: Mechanism of bacteriorhodopsin's light-driven proton pump. *Annu. Rev. Biophys. Biophys. Chem.* 20: 491-518.

Needleman, R., M. Chang, B. Ni, G. Váró, J. Fornes, S.H. White and J.K. Lanyi. 1991. Properties of asp212-asn bacteriorhodopsin suggest that asp212 and asp85 both participate in a counterion and proton acceptor complex near the Schiff base. *J. Biol. Chem.*

Otto, H., T. Marti, M. Holz, T. Mogi, L.J. Stern, F. Engel, H.G. Khorana and M.P. Heyn. 1990. Substitution of amino acids Asp-85, Asp-212, and Arg-82 in bacteriorhodopsin affects the proton release phase of the pump and the pK of the Schiff base. *Proc. Natl. Acad. Sci. USA* 87: 1018-1022.

Simmeth, R. and G.W. Rayfield. 1990. Evidence that the photoelectric response of bacteriorhodopsin occurs in less than 5 picoseconds. *Biophys. J.* 57: 1099-1101.

Trissl, H.W. 1990. Photoelectric measurements of purple membranes. *Photochem. Photobiol.* 51: 793-818.

Váró, G. and L. Keszthelyi. 1983. Photoelectric signals from dried oriented purple membranes of Halobacterium halobium. *Biophys. J.* 43: 47-51.

Váró, G. and J.K. Lanyi. 1990. Protonation and deprotonation of the M, N, and O intermediates during the bacteriorhodopsin photocycle. *Biochemistry* 29: 6858-6865.

Váró, G. and J.K. Lanyi. 1991. Thermodynamics and energy coupling in the bacteriorhodopsin photocycle. *Biochemistry* 30: 5016-5022.

Electric signals and the photocycle of bacteriorhodopsin

András Dér, Rudolf Tóth-Boconádi and Sándor Száraz

Institute of Biophysics, Biological Research Centre of the Hungarian Academy of Sciences H-6701 Szeged, P.O.B. 521, Hungary

INTRODUCTION

Elucidation of the bacteriorhodopsin (bR) photocycle is one of the most important problems in the investigation of the intramolecular proton pump of bR. The first photocycle model was suggested by Lozier et al. (1975). The linear, unidirectional scheme they have given proved to be a good approximation, but further studies have shown that the reaction pathway is more complicated. Recently Nagle (1991a) have proved that a unique reaction scheme can be determined by analyzing spectroscopical data, if plausible assumptions (first-order reactions, temperature-independent spectra, etc.) are accepted. Despite, latest analyses of data obtained by either absorption kinetic or resonance Raman experiments alone have not lead to a single best model with full confidence (Lozier et al., 1992; Nagle 1991b). The photocycle of bR, therefore, is still a matter of debate, and some authors suggest that combined results of different types of experiments should be the object of future evaluation procedures (Lozier et al., 1992). In this paper we present a simple theory, how photoelectric signals might contribute to the solution of this problem.

Electric signals have been measured by a number of investigators (see Trissl, 1990). Keszthelyi & Ormos (1980), then Fahr et al. (1981) presented essentially the same theory for the evaluation of electric signals based on the unidirectional, unbranched model. Our goal was to present a generalized procedure that allows evaluation of the electric signals in case of more complex photocycle kinetics.

THEORY

Evaluation of electric signals is based on the observation that absorption kinetic and photoelectric signals of bR are coupled (Keszthelyi & Ormos, 1980), implying that optical and electric changes in bR reflect the same molecular events, namely transitions between the conformational states of the protein (see also Müller et al., 1991). Denotation of the intermediates based on spectroscopic observations, therefore, can be adapted to describe photoelectric changes as well.

Let us consider a single reaction in the scheme of a potential photocycle model. If X_n and X_m are two of the intermediates, an equilibrium reaction is assumed as follows:

$$X_n \rightleftharpoons X_m \tag{1}$$

The "flow of matter" from X_n to X_m expressed as

$$-\frac{d[X_n]}{dt} = k_{nm}[X_n](t) - k_{mn}[X_m](t) \tag{2}$$

is assumed to accompany with a flow of charge, that can be observed as a current (i_m) in an appropriate measuring circuit. On the basis of Keszthelyi & Ormos (1989), this current component is proportional to

$$\mu_{nm} = \frac{Q_{nm} d_{nm}}{e_{nm}} f_{nm} , \qquad (3)$$

where Q_{nm} is the charge that moves d_{nm} distance in a segment of the bR molecule characterized by a dielectric constant of enm. The factor f_{nm} involves all the other parameters which influence the size of i_{nm}, such as the orientational function describing the degree of orientation of bR molecules, sample geometry (e.g. the distance between measuring electrodes), and takes into account the screening effect of the electrolyte. With the assumption of $\mu_{nm}=\mu_{mn}$ and using (2) and (3):

$$i_{nm}(t) = \mu_{nm}(k_{nm}[X_n](t) - k_{mn}[X_m](t)) , \qquad (4)$$

while the overall current measured in the outer circuit can be expressed as:

$$I(t) = \sum_{n>m} i_{nm}(t). \qquad (5)$$

Fitting of absorption kinetic data measured at, at least, 3 temperatures with a particular model, as it has been done in (Lozier et al., 1992), provides us with $[X_{nm}](t)$ and k_{nm} values for each intermediate, transition and temperature as output values, thus determination of μ_{nm} (using (4) and (5)) is reduced to a multilinear regression problem, i.e. the linear combination of the $k_{nm}[X_n](t) - k_{mn}[X_m](t)$ differences should be fitted to the measured $I(t)$. For practical reasons

$$U(t) = \int_0^t I(\tau) d\tau \qquad (6)$$

is fitted with the same type of integral functions of the differences in (2), in order not to underestimate the weight of slow components. Integrating (2)

$$\alpha_i = k_{nm} \int_0^\infty [X_n](t)dt - k_{mn} \int_0^\infty [X_m](t)dt \qquad (7)$$

gives the relative number of molecules passing through the i-th branch of the photocycle which involves the n-m transition. Normalizing α_i-s by $\sum_i \alpha_i = 1$, and using (4), (5) and (6), it can be concluded that $U(\infty) = \sum_{n>m} \mu_{nm}/\sum_i i$, while $U(0)=0$, so $U(t)$ is the quantity which is properly scaled in terms of μ_{nm}-s.

Imposing plausible criteria on the model-dependent μ_{nm}-s, an order of preference can be made among the different models tested. For a good model, we require an approximate temperature-independency of the μ_{nm} values. In practice, a global fitting of the integrated photoelectric signals measured at different temperatures is performed, with the restriction of temperature-independency for each μ_{nm}, then the goodness-of-fit for the electric and optical data can be used together to decide between different schemes. Comparison of two photocycle models is presented in Results and Discussion as an example for the application of the above procedure.

METHODS

Photoelectric and absorption kinetic signals were measured on an oriented suspension of purple membranes immobilized in gel, from 400 to 675 nm by 25 nm steps and three different temperatures (13°C, 25°C and 31°C). Incubating solution contained 50 mM Na_2SO_4 and 50 mM MES at pH=7. Sample preparation procedure and the measuring set-up have been described by Dér et al. (1985) in detail. Data for all temperatures were fitted simultaneously with different photocycle models assuming that the visible spectra of the intermediates are temperature-invariant. The model-fitting program was written by one of us (S. Sz.) using the theoretical background described in (Nagle, 1991) although different numerical procedures were implemented.

RESULTS AND DISCUSSION

As an example we present the comparison of two entirely different kinetic models, both of which fitted experimental data fairly well:

$$K \to L \rightleftharpoons X \rightleftharpoons M \rightleftharpoons Y \rightleftharpoons O \to bR \tag{8}$$

$$K \to L_1 \rightleftharpoons M_1 \to O \to bR \atop L_2 \rightleftharpoons M_2 \tag{9}$$

Scheme (8) looks almost equivalent to models have been thoroughly discussed in recent papers (Lozier et al., 1992, Váró & Lanyi, 1991), while model (9) contains the same intermediates arranged in a branched scheme.

Fitting of models (8) and (9) to the optical data set was carried out as the first step. The difference absorption spectra of the intermediates (Fig.1a) in model (8) show that X is an L-like intermediate, but with distinct differences as compared to its precursor, while Y must be considered as an M-like one. The relatively small extinction in the red region of the "O" difference spectrum suggests that here we might observe rather an equilibrium mixture of N and O (NO), which would be consistent with (Chernavskii et al., (1989). The program provided negligible reaction rates to the X→L and Y→M reactions.

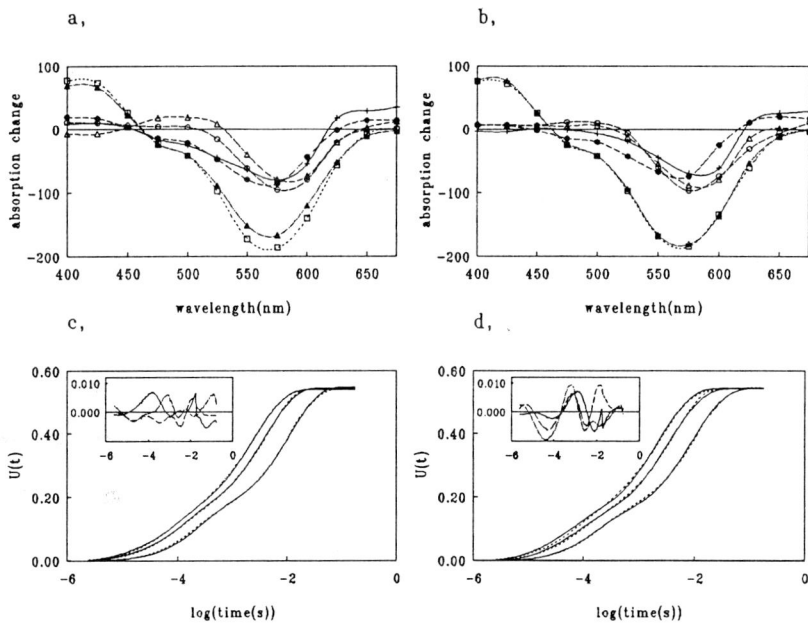

Fig. 1: a, the calculated difference spectra of the intermediates using the assignments of model (10); b, same as a, but for model (11) (+-K, ▲-L_1, ○-L_2, □-M_1, △-M_2, ○-NO); c, electric signals measured at 13, 25 and 31 °C (··) and the fit using model (10) (-); d, same as c, but model (11) was used. (In c, and d, the inserts show the difference of the fitted line and the measured data. All vertical axes are in arbitrary units.)

Taking into account the above arguments, the symbols in (8) can be rewritten as

$$K \to L_1 \to L_2 \rightleftharpoons M_1 \to M_2 \rightleftharpoons (NO) \to bR \tag{10}$$

Similar observations can be made for the difference spectra of model (9) (Fig. 1b), and the program eliminated the M_2-L_2 reaction, so the "effective scheme" is:

$$K \to L_1 \rightleftarrows M_1 \to O \to bR$$
$$\searrow L_2 \to M_2 \nearrow \qquad (11)$$

Models (10) and (11) contain the same number of parameters (intermediates and true rate constants), and fit optical data almost equally well (χ^2=498.8 for model (10) and 473.3 for model (11)). The average correlation coefficient was above 0.95 in both cases. Calculated difference spectra are shown in Fig. 1a and b. Evaluation of the electric signals was carried out as described in Theory, using the results of model-fitting to the optical data as input parameters. (Electric current traces (I(t)) were integrated from 150 ns, in order to ignore the fast negative component associated with the BR-K transition.) Figs. 1c and d show that model (10) can fit electrical data better than model (11) does (χ^2=6.49x10^{-2} for model (10) and 9.79x10^{-2} for model (11)). This difference is more significant than that for the optical data. The μ_{mn} values got with model (10) (see Table 1) suggest that major charge displacements occur only in the last two transitions of the cycle (M→NO and NO→BR), which is consistent with the results of e.g. Keszthelyi & Ormos (1980).

Table 1: μ_{nm} values (in arbitrary units) calculated from model (10):

	K-L_1	L_1-L_2	L_2-M_1	M_1-M_2	M_2-NO	NO-bR
μ_{nm}:	-2.14x10^{-4}	7.79x10^{-2}	1.47x10^{-2}	7.44x10^{-2}	1.55x10^{-1}	1.75x10^{-1}

We must emphasize, however, that model (10) may well turn out not to be the best one, as the above comparison is related only to two particular models. There are some experimental evidences which argue with the possible existence of a parallel pathway in the bR photocycle (suggested originally by Sherman et al., 1979). In order to derive further conclusions, analysis of a more complete data set (including pH-dependency, as in Lozier et al., 1992) by using a number of other models is planned.

REFERENCES

Chernavskii, D.S., Chizhov, I.V., Lozier, R.H., Murina, T.M., Prokhorov, A.M., and Zubov, B.V. (1989): Kinetic model of bacteriorhodopsin photocycle: Pathway from M state to bR. Photochem. Photobiol. 49: 649-653.

Dér, A., Hargittai, P., and Simon, J. (1985): Time-resolved photoelectric and absorption signals from oriented purple membranes immobilized in gel. J. Biochem. Biophys. Meth. 10:295-300.

Fahr, A., Lauger, P., and Bamberg, E. (1981): Photocurrent kinetics of purple-membrane sheets bound to planar bilayer membranes. J. Membrane Biol. 60:51-62.

Keszthelyi, L., and Ormos, P. (1980): Electrical signals associated with the photocycle of bacteriorhodopsin. FEBS Lett. 109: 189-193.

Keszthelyi, L., Ormos, P. (1989): Protein electric response signals from dielectrically polarized systems. J. Membrane Biol. 109:193-200.

Lozier, R.H., Bogomolni, R.A., and Stoeckenius, W. (1975): Bacteriorhodopsin: A light-driven proton pump in Halobacterium halobium. Biophys. J. 15:955-962.

Lozier, R.H., Xie, A., Hofrichter, J., and Clore, M. (1992): Reversible steps in the bacteriorhodopsin photocycle. Proc. Natl. Acad. Sci. USA 89: 3610-3614.

Müller, K.-H., Butt, H.J., Bamberg, E., Fendler, K., Hess, B., Siebert, F., and Engelhard, M. (1991): The reaction cycle of bacteriorhodopsin: an analysis using visible absorption, photocurrent and infrared techniques. Eur. Biophys. J. 19: 241-251.

Nagle, J.F. (1991a): Solving complex photocycle kinetics. Biophys. J. 54: 476-487.

Nagle, J.F. (1991b): Photocycle kinetics: Analysis of Raman data from bacteriorhodopsin. Photochem. Photobiol. 54: 897-903.

Sherman, W.V., Eicke, R.R., Stafford, S.R., and Wasacz, F.M. (1979): Branching in the bacteriorhodopsin photochemical cycle. Photochem. Photobiol. 30:727-729.

Trissl, H.-W. (1990): Photoelectric measurements of purple membranes. Photochem. Photobiol. 51: 793-818.

Váró, Gy, Lanyi, J.K. (1991): Thermodynamics and energy coupling in the bacteriorhodopsin photocycle. Biochemistry 30:5016-5022.

Low pH photovoltage kinetics of bacteriorhodopsin with replacements of Asp-96, -85, -212 and Arg-82

Stephan Moltke, Maarten P. Heyn, Mark P. Krebs*, Ramin Mollaaghababa* and H. Gobind Khorana*

*Biophysics Group, Freie Universität Berlin, Arnimallee 14, W-1000 Berlin 33, Germany, and *Departments of Biology and Chemistry, Massachusetts Institute of Technology, Cambridge, MA 02139, USA*

INTRODUCTION

At acid pH the chromophore of Bacteriorhodopsin (BR) in purple membranes (PM) shows a pronounced red shift of its absorption maximum from 568nm to 605nm. This form is called the Blue Membrane (BM). Its photocycle is also changed: the M-intermediate, which is thought to be essential for vectorial proton release and uptake, is no longer formed (Váró und Lanyi, 1989). Nevertheless, charge movements take place inside the protein (Drachev et al., 1981). We have investigated the photovoltage kinetics of BR from strain S9 and various single-site mutants expressed in Halobacteria (Krebs et al., 1991) at low pH.

Purple or blue membranes were adsorbed to lipid impregnated polyethylene sheets in 150mM KCl, 3mM Trizma, 3mM HEPES, 3mM Acetate buffer at pH 7, 25°C. Titrations to low or high pH were carried out with H_2SO_4 and KOH, respectively. H_2SO_4 is used to prevent formation of the acid purple form. Voltage data are acquired simultaneously in 2 channels (64kb memory each) with sampling times of 50ns and 1–2ms covering a total time scale from 50ns to almost 200s. The analog bandwidth of the system is ~5 MHz and the system discharges with a bi-exponential characteristic at 1–10 s, depending on the shunt resistance. Data were analyzed by fitting sums of exponentials with gaussian distributed rate constants (Holz et al., 1988). Signal-to-noise ratios were 100–200.

BLUE MEMBRANE

Three representative data traces from PM at various pH values are shown in Fig.1. The sign conventions are the following: a positive phase, i.e. exponential function with positive amplitude, corresponds to either a movement of a positive charge in the direction to the extracellular side or to that of a negative charge in the opposite direction. At pH 7 five steps can be distinguished: a time-unresolved negative movement ($\tau_1 = 0.2\mu s$) is followed by a very small positive phase ($\tau_2 \sim 1\mu s$). The three major steps occur at $\tau_3 = 30\mu s$ (+30%), $\tau_4 = 800\mu s$ (+12%) and $\tau_5 = 12ms$

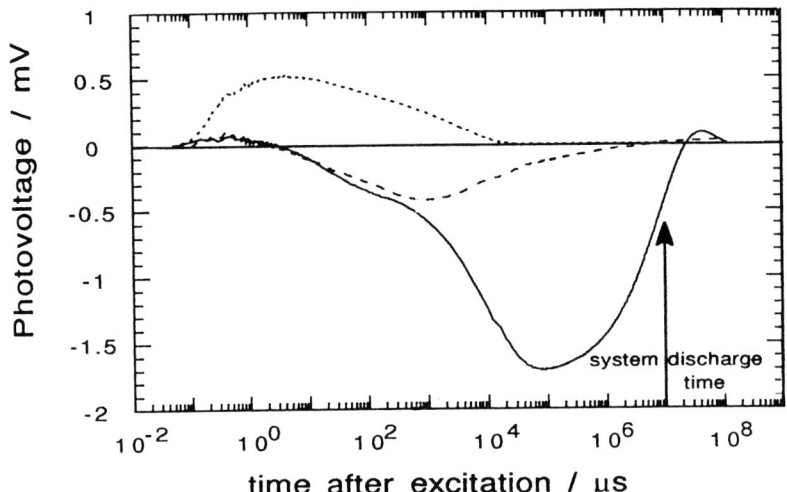

Figure 1: Photovoltage traces from Bacteriorhodopsin at pH 7 (———), pH 12 (- - -) and pH 0.6 (- - - -), for sign conventions see text. The traces at pH 7 and pH 0.6 are from the same titration and drawn to scale, the pH 12 signal comes from a different series.

(+55%) [1]. Above pH 8 a sixth, pH-dependent phase separates from τ_5, while the total amplitude is conserved.

In the blue form (pH<2) the fast negative amplitude increases dramatically and a second negative phase is observed. A triphasic compensatory movement completely discharges the system with rate constants very much faster than the passive system discharge function: $\tau_1 = 0.2\mu s$ (-45%), $\tau_2 = 1\mu s$ (-5%), $\tau_3 = 30\mu s$ (+13%), $\tau_4 = 100\mu s$ (+7%), $\tau_5 = 3.5ms$ (+30%). Therefore, the charge(s) must move in one direction and back again. The kinetics of the positive phases are slightly pH-dependent with slopes of the $\log \tau$ over pH plot between 0.1 and 0.3, suggesting that they are associated with H^+ translocation.

Comparing the absolute amplitudes from one titration we see that the total movement in one direction, i.e. A_1+A_2, lies between $A_3^{pH 7}$ (proton release to the extracellular side) and $A_5^{pH 7}$ (proton reuptake from the cytoplasm). If we want to identify the blue charge transport with $A_5^{pH 7}$, an apparent decrease of the amplitude must be explained. Several reasons can be suggested:

- the shift of the absorption maximum, because the excitatory flash always has $\lambda = 580nm$,
- a different *cis* : *all-trans* ratio in the blue form,
- a weakened adsorption of the membranes to the supporting sheet, which is evidenced by the fact that adsorption in a solution at pH 2 leads to very small absolute amplitude.

There do not seem to be any good reasons for an apparent increase. Therefore, we suggest that the charge movement in BM arises from proton transfer along the cytoplasmic pathway of the protein ($A_5^{pH 7}$): taking into account the sign of the blue signal, it must be first release and then uptake on the cytoplasmic side of the membrane. Because the Schiff base no longer deprotonates under these conditions, the question arises which group's proton is seen.

---

[1] Relative amplitudes are calculated in relation to the total charge movement, i.e. the sum of the moduli of all amplitudes.

At very high pH>12 a positive voltage signal is exactly balanced by an active discharge slightly faster than the system discharge. Argueing analogously this indicates proton release and subsequent uptake on the extracellular side, perhaps driven by the gradient resulting from the released protons.

Interestingly, each of the single-site mutants shows an electric signal at low pH, which is qualitatively identical to that of BM. If not indicated otherwise, the descriptions given below refer to these low pH conditions.

SINGLE-SITE MUTANTS

Replacements at position 96. Substituting Asp96 with Glu results in no drastic change below pH 2 except a slowing down of $\tau_4 = 800\mu s$ and $\tau_5 = 15$ms. Removing the carboxyl group leads to a dramatic loss of amplitude for A_3 and A_4:

D96N: $A_3 = 3\%$ ($\tau_3 = 30\mu s$), $A_4 = 7\%$ ($\tau_4 = 500\mu s$), $A_5 = 40\%$ ($\tau_5 = 15$ms)
D96A: $A_3 = 2\%$ ($\tau_3 = 60\mu s$), A_4 = not detectable, $A_5 = 50\%$ ($\tau_5 = 16$ms).

The last step in both is as slow as with D96E. Obviously D96 is involved in the transfer corresponding to step 3 and 4. Since D96 presumably lies on the cytoplasmic pathway, this is in good agreement with the interpretation given above.

Removing the protonable group at positions 82, 85 or 212. The only difference to the signal of BM in the mutants R82Q, R82A, D85N, D85A and D212N at low pH is the higher apparent pK of the transition to the 'blue signal' (between 3 and 5, cf. the traces for D212N in Fig.2). Also, the positive decay is just biphasic with $\tau_3 \sim 150\mu s$ ($A_3 \sim 20\%$), $A_3 = 0$ and $\tau_5 \sim 3$ms ($A_5 \sim 25\%$). Although the active discharge has one phase missing, the mean amplitude $A_5 = 25 \pm 5\%$ is not greater than that of BM, contrary to D96N and D96A. The slowing down of τ_3 may be due to the higher pH of the transitions. Contrary to this, Needleman et al. (1991) find an inverted purple-to-blue transition around pH 7 with no blue form at low pH for the absorption spectrum and photocycle of D212N (for our high pH results see below).

Replacement of Asp85 or ASP212 with Glu. Each of these single-site mutants has an effect comparable to that of removing the carboxyl at 96: the positive decay is almost monophasic with $\tau_5 = 17$ms ($A_5 = 43\%$) in D85E, and $\tau_5 = 50$ms ($A_5 = 50\%$) in D212E. This fact might be explained by assuming that either replacement lowers the pK of the internal proton source so much that reprotonation of this group is now rate limiting for proton re-uptake.

The photovoltage at very high pH. At pH>10 all of the mutants discussed above show a photovoltage signal similar to that of PM at pH 12 (cf. Fig.1) with the notable exception of D212N (see Fig.2). Titration of this mutant from low to high pH leads from the electrically blue form via an intermediate state again to the pure blue signal with $\tau_3 = 10\mu s$ ($A_3 = 15\%$) and $\tau_5 = 3$ms ($A_5 = 35\%$), which is in good agreement with Needleman et al. (1991). If we interpret the high pH kinetics as showing transient release and uptake of a proton on the extracellular side, we must conclude that the absence of the carboxyl at position 212 almost perfectly blocks the release pathway to the extracellular side. Because D85N and D85A do show the usual high pH behaviour, this fact hints at a fundamental difference in the roles of D85 and D212 for proton release.

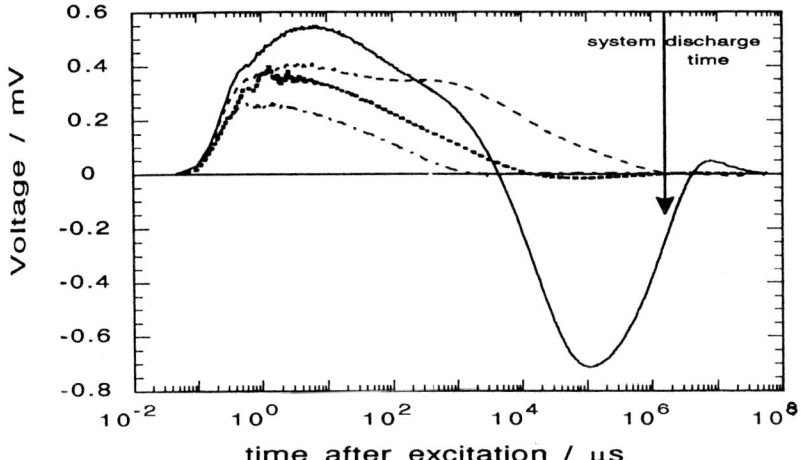

Figure 2: Photovoltages from D212N at pH 2.8 (----), pH 5.1 (———), pH 9 (- - -) and pH 11 (— · — · —·).

FINAL REMARKS

The above interpretations are based on the following assumptions:
- the signals arise from proton movement (movement solely of charged amino acid residues can be excluded due to the great absolute amplitudes),
- the high and low pH signals do not come from translocations across the whole membrane,
- even though the Schiff base does not deprotonate, the charge transfer originates near the middle of the membrane,
- effects due to structural changes can be neglected.

REFERENCES

Drachev, L. A., Kalamkarov, G. R., Kaulen, A. D., Ostrovsky, M., and Skulachev, V. P. (1981). Fast stages of photoelectric Processes in biological membranes: I. Bacteriorhodopsin. *Eur.J.Biochem.* **117**: 461–470.

Holz, M., Lindau, M., and Heyn, M. P. (1988). Distributed kinetics of the charge movements in Bacteriorhodopsin: evidence for conformational substates. *Biophys.J.* **53**: 623–633.

Krebs, M. P., Hauß, T., Heyn, M. P., RajBhandary, U. L., and Khorana, H. G. (1991). Expression of the bacterioopsin gene in *Halobacterium halobium* using a multicopy plasmid. *Proc.Natl.Acad.Sci.USA* **88**: 859–863.

Needleman, R., Chang, M., Ni, B., Váró, G., Fornés, J., White, S. H., and Lanyi, J. K. (1991). Properties of $Asp^{212} \rightarrow Asn$ bacteriorhodopsin suggest that Asp^{212} and Asp^{85} both participate in a counterion and proton acceptor complex near the Schiff base. *J.Biol.Chem.* **266**: 11478–11484.

Váró, G. and Lanyi, J. K. (1989). Photoreactions of bacteriorhodopsin at acid pH. *Biophys.J.* **56**: 1143–1151.

Proton movement and surface charge in bacteriorhodopsin detected by selectively attached pH-indicators[1]

Peter Scherrer, Ulrike Alexiev, Harald Otto, Maarten P. Heyn, Thomas Marti* and Gobind H. Khorana*

Biophysics Group, Department of Physics, Freie Universität Berlin, Arnimallee 14, D-1000 Berlin 33, Germany and *Department of Biology and Chemistry, Massachusetts Institute of Technology, Cambridge, Massachusetts 02139, USA

ABSTRACT

The light-induced proton movement in bacteriorhodopsin (bR) micelles was monitored with the optical pH-indicator dye fluorescein attached to various sites on the protein surface. This was achieved by covalent linkage of 5-iodoacetamidofluorescein (IAF) to cysteine residues introduced by site-directed mutagenesis. The proton release time measured with fluorescein in position 130 (V130C-AF) on the extracellular side was 22 ± 4 μs compared to 62 ± 5 μs in position 35 (S35C-AF) on the cytoplasmic side. The latter time was still clearly faster than that detected by pyranine ($124 \pm 9 \mu$s) in the bulk medium. Only the light-induced protonation changes detected with fluorescein bound to the cytoplasmic side were dependent on the pK of the lipid headgroups in the mixed micelles. These results indicate proton movement along the micellular surface.

The apparent pK of the bound fluorescein was dependent on the salt concentration in the medium. Using the Gouy-Chapman equation, the surface charge density could be determined as -2.5 ± 0.2 charges/bR for the cytoplasmic and -1.8 ± 0.2 charges/bR for the extracellular side.

INTRODUCTION

Flash-induced proton release and uptake in purple membrane (pm) suspension were studied using pH-indicators like 7-hydroxy-coumarin (Lozier et al. 1976; Scherrer et al. 1981), pyranine (Drachev et al., 1984) or phenol red (Varo and Lanyi, 1990) to detect the proton in the aqueous medium. To measure the proton appearance on the surface of the protein, Heberle and Dencher (1990) linked the indicator to ϵ-amino groups of lysines on the pm surface using a succinimidylester derivative of fluorescein. In our work we developed a method to monitor the light-induced proton movement at various selected positions on the extracellular and

---
[1] This article combines the results of two contributions.

cytoplasmic side of the protein. We correlate these data with the absorbance changes of the chromophore in the course of the photocycle.

The lack of the amino acid cysteine in bacteriorhodopsin provided the unique possibility to introduce this amino acid into the protein by site-directed mutagenesis at any desired position and to use its reactive sulfhydryl group to attach our indicator. As a proton indicator we chose fluorescein, with the highly pH-dependent absorbance at 490nm (Fothergill, 1964), and linked its iodo-acetamido derivative to the cysteine (Fig. 1).

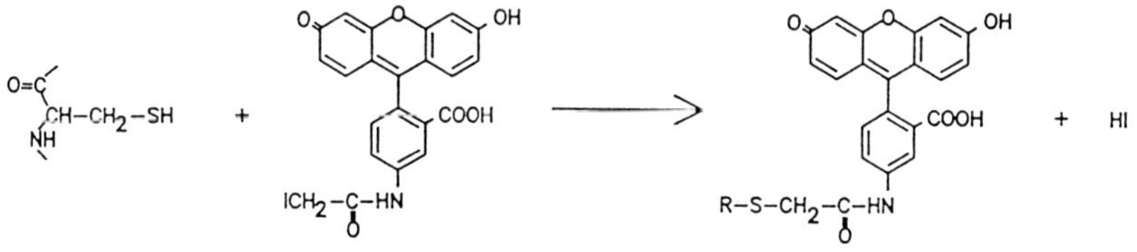

CYSTEINE   IAF   C-AF

Fig.1 Reaction of the iodoacetamido derivative of fluorescein (IAF) with cysteine at pH 8.0.

This system allowed us to monitor the appearance of the proton at various positions on the protein surface and to test the effects of different lipid headgroups (phosphatidylcholine and phosphatidic acid) on the proton movement over the micelle surface. Furthermore, the dependency of the bound fluorescein pK on salt concentration allowed to calculate the surface charge density on either side of bR.

MATERIAL AND METHODS

DMPC: L-α-Phosphatidylcholine, dimyristoyl; DMPA: L-α-Phosphatidic acid, dimyristoyl.
Bacterio-opsin mutants:
Construction and production of bacterio-opsin mutants were as described previously (Nassal et al., 1987; Karnik et al., 1987; Braiman et al., 1987).
Regeneration:
The isolated freeze dried bR mutant protein (0.2 μmole) was dissolved in 1ml 1% CHAPS; 1% DMPC; 50 mM NaCl; 1mM DTT; 10 mM $NaPO_4$-buffer pH 6.2 and regenerated with 0.25 μmole retinal and 150 mM KCl. Excess lipids, detergents and buffer were removed by chromatography on a Sephadex G-25 column pre-equilibrated and eluted with 0.1% CHAPS; 0.0025% DMPC; 150 mM KCl.
IAF-labelling:
0.1 μmole bR in 0.1% CHAPS; 0.0025% DMPC; 150 mM KCl; 4 μM EDTA; 0.1 mM DTT; 50 mM Tris-buffer pH 8.0 were reacted with 1 μmole 5-IAF for 30 minutes under argon. The reaction was stopped by addition of 20 μmole glutathione; 20 μmole DTT and the excess reagents were removed by chromatography on Sephadex G-25 in 0.1% CHAPS; 0.0025% DMPC; 150 mM KCl.
Flash spectroscopy was performed as been described elsewhere (Otto et al. 1989)

RESULTS AND DISCUSSION

Regeneration

The DMPC concentration needed for refolding and regeneration of bR was dependent on the SDS concentration present, due to the earlier protein isolation procedure, since mixed SDS-DMPC micelles are formed prior to bR-DMPC-CHAPS micelles. The SDS-DMPC-micelles did not precipitate under these conditions. The SDS-DMPC and excess DMPC-CHAPS micelles were separated by hydrophobic interaction on G-25 Sephadex resin.

bR / lipid / detergent ratio

Since the function of membrane proteins in mixed micelles is often greatly affected by the protein/lipid/detergent ratio, it is important to characterize the mixed micelle systems first, by searching for conditions with a) sufficient stability of the protein and b) the least effect on the photocycle caused by small changes in the protein/lipid/detergent ratio. The ebR/lipid/detergent ratio was varied and the kinetics of the M-intermediate analyzed (Fig. 2). An increase in the detergent and/or in the lipid concentration resulted in a slower M-decay. The same protein/lipid/detergent ratio with different final concentrations showed the same decay kinetics.

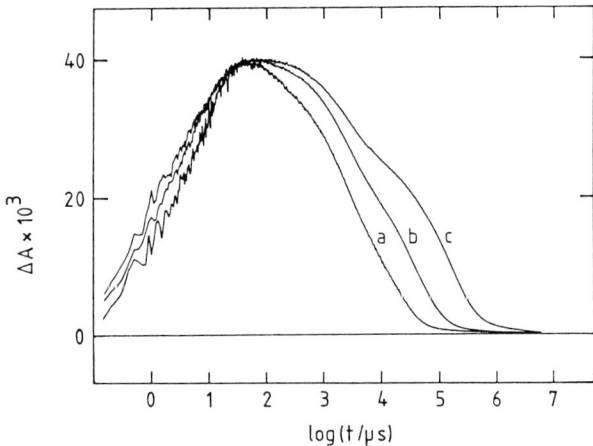

Fig.2 Time course of the flash-induced absorbance changes of bR at 410nm (M-intermediate), 22°C, pH7.3 in
a) 0.1%CHAPS, 0.0025%DMPC 4-10 μM bR
(bR/DMPC/CHAPS ratio: 1/8/330 - 1/3/125)
b) 1%CHAPS, 0.025%DMPC, 9.5 μM bR (1/35/1400)
c) 1%CHAPS, 0.025%DMPC, 4 μM bR (1/80/3300)

In 0.1% CHAPS, 0.0025% DMPC, 150 mM KCl a change in protein concentration between 4 and 10 μM (this corresponds to a bR/lipid/detergent ratio of 1/8/330 to 1/3/125) did not cause any change in the kinetics of the M-intermediate. Moreover, the kinetics for the M-intermediate was most similar to that of native bR (Fig.2, trace a). Under these conditions we found good stability for the reconstituted micelles.

Proton release detected with bound fluorescein

Iodoacetamidofluorescein was reacted with regenerated bR containing a cysteine residue either in position 35 or 130. Within 30 minutes reaction time approx. one fluorescein molecule was incorporated/bR. In the absence of cysteine residues, less than 3% IAF were bound to bR within 60 minutes. The dye was covalently attached to bR as confirmed by the fluorescence of the bR-band in SDS-gels. The extent of labelling was estimated based on the pH dependent absorbance change of fluorescein at 490nm. This pH sensitive absorbance band was used to probe for light-induced protonation changes.

Figures 3 & 4 show the light-induced absorbance changes at 410nm for the M-intermediate and the proton release and uptake measured with pyranine(DIFPY) at 450nm and the bound fluorescein (DIFAF) at 490nm. DIFPY is the difference of the absorbance changes measured at 450nm in the absence and presence of pyranine. DIFAF is the difference of the absorbance changes at 490nm for an IAF-labelled sample in the absence and presence of buffer (10 mM MOPS), in order to correct for possible effects due to changes in the charges near the label and/or in the surface potential. The proton concentration changes in the bulk medium were recorded with the negatively charged pyranine (characterized by Kano & Fendler, 1978). The negative charges prevented interactions with the negatively charged membrane surface. The kinetics for the M - intermediate did not change significantly compared to ebR for either mutant protein even after labelling with IAF. Fluorescein in position 130 on the extracellular side detected the proton much faster (22 ± 4µs) than pyranine (124± 9 µs) in the bulk phase (Fig.3). This is in agreement with the measurements by Heberle & Dencher (1990). However fluorescein attached to the cytoplasmic side in position 35, detected a light-induced proton release approximately 3 times slower (62±5 µs) than in position 130 (Fig.3 & 4), but still clearly faster than measured by pyranine in the bulk phase. These results

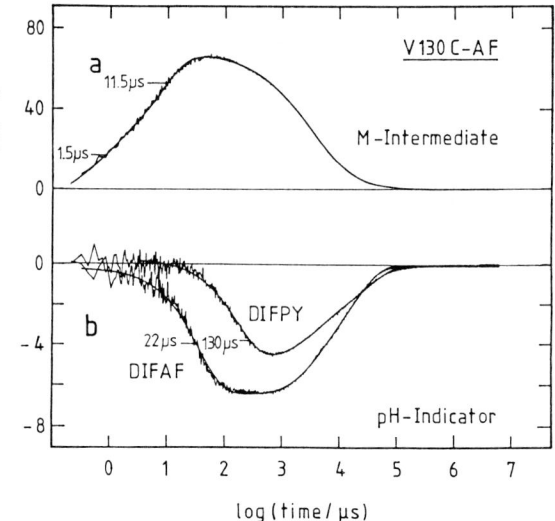

Fig.3
a) Flash-induced absorbance changes of V130C-AF measured at 410nm (M-Intermediate).
b) Kinetics of proton release and uptake as measured by bound fluorescein at 490nm (DIFAF; V130C-AF) and by pyranine at 450nm (DIFPY) in the aqueous bulk phase at pH 7.3 and 22°C.

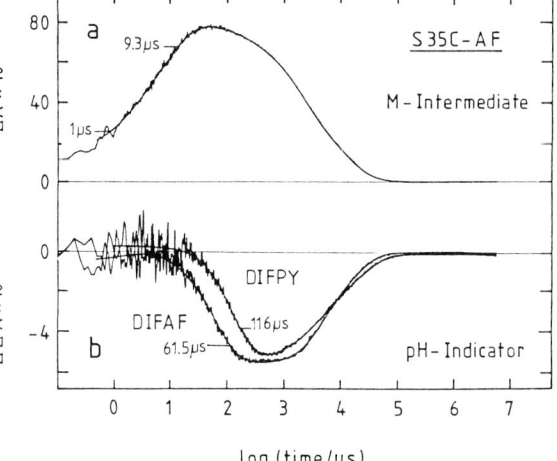

Fig.4 Kinetics of the M-intermediate and the dye signals of S35C-AF measured as described in Fig.3 .

indicate a delay in proton transfer from the protein surface to the bulk phase with a faster proton movement along the mixed micelle surface than from the protein surface to the bulk medium.

A proper understanding of the proton transfer from surface to bulk is one of the most challenging problems in modern bioenergetics. A number of experiments with model membrane systems performed by Gutman and coworkers (Gutman and Nachliel, 1985; Nachliel and Gutman, 1988) indicated that protons were retarded by the lipid headgroups depending on their pK. Groups with higher pK's (phosphatidylserine, pK 4.6) retained the proton much longer ($2\mu s$) than groups with lower pK's (phosphatidylcholine, pK 2.2; 30 ns). As a consequence protons diffuse over longer distances in the presence of low pK lipid headgroups which allow many hopping steps and over shorter distances with high pK headgroups which delay the proton and reduce its mobility. If proton migration along the micellular surface is a major pathway for protonation of the dye on the cytoplasmic side, one would expect that addition of lipids with low pK headgroups will lead to an increase in the amplitude of the protonated dye signal, whereas addition of lipids with high pK headgroups will lead to a reduction. With the dye on the extracellular side, near the source of the released proton, no proton migration along the micellular surface is required to reach the dye and no effect caused by the pK of lipid headgroups is expected. We added to the mixed micelles DMPC with a pK of 2.2 and DMPA with a much higher pK of 8.0 and measured the proton release signal with fluorescein in position 35 and 130. While the lipid addition had no effect on the proton release signal in position 130 on the release side (data not shown), the predicted increase and decrease of the amplitude of the proton signal in the position 35 was detected (Fig. 5). Our results are in good agreement with the Gutman findings and demonstrated that protons move along the mixed micellular surface.

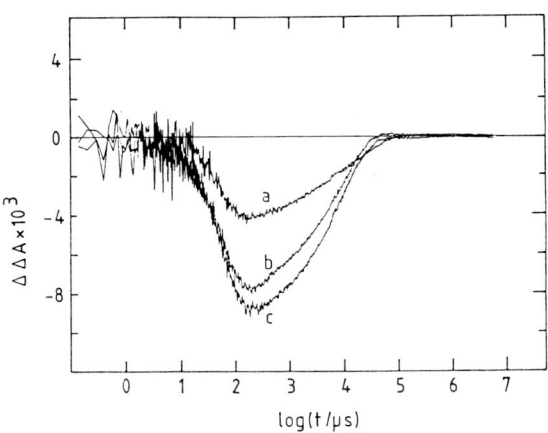

Fig.5 Difference of the flash-induced absorbance changes at 490nm of S35C-AF between an unbuffered and buffered (10 mM MOPS) sample in 150mM KCl, pH 7.3, 22°C and
a) 0.1% CHAPS, 0.002% DMPA
b) 0.1% CHAPS,
c) 0.1% CHAPS, 0.05% DMPC .

Surface charge

When fluorescein was bound to the protein a change in its pK of up to almost one unit was detected in 150 mM KCl. Such effects can be caused by the surface potential and/or by interactions with nearby charges. If the pK change is due to the surface potential then the apparent pK will be dependent on the salt concentration. The pH-titration of V130C-AF in various salt concentrations (Fig. 6) shows that the fluorescein pK is clearly salt dependent. By using the Gouy-Chapman equation the surface charge density (σ) was calculated. The relevant equations are:

$$\sinh e\psi/2\ kT = A\ \sigma\ C^{-1/2}$$

$$pK_t - pK_a = e\psi/2.3\ kT$$

ψ is the surface potential, C the salt (KCl) concentration, A equals 134.6 $M^{1/2}$ at 22°C and pK_a is the apparent pK at a given salt concentration. The true pK (pK_t) was calculated with a transformed Gouy-Chapman equation introduced by Koutalos et. al. (1990).

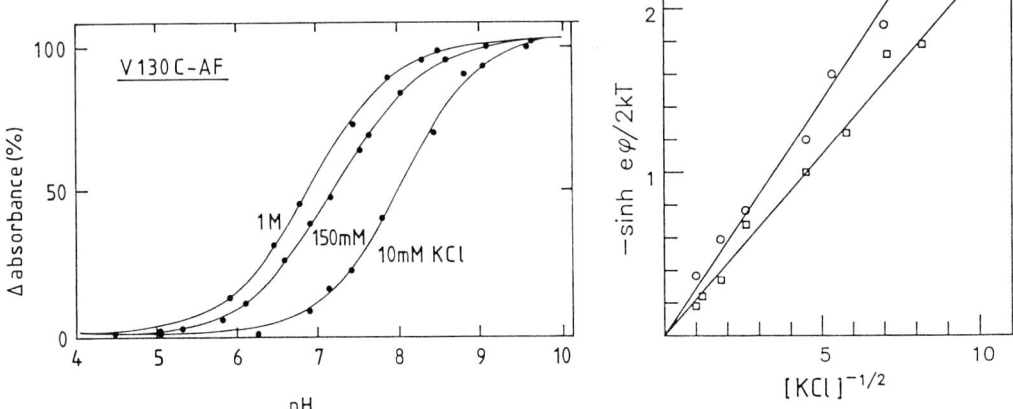

Fig.6 Titration data and fitted curves of V130C-AF at 490nm in 10 mM, 150 mM and 1 M KCl.

Fig.7 Gouy-Chapman plot of V130C-AF (□) and A160C-AF (o)

The Gouy-Chapman plots for V130C-AF (extracellular) and A160C-AF (cytoplasmic) are presented in Figure 7, showing that the cytoplasmic side of the protein is more negatively charged. For the extracellular side we calculated 1.8 - 1.9 negative elementary charges/bR, based on the V130C-AF titration data and assuming a surface area for the bR molecule of about 1150 $Å^2$. For the cytoplasmic side with the A160C-AF data and the same surface area we obtained 2.4 - 2.5 negative charges/bR. To verify the validity of our approach we used the mutant R134C-AF with a charge change on the extracellular side. Here the positive charge of arginine was removed and with the fluorescein label in this position we estimated a charge of -3.0, as expected a change of about one unit.

CONCLUSIONS

1) Good protein stability and reproducible flash-induced absorbance changes with bR-micelles were obtained in 0.1% CHAPS; 0.0025% DMPC; 150 mM KCl.

2) The pumped proton was detected later by fluorescein on the cytoplasmic side than on the extracellular, the release side, but

clearly earlier than measured by pyranine in the aqueous bulk phase. The pK of the lipid headgroups affected the amplitude of the measured proton signal only with fluorescein on the cytoplasmic side. These results suggest that protons move faster along micellular surfaces than from the micelle surface to the aqueous bulk phase.

3) The surface charge was determined separately on the cytoplasmic (-2.5 ± 0.2 charges/bR) and on the extracellular side (-1.8 ± 0.2 charges/bR) assuming a surface area of 1150 $Å^2$ for either side.

REFERENCES

Braiman M.S., Stern L.J., Chao B.H. and Khorana H.G. (1987) J. Biol. Chem. 262: 9271 - 9276.
Fothergill, J.E. (1964): in Fluorescent Protein Tracing (Nairn, R.C. Ed.) pp 34-59, E.&S. Livingstone Ltd., Edinburgh and London.
Gutman, M. and Nachliel, E. (1985) Biochemistry 24: 2941-2946.
Heberle, J. and Dencher, N.A. (1990) FEBS Lett. 277: 277-280.
Kano, K. and Fendler, J. H. (1978) Biochim. Biophys. Acta. 509: 289-299.
Karnik, S.S., Nassal M., Doi T., Jay E., Sgaramella V. and Khorana H.G. (1987) J. Biol. Chem. 262: 9255 -9263.
Koutalos, Y., Ebrey, T.G., Gilson, H.R. and Honig, B. (1990) Biophys. J. 58: 493-501
Nachliel, E. and Gutman, M. (1988) J. Am. Chem. Soc. 110: 2629-2635.
Nassal, M., Mogi, T., Karnik, S.S. and Khorana, H.G.(1987) J. Biol. Chem. 262: 9264 - 9270.
Otto, H., Marti, T., Holz, M., Mogi, T., Lindau, M., Khorana, H.G. and Heyn M.P. (1989) Proc. Natl. Acad. Sci. USA 86: 9228 - 9232.
Scherrer, P., Packer, L. and Seltzer, S.,(1981) Arch. Biochem. & Biophys. 212: 589-601.
Varo G. and Lanyi J.K. (1990) Biochemistry 29: 6850-6865.

Active and passive proton transfer steps through bacteriorhodopsin are controlled by a light-triggered hydrophobic gate

Norbert A. Dencher[1], Joachim Heberle[1], Georg Büldt[2], Hans-Dieter Höltje[3] and Monika Höltje[3]

[1] Han-Meitner-Institut, BENSC-N1, Glienicker Strasse 100, W-1000 Berlin 39, Germany; [2] Department of Physics/Biophysics, Arnimallee 14 and [3] Department of Pharmacy, Free University, Königin-Luise-Strasse 2 + 4, W-1000 Berlin 33, Germany

We are still far away from understanding the details of each step in light-energized proton pumping by bacteriorhodopsin (BR). However, due to the recent vast increase in structural and functional knowledge about BR, we can emphasize key elements and reaction steps inherent in vectorial transmembrane proton transfer, i.e., the involvement of conformational alterations of both the chromophore and the protein, and the participation of water molecules as well as of specific amino acids such as Asp85 and Asp96.

THE PROTON PUMPING MECHANISM OF BACTERIORHODOPSIN: STRUCTURE-FUNCTION RELATIONSHIP

Figure 1 depicts a computer graphics model of the functionally important part of BR, constructed from the coordinates provided by Henderson et al. (1990). Only the functional relevant amino acid side chains, polar hydrogens involved in hydrogen bonding, water molecules, and the chromophore are shown. These water molecules together with exchangeable hydrogens are suggested to be necessary elements of the active centre in BR, formed by the chromophore retinal, the protonated Schiff's base linkage to lysine-216 of the protein moiety, and the amino acids composing the binding pocket. In addition, hydrogen-bonded chains of water molecules provide the conductance pathway for protons connecting the active centre with both surfaces of the membrane. The structure was calculated with the program SYBYL (Tripos Assoc. Inc., St. Louis; general details are described elsewhere by Höltje & Höltje (1991)) without, on purpose, any optimization of the geometry as well as changes in the conformation of side chains. Only Arg82 was slightly displaced to relieve steric contacts and to allow formation of an hydrogen bond with Glu204.

(i) Participation of water molecules in proton transfer. A very relevant result for the structure-function relationship of BR is the localization of 4 $\pm$ 1 water molecules and of some exchangeable hydrogens at the retinal Schiff's base end in the projected map at 15 % rel. humidity. These water molecules are strongly associated with BR, since they are still present in part even at 0% rel. humidity and can only be removed by the application of vacuum (Dencher et al., 1988; Papadopoulos et al., 1990). From resonance Raman spectroscopy, tightly bound water molecules at the chromophore site have been suggested (Hildebrand & Stockburger, 1984), which should serve as proton donors/acceptors for the Schiff's base nitrogen and stabilize the ion-pair structure between the protonated positively charged Schiff's base nitrogen and its negative counterion(s) in the protein. It is very probable that at least one of the four water molecules localized by our neutron diffraction study at 15 % rel. humidity (Dencher et al., 1988; Papadopoulos et al., 1990) is in contact with the Schiff's base. In fact, one of the ten predicted water molecules in our modelled structure was positioned in the calculation in the direct vicinity of the Schiff's base nitrogen, between the side chains of Asp212 and Asp85 (Fig. 1). The Schiff's base nitrogen reversibly deprotonates during the

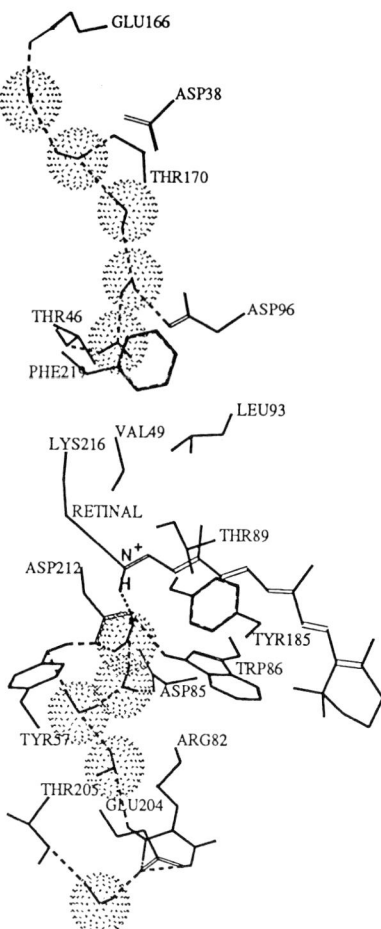

Fig. 1. Computer graphics model of the pathway for protons across bacteriorhodopsin, emphasizing two chains of hydrogen-bonded water molecules. Only side chains and explicit polar hydrogens involved in hydrogen-bonding are plotted. The position of water molecules derived from the program GRIN/GRID (Molecular Discovery Ltd., Oxford, UK) are marked by dotted spheres, reflecting the van der Waals contact radii. The bottom of the structure represents the extracellular surface of BR towards which the proton is ejected.

photocycle (Lewis et al., 1974) and is in an environment concomitantly undergoing structural alterations. Therefore, one or several water molecules at this site could be directly involved in the processes of light-energized vectorial proton transport. The value of 4 water molecules localized in the projected structure of the Schiff's base was obtained for relatively dry PM samples, i.e., at 15 % relative humidity. Obviously, the number of water molecules at 100 % relative humidity, where BR is active as proton pump, has to be known. Because of lack of measuring time with neutrons, the necessary measurements at 40 - 95 % relative humidity cannot be performed in the near future. Therefore, we rely on the value of *10 water molecules* positioned by the program GRIN/GRID across the BR structure in Fig. 1. These *water molecules and the exchangeable hydrogens might be components of the proton conductance pathway between the active centre and both surfaces of BR.*

(ii) Transient encounter of the Schiff's base nitrogen and Asp85. Upon photon absorption by the chromophore, the Schiff's base nitrogen deprotonates (Lewis et al., 1974) and a proton appears at the extracellular surface of BR, on the time scale of the L to M transition ($\tau = 60$ μs at 22 °C; Heberle & Dencher, 1992). It is generally believed that Asp85 is the primary acceptor of the

Schiff's base proton. However, *in the ground-state structure of BR (Fig. 1) the proton of the Schiff's base is forming a hydrogen-bond with Asp212. Asp85 is too far away*, even a change in the conformation of its side chain would not allow formation of a hydrogen-bond to the Schiff's base. We therefore *propose that during the BR_{568} to M_{412} transition, Asp85 and the Schiff's base transiently get closer, leading to the formation of a hydrogen-bond and subsequent transfer of the proton to Asp85*. This transient approach is *induced either by the observed movement of the Schiff's base linkage of the chromophore (Hauß et al., 1991) or by the detected alterations in the protein moiety* (Dencher et al., 1988, 1989, 1991; Koch et al., 1991). Neutron diffraction and time-resolved X-ray diffraction experiments have for the first time established *significant reversible structural changes in bacteriorhodopsin*, without loss of crystalline order, during the light-induced transition from the BR_{568} ground state to the M_{412} intermediate of the photocycle. The difference density map (M_{412} - BR_{568}) displays strong peaks at helix G and F and between helix D and E/F, caused by a shift of the projected density in the neighborhood of the cyclohexene ring and at the Schiff's base end of the chromophore retinal during M_{412} formation. These intensity changes in the resolution range 60-7 Å are indicative of *alterations in the tertiary structure*, such as a small shift or a 1-2° tilt of helices, e.g. of helix G. Also positional changes of four or five amino acids over distances of 3 - 5 Å would be in line with these data (Dencher et al., 1988, 1989, 1991; Koch et al., 1991). The X-ray synchrotron data suggest that after the *light-induced structural changes generated during the BR_{568} to M_{412} transition, BR relaxes to its original conformation during the N_{550} to BR_{568} transition* (Koch et al., 1991). Retinal undergoes a transient isomerization from all-*trans* in the BR_{568} ground state to 13-*cis* in the M_{412} intermediate during the photocycle of BR. Since it was not yet clear, if and how this isomerization is related to the observed structural changes in the protein moiety of BR, we have employed neutron diffraction to visualize light-triggered positional alterations of the chromophore (Hauß et al., 1991). One perprotonated and two specifically deuterated retinals were used to show possible movement of the ring (D11-retinal) and of the Schiff's base portion (D5-retinal). Structural data were recorded at 90 K on PM in the light-adapted ground state and in the trapped M state. According to the data, the cyclohexene ring does not change its position during the BR_{568} to M_{412} transition (less than 0.7 Å, as compared to the error of 1 Å inherent in determination of the label position; Hauß et al., 1991). In contrast, the D5 label position (close to the Schiff's base) shifts by 1.4 Å towards the ring position (Hauß et al., 1991; T. Hauß, personal communication). This *alteration of the D5 label position between BR_{568} and M_{412} suggests a tilt of the retinal*, which seems to be fixed with its ring in the protein pocket, *increasing the angle between the membrane plane and the polyene chain by about 11°* (T. Hauß, personal communication). This positional change of the chromophore might trigger the alterations in the tertiary structure as well as the active proton translocation across bacteriorhodopsin. According to our X-ray (Büldt et al., 1991) and neutron diffraction data (Hauß et al., 1991), the *longitudinal axis of the polyene chain makes an angle of about 24-29° with respect to the membrane plane in the BR_{568} ground state that increases by about 11° in the M_{412} state*.

The importance of adequate distances between the Schiff's base proton and the primary proton acceptor, as well as between the primary and the secondary proton acceptor, is demonstrated by the observation that replacement of Asp85 by Glu in the Asp85Glu mutant BR (resulting in an elongation of the side chain by only one C-C bond and maybe in a slight pK-change) leads to an 50 times faster formation of M_{412} (deprotonation of the Schiff's base). On the other hand, instead of following this acceleration, the appearance of the proton at the extracellular BR surface is strongly delayed as compared to wild-type BR (Dencher et al., 1991).

Movement of the Schiff's base linkage of the chromophore and/or alterations in the tertiary structure of the protein could be the driving force for the active step(s) of proton transfer. It has been suggested (Scheiner & Hillenbrand, 1985) that alteration of the H-bond geometry can result in transfer of a proton by reversing the order of pKs of two residues. Therefore, conformational changes that affect the H-bond geometry would lead to pK shifts, which in turn induce active proton translocation.

(iii) A hydrophobic gate controls vectorial proton transfer. It is evident from Fig. 1 that the proton conducting pathway is interrupted for about seven angstroms . Between Thr46 and Thr89, just above the Schiff's base, the hydrophobic amino acids Phe 219, Leu93, and Val49 are gathered, leaving no space for water molecules. Here, the hydrogen-bonding system is interrupted. This explains the experimental finding that the proton pathway through BR, represented by the network of hydrogen bonds formed from the amino acid side chains and the water molecules indicated in Fig. 1, is completely blocked in the ground state of BR (Burghaus & Dencher, 1989). Obviously, it is essential that the proton pathway is closed as long as BR is not active. This property prevents

uncontrolled proton backflow via this specific pathway, which would otherwise lead to undesired collapse of any electrochemical proton gradient across the membrane. *Only during the short period of conformational change resulting in an "opening" of this hydrophobic gate, i.e., in a transient populating with 2-3 water molecules, can protons surmount the barrier in the proton conducting pathway.* Furthermore, this only *transiently passable hydrophobic gate can explain and is required for the vectoriallity of this proton pump.* In the first part of the pumping cycle, the Schiff's base proton protonates Asp85, and via a "domino effect" a proton is ejected nearly simultaneously from the extracellular surface of BR (Heberle & Dencher, 1992). In the second part of the cycle, via the transiently formed network of hydrogen bonds in this gate, the Schiff's base is reprotonated from Asp96, which in turn is later reprotonated by uptake of a cytoplasmic proton. *During the N_{550} to BR_{568} transition three important molecular events occur: (i) conformational relaxation of the protein, (ii) reisomerization of the chromophore retinal, and (iii) uptake of a cytoplasmic proton by the protein.*

ACKNOWLEDGEMENTS. N.A.D. thanks Dr. Thomas Hauß (Dept. Physics, Free University Berlin, Biophysics Group M.P. Heyn) for successfully conducting with him the neutron diffraction measurements at the ILL in Grenoble, resulting in the localization of light-induced positional changes of retinal (and in a few, because of lack of time, enjoyable common cultural activities, e.g., attendance at two Grenoble jazz festivals).

REFERENCES

Burghaus, P.A. & N.A. Dencher (1989): The chromophore retinal hinders passive proton/hydroxide ion translocation through bacteriorhodopsin. *Arch. Biochem. Biophys.* 275, 395-409.

Büldt, G., K. Konno, K. Nakanishi, H.-J. Plöhn, B.N. Rao & N.A. Dencher (1991): Heavy-atom labelled retinal analogues located in bacteriorhodopsin by X-ray diffraction. *Photochem. Photobiol.* 54, 873-879.

Dencher, N.A., D. Dresselhaus, G. Maret, G. Papadopoulos, G. Zaccai & G. Büldt (1988): Light-induced structural changes in bacteriorhodopsin and topography of water molecules in the purple membrane studied by neutron diffraction and magnetic birefringence. *Proceedings of the Yamada Conference* XXI, pp. 109-115.

Dencher, N.A., D. Dresselhaus, G. Zaccai & G. Büldt (1989): Structural changes in bacteriorhodopsin during proton translocation revealed by neutron diffraction. *Proc. Natl. Acad. Sci. USA* 86, 7876-7879.

Dencher, N.A., Heberle, J., Bark, C., Koch, M.H.J., Rapp, G., Oesterhelt, D., Bartels, K. & Büldt, G. (1991): Proton translocation and conformational changes during the bacteriorhodopsin photocycle: Time-resolved studies with membrane-bound optical probes and X-ray diffraction. *Photochem. Photobiol.* 54, 881-887.

Hauß, T., Heyn, M.P., Büldt, G. & Dencher, N.A. (1991): Movement of retinal in bacteriorhodopsin during the BR_{570} to M_{411} transition: A neutron diffraction study. *Jahrestagung DG Biophysik, Homburg*, abstract P32.

Heberle, J. & Dencher, N.A. (1992): Surface-bound optical probes monitor proton translocation and surface potential changes during the bacteriorhodopsin photocycle. *Proc. Natl. Acad. Sci. USA* 89.

Henderson, R., J.M. Baldwin, T.A. Ceska, F. Zemlin, E. Beckmann & K.H. Downing (1990): Model for the structure of bacteriorhodopsin based on high-resolution electron cryo-microscopy. *J. Mol. Biol.* 213; 899-929.

Hildebrandt, P. & Stockburger, M. (1984): Role of water in bacteriorhodopsin's chromophore: resonance Raman study. *Biochemistry* 23, 5539-5548.

Höltje, M. & Höltje, H.-D. (1991): Molecular modelling study on the negative inotropic potencies of 1,4-dihydropyridines. *Pharm. Pharmacol. Lett.* 1, 19-13.

Koch, M.H.J., N.A. Dencher, D. Oesterhelt, H.-J. Plöhn, G. Rapp & G. Büldt (1991): Time-resolved X-ray diffraction study of structural changes associated with the photocycle of bacteriorhodopsin. *EMBO J.* 10, 521-526.

Lewis, A., J. Spoonhower, R.A. Bogomolni, R.H. Lozier & W. Stoeckenius (1974): Tunable laser resonance Raman spectroscopy of bacteriorhodopsin. *Proc. Natl. Acad. Sci. USA* 71, 4462-4466.

Papadopoulos, G., N.A. Dencher, G. Zaccai & G. Büldt (1990) Water molecules and exchangeable hydrogen ions at the active centre of bacteriorhodopsin localized by neutron diffraction. *J. Mol. Biol.* 214, 15-19.

Scheiner, S. & Hillenbrand, E.A. (1985): Modification of pK values caused by change in H-bond geometry. *Proc. Natl. Acad. Sci. USA* 82, 2741-2745.

Bacteriorhodopsin pump activity at reduced humidity

Gerd Thiedemann, Joachim Heberle and Norbert A. Dencher

Hahn-Meitner-Institut, Glienicker Strasse 100, W-1000 Berlin 39, Germany

INTRODUCTION

One successful approach to elucidate the molecular mechanism of an transport protein such as bacteriorhodopsin (BR) is to study its structure and function in response to alterations in the environment. Water is of fundamental importance for various functional properties of BR as has been previously demonstrated. The kinetics of spectroscopic intermediates as well as of displacement currents upon photoexcitation of BR are strongly affected by the water content of the sample [1, 2, 3]. From resonance Raman spectroscopy, water molecules at the chromophore site have been suggested [4] and by neutron diffraction tightly bound water molecules at the Schiff's base could be visualized [5]. The effect of water directly on the proton release and reuptake events was unknown to date, however. Therefore, in order to get information on the participation of water molecules both in the active steps of H^+-translocation and in the proton pathway, we have measured the dependency of the kinetics of the photocycle as well as of H^+-ejection and H^+-uptake with a time resolution of 20 ns on the water content of purple membranes (PM).

MATERIALS AND METHODS

Transient absorption spectrometer

A transient absorption spectrometer (fig. 1) was constructed in order to monitor time resolved absorbance changes of PM suspensions or PM films with high sensitivity and temporal resolution. The pulse of a Q-switched, frequency-doubled Nd:Yag solid state laser (details in fig. 1) is used to excite the sample with a pulse energy of 0.8 mJ. Monochromators, in front of and behind the sample, as well as appropriate interference filters select the wavelength λ_m of the analyzing light from a tungsten-halogen lamp (electrical power 100 W). To prevent artefacts due to rotational diffusion of PM-patches in suspension, the polarisation of the analyzing light is adjusted to the magic angle.

The anode signal of a standard side-on photomultiplier tube is preamplified and feed to the plus-input of a transient recorder with differential amplifier. The minus-input is connected to a computer-controlled digital-analog-converter (DAC) in order to compensate the DC-component. Laser pulse and transient recorder are synchronized within a jitter of $\delta t < 1$ ns. The transient

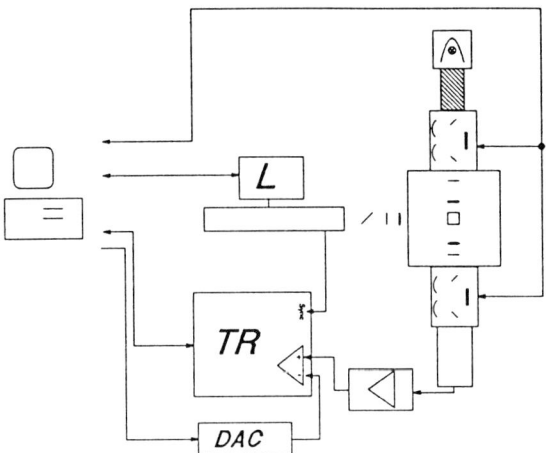

Figure 1: Sketch of the apparatus for the time-resolved absorption measurements. (L) pulsed laser (wavelength: $\lambda_{ex} = 532\ nm$, duration: $t = 8\ ns$, energy: $E = 120\ mJ$); (TR) transient recorder (sampling rate: $100\ MHz \ldots 2.5\ kHz$, resolution: 10 bit, memory: 512 kword); time constant of detector and electronics: $\tau \approx 20\ ns$.

Table 1: Samples

pm: PM suspension in deionized water			
pmP: pH-indicator pyranin added		pmN: pH-indicator neutral red added	
pmPi: imidazole buffer added	pmPp: phosphate buffer added	pmNi: imidazole buffer added	pmNp: phosphate buffer added

recorder (details in fig. 1) suits well our need of measuring small absorbance changes containing time constants that are distributed over up to 8 orders of magnitude. The high density of sampling points leads to a high signal-to-noise ratio when neighbouring points are averaged. A microcomputer controls the various devices, calculates and displays the change of absorbance, and allows averaging of numerous traces for further improvement of the signal-to-noise ratio.

Sample preparation

Seven kinds of BR-suspensions (concentration: $13.3\ mg/ml$ BR) were prepared containing different pH-indicating dyes and pH-buffers, respectively (details in table 1). The ratio of molecules in suspension was optimized in initial measurments and adjusted to a molar ratio of $BR : pH\text{-}indicator : buffer = 1 : 1 : 2$ for the experiments presented.

The pH of all suspensions was adjusted to pH 7.5. Seven glass slides ($13 \times 25\ mm^2$) were covered with the respective suspensions, giving allways a BR-surface density of $425\ \mu g/cm^2$ and quickly dried under a stream of nitrogen. These samples were thereafter equilibrated in a closed chamber for at least $15\ h$ at each relative humidity (r.h.) in the chamber. A defined r.h. in the chamber was achieved with a beaker containing water or certain saturated salt solutions at a defined temperature

(water: 100 per cent r.h., KCl: 85 per cent r.h., NaCl: 75 per cent r.h.).
The slides were then placed diagonally in closed $10 \times 10 \times 45$ mm^3 cuvettes containing a reservoir filled with the respective solution to keep the adjusted r.h.
At each relative humidity absorption spectra and time-resolved absorbance changes of the light adapted samples were recorded. The spectra obtained from samples containing pH-indicators were used to calculate the actual pH of the sample. Time-resolved absorbance changes were detected at $\lambda_m = 407$ nm, 457 nm, 578 nm, and 656 nm. The samples were thermostated at $T = 22°C$.

RESULTS

Static pH in PM-stacks at reduced relative humidity

The ability to measure the pH in PM-stacks is important not only for this study, but also for all investigations of PM on solid supports at reduced humidities, e.g., diffraction studies or FTIR investigations. The pH of the intermembrane space of the films was calculated from spectra of the samples containing pyranine, i.e., pmP, $pmPi$, $pmPp$. It was found to be pH 7.0 at 100 per cent r.h. and to decrease to pH 6.8 at 85 and 75 per cent r.h., respectively. The buffered samples ($pmPi$, $pmPp$) had the same pH as the unbuffered ones (pmP).

Time resolved photocycle and pumping dynamics in PM-films at reduced relative humidity

The BR-photocycle intermediate M was measured using the transient absorption spectrometer (fig. 1) at $\lambda_m = 407$ nm with samples containing PM only, i.e., pm (fig. 2, upper traces). The overall decay of the M-intermediate is accellerated by a factor of 2 in the 100 per cent r.h. film sample ($\tau_{decay} \approx 2$ ms) as compared with a PM-suspension (data not shown), whereas the time constants[1] for the rise are found to be similar ($\tau_{rise} \approx 50 \mu s$). When reducing r.h. from 100 to 85 per cent the rise time of the M-intermediate in the film samples is nearly unchanged. The M-intermediate decay time is slightly slowed down ($\tau_{decay} \approx 5$ ms) at 85 per cent r.h.. However, at 75 per cent r.h., both, the rise and decay time of PM-films are increased ($\tau_{rise} \approx 0.10$ ms, $\tau_{decay} \approx 14$ ms). In addition, a very slow decay component develops.
The O-intermediate was observed in all samples, i.e., pm, pmP, $pmPi$, $pmPp$, pmN, $pmNi$, $pmNp$ at $\lambda = 656$ nm and at a r.h. of 100 and 85 per cent. At 75 per cent r.h., however, no absorption changes reflecting the O-intermediate were detected.
The transient changes of the pH-indicator-signal were derived by subtracting corresponding traces at $\lambda_m = 457$ nm of samples with and without pH-buffer:

$$\text{pH-indicator-signal} = pm\ I - pm\ I\ b\ , \text{where } I \in \{P, N\}, b \in \{i, p\}$$

The use of the pH-indicator neutral red instead of pyranine results in a similar, but smaller difference pH-indicator transient (data not shown), due to its smaller change of extinction coefficent ϵ per pH unit at $\lambda_m = 457$ nm. Both, pH-indicators and pH-buffers do not affect the photocycle in the film, i.e., absorbance changes of the $pm\ I$ and $pm\ I\ b$-samples do not differ from the pm-sample in wavelength regions where the respective dye does not absorb.
The transient change of the pH-indicator signal (fig. 2, lower traces) is proportional to the change of protonation of the pH-dye. A signal is observed at every investigated r.h., even at 75 per cent r.h., reflecting the release and reuptake of a proton by the PM. The kinetics follow closely the absorbance change at $\lambda_m = 407$ nm. It should be noted, that the pH-dye is located close to the PM-surface. However, from these results one cannot decide from which side of the PM the

---

[1] All time constants τ are dominating points of inflection of absorbance traces over the logarithmic time axis

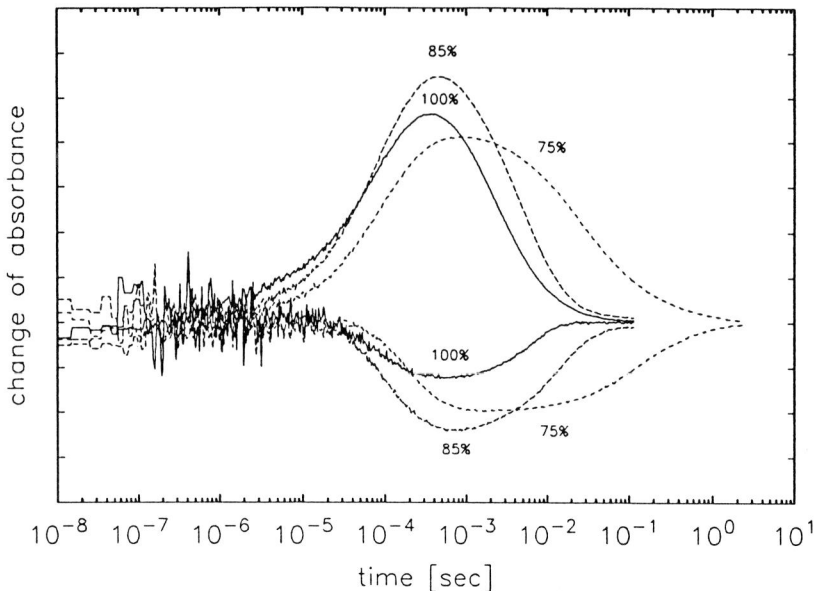

Figure 2: Upper traces: BR-photocycle intermediate M at 100 %, 85 % and 75 % relative humiditiy (absorbance change at $\lambda_m = 407\ nm$). Lower traces: respective absorbance change of pH-indicator pyranine at $\lambda_m = 457\ nm$, proportional to its change of protonation. (A PM-film containing pyranine and additional phosphate buffer served as reference.)

reuptake of the proton takes place. Future experiments will go to even lower r.h., to clarify how many water molecules are necessary for the function of BR.

References

[1] Korenstein, R. & Hess, B. (1977): Hydration effects on the photocycle of bacteriorhodopsin in thin layers of purple membrane. *Nature* **270**, 184-1186.

[2] Váró, G. & Keszthelyi, L. (1983): Arrhenius parameters of the bacteriorhodopsin photocycle in dried oriented samples. *Biophys. J.* **43**, 47-51.

[3] Váró, G. & Lanyi, J.K. (1991): Distortions in the photocycle of bacteriorhodopsin at moderate dehydration. *Biophys. J.* **59**, 313-322.

[4] Hildebrand,P. & Stockburger, M. (1984): Role of Water in Bacteriorhodopsins Chromophore: Resonance Raman Study. *Biochemistry* **23**, 5539-5548.

[5] Papadopoulos, G., Dencher, N.A., Zaccaï, G. & Büldt,G. (1990): Water molecules and exchangeable hydrogen ion at the active centre of bacteriorhodopsin localized by neutron diffraction. *J. Mol. Biol.* **214**, 14-19

The surface of the purple membrane: a transient pool for protons?

Joachim Heberle and Norbert A. Dencher

Hahn-Meitner-Institut, BENSC-N1, Glienicker Strasse 100, W-1000 Berlin 39, Germany

INTRODUCTION

Protons play a fundamental role in bioenergetics. H^+-pumps generate H^+-gradients across cell membranes. These H^+-gradients are utilized by ATPsynthases to produce the ubiquitous energy storage ATP. But still certain aspects of Mitchell's chemiosmotic hypothesis are vehemently discussed (see, e.g., *Prats et al., 1987* and *Kasianowicz et al., 1987*). In particular, a central question is not answered yet: Can the membrane surface act as a storage for H^+? Such a H^+-pool may be used to drive ATP-synthesis without the need of large pH-differences between the bulk phases on both sides of the membrane.

In this article we will provide experimental evidence which lead us to propose a **transient H^+-pool** along the surface of the purple membrane (PM) of *halobacteria*. We have employed flash absorption spectroscopy in combination with pH indicators *(Dencher et al., 1991)* in order to monitor H^+-transfer along the membrane not only in time but also - by site-specific positioning of various pH indicators - in space. This system (PM with the pH indicators) is then subjected to different physico-chemical alterations to demonstrate the necessity of postulating a transient H^+-diffusion barrier at the PM.

PROTON TRANSFER ACROSS AND ALONG THE PURPLE MEMBRANE

Bacteriorhodopsin (BR) residing in the PM was labeled with the pH indicator fluorescein by conjugation to Lys-129 at the extracellular side of BR as described in *Heberle & Dencher (1990, 1992)*. To study H^+-transfer reactions of BR in PM, we applied transient absorption spectroscopy with a time-resolution of 20 ns (a detailed description of the laser flash apparatus will be given in *Thiedemann et al., this book*). Representative results are depicted in Fig. 1. The upper trace corresponds to formation and decay of the photocycle intermediate M. The lower traces in Fig. 1 represent absorbance changes detected by the pH indicator fluorescein covalently bound to BR, and pyranine residing in the surrounding aqueous bulk phase. Due to the H^+-pumping activity of BR, the pH is transiently lowered in the environment of the respective pH indicator which is indicated by a decrease in absorbance of both pH probes. However, *fluorescein bound to the surface of BR responds to the pH decrease about 1 order of magnitude faster than pyranine*. Subsequently, fluorescein indicates the reestablishment of the original pH by two relaxation processes whereas for pyranine only one exponential is

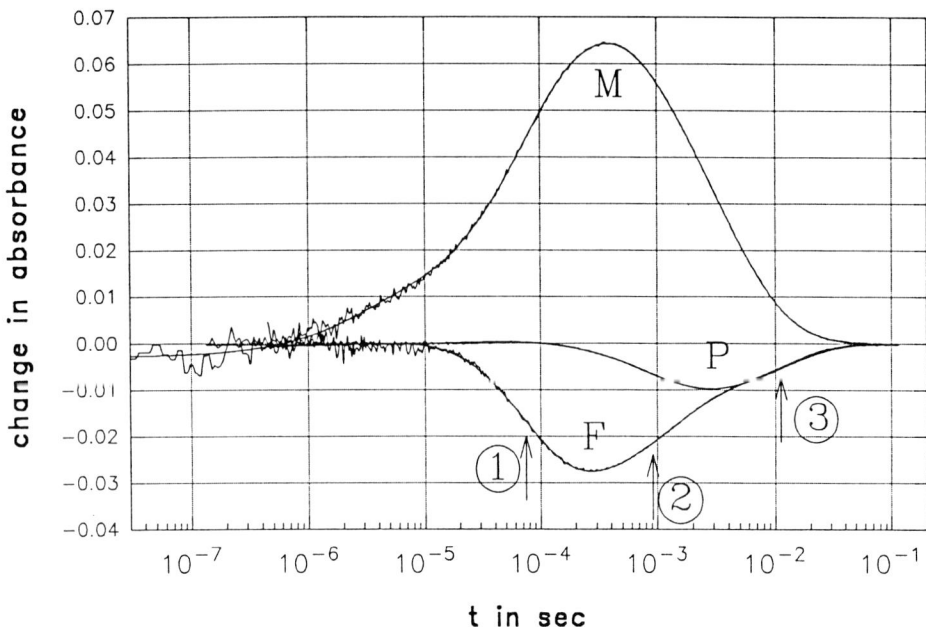

Fig. 1: Comparison of the time course of pH-change with rise and decay of the M intermediate. M (c_{BR} = 11 μM) was monitored at 409 nm while BR bound fluorescein (c_F = 6 μM) and pyranine in the aqueous bulk phase (c_P = 50 μM) were measured at 489 nm and 457 nm, respectively. T = 22 °C; pH = 7.5; 150 mM KCl.

Smooth lines result from fitting the data to sums of exponentials (at longer times the data traces are indistinguishable from the fitting curve). Arrows correspond to the H^+-transfer steps detected by the pH indicators. The absorbance change of fluorescein (F) is expanded by a factor of 4 to illustrate the common features with the response of pyranine (P).

sufficient. The molecular scenario underlying the absorbance changes of both pH indicators is described as follows: (1) Light-induced H^+-release from the interior of the protein to the extracellular surface of BR which can only be detected by surface bound fluorescein. The pumped H^+ appears at the membrane surface concomitantly with the rise of the photocycle intermediate M. (2) The established H^+-gradient between membrane surface and aqueous bulk dissipates, concurrently indicated by a partial increase in absorbance of fluorescein and a decrease in absorbance of pyranine (surface/bulk transfer of H^+). At this point, it is important to emphasize that *a pH indicator that resides in the aqueous bulk does not detect the kinetics of H^+-pumping within the protein but the subsequent H^+-transfer from the surface into the bulk*. (3) H^+-reuptake by BR is revealed by the complete increase in absorbance of fluorescein and pyranine, respectively.

In this article, we will focus on the surface/bulk transfer of H^+ (2 in Fig. 1). The difference in time between the initial response of fluorescein and pyranine is associated to the dwell time for H^+ at the PM surface of several hundred microseconds. This dwell time also manifests when the stoichiometry of H^+ pumped per photoactivated BR is calculated. We confirm earlier results of 1 H^+/BR when pyranine (as a pH indicator of the aqueous bulk medium) is titrated. However, the **apparent** stoichiometry calculated from the absorbance changes of surface bound fluorescein is 3-4 H^+/BR - the result of the transient dwell time of H^+ at the PM surface!

By increasing the concentration of (the aqueous bulk pH indicator) pyranine, the average distance between pyranine and the PM surface is reduced. Thus, the time for the protonation reaction of pyranine might be decreased, too. We varied the pyranine concentration from 5 to 150 µM (data not shown) but the time constant for the protonation of pyranine is only slightly accelerated. This acceleration can be attributed to a 'buffer effect' (see below) of pyranine itself. The deprotonation reaction of pyranine corresponding to the H^+-reuptake of BR, is not influenced.

Increasing concentrations of cations like K^+ will lead to a narrowing of the diffuse double-layer along the PM surface. Consequently, the PM-pyranine distance decreases. However, in experiments where the K^+-concentration was varied between 0.01 and 500 mM, no significant acceleration in the 'response-time' of pyranine is observed.

The extraordinary fast diffusion of H^+ in water is hampered by the presence of fixed buffers along biological membranes (amino acids, lipids, etc.). Water soluble buffers compensate this effect due to their mobility *(Junge & McLaughlin, 1987)*. As a result, the apparent rate of H^+-diffusion can be accelerated by increasing the concentration of mobile buffers *(Grzesiek & Dencher, 1986; Gutman & Nachliel, 1990)*. This 'buffer effect' on the surface/bulk H^+-transfer reaction is reflected both in the signals of pyranine and of surface bound fluorescein. The H^+-transfer from the surface into the aqueous bulk is accelerated by a factor of 3 at least when 1 mM imidazole is added to the PM suspension. The H^+-release reaction from the interior of BR to the PM surface, which can only be detected by surface bound fluorescein, is only slightly affected by the presence of imidazole. The activation energy of this diffusion process [E_a = 35 kJ/mol, *(Heberle & Dencher, 1992)*] suggests that not only water molecules are involved. Protonable amino acids like Asp 85 *(Heberle et al., 1991)* participate in the H^+-extrusion pathway of BR.

The experiments described above lead us to the conclusion that at least a **transient** H^+-pool at the surface of the PM exists. This shall be emphasized by the following calculation. The effective H^+-diffusion coefficient in a PM suspension is $D_{H^+}^{eff} = 3.4 \times 10^{-7}$ cm^2/s *(Heberle & Dencher, 1992)*. The time constant for the surface/bulk transfer is about 800 µs at 22 °C. With $x^2 = 2Dt$, the distance from the PM surface to pyranine residing in the bulk water phase would be more than 200 nm. This value is far too high for being acceptable. Hence, a different physico-chemical mechanism has to be involved.

But what is the origin of such a transient H^+-pool? It was pointed out in several investigations that the water molecules along biological membranes are highly structured (see *Tocanne & Tessie, 1990; Cevc, 1990*, and literature cited therein). To approximate surface and bulk water structure we have performed measurements on PM in ice *(Heberle & Dencher, 1990)*. There, pyranine exhibits the same response as surface bound fluorescein (detection of the released H^+ concomitant with the rise of M). As there is no faster H^+-diffusion in ice than in water *(Pines & Huppert, 1985)* - in contrast to *Eigen & deMaeyer (1958)* - it is concluded that H^+-transfer from the PM surface into the bulk will not limit the protonation rate of pyranine in ice. We suggest that the water structure along the PM may be responsible for the dwell time of H^+ at the surface. To clarify this point, measurements with the pH jump method *(Gutman, 1984)* are currently in progress.

The above calculation for H^+-diffusion in PM suspension resulted in an average distance the pumped H^+ can diffuse (about 200 nm) during the dwell time at the surface. This distance is in the order of magnitude of the distance from the center of a PM patch to the H^+-ATP-synthase within the cell membrane. 'Localized' coupling of the H^+-generator (BR) and the H^+-consumer (ATP-synthase) is therefore not excluded, at least. We will investigate this problem by co-reconstitution of BR and H^+-ATPsynthase in lipid vesicles. By site-directed positioning of pH indicators we will be able to determine H^+-transfer steps both spatially and temporally.

REFERENCES

Cevc, G. (1990): Membrane electrostatics. Biochim. Biophys. Acta 1031: 311-382.

Dencher, N.A., Heberle, J., Bark, C., Koch, M.H.J., Rapp, G., Oesterhelt, D., Bartels, K., and Büldt, G. (1991): Proton translocation and conformational changes during the bacteriorhodopsin photocycle: Time-resolved studies with membrane-bound optical probes and X-ray diffraction. Photochem. Photobiol. 54: 881-887.

Eigen, M., and deMaeyer, L. (1958): Self-dissociation and protonic charge transport in water and ice. Proc. Roy. Soc. London 247: 505-533.

Grzesiek, S., and Dencher, N.A. (1986): Time-course and stoichiometry of light-induced proton release and uptake during the photocycle of bacteriorhodopsin. FEBS Lett. 208: 337-342.

Gutman, M. (1984): The pH jump: Probing of macromolecules and solutions by a laser induced, ultrashort proton pulse. Theory and applications in Biochemistry. Method. Biochem. Anal. 30: 1-105.

Gutman, M., and Nachliel, E. (1990): The dynamic aspects of proton transfer processes. Biochim. Biophys. Acta 1015: 391-414.

Heberle, J., and Dencher, N.A. (1990): Bacteriorhodopsin in ice: Accelerated proton transfer from the purple membrane surface. FEBS Lett. 277: 277-280.

Heberle, J., Oesterhelt, D., and Dencher, N.A. (1991): Replacement of Asp-85 by Glu in bacteriorhodopsin kinetically decouples Schiff's Base deprotonation and H^+-ejection. Biophys. J. 59: 327a.

Heberle, J., and Dencher, N.A. (1992): Surface-bound optical probes monitor proton translocation and surface potential changes during the bacteriorhodopsin photocycle. Proc. Natl. Acad. Sci. 89: (in press).

Junge, W., and McLaughlin, S. (1987): The role of fixed and mobile buffers in the kinetics of proton movement. Biochim. Biophys. Acta 890: 1-5.

Kasianowicz, J., Benz, R., Gutman, M., and McLaughlin, S. (1987): Reply to: Lateral Diffusion of Protons along Phospholipid Monolayers. J. Membrane Biol. 99: 227.

Pines, E., and Huppert, D. (1985): Kinetics of proton transfer in ice via the pH-jump method: Evaluation of the proton diffusion rate in polycrystalline doped ice. Chem. Phys. Lett. 116: 295-301.

Prats, M., Tocanne, J.F., and Teissie, J. (1987): Lateral Diffusion of Protons along Phospholipid Monolayers. J. Membrane Biol. 99: 225-226.

Tocanne, J.-F., and Tessie, J. (1990): Ionization of phospholipids and phospholipid-supported interfacial lateral diffusion of protons in membrane model systems. Biochim. Biophys. Acta 1031: 111-142.

Fluorescence studies on the surface potential of deionized forms of purple and bleached membranes*

Francesc Sepulcre and Esteve Padrós

Unitat de Biofísica, Departamento de Bioquímica i de Biologia Molecular, Facultat de Medicina, Universitat Autònoma de Barcelona, 08193 Bellaterra, Barcelona, Spain

INTRODUCTION

Native purple membrane possess a net negative surface potential, due to the presence of several negatively charged groups and about 5 bound cations (for recent review, see Jonas *et al.*, 1990). It is known that bleaching of bacteriorhodopsin produces conformational changes involving tertiary structure without altering noticeably its secondary structure (Becher & Cassim, 1977). All these changes are reversed upon regeneration with retinal, but there are some results which show that removal of Ca^{2+} and Mg^{2+} of the bleached membrane causes irreversible conformational changes. In other words, fully native purple membrane cannot be restored by pigment and/or cation addition to bleached deionized membrane (Duñach *et al.*, 1986; Chang *et al.*, 1988). There are many experiments supporting the specific binding of divalent cation to purple membrane (Chang *et al.*, 1986; Duñach *et al.*, 1987; 1988). Therefore, these irreversible conformational changes in bleached deionized membrane suggest that these sites are modified.

In this work we examine the role of the retinal chromophore in cation binding by comparing the surface potential values of purple and bleached membranes as well as their deionized forms, as a function of pH. For the surface potential measures we have used the fluorescent probe 1-anilinonaphatalene-8-sulfonic acid (ANS)[1]. This probe binds noncovalently to membrane proteins as well as lipids and its fluorescence is extremely sensitive to the changes in the probe environment (Slavik, 1982).

---

*This work was supported by *DGICYT* (PB89-0301) and *CIRIT* (AR91-264)

[1] Abbreviations: ANS, 1-anilinonaphatalene-8-sulfonic acid; BR, bacteriorhodopsin

MATERIALS AND METHODS

The purple membrane was isolated from *H. halobium* strain S9 (Oesterhelt & Stoeckenius, 1974). Deionized membranes were prepared by passage of purple or bleached membrane suspensions through a well-washed cation-exchange Dowex AG-50W column. A separate aliquot of deionized membrane samples was always used for pH measurements in order to avoid contamination by KCl leaks from the electrode (Duñach *et al.*, 1988). Bleaching of the purple membrane was effected by its illumination in the presence of 1M hydroxylamine, 4M NaCl at room temperature with a 150 W lamp. Fluorescence measurements were performed on a Perkin-Elmer 650-40 or a SLM Aminco 8000 fluorescence spectrophotometer at room temperature, with emission and excitation slits at 8 nm. ANS was excited at 370 nm and the emission observed at 464 nm. The membrane surface potential was calculated according to the Boltzman equation (Castle & Hubbell, 1976):

$$\Delta\Theta = (RT)/(ZF)\ln(P_1/P_2)$$

where R, T, Z and F are, respectively, the universal gas constant, the absolute temperature, the ion charge and the Faraday constant. P_1 and P_2 are the partition coefficients of ANS and were calculated from

$$P_i = \text{(bound ANS)/(free ANS)} = (I-I_0)/(I_m-I)$$

where I is the fluorescence intensity at a given pH, I_0 is the intensity in the absence of protein (ANS in water) and I_m is the maximun intensity, corresponding to all of the ANS bound to the protein (Flanagan & Hesketh, 1973). For the determination of I_m we have compared several graphical procedures, in order to avoid the errors produced in the double reciprocal plot representation [1/I against 1/[Prot] at a fixed ligand concentration] (Zierler, 1977). These graphical methods are: $\cosh^{-1}(1+I)$, $\sinh^{-1}(1/I)$ and $1/\sinh^{-1}(I)$ against 1/[BR] at a fixed ANS concentration (Rajkowski, 1990), and I against [ANS] at a fixed BR concentration (Zierler, 1977). A different I_m was obtained for purple, blue (deionized), bleached and bleached deionized membranes.

RESULTS AND DISCUSSION

The fluorescent probe ANS has been commonly used in biomembrane studies because its fluorescence intensity depends pronouncedly on the nature of the solvent, especially on its polarity. Because quantum yield of ANS in water is very low (Slavik, 1982), only the bound ANS contributes to the observed fluorescence signal. On the other hand, as the ANS negative charge remains unaffected within the pH range used, the changes in the fluorescence signal only reflect changes in the membrane surface potential.

One of the difficulties inherent with this method is the calculation of the maximum fluorescence intensity value, I_m. This point has been adressed by several authors (Zierler, 1977; Panjehshahin *et al.*, 1989; Rajkowski, 1990), which have pointed out that double reciprocal plots can develop upward concave curvature as the protein concentration increases. Therefore, a linear extrapolation of these plots can give overestimated I_m values. In order to avoid this problem, we have compared the I_m values obtained from the graphical procedures described in Materials and Methods. There are no significant differences between these values, with the exception of the plots of $1/I$ and $1/\sinh^{-1}(I)$ against $1/[BR]$, which gave different values, in agreement with the above authors. Therefore, the surface potential was estimated from the I_m value obtained from extrapolation of the plot of $\cosh^{-1}(1+I)$ against $1/[BR]$ (Rajkowski, 1990).

At pH 3.5 all the membrane samples have the ANS fluorescence maximum intensity located at the same position (about 464 nm), which is similar to the reported values for ANS bound to biological membranes (Slavik, 1982). Therefore, the environment of the ANS-bound molecules in each of these forms must be similar, despite the presence or absence of retinal and cations. Similarly, the fact that the fluorescence intensity corresponding to the bleached membrane is higher than the purple or blue samples (results not shown), can be explained by energy transfer between the ANS chromophore and the retinal.

Fig. 1A presents the plots corresponding to purple and bleached membranes, and Fig. 1B presents the plots corresponding to blue (deionized) and bleached deionized membranes. For purple and blue membranes the results are in agreement with previous findings using ESR techniques (Duñach et al., 1988). The plot corresponding to the bleached membrane is similar to that of the purple membrane in the whole pH interval, in agreement with the data of Ehrenberg & Meiri (1983) at neutral pH. The surface potential for bleached deionized membrane between pH 3.5 and 6.0 has a behaviour like the purple or the bleached membranes, and is less negative than the blue membrane. Above pH 6.0, the surface potential continues to decrease until the

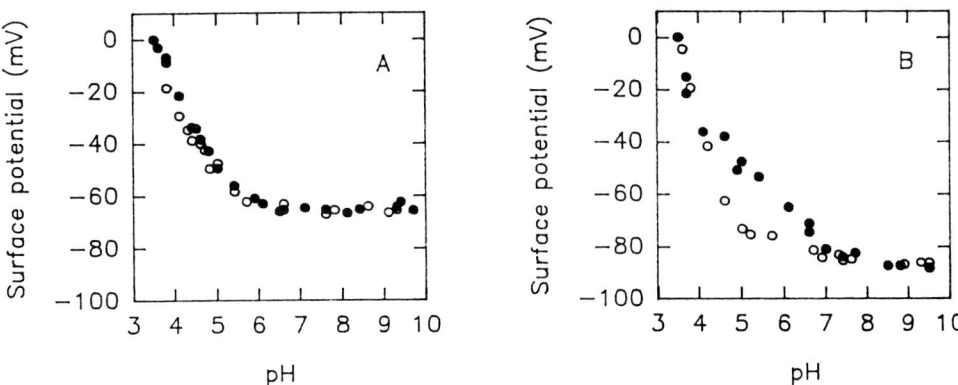

Fig 1. Influence of bleaching and/or cation removal on purple membrane surface potential. Protein concentration, 10 μM; ANS concentration, $7.4 \cdot 10^{-6}$ M. Surface potential value at pH 3.5 was taken as an arbitrary origin of the surface potential changes. (A) purple (O) and bleached (●) membranes. (B) blue (O) and bleached deionized (●) membranes.

value corresponding to the blue membrane is reached. This behaviour strongly suggest that some of the ionizable groups responsible of cation binding are still protonated in the bleached deionized membrane between pH 5.5 and 7.5, whereas in the blue membrane they are deprotonated. Thus, it is likely that retinal deprivation produces an increase in the pK_a of these groups.

The present work gives support to the idea that changes in the retinal environment can alter the surface potential through changes in the geometry and the properties of the cation binding sites. It also gives further suport to the presence of specific cation binding sites in the native purple membrane.

REFERENCES

Becher, B., and Cassim, J.Y. (1977): Effects of bleaching and regeneration on the purple membrane structure of *Halobacterium halobium*. *Biophys. J.*, 19, 285-297.

Castle, J.D., and Hubbell, W.L. (1976): Estimation of membrane surface potential and charge density from the phase equilibrium of a paramagnetic amphiphile. *Biochemistry* 15, 4818-4831.

Chang, C-M., Jonas, R., Melchiore, S., Govindjee, R., and Ebrey, T.G. (1986): Mechanism and role of divalent cation binding of bacteriorhodopsin. *Biophys. J.*, 49, 731-739.

Chang, C-M., Jonas, R., Govindjee, R., and Ebrey, T.G. (1988): Regeneration of blue and purple membranes from deionized bleached membranes of *Halobacterium halobium*. *Photochem. Photobiol.* 47, 261-265.

Duñach, M., Seigneuret, M., Rigaud, J.-L., and Padrós, E. (1986): The relationship between the chromophore moiety and the cation binding sites in bacteriorhodopsin. *Bioscience Rep.* 6, 961-966.

Duñach, M., Seigneuret, M., Rigaud, J.-L., and Padrós, E. (1987): Characterization of the cation binding sites of the purple membrane. Electron spin resonance and flash photolysis studies. *Biochemistry* 26, 1179-1186.

Duñach, M., Seigneuret, M., Rigaud, J.-L., and Padrós, E. (1988): Influence of cations on the blue to purple transition of bacteriorhodopsin. *J. Biol. Chem.* 263, 17378-17384.

Ehrenberg, B., and Meiri, Z. (1983): The bleaching of purple membranes does not change their surface potential. *FEBS Lett.* 164, 63-66.

Flanagan, M.T., and Hesketh, T.R. (1973): Electrostatic interactions in the binding of fluorescent probes to lipid membranes. *Biochim. Biophys. Acta*, 298, 535-545.

Jonas, R., Koutalos, Y., and Ebrey, T.G. (1990): Purple membrane: surface charge density and the multiple effect of pH and cations. *Photochem. Photobiol.*, 52, 1163-1177.

Oesterhelt, D., and Stoeckenius, W. (1974): Isolation of the cell membrane of *Halobacterium halobium* and its fractionation into red and purple membranes. *Methods Enzymol.*, 31, 667-678.

Panjehshahin, M.R., Bowmer, C.J., and Yates, M.S. (1989): A pitfall in the use of double-reciprocal plots to estimate the intrinsic molar fluorescence of ligands bound to albumin. *Biochem. Pharmacol.* 38, 155-159.

Rajkowski, K. M. (1990): Comparison of graphical procedures for estimating the intrinsic molar fluorescence of protein-bound drugs for drug-binding studies. *Biochem. Pharmacol.* 39, 895-900.

Slavik, J. (1982): Anilinonaphthalene sulfonate as a probe of membrane composition and function. *Biochim. Biophys. Acta* 694, 1-25.

Zierler, K. (1977): An error in interpretation of double-reciprocal plots and Scatchard plots in studies of binding of fluorescent probes to proteins, and alternative proposals for determining binding parameters. *Biophys. Struct. Mechanism* 3, 275-289.

Rhodopsins : from ion pumps to specialized photoreceptors

Vladimir P. Skulachev

Department of Bioenergetics, A.N. Belozersky Institute of Physico-Chemical Biology, Moscow State University, Moscow 119899, Russia

BACTERIORHODOPSIN: THE MECHANISM OF H^+ PUMPING

Proton pumping by bacteriorhodopsin (bR) involves two main steps - the transfer of H^+ ions (i) from the protonated Schiff base (localized in the middle part of this membrane) to the outer aqueous space, and (ii) from the cytosol to the deprotonated Schiff base. In these two steps the H^+ traverses a rather hydrophilic and a very hydrophobic region of the bR molecule, respectively. Between the outer membrane surface and the Schiff base, there are four charged amino acids and no valine, leucine and isoleucine. On the other hand, five leucines, valine and only one charged amino acid (Asp-96) seem to be localized between the Schiff base and the cytoplasmic membrane surface (Henderson et al., 1990). This means that the dielectric constant in the regions of the outward H^+-conducting pathway is much higher than that of the inward pathway, accounting for the fact that the contribution of the outward H^+ pathway to the light-induced $\Delta\Psi$ generation by bR is as little as 20% whereas that of the inward one is equal to 80% (Drachev et al., 1978). It was suggested (Skulachev, 1992) that H^+ movement through the hydrophobic part of the bR is organized in such a way that a cleft is formed in the hydrophobic part of the bR molecule so that Asp-96 carboxylate becomes accessible to solutes (Fig. 1). The light-induced conformation change, bR --> bR*, resulting in the cleft formation, is assumed to disrupt a hydrogen bond between (Asp-96)-COOH and nucleophilic group X, allowing for H^+ transfer from Asp-96 to the Schiff base. Regeneration of (Asp-96)-COOH ··· X facilitates formation of the bR ground state (bR* --> bR). The formation of the cleft explains why light sensitizes bR to hydroxylamine.

BACTERIORHODOPSIN: THE BRIGHT LIGHT PHOTOSENSOR

As it has recently been shown (Bibikov et al., 1991), the *H.halobium* mutant lacking bR, halorhodopsin (hR), and sensory rhodosins (sR) I and II, is "blind" whereas its transformant containing the bR gene becomes photosensitive. The transformant shows lower sensitivity to dim light and higher sensitivity to bright light than a sR-containing mutant possessing no bR and hR. Effects of agents that change

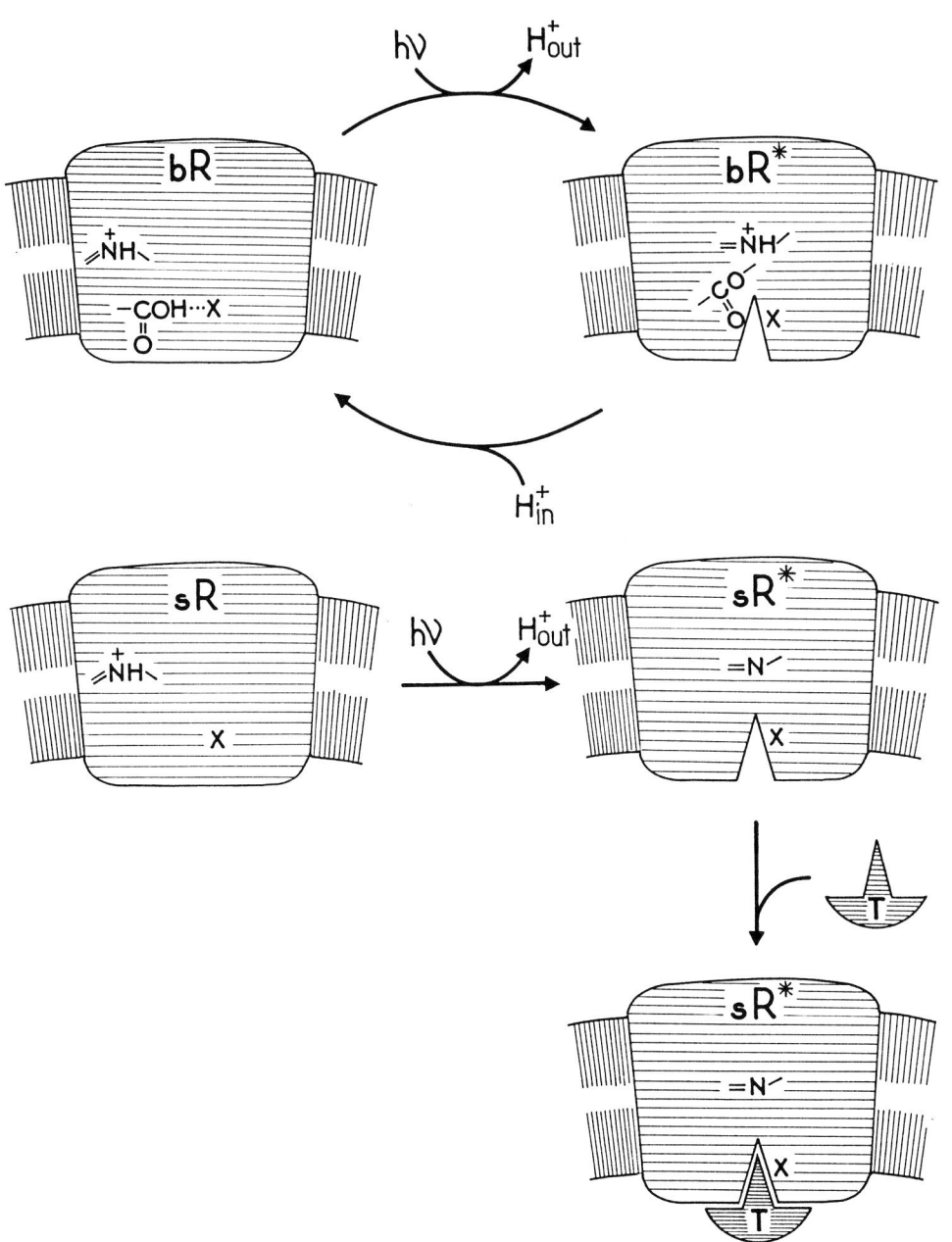

Fig. 1. bR-mediated H^+ pumping and sR-mediated photoreception. bR^* and sR^*, the photocycle intermediates (N for bR and M for sR). $-\underset{\underset{O}{\|}}{C}OH$, Asp-96 carboxylic group which in bR forms a hydrogen bond with nucleophilic group X. T, the first protein of the signal amplification cascade (in the visual system, transducin).

the $\Delta\bar{\mu}H$ level clearly indicate that phototaxis in the transformant is mediated by a $\Delta\bar{\mu}H$ receptor ("protometer") which monitors the $\Delta\bar{\mu}H$ level generated by bR and other H^+ pumps in *H.halobium*.

SENSORY RHODOPSINS: THE DIM LIGHT PHOTOSENSORS

It is assumed that the initial steps of sR-linked photoreception are similar to those of bR-linked photoreception up to the stage of the conformation change (sR^* formation). sR^*, in contrast to bR^*, is rather long-lived, to have time to be recognized by the first protein of the amplification cascade:

dim light $\longrightarrow$ sR $\longrightarrow$ sR^* $\longrightarrow$ amplification cascade $\searrow \cdots \longrightarrow$ flagellum
bright light $\longrightarrow$ bR $\xrightarrow{bR^*}$ $\Delta\bar{\mu}H$ $\longrightarrow$ "protometer" $\nearrow$

Due to the amplification cascade, the sR systems are much more sensitive than bR (Bibikov et al., 1991). $\Delta\bar{\mu}H$ is not involved in sR-mediated photoreception (Oesterheld and Marwan, 1987). Apparently, when sR evolved from bR a mutation occur in the inward H^+-conducting pathway, the event resulting in the inhibition of the terminal steps of the photocycle. In fact, Asp-96 is absent from sR (Fig.1).

VISUAL RHODOPSIN

Visual rhodopsin (R) seems to operate like sR at least in dim light. On the other hand, some features resembling bR-mediated photoreception are also inherent in its mechanism. Thus, operation of the amplification cascade leads to a change in the membrane potential ($\Delta\Psi$) which is monitored by a $\Delta\Psi$-sensitive system ("voltmeter"). It should be stressed that the *H.halobium* "protometer" deals, first of all, with the $\Delta\Psi$ constituent of $\Delta\bar{\mu}H$ since ΔpH changes always proceed much more slowly than those of $\Delta\Psi$.

Another apparent similarity of R and bR photoreception is that R, like bR, is competent for photopotential generation. I mean formation of the early receptor potential (ERP). As was found in our group (Drachev et al., 1981), the kinetics of the initial stages of $\Delta\Psi$ formation which accompany the laser flash-induced single turnover of bR and R are similar, resembling very much the ERP. A saturating flash was shown to induce $\Delta\Psi$ up to 30 and 80 mV in the case of R and bR, respectively. The direction of ERP coincided with that of the late receptor potential (LRP), when develops due to the operation of the amplification cascade (hyperpolarization of the plasma membrane in vertebrates and its depolarization in invertebrates). This means that a strong and sudden increase in the light intensity might induce the excitation of the photoreceptor cell by two entirely differing mechanisms triggered by the R molecules localized in the plasma membrane: one, via the cascade, and another, directly initiated by ERP, without the cascade (it should be mentioned that in cones, responsible for bright light reception, it is the plasma membrane which contains all the R pool). The direct mechanism might have some advantages under high back-ground illumination, when the cascade is saturated. It should be noted that in invertebrates (i) excitation is induced by the $\Delta\Psi$ decrease, and (ii) ERP is oppositely directed to that in vertebrates, being linked to absorption of the second photon (the photoinduced R^* --> R transition):

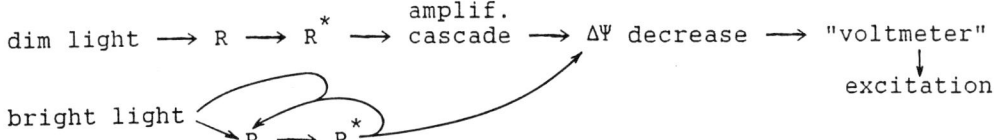

Thus in invertebrates, ERP and the cascade-linked LRP represent two different photoreactions i.e. (i) $R + h\nu_1 \rightarrow R^* + LRP$ and (ii) $R + h\nu_2 \rightarrow R + ERP$. Due to difference in the R and R^* spectra, ERP and LRP can be easily distinguished. As shown by Pack and Lidington (1974), the *Drosophila* norp A mutant, lacking LRP because of a lesion in the cascade, exhibited a positive waveform in the electroretinogram in response to a bright orange light flash ($R^* \rightarrow R$ transition), added after blue pre-illumination causing $R \rightarrow R^*$ transition. This event proved to be a summation of retina and lamina components related to ERP (Minke and Kirschfeld, 1980; Stephenson and Pack, 1980; Gagne et al., 1989). Thus it seems probable that at least in invertebrates, ERP can participate in photoreception at high light intensities in agreement with the concept developed in our group (Skulachev, 1982).

REFERENCES

Bibikov,S.I., Grishanin,R.N., Marwan,W., Oesterhelt,D. and Skulachev,V.P. (1991): The proton pump bacteriorhodopsin is a photoreceptor for signal transduction in *Halobacterium halobium*. FEBS Lett. 295, 223-226.

Drachev,L.A., Kalamkarov,G.R., Kaulen,A.D., Ostrovsky,M.A. and Skulachev,V.P. (1981): Fast stages of photoelectric processes in biological membranes. II. Visual rhodopsin. Eur.J.Biochem. 117, 471-481.

Drachev,L.A., Kaulen,A.D. and Skulachev,V.P. (1978): Time resolution of the intermediate steps in the bacteriorhodopsin-linked electrogenesis. FEBS Lett. 87, 161-167.

Gagne,S., Roebroek,J.G.H. and Stavenga,D.G. (1989): Enigma of early receptor potential in fly eyes. Vision Res. 29, 1663-1670.

Henderson,R, Baldwin,J.M., Ceska,T.A., Zemlin.F., Beckmann,F. and Downing,K.H. (1990): Model for the structure of bacteriorhodopsin based on high-resolution electron cryo-microscopy. J.Mol.Biol. 213, 899-929.

Minke,B., Hochstein,S. and Hillman,P. (1973): Early receptor potential evidence for the existence of two thermally stable states in the barnacle visual pigment. J.Gen.Physiol. 62, 87-104.

Oesterhelt,D. and Marwan,W. (1987) Change in membrane potential is not a competent of photophobic transduction chain in *Halobacterium halobium*. J.Bacteriol. 169, 3515-3520.

Pak,W.L. and Lidington,K.J. (1974): Fast electrical potential from a long-lived, long-wawelength photoproduct of fly visual pigment. J.Gen.Physiol. 63, 740-756.

Skulachev,V.P. (1982): The dual role of rhodopsin in vision: light-driven charge translocation and formation of long-lived photoproducts. FEBS Lett. 146, 244-254.

Skulachev,V.P. (1992): Chemiosmotic system in bioenergetics: H^+- and Na^+-cycles. Biosci.Reports (accepted).

Stephenson,R.S. and Pak,W.L. (1980): Heterogenic components of a fast potential in *Drosophila* compound eye and their relation to visual pigment conversion. J.Gen.Physiol. 75, 353-379.

FT-IR difference spectroscopy of halorhodopsin in the presence of different anions

Timothy J. Walter and Mark S. Braiman

Department of Biochemistry, University of Virginia Health Sciences Center, Charlottesville, VA 22908, USA

An important question regarding halorhodopsin (hR) is the location and type of residues which serve as transient chloride binding sites in the transmembrane domain. Several arginine residues have been proposed as the most likely candidates [Lanyi, 1990]. We present here Fourier-transform infrared (FTIR) spectroscopic evidence consistent with this idea. FTIR difference spectra corresponding to the hR→hL photoreaction were obtained by kinetically trapping the hL intermediate at low temperature, and also by collecting 5-ms time-resolved spectra at room temperature following initiation of the photocycle with a laser flash. This second approach has been applied previously to bR [Braiman et al., 1990] but not hR.

A difference band near 1690 cm^{-1} in the resulting hR→hL spectra can be assigned partly to arginine based on (i) its similarity to the characteristic group frequency of the guanidino $CN_3H_5^+$ antisymmetric stretch mode; (ii) a counterion dependence for the frequency of the *positive* component that is paralleled in salts of the model compound phenylguanidine; and (iii) similarity of the frequency downshifts observed for the ~1690- cm^{-1} bands in hR and in phenylguanidinium salts when their protons are exchanged for deuterons. Our results suggest that the hR→hL difference band near 1690 cm^{-1} arises in part due to transient displacement of an internal carboxylate (Asp or Glu) counterion from an arginine residue by a chloride ion.

MATERIALS AND METHODS

Purification and reconstitution. Halorhodopsin was purified from *H. halobium* strain OD2W (generously provided by Janos Lanyi) essentially as described earlier [Duschl et al., 1988]. The purified hR, with $A_{280}/A_{570} \approx 2$, was reconstituted into native *H. halobium* lipids, at a lipid:protein ratio of 5:1 (w/w). For FTIR spectroscopy, pelleted samples (in 25 mM Tris-Cl pH 7.2 containing saturated KCl or KI) were squeezed between sealed CaF$_2$ windows for both low-temperature and time-resolved measurements.

FTIR spectroscopy. Static FTIR difference spectra of the hR→hL photoreaction were measured at 250 K as described previously [Rothschild et al., 1988]. Time-resolved difference spectra were obtained at 293 K, using a stroboscopic technique [Braiman et al., 1990].

Preparation of phenylguanidine salts. Phenylguanidium carbonate (ICN Biomedicals) was dissolved in water (100 mg/mL) and titrated with HCl, HBr, or HI until liberation of CO$_2$ was complete. The resulting salts were then lyophilized. To make the deuterium-exchanged sample, the chloride salt was alternately dissolved in D_2O and lyophilized (3× total). After the final lyophilization of each sample, several mg of the oily residue were pressed to a thickness of ~10 μm between CaF_2 windows for FTIR spectroscopy.

RESULTS AND DISCUSSION

Assignment of difference bands near 1690 cm^{-1} to arginine. Figure 1 presents a comparison of hR→hL difference spectra obtained in the presence of Cl^- and I^-. The spectra are generally similar to the 250 K static hR→hL difference spectrum obtained previously at 4-cm^{-1} resolution [Rothschild et al., 1988]. However, there are several features with an interesting counterion dependence. A portion of the positive band at 1689 cm^{-1} in the Cl^- spectrum (Fig. 1A) is downshifted to 1680 cm^{-1} in I^- (Fig. 1B); the negative peak of the same difference band shifts by only ~2 cm^{-1}. These results are reproduced in the 293 K time-resolved hR→hL difference spectrum (Fig. 1C,D), although with the poorer spectral resolution the negative band of the hR·I^- spectrum is less pronounced and the splitting of the positive band is not resolved despite the clear downshift of its peak.

The only strong characteristic group vibrations in the 1680–95 cm^{-1} region that arise from chemical moieties found in hR are: (i) the peptide backbone's amide I modes; (ii) the $CN_2H_5^+$ antisymmetric stretches of arginine guanido groups, (iii) the carbonyl stretches of asparagine and glutamine side chains [Venyaminov & Kalnin, 1990]. A portion of the intensity in the 1695/1689 cm^{-1} difference band can be assigned to arginine, based on shifts observed when the hR→hL spectrum is measured in deuterated buffer. Most of the difference band intensity appears to downshift by ~90 cm^{-1}, resulting in new 1608 cm^{-1} negative and 1600 cm^{-1} positive bands (Fig. 1E). Asparagine and glutamine carbonyl stretch modes are expected to downshift only ~35 cm^{-1} upon deuteration [Chirgadze et al., 1975; Venyaminov & Kalnin, 1990]. Even smaller shifts are generally expected for deuterium-exchanged amide I vibrations [Susi, 1972]. However, the $CN_3H_5^+$ antisymmetric stretch of an arginine side chain has been shown to exhibit a considerably larger deuteration-induced shift than amide carbonyls ($\nu_{CN_3H_5^+} \simeq 1673$ cm^{-1}, $\nu_{CN_3D_5^+} \simeq 1608$ cm^{-1}) [Chirgadze et al., 1975; Venyaminov & Kalnin, 1990]. A crucial consideration in our assignment of hR→hL difference bands around 1690 cm^{-1} to arginine is that no strong characteristic group vibration of any other amino acid, nor of the retinal chromophore itself, can explain the downshifted peaks observed near 1605 cm^{-1} upon deuteration.

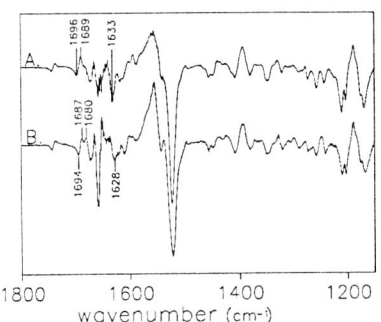
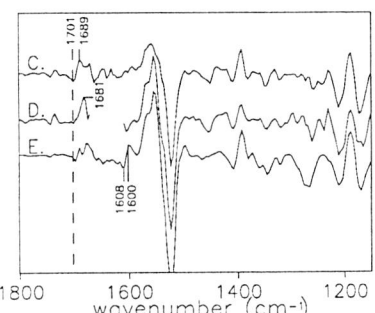

Figure 1: FTIR spectra of the hR→hL photoreaction. **A**, static difference spectrum, obtained at 250 K by using alternating 10-min periods in the light and dark, of a sample containing saturated KCl. **B**, same as **A** but with KI instead of KCl. **C** and **D**, same as **A** and **B**, but obtained with 8-cm^{-1} resolution at 293 K using a stroboscopic time-resolved method. These data were collected during a time window extending from 0.2–5.3 ms after each 530-nm flash. (The amide I region of **D** is blanked because of large noise peaks.) **E**, same as **C** but using deuterated buffer.

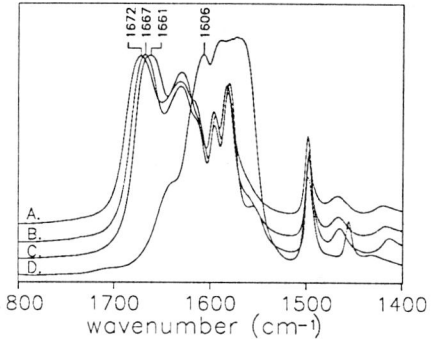

Figure 2: FTIR spectra of salts of phenylguanidine: **A**, chloride; **B**, bromide; **C**, iodide; **D**, chloride salt after deuterium exchange.

The counterion- and deuteration-induced shifts observed in the model compound phenylguanidine (Fig. 2) further support this assignment. The most intense band is due to the $CN_3H_5^+$ antisymmetric stretch, with frequencies of 1672, 1667, and 1661 cm^{-1} for Cl$^-$, Br$^-$, and I$^-$ salts, respectively (Fig. 2A–C). Finally, Fig. 2D demonstrates the 66- cm^{-1} deuteration-induced downshift of the $CN_3H_5^+$ antisymmetric stretch.

We interpret the negative band at 1696 cm^{-1} in Fig. 1A–D as arising principally from the arginine $CN_3H_5^+$ antisymmetric stretch when it is bound to an internal carboxylate (Asp or Glu). First, the deuteration-induced shift of this negative band to 1607 cm^{-1} (Figure 2) supports its assignment as an arginine $CN_3H_5^+$ vibration. Second, the frequency observed for this vibration is not significantly dependent on whether the external anion is Cl$^-$ or I$^-$ (Figure 1A,B). Finally, its higher frequency (relative to the positive component of the difference band) is consistent with the 8- cm^{-1} -higher frequency we observe for phenylguanidium acetate (data not shown) than for the chloride salt.

Counterion effects on the retinylidene Schiff base C=NH$^+$ stretch frequency. The C=NH stretching mode of the hR·Cl$^-$ chromophore has a resonance Raman frequency of 1633 cm^{-1} [Smith et al., 1984]. This low Schiff base frequency (compared to 1640 cm^{-1} in bR) was the basis for the previous conclusion that interaction of halide ions with the Schiff base in hR is quite weak. Nevertheless, other resonance Raman studies demonstrated an effect of simple vs. polyatomic anions on the Schiff base frequency [Pande et al., 1989], supporting a prior conclusion [Steiner et al., 1984] that there is indeed a binding site for external anions near the Schiff base nitrogen. However, no shift in the C=NH frequency was observed between hR·Cl$^-$ and hR·Br$^-$, whereas with model compound retinylidene Schiff bases, there is a 5- cm^{-1} difference between Cl$^-$ and Br$^-$ salts [Blatz & Mohler, 1975]. This result suggested that simple halide ions do not bind *directly* to the protonated Schiff base nitrogen [Pande et al., 1989].

In contrast to the resonance Raman work, our data show a distinct 5- cm^{-1} downshift of the C=N frequency upon exchange of Cl$^-$ for I$^-$. The FTIR difference spectra in Fig. 1 thus suggest the presence of a halide binding site which is near the retinylidene Schiff base and is occupied in the hR state (as opposed to the aforementioned arginine site which is occupied in the hL state).

Mechanistic interpretation. We propose the following model for changes in anion binding during the hR→hL photoreaction. In the starting (hR) state, there is a chloride bound near the Schiff base nitrogen, which has been previously designated as site I [Steiner et al., 1984]. The binding strength of this site, and its precise distance from the Schiff base, cannot be inferred directly from our data. However, as discussed above, results from other investigators argue that it is not within hydrogen-bonding distance of the Schiff base nitrogen.

During the transition to hL, the chloride moves to form a hydrogen-bonded salt bridge with an arginine. This arginine is likely to be situated within a relatively hydrophobic region of the protein, because our model compound studies indicate that arginines solvated by ethanol or water do not show anion-dependent shifts (data not shown).

In our model, this arginine binding site is unoccupied by halide in the hR state, but is stabilized by formation of a salt bridge with an internal Asp or Glu carboxylate, which is displaced during the hR→hL transition. The absence of a COOH protonation signal in the hR→hL spectrum (Fig. 1; see also [Rothschild et al., 1988]) suggests that this displacement does not lead to protonation of the carboxylate group. One intriguing possibility is that the Schiff base and a nearby arginine residue simply exchange anions during the hR→hL reaction.

ACKNOWLEDGEMENTS

MSB is a Lucille P. Markey Scholar and this work was supported by a grant from the Lucille P. Markey Charitable Trust, as well as by NIH Biomedical Research Support Grant 5-S07RR05431-30. TJW is a Howard Hughes Medical Institute Predoctoral Fellow.

REFERENCES

Blatz, P. E. and J. H. Mohler (1975): Effect of selected anions and the solvents on the electronic absorption, nuclear magnetic resonance, and the infrared spectra of the N-retinylidene-n-butylammonium cation. *Biochemistry* 14, 2304–2309.

Braiman, M. S., O. Bousché and K. J. Rothschild (1990): Protein dynamics in the bacteriorhodopsin photocycle: Submillisecond Fourier transform infrared spectra of the L, M, and N photointermediates. *Proc. Natl. Acad. Sci. USA* 88, 2388–2392.

Chirgadze, Y. N., O. V. Fedorov and N. P. Trushina (1975): Estimation of amino acid residue side-chain absorption in the infrared spectra of protein solutions in heavy water. *Biopolymers* 14, 679–694.

Duschl, A., M. A. McCloskey and J. K. Lanyi (1988): Functional reconstitution of halorhodopsin. Properties of halorhodopsin-containing proteoliposomes. *J. Biol. Chem.* 263, 17016–17022.

Lanyi, J. K. (1990): Halorhodopsin, a light-driven electrogenic chloride-transport system. *Physiol. Rev.* 70, 319–330.

Pande, C., J. K. Lanyi and R. H. Callender (1989): Effects of various anions on the Raman spectrum of halorhodopsin. *Biophys. J.* 55, 425–31.

Rothschild, K. J., O. Bousché, M. S. Braiman, C. A. Hasselbacher and J. L. Spudich (1988): Fourier transform infrared study of the halorhodopsin chloride pump. *Biochemistry* 27, 2420–2424.

Smith, S. O., M. J. Marvin, R. A. Bogomolni and R. Mathies (1984): Structure of the retinal chromophore in the hR_{578} form of halorhodopsin. *J. Biol. Chem.* 259, 12326–12329.

Steiner, M., D. Oesterhelt, M. Ariki and J. K. Lanyi (1984): Halide binding by the purified halorhodopsin chromoprotein. I. Effects on the chromophore. *J. Biol. Chem.* 259, 2179–2184.

Susi, H. (1972): Infrared spectroscopy — conformation. *Methods Enzymol.* 26, 455–472.

Venyaminov, S. Y. and N. N. Kalnin (1990): Quantitative IR spectrophotometry of peptide compounds in water (H_2O) solutions. I. Spectral parameters of amino acid residue absorption bands. *Biopolymers* 30, 1243–1257.

Absorbance change of a C-50 carotenoid, bacterioruberin related to Cl⁻ translocation in a Cl⁻ pump, halorhodopsin

H. Tomioka[1], N. Kamo[2], K. Fujikawa[2] and H. Sasabe[1]

[1] Nano-Photonics, Frontier Research Program, Riken Institute, Wako 351-01, Japan. [2] Faculty of Pharmaceutical Sciences, Hokkaido University, Sapporo 060, Japan

Halophilic archaebacterium, *Halobacterium halobium* cells have four retinal proteins in the cell membrane: bacteriorhodopsin, halorhodopsin, sensory rhodopsin and phoborhodopsin (Oesterhelt & Stoeckenius, 1971; Mukohata et al., 1980; Spudich & Bogomolni, 1984; Tomioka et al. 1986). Halorhodopsin (hR) is a light-driven Cl⁻ pump (Schobert & Lanyi, 1982). Cl⁻ is translocated across the cell membrane from the outside to the cytosol. The Cl⁻ pumping is coupled with a cyclic photoreaction of hR (Ogurusu, 1981; Bogomolni & Spudich, 1982; Tsuda et al., 1982). hR (λmax=578 nm) is photoexcited and converted into a photointermediate absorbing maximally 520 nm (hR520) in msec time window. And then hR520 reverts with half-time of several msec to the original pigment, hR578.

A wild-type *Halobacterium halobium* cells are orange to red colored due to the presence of carotenoids. A major red pigment in the cell is a C-50 carotenoid, bacterioruberin (Rub). Roles of Rub have been investigated with Rub⁺ strain and Rub⁻ strain. It has been suggested that the carotenoids could protect the cell from damage by the intense light radiation (Dundas, 1977). However, Hescox & Carlberg (1972) and Sharma et al.(1984) reported that the carotenoid pigments do not exert the protective effect against UV light. And they also reported the carotenoids function as an accessory pigment in photoreactivation of the halobacteria from UV damage. However, recent works suggest that the carotenoids have no influence on the photoreactivation (Iwasa et al., 1988; Eker et al., 1991). The role of the carotenoids remain obscure. We report here a clear effect of a C-50 carotenoid, bacterioruberin on halorhodopsin and absorbance change of the carotenoid coupled with the photocycle of halorhodopsin.

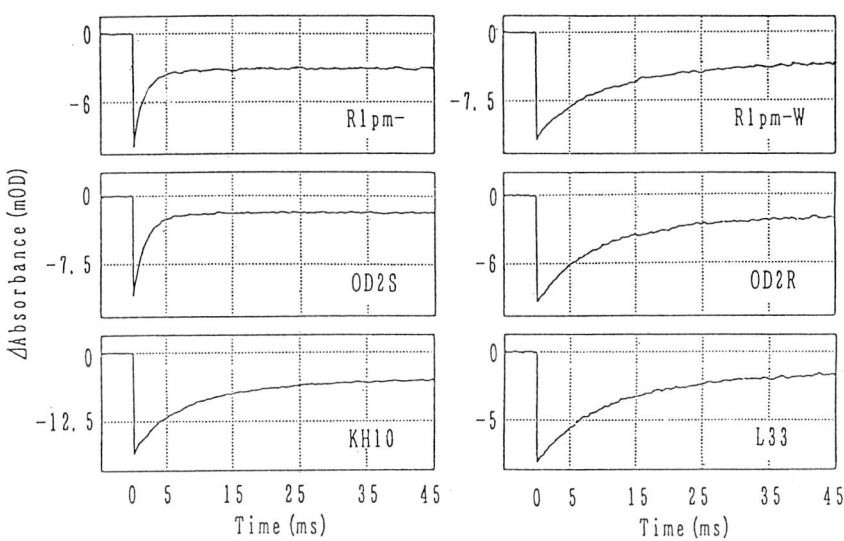

Fig.1. Flash-induced absorbance changes at 590 nm in membrane suspension from 6 bR⁻hR⁻sR⁻ strains. The membrane was suspended in 4 M NaCl, 10 mM PIPES (pH7.0).

In Fig. 1 we compare the transient absorbance changes at 590 nm after red flash (>620 nm) for 6 bR⁻hR⁺sR⁺ strains (R1pm⁻, R1pm⁻W, OD2S, OD2R, L33 and KH-10). These flash-induced absorbance changes were divided into two groups: the first contained R1pm- and OD2S; the second contained KH-10, R1pm⁻W, OD2R and L33. The relaxation of the absorbance changes of the former was divided into two kinetic components with half-times of ≈1 msec and 650 msec. That of the latter was also divided into two components with half-times of ≈6 msec and 650 msec. The slow component is assigned to the photochmical activity of sR and the fast component is attributed to hR (Bogomolni & Spudich, 1982; Tsuda et al., 1982; Lanyi & Schobert, 1983). But half-time of the fast component is different between the two groups. Visible observation of the 6 strains suggests that red colored strains have faster hR with a half-time of ≈1 msec and white colored strains have slower hR with a half-time of ≈6 msec. We invesigated the existence of four carotenoids in the 6 strains (Table 1). The presence of bacterioruberin is correlated well with the faster hR. These results strongly suggests that bacterioruberin accelerates the rate of hR photocycle. The Cl^- dependence of the amount of photocycling hR is also affected by Rub (data not shown). Half-maximal photocycling hR is at 4 mM Cl^- in a Rub⁺ strain and at 20 mM Cl^- in a Rub⁻ strain, suggesting that Rub interacts with hR and increases the affinity of hR for Cl^-.

Strain	Half time (ms)	Bacterio-ruberin	β-Carotene	Lycopene	Retinal
R1pm-	1.1	+	—	—	+
OD2S	1.5	+	—	—	+
R1pm-W	5.8	—	+	+	+
KH10	5.5	—	+	+	+
OD2R	5.5	—	—	—	—
L33	6.6	—	—	—	+

Table. Half-times of hR578 recovery and the presence of four carotenoid pigments in the 6 bR$^-$hR$^+$sR$^+$ strains. The carotenoid pigments were separated with thin layer chromatography. Each carotenoids identified with the Rf value.

The Red flash (>620 nm) which photoexcites hR but not Rub (λ max=500 nm) brought about the hR absorbance change and unidentified change in the range of 440-560 nm. The unknown absorbance change was detected only in the Rub$^+$ strain. The Rub absorbance change has a positive band in the 500-560 nm region and a negative band in the 440-500 nm. The positive band has two peaks: one is a large peak at 540 nm and the other is a small peak at 510 nm. The negative band also has two peaks: a large one at 485 nm and a small one at 460 nm. The Rub absorbance change was not observed in Cl$^-$-free hR membrane and depended upon Cl$^-$ in a similar manner of hR. The Rub absorbance change was not affected with a protonophore and an ionophore. These results suggest that the Rub absorbance change does not depend on the membrane potential but depend on Cl$^-$.

We considered that the Rub absorbance change might be originated from the local change accompanied with Cl$^-$ translocation in hR. If this is correct, the Rub absorbance change should be caused by an addition of Cl$^-$ to Cl$^-$-free hR membrane. hR membrane was washed with potassium citrate and Cl$^-$ was removed. The addition of Cl$^-$ to the Cl$^-$-free hR induced a Rub absorbance change, which agreed with the Rub absorbance change induced by the red flash. These results strongly suggest that the Rub absorbance change is caused by Cl$^-$ translocation in hR.

REFERENCES

Bogomolni, R.A. & Spudich, J.L. (1982). Identification of a third rhodopsin-like pigment in phototactic *Halobacterium halobium*. *Proceedings of the National Academy of Sciences of the United States of America* 79, 6250-6254.

Dundas, I.E.D. (1977) Physiology of Halobacteriaceae. *Advances in Microbial Physiology* 15, 85-120

Eker, A.P.M., Formenoy, L. & Wit, L.E.A. (1991) Photoreactivation in the extreme halophilic archaebacterium *Halobacterium halobium*. *Photochemistry and Photobiology* 53, 643-651

Hescox, M.A. & Carlberg, D.M. (1972) Photoreactivation in *Halobacterium cutirubrum*. *Canadian Journal of Microbiology* 18, 981-985

Iwasa, T., Tokutomi, S. & Tokunaga, F. (1988) Photoreactivation of *Halobacterium halobium:* action spectrum and role of pigmentaion. *Photochemistry and Photobiology* 47, 267-270

Lanyi, J.K. & Schobert, B. (1983). Effects of chloride and pH on the chromophore and photochemical cycling of halorhodopsin. *Biochemistry* 22, 2763-2769.

Mukohata, Y., Matsuno-Yagi, A. & Kaji, Y. (1980). Light-induced proton uptake and ATP synthesis by bacteriorhodopsin-depleted *Halobacterium*. In *Saline Environment*, pp. 31-37. Edited by H. Morishita & M. Masui. Tokyo: Business Center for Academic Society of Japan.

Oesterhelt, D. & Stoeckenius, W. (1971). Rhodopsin-like protein from the purple membrane of *Halobacterium halobium*. *Nature New Biology* 233, 149-152.

Ogurusu, T., Maeda, A., Sasaki, N. and Yoshizawa, T. (1981) Light-induced reaction of halorhodopsin prepared under low salt conditions. *Journal of Biochemistry* 90, 1267-1273

Schobert, B., & Lanyi, J.K. (1982). Halorhodopsin is a light-driven chloride pump. *Journal of Biological Chemistry* 257, 10306-10316.

Sharma, N., Hepburn, D. & Fitt P.S. (1984) Photoreactivation in pigmented and non-pigmented extreme halophiles. *Biochimica et Biophysica Acta* 799, 135-142.

Spudich, J.L. & Bogomolni, R.A. (1984) Mechanism of colour discrimination by a bacterial sensory rhodopsin. *Nature (London)* 312, 509-513

Tomioka, H., Takahashi, T., Kamo, N. & Kobatake, Y. (1986). Flash spectrophotometric identification of a fourth rhodopsin-like pigment in *Halobacterium halobium*. *Biochemical and Biophysical Research Communications* 139, 389-395.

Tsuda, M., Hazemoto, N., Kondo, M., Kamo, N., Kobatake, Y. & Terayama, Y. (1982). Two photocycles in *Halobacterium halobium* that lacks bacteriorhodopsin. *Biochemical and Biophysical Research Communications* 108, 970-976.

VI. Structure and function of retinal

VI. Structure et fonction du rétinal

The triggering process of visual transduction

Arh-Hwang Chen[1], Fadila Derguini, Paul Franklin, Shunghua Hu,
Koji Nakanishi[2], Beatriz Ruiz Silvo and Jun Wang

[1] Department of Chemistry, Columbia University, New York, NY 10027, USA and [2] National Kaohsiung Normal Universisty, Kaohsiung, Taiwan

Despite rapid advances in various areas related to visual transduction, clarification of the classical problem, the triggering process itself, has remained unsolved. In the following we address these points and conclude that triggering of the enzymatic cascade requires complete cis-trans isomerization of the 11-ene involving the entire polyene moiety.
Bathorhodopsin was believed to be the primary photoproduct for many years until photorhodopsin (λmax 570 nm), which decays to bathorhodopsin in ca. 40 ps, was detected using a very low intensity laser (Shichida et al., 1984). Recent femtosecond spectroscopy has shown that photorhodopsin is formed from the excited state of rhodopsin in less than 200 fs (Schoenlein at al., 1991); however, according to Yan et al. (1991), rhodopsin excited state decays in ca. 3 ps. Although no spectroscopic measurements have yet been performed with photorhodopsin, it is undoubtedly less relaxed than bathorhodopsin, both in terms of the chromophore and the surrounding protein conformations. The largest factor contributing to its red-shifted λmax of 570 nm (bathorhodopsin 543 nm), is most likely the distorted 11, 12-double bond.
Opsin comfortably accommodates the 11-cis, but not the all-trans chromophore, through the fixed ionone ring binding site at one end and the lysine binding site at the other. Photoisomerization of the 11-cis-ene to trans results in storage of internal energy resulting from strain between the distorted all-trans chromophore and the protein; the latter thermally relaxes through a series of intermediates to the stage of unprotonated meta-II, which interacts with the G protein, transducin, thus initiating the enzymatic cascade. Although spectroscopy has shown that photorhodopsin is the primary photoproduct, it is not known whether its formation alone can trigger subsequent conformational changes leading to enzymatic activation. Two mechanisms have been considered for visual triggering (Fig. 1). One is cis-trans isomerization (CTI) stemming from Wald's establishment of 11,12-double bond isomerization. The second is "sudden polarization" (SP) first applied as a theoretical model to retinals (Salem and Bruckman, 1975; Salem 1979). However, experimental evidence to differentiate the two and characterize the structural requirements leading to the process of visual transduction has been lacking.
According to the SP mechanism, based on *ab initio* calculations, absorption of a photon leads to charge separation and creation of a dipole in the zwitterionic singlet state. In the pentadienyl PSB model (Fig. 1), this charge separation leads to a new excited state dipolar species. Calculations showed that charge separation is negligible at low C-12/C-11 torsional angles but suddenly increases in the narrow region of 89°/90°/91°, hence the term "sudden polarization". More recent calculations (Albert and Ramasesha, 1990) suggest that polarization can occur over a wider torsional angle range from 40° to 130°. Therefore, the polarization may not be "sudden," but it is a phenomenon likely to be occurring during photoexcitation accompanied by a 90° torsional angle twist. The SP mechanism proposes that the resulting electric signal could act as a trigger and have a direct electrostatic interaction, or an indirect interaction with the protein leading to conformational changes.

Two possible triggering processes in visual transduction
i) cis-trans isomerization
11
12
200 fs
rhodopsin
500 nm
photo-:570 nm
highly distorted "all-trans"
meta-II, 380 nm
ii) "sudden polarization" of polyene in excited state
(+0.17)
7
12
(+0.83)
R
hν
(calcd. for 12-s-cis)
(+1.11)
90°
(-0.11)
excited state
Δμ 33 Debyes
Δq
88 90 92
θ°
charge separatio
Salem, Bruckman, Nature 258 525 (1975)

Fig. 1. The pi orbitals in photorhodopsin denote the highly distorted double bond. In (ii), the pentaene is shown as 12-s-trans form (in line with current picture) rather than in the original 12-s-cis form. Charge suddenly develops at distortional angles close to 9°.

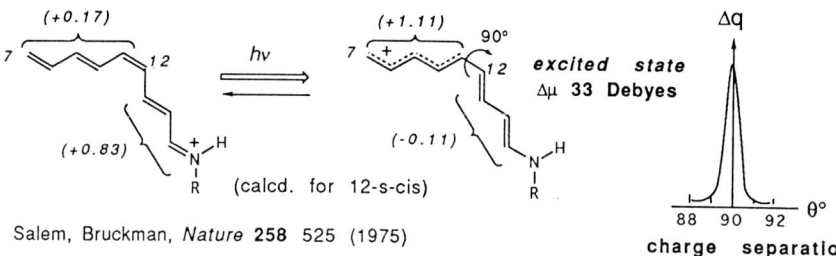

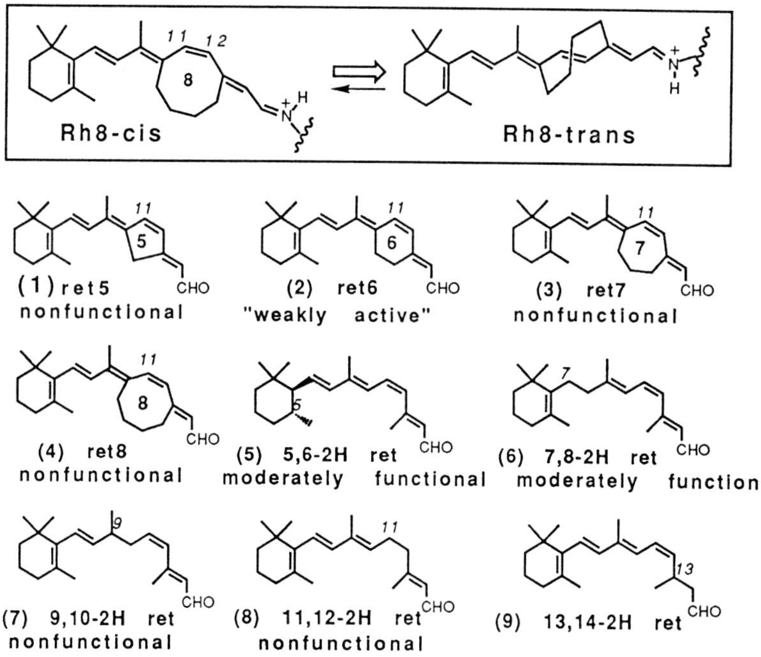

11 12
8
Rh8-cis
Rh8-trans
(1) ret5
nonfunctional
(2) ret6
"weakly active"
(3) ret7
nonfunctional
(4) ret8
nonfunctional
(5) 5,6-2H ret
moderately functional
(6) 7,8-2H ret
moderately function
(7) 9,10-2H ret
nonfunctional
(8) 11,12-2H ret
nonfunctional
(9) 13,14-2H ret
CHO

Fig. 2. Retinal analogs with fixed 11-ene and dihydroretinals; activity as measured by phosphodiesterase activation and kinase phosphorylation are also given.

Retinal analog with 11-cis-ene fixed by 8-membered ring (ret8). Ret8 (Fig. 2) was prepared (Caldwell et al., in prepartaion) because the flexibile 8-membered ring might be able to accomodate a trans bond and thus give rise to a bathorhodopsin stable at room temperature. The rigid ret5 gave Rh5 but this was photochemically inert (Kandori et al., 1989). Ret6 yields Rh6, λmax 510-512 nm, the weak activity of which probably arises from structural changes caused by irradiation of this photounstable pigment (de Grip et al., 1990; Bhattacharya et al., 1992). Ret7, the first of the 11-cis-ene fixed analogs, gave a nonbleachable and inactive rhodopsin (Rh7)(Zankel, et al.,1990). However, it did give rise to a primary photoproduct absorbing at 580 nm (photorhodopsin), but failed to give bathorhodpsin and reverted to the original pigment, λmax 490 nm (Kandori et al., 1989). In the case of Rh8, a photorhodopsin intermediate formed in 15 ps, λ_{max} 585 nm; this relaxed to a batho-like intermediate in 1 ns, λ_{max} 577 nm, which thermally reverted to the initial pigment after 50 ns (Mizukami et al., 1992). Since the chromophore in native rhodopsin adopts a distorted all-trans configuration in photo and bathorhodopsin, the fact that Rh8 yields these two intermediates clearly demonstrates that the more flexible 8-membered ring must have isomerized to the all-trans form (Fig.2).

Bioassays with MeRh8 and other rhodopsin analogs (Fig. 2).
A major difficulty encountered in attempts to make vesicles of the 502 nm Rh8 pigment was formation of a 425 nm species, a nonspecific Schiff base formed with thermally denatured opsin a crude enzyme extract, unbleached ROS, phosphatidylethanolamine vesicles, etc.
All attempts to prepare Rh8-containing vesicles failed until we used methylated opsin (Me-opsin) in which all non-active-site lysines are dimethylated (Longstaff & Rando, 1985). Incubation of Me-opsin with ret8 provided only the 502 nm pigment, MeRh8. Importantly, the similarity in the UV/VIS and CD spectra of MeRh8 and native rhodopsin showed that the overall environment of the chromophores in both pigments were quite similar.
The results of the two assays are summarized in Fig. 2. After careful experiments, it was concluded that the crucial MeRh8 is incapable of activating the enzymatic cascade.
The bioassay results of the series of dihydroretinals are also shown in Fig. 2. The general trend seen in the assay indicates diminished activity upon saturation of the 5,6 and 7,8 double bonds, followed by complete loss of activity upon saturation of the 9,10 and 11,12 double bonds.

Discussion.
Upon replacement of double bonds with C-C single bonds of the polyene, the rigidity of the chromophore is lost, the protein has additional relaxation modes available after the 11-ene isomerization, and thus can bypass the necessary relaxational pathway leading to activity. The 9-methyl plays an integral role in both binding and activation (Ganter et al., 1989). Thus the GTPase of the 9-desmethyl pigment was only 8% of that of rhodopsin, while the intermediate corresponding to meta-II was still protonated. In the case of the nonfunctional 9,10-dihydro pigment, the 9-methyl is no longer held sufficiently rigid to induce proper conformational changes in the rhodopsin after photoisomerization. Namely, 9-methyl is crucially involved in relaxation of the protein conformation through hydrophobic bonding. Similar to the role of the 9-methyl in the vertebrate photosensor, the 13-methyl in sensory rhodopsin I (SR-I) from *H. halobium* is also critical in converting photon absorption into protein conformational changes (Yan et al.). The inactivity of 11,12 dihydro-retinal is not surprising and agrees with earlier *in vivo* results where its intraperitoneal injection into vitamin A deficient rats did not lead to recovery of vision (Crouch et al., 1981).

Rh7 and Rh8 give photorhodopsin as the primary photoproduct; Rh8 yields one further intermediate corresponding to bathorhodopsin. Their photoexcited states involve charge polarization (or sudden polarization) of the chromophore, yet the pigments exert no activity. If only charge polarization were necessary for triggering subsequent events, Rh8 certainly should have been active since the environment of both the native and ret8 chromophores are very similar, both in the original pigment and in the primary photoproduct: (i) the UV/VIS and CD spectra of rhodopsin and MeRh8 are similar; and (ii) the maxima of photorhodopsin derived from native and MeRh8 are 570 nm and 585 nm, respectively. The similarity in the maxima of the two photorhodopsins allows one to hypothesize that the shape of the trans-chromophore in native photorhodopsin could be very close to that in MeRh8 photorhodopsin; i.e., in both systems, the torsional angles must be greater than 90° (transoid), thus fulfilling the theoretical requirement for charge (sudden) polarization to occur. The ring strain in the Rh8 bathorhodopsin forces it to revert to its original conformation, together with the protein. It appears that a full isomerization of 11-cis to the extended all-trans-retinal is necessary to induce conformational changes in the opsin leading to activation. This change should be such that the PSB proton in meta-I must be delivered to the designated proton acceptor in meta-II (Longstaff et al., 1986); recall that the meta-II equivalent of the inactive 9-desmethyl-Rh still retained the PSB (Ganter et al., 1989). Experiments with the dihydro series also show that occurrence of a satisfactory conformational change of the protein requires the assistance of the full and rigid delocalized pi system. It is concluded that although charge polarization or "sudden polarization" occurs in the excited state, this itself is insufficient, and that the triggering process of visual transduction requires complete cis-trans isomerization of the chromophoric 11-ene to the all-trans form involving the entire pi system. (The studies were supported by GM 36564).

REFERENCES AND NOTES

Akita,H., Tanis, S., Adams, M., Balogh-Nair, V. & Nakanishi, K. (1980): Nonbleachable rhodopsin retaining the full natural chromophore. J. Am. Chem. Soc., 102, 6370.

Albert, I. D. L. and Ramasesha, S. (1990): Sudden polarization in push-pull polyene: A model exact study. J. Phys. Chem. 94, 65406543.

Bhattacharya, S., Ridge, K. D., Knox, B. E., Khorana, H. G. (1992): Light stable rhodopsin. J. Biol. Chem. 267, 6763-6769.

de Grip, W. J., van Oostrom, J., Bovee-Geurts, P.H. M. , van der Steen, R. , van Amsterdam, L. J. P. , Groesbeek, M., Lugtenburg, J. (1990): 10,20-Methanorhodopsins: (7E,9E,13E)-10,20-methanorhodopsin and (7E,9Z,13Z)-10-20-methanorhodopsin. Eur. J. Biochem, 191, 211-220.

Kandori, H., Matuoka, S., Shichida, Y., Yoshizawa, T., Ito, M., Tsukida, K., Balogh-Nair, V., Nakanishi, K. (1989): Mechanism of isomerization of rhodopsin studied by use of 11-cis-locked rhodopsin analogs excited with a picosecond laser pulse. Biochemistry 28, 6460 6467.

Liebman, P. A., Evanczsuk, A. T. (1982): Real time assay of rod disk membrane cGMP phosphodiesterase and its controller enzymes. Methods in Enzymol. 81, 532-542.

Longstaff, C., Rando, R. R.(1985): Methylation of the active-site lysine of rhodopsin. Biochemistry 24, 8137-8145.

Longstaff, C., Calhoon, R. D., Rando, R. R. (1986): Deprotonation of the Schiff base of rhodopsin is obligate in the activation of the G protein. Proc. Natl. Acad. Sci. USA 83, 4209-4213.

Mathies, R., Stryer, L. (1976): Retinal has a highly dipolar vertically excited singlet state: implications for vision. Proc. Nat. Acad. Sci. USA 73, 2169-2173.

Mizukami, T., Kandori, H., Shichida, Y., Cheng, A.-H, Derguini, F., Nakanishi, K. & Yoshizawa, T. (1992): Photoisomerization mechanism of the rhodopsin chromophore: picosecond photolysis of pigment containing 11-cis-locked-8-membered retinal (submitted).

Ganter, U. M. , Schmid, E. D., Perez-Sala, D., Rando, R. R., Siebert, F. (1989): Removal of the 9-methyl group of retinal inhibits signal transduction in the visual process. A Fourier transform infrared and biochemical investigation. Biochemistry 28, 5954 5962.

Salem, L., Bruckmann, P. (1975): Conversion of a photon to an electrical signal by sudden polarization in the N-retinylidene visual chromophore. Nature 258, 526-528.

Salem, H.(1979): The sudden polarization effect and its possible role in vision. Acc. Chem. Res. 12, 87-92 (1979).

Schaffer, A. M., Yamaoka, T., Becker, R. S. (1975): Visual pigments - V. Ground and Excited state acid dissociation constants of protonated all-trans retinal Schiff base and correlation with theory. Photochemistry and Photobiology 21, 297-301 .

Shichida, Y., Matuoka, S., Yoshizawa, T. (1984): Formation of photorhodopsin, a precursor of bathorhodopsin, detected by picosecond laser photolysis at room temperature. Photobiochem. Photobiophys. 7, 221-228

Schoenlein, W., Peteanu, L. A., Mathies, R. A., Shank, C. V. (1991): The first step in vision: Femtosecond isomerization of rhodopsin. Science 254, 412-415.

Yan, M., Manor, D., Weng, G., Chao, H., Rothberg, H., Jedu, T. M. , Alfano, R. R. , Callender, R. R. (1991): Ultrafast spectroscopy of the visual pigment rhodopsin. Proc. Natl. Acad. Sci. U. S. A. 88, 9809-9812.

Yan, B.,Nakanishi, K., Spudich, J. L. (1991): Mechanism of activation of sensory rhodopsin I: Evidence for a steric trigger. Proc. Natl. Acad. Sci. USA 88, 9412-9416.

Zankel, T., Ok, H., Johnson, R., Chang, C. W., Sekiya, N., Naoki, H., Yoshihara, K. & Nakanishi, K. (1990): Bovine rhodopsin with 11-cis-locked retinal chromophore neither activates rhodopsin kinase nor undergoes conformation change upon irradiation. J. Am. Chem. Soc., 112, 5387-5388.

Evidence for a curved retinal in BR from solid-state ^{2}H-NMR

Anne S. Ulrich[1], Ingrid Wallat[2], Maarten P. Heyn[2] and Anthony Watts[1]*

(1) Department of Biochemistry, University of Oxford, South Parks Road, Oxford OX1 3QU, UK. (2) Department of Physics, Freie Universität Berlin, Arnimallee 14, 1000 Berlin 33, Germany

* Correspondence author

A strategy to solve the structure of retinal in bacteriorhodopsin (BR) is based on the determination of its chemical bond vectors. Individual methyl-groups on the chromophore are specifically deuterium-labelled and examined by solid-state deuterium (^{2}H-)NMR measurements of uniaxially oriented purple membrane (PM) patches. From the spectral quadrupole splitting, the angle between the deuteromethyl-group and the sample normal can be calculated and thus the orientation of the molecular segment is found. Bond-angles have been determined for the three methyl-groups on the cyclohexene ring of retinal (Ulrich et al., 1992), and here we report recently acquired ^{2}H-NMR data on the methyl-group attached to carbon C_{13}, near the Schiff base. By determining several such geometrical constraints along the whole chromophore it is possible to characterize its complete structure, if the intramolecular flexibility of the molecule is taken into account. The resulting picture of the orientation and conformation of retinal within BR is consistent with other structural information on the chromophore, which confirms the validity of this non-perturbing solid-state NMR approach. A new feature of the retinal structure, however, emerges from these ^{2}H-NMR experiments, namely that there is a distinct curvature within the plane of the polyene-chain.

For the analysis of ^{2}H-NMR spectra from immobilized samples, there exists a simple relationship between the spectral quadrupole splitting Δv_Q and the angle θ of the deuteromethyl-group bond vector relative to the magnetic field direction (Seelig, 1977):

$$\Delta v_Q = (3\cos^2\theta - 1) \cdot 40 \text{ kHz}$$

Practically, the oriented sample is aligned with its normal parallel to the spectrometer magnetic field. All bond vectors which make a constant angle θ with the sample normal, now make the same well-defined angle with the spectrometer field, and the ^{2}H-NMR spectrum of a deuterium-labelled methyl-group simply consists of one pair of resonances separated by a quadrupole splitting Δv_Q. A certain degree of spectral linebroadening is associated with the quality of membrane alignment in the macroscopically oriented sample, where the PM patches have been deposited on small glass supports by controlled evaporation from a concentrated suspension (Seiff et al., 1987). The corresponding mosaic spread can be quantified by lineshape simulation of the ^{2}H-NMR spectrum, which further provides an independent way of analyzing the spectral features.

Figure 1 shows the ^{2}H-NMR spectrum from dark-adapted BR (110 mg) in uniaxially oriented PM patches, containing retinal that is deuterium-labelled at the methyl-group on carbon C_{13} (Heyn et al., 1988). Two broad resonances are seen with a quadrupole splitting of 46 kHz and a central sharp line due to HDO. Using the above equation, an angle $\theta = 32°$ is calculated for the deuteromethyl-group relative to the membrane normal. The good fit of the lineshape simulation shown superimposed in Fig. 1 confirms this evaluation. The fitted linewidth provides

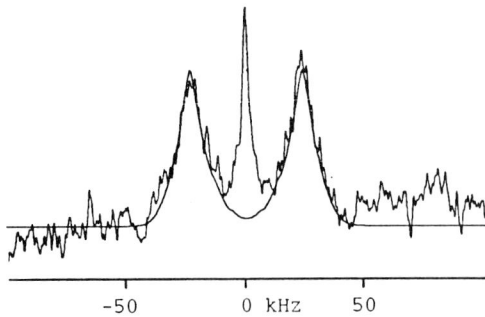

Figure 1: ^{2}H-NMR spectrum from dark-adapted BR (110 mg in oriented PM patches), containing retinal with a labelled deuteromethyl-group on carbon C_{13}. A lineshape simulation is shown superimposed over the experimental spectrum. The oriented sample was measured at room temperature with its normal parallel to the magnetic field direction, at a deuterium resonance frequency of 61 MHz using a quadrupole-echo pulse sequence with a $\pi/2$ pulse width of 6µs, echo delay times around 30 µs and a repetition time of 200 ms.

an estimate for the mosaic spread of the macroscopically oriented sample of around ±10°, a value which is close to the mosaic spread found by independent ^{31}P-NMR measurements of the membrane phospholipids in an oriented PM sample prepared in the same way (Ulrich et al., 1992). It is thus noted that there is no necessity to introduce any additional linebroadening in the simulation procedure to account for the presence of both all-*trans* and 13-*cis* retinal in the dark-adapted BR sample. This mixture does not seem to produce shifted spectral components, which means that the thermal equilibrium of isomerization of the Schiff base terminus around the $C_{13}=C_{14}$ double bond (and, simultaneously, the $C_{15}=N$ bond) has no effect on the orientation of the methyl-group on C_{13} and consequently neither on any other bond further down the molecule. During the light-induced isomerization in the photocycle of BR, on the other hand, some structural changes are expected to occur along the whole chromophore, which might be revealed by ^{2}H-NMR measurements of trapped intermediates.

The bond angle of 32° for the C_{13} methyl-group on retinal provides information about the orientation of the polyene chain in BR. Three further methyl-groups have been measured on the cyclohexene ring (Ulrich et al., 1992), namely the two groups on carbon C_1 (95° and 75°) and the one on C_5 (46°). Figure 2 illustrates how all these orientations are accommodated in space by the proposed structure of retinal within BR. This geometry was obtained by modelling the molecular framework to the set of geometrical constraints, i.e. to the known methyl-group orientations. While the three-dimensional orientation of the cyclohexene ring is defined by three methyl-groups, the polyene chain is represented by a single bond angle only. The molecular modelling approach is therefore concerned with fitting possible structures of the polyene-chain to the fixed cyclohexene ring, which is aligned roughly vertical within the membrane as seen in Fig.2. The intramolecular flexibility of retinal is taken into account by allowing for a complete rotation around the C_6-C_7 bond between the ring and the chain and by permitting a limited skew around C_1-C_6 within the puckered cyclohexene ring.

To evaluate the retinal conformation around the C_6-C_7 bond, i.e. of the polyene chain relative to the ring, it is constructive to focus on the two methyl-groups on C_5 and C_{13}, which are at either end of the conjugated system of double bonds. The two respective methyl-group orientations of 46° and 32° are roughly parallel, from which it is evident that the chromophore has a 6-S-*trans* conformation rather than 6-S-*cis*. The fact that the two methyl-groups are not exactly parallel may simply appear to be a consequence of the intramolecular flexibility, particularly of the rotational freedom around C_6-C_7. A second reason for the difference in bond angles, however, could be an in-plane curvature within the conjugated system of the polyene chain, or even a slight twist within this plane, which would relieve the steric crowding of the methyl-groups along the chain. Such a distortion is generally found in the structure of the crystalline compound (Santasiero et al., 1990) and in other retinal containing proteins (Newcomer et al., 1984; McRee et al., 1989).

Molecular modelling was used to assess the contributions of the various modes of flexibility and to determine theoretically the rotation around C_6-C_7 and the skew around C_1-C_6 that would be necessary to satisfy the geometric constraints. It is found that the retinal molecule would have to be very severely distorted away from its known

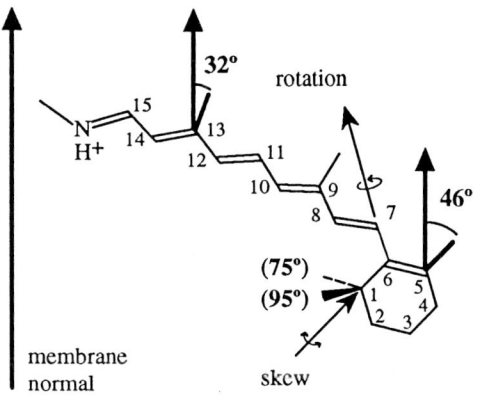

Figure 2: Orientation and conformation of retinal in BR as modelled from the methyl-group bond angles determined by solid-state ^{2}H-NMR spectroscopy. The bond vector on C_{13} was measured from the spectrum shown in Fig. 1 and used to characterize the structure of the polyene chain relative to the cyclohexene ring orientation. The difference between this bond angle (32°) and that of the methyl-group on C_5 (46°) indicates that the chromophore possesses a curvature within the plane of the conjugated system.

structure within the protein binding pocket, unless an appropriate curvature of the polyene chain is included in the modelling. The chromophore long axis has been shown to be tilted by about 67° relative to the membrane normal, and it lies with its conjugated plane approximately perpendicular to the membrane surface (Heyn et al., 1977; Earnest et al., 1986; Lin & Mathies, 1989). Therefore, it would not be reasonable to rotate the polyene chain too far away from a planar 6-S-*trans* conformation because not only would this move the conjugated plane away from its known perpendicular alignment within the membrane, but it would also yield a tilt angle for the polyene chain that is much too steep.

In a previous ^{2}H-NMR analysis of the cyclohexene ring structure in BR (Ulrich *et al.*, 1992), we estimated the local tilt of the chromophore long-axis at the ring, which was considerably steeper than the average tilt angle (≈67°) known from spectroscopic studies. This observation had lead to the initial prediction of a curved polyene chain of retinal in BR. Together with the new results, a more detailed picture of the chromophore structure has now emerged, from which an end-to-end tilt angle can be estimated. This end-to-end tilt angle of retinal in BR is still somewhat steeper than the orientation of the electronic transition moment, which is the property detected by optical spectroscopy. This difference, however, is readily explained by a deviation of the electronic transition moment from the long-axis of the conjugated system of double bonds (Shang *et al.*, 1991). Since there are three variable modes of flexibility, it is not feasible at present (without knowing the bond angle of the methyl-group on C_9) to quantify the contribution of each and to propose an exact value of the end-to-end tilt angle. Nevertheless, it has been possible to establish that an appreciable curvature of the chromophore constitutes a critical feature of its detailed structure within BR.

We would like to acknowledge Norman Gregory for his continued patience with the spectrometer, and SERC and the CEC for financial support.

References:
Earnest, T. N., Roepe, P., Braiman, M. S., Gillespie, K. J. & Rothschild, J. (1986) *Biochem.* **25**, 7793-7798
Heyn, M. P., Cherry, R. J. & Müller (1977) *J. Mol. Biol.* **117**, 607-620
Heyn, M. P., Westerhausen, J., Wallat, I. & Seiff, F. (1988) *Proc. Natl. Sci. USA* **85**, 2146-2150
Lin, S. W. & Mathies, R. A. (1989) *Biophys. J.* **56**, 653-660
McRee, D., E., Tainer, J., a., Meyer, T., E., Van Beeumen, J., Cusanovich, M., A. & Getzoff, E., D. (1989) *Proc. Natl. Acad. Sci. USA* **86**, 6533-6537
Newcomer, M. E., Jones, T., A., Aqvist, J., Sundelin, J., Eriksson, U., Rask, L. & Petersen, P., A. (1984) *EMBO J.* **3**, 1451-1454
Santarsiero, B.D., James, M.N.G., Mahendran, M. & Childs, R.F. (1990) *J. Am. Soc.* **112**, 9416-9418
Shang, Q.-y., Dou, X. & Hudson, B. S. (1991) *Nature* **352**, 703-705
Seelig, J. (1977) *Q. Rev. Biophys.* **10**, 353-418
Seiff, F., Westerhausen, J., Wallat, I. & Heyn, M. P. (1987) in *Receptors and Ion Channels* (Ovchinnikov, Y. A. & Hucho, F., Eds.) pp 255-263, de Gruyter, Berlin
Ulrich, A. S., Heyn, M. P. & Watts, A. (1992) submitted to *Biochemistry*

^{19}F-NMR in studies of fluorinated visual pigment analogs.
A method for detecting neighbouring groups or empty space in a binding site

Robert S.H. Liu and Leticia U. Colmenares

Department of Chemistry, University of Hawaii, Honolulu, HI 96822, USA

Abstract: ^{19}F-NMR data of six chain labelled and two ring labelled rhodopsin analogs are presented along with those of the corresponding protonated Schiff bases (PSB). The fluorine opsin shift values (FOS) are examined in relation to the shape of the binding cavity. It is suggested that van der Waals interaction is the principal controlling factor that determines the chemical shifts of the pigment analogs. The method also provides a direct way for demonstrating ring-chain restricted rotation for a substrate imbedded in the protein cavity.

Introduction

Nuclear magnetic resonance spectroscopy in conjunction with the use of isotopically substituted retinal isomers has been fruitfully applied to probing structural information of visual pigments and other retinal binding proteins. As a result, information such as the protonated nature of the visual chromophore, its configurational and conformational information and the extent of charge delocalization and the associated question of location of the second point charge has been firmly established (Mateescu et al., 1983; Mollevanger et al., 1987; Smith et al., 1990).

The use of a fluorine atom as a reporting group for studies of visual pigments, while suffering from the obvious drawback in having to deal with modified retinals and rhodopsins, has the advantage over the isotopic reporting groups in occupying space beyond the immediate regions defined by the retinyl chromophore. Hence, potentially ^{19}F-NMR studies of F-labelled rhodopsins could become a complementary NMR technique for probing information beyond those already elegantly demonstrated in studies with isotopically labelled retinals.

Results & Discussion

The program of studying fluorine labelled visual pigment analogs began more than a decade ago when the synthesis of several chain labelled retinal and rhodopsin analogs was reported by Asato et al. (1978). Preparation of trifluoromethylated retinals and visual pigment analogs followed after successful use of trifluoroacetone as a labelling reagent through secondary amine catalyzed condensation reactions (Mead et al., 1985).

The first ^{19}F-NMR work on a visual pigment appeared in 1981 (Liu et al., 1981). The excessive linewidth (approx. 1 ppm) associated with such membrane proteins, soluble only in detergent micelles, impeded rapid measurements of other analogs. Only recently, after having mastered the procedure for concentrating protein samples and improved local ^{19}F-NMR capability, we were able to do such studies routinely and have completed the study of isomers of eight labelled analogs (**1-8**) (Colmenares et al., 1991; Colmenares & Liu; 1992; Colmenares & Liu, submitted). The relevant chemical shift data are listed in Table 1.

11-cis-retinal

1: 14F
2: 12F
3: 10F
4: 8F
5: 19,19,19F$_3$
6: 20,20,20F$_3$

7: X = Y = CF$_3$
8: X = CF$_3$; Y = F

Table 1. F-19 NMR of fluorinated rhodopsins[a]

Compound	F-19 chemical shift, δ (ppm)		
	PSB	Pigment	FOS[b]
14F, 11-cis, 1	-125.5	-117.1	8.4
9-cis	-131.4	-123.5	7.9
12F, 11-cis, 2	-107.8	-94.2	13.6
9-cis	-120.9	-114.3	6.6
9,11-dicis	-108.7	-110.8	-2.1
10F, 11-cis, 3	-112.2	-107.7	4.5
9-cis	-119.7	-114.9	4.8
8F, 11-cis, 4	-116.8	-115.0	1.8
9-cis	-105.9	-99.3	6.6
9-CF$_3$, 11-cis, 5	-58.2	-53.8	4.4
9-cis	-64.6	-60.2	4.4
13-CF$_3$, 9-cis, 6	-61.4	-57.8	3.6
o,o-CF$_3$, 11-cis, 7	-58.5	-55.2	3.3
	-58.5	-60.7	-2.2
o-F,o-CF$_3$, 11-cis, 8	-59.6	-61.6	-2.0
	-110.5	-114.5	-4.0

a. CF$_3$CCl$_3$ as external standard. b. Pigment minus PSB.

For the chain labelled compounds (1-6), most of the pigment data fall within the range of ~4-8 ppm downfield from those of the corresponding PSB (fluorine opsin shift, FOS). This range corresponds to what is commonly known as the hydrophobic shift for protein bound species or increased van der Waals interaction (Gerig, 1989). Several exceptional cases, exhibiting values considerably outside this range, are discussed separately below.

We believe that all these cases can be accounted for largely by the different extent of local van der Waals interaction. For 12F-rhodopsin (2), two of the three values for its isomers are abnormal. A substituent at this position is known to interact with the protein residues. Thus, the extremely low yield of pigments for 12-methyl- and 12-chloro-rhodopsin was attributed by Liu et al., (1984) to steric crowding of the 12-substituent with the suspected counterion. This interaction has recently been more clearly illustrated in a molecular modeling study (Liu & Mirzadegan, 1988). The 12F-substituent, while smaller than the methyl and chloro substituents, must still be projecting into a crowded protein region. The increased downfield shift is, therefore, consistent with increased vdW interactions plus induced polarization of the 12F substituent by the negatively charged carboxylate. The same substituent for the 9-cis isomer is instead more buried in the binding site, thus not in close contact with protein residues. For 9,11-dicis-rhodopsin, the doubly twisted dicis geometry makes the F-substituent orient in a direction opposite to that of the other isomers, thus experiencing a different magnetic environment. Furthermore, molecular modeling studies (Mirzadegan & Liu, 1991) show that the F-substituent is located close to Trp-265 to the center of the modeled binding site. It should occupy a more open space (reduced vdW interactions), perhaps additional ring current effect giving an upfield FOS shift of 2.1 ppm.

The 11-cis isomer of 8F-rhodopsin exhibits the least downfield shift (FOS = 1.8 ppm). It is accountable by a large space suspected to exist around the 8F substituent, as revealed in the molecular modelling study, making the extent of van der Waals interactions more akin to that in solution. It is also in agreement with the reported analog studies (Liu & Asato, 1990) where several 5-substituted pigment analogs were successfully prepared, implying a large space available around the 5-methyl and the adjacent 8F region. It should be noted that this effect must be highly dependent on the shape of the polyene chromophore because the related 9-cis isomer exhibits a normal FOS value. For the latter, the increased F-8, H-11 steric interaction owing to the 9-cis linkage could possibly alter the conformation nearby this region. Thus, a twist at the 8,9-bond orients F-8 toward a different protein environment as that of the 11-cis pigment.

9-Cis-10F-rhodopsin, known to exhibit unusually inert photochemical behavior (Liu et al., 1986), does not show an abnormal FOS value. The molecular modeling study reveals its close proximity to Tyr-268 which could provide the postulated H-bonding for the altered photochemical activity. The apparent "negative" nmr result could be a reflection of the low sensitivity of F-chemical shifts toward polar effects (see below). Also, a photochemical reaction is a dynamic behavior which could involve specific protein perturbation after light excitation. Such a situation would not be detectable in a static nmr study.

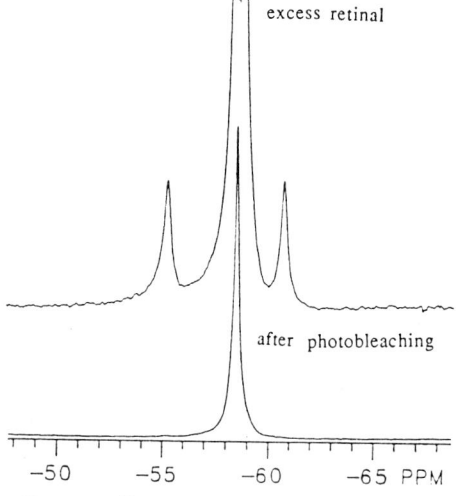

Figure 1. ^{19}F-NMR spectrum of the pigment from o,o-bis-CF$_3$-phenylretinal

^{19}F-NMR spectra of the two doubly ring labelled (**7, 8**) phenylrhodopsins show unusual protein substrate interactions. The spectrum of o,o-bis-CF$_3$ labelled analog, in the presence of excess retinal, is shown in Figure 1 as a representative example. Pigment formation was revealed by the appearance of two new peaks of equal intensity on either side of the main band. That these peaks are associated with the pigment is reflected by their disappearance upon irradiation with orange light (>460 nm).

In solution, only a single F-signal is present for the labelled retinyl chromophore. When protein bound, two peaks appeared. Clearly, the ring-chain rotational equilibraion is stalled with the two CF$_3$ groups now facing different magnetic environment. In fact, one may further deduce that the upfield signal (negative FOS) corresponds to the CF$_3$ occupying the same position as the CH$_3$-5 in the parent retinal. This is because of the open space nearby this region as discussed above. That the signal of 8F-rhodopsin also exhibits a high field shift (FOS = 1.8 ppm rather than 4-8 ppm for pigments of normal hydrophobic shifts) is also consistent with this notion because 8F should be located in the same open region of the binding cavity to face a more "solution like" environment. The low field signal, therefore, corresponds to the CF$_3$ occupying the 1,1-CH$_3$ position of the parent retinal. The presence of a recognition site for the latter group in opsin is well known. Close interaction of the CF$_3$ group with the nearby amino acid residues probably caused this increased protein (down field) shift. It is interesting to note that the low field signal exhibited a broader line width (88 vs 72 Hz) consistent with a more crowded environment.

For the o,o-F,CF$_3$-phenyl sample (**8**), the pigment showed two peaks of 3 : 1 in area, both at higher field than the corresponding signals in the free chromophore. The appearance of only one set of such signals indicates that the pigment is conformationally homogeneous. The high field shift of the CF$_3$ group is consistent with the notion that it occupies the more open 5-methyl position. The similar shift of the o-F-signal (rather than the down field shift for the bis-CF$_3$) is consistent with the smaller size of the substituent, no longer in close contact with the dimethyl recognition site. This interpretation of F-shift implies that opsin has preferentially picked up the less stable but twisted ring/chain conformer, presumably being more compatible with the shape of the binding site.

We believe that the examples presented above clearly demonstrated the value of the ^{19}F-NMR method for probing information of the binding cavity. During the course of this work, it became obvious to us that among factors known to perturb F-chemical shifts (solvent effect, ring current, H-bonding, local point charges, etc.), vdW interactions is by far the dominant term. It accounts for all the abnormal shifts either to a low field due to projection of the F-label into a crowded region or to a high field due to projection into a void as well as the 4-8 ppm shifts for the bound substrates. These findings confirm those of early workers who led the way in employing the same techniques for studies of other proteins and labelled substrates (Hull and Sykes, 1976; Gerig, 1989). The literature is also replete with examples of small organic molecules where severe steric crowding has induced large down field shift of a nearby F-atom. The most dramatic example is perhaps the 27 ppm down field shift of a F-atom periparallel to a t-butyl group in a substituted naphthalene (Gribble et al., 1991). What has not been stressed in the literature is the effect of moving groups away from a F-label, to cause upfield shifts. Thus, we muse at the thought whether ^{19}F-NMR is potentially a superb method for measuring nothing, i.e., empty space in a protein cavity.

Acknowledgment: This research work was supported by a grant from The U. S. Department of Health and Human Services (DK-17806). Other early contributors to the development of our F-nmr program are A. E. Asato, D. Mead and M. Denny.

References:

Asato, A. E.; Matsumoto, H.; Denny, M.; Liu, R. S. H. (1978): Fluorinated Rhodopsin Analogues from 10-Fluoro- and 14-Fluororetinal. J. Am. Chem. Soc. 100, 5957-5960.

Colmenares, L. U.; Asato, A. E.; Denny, M.; Mead, D; Zingoni, J. P.; Liu, R. S. H. (1975): NMR Studies of Fluorinated Visual Pigment Analogs. Biochem. Biophys. Res. Com. 179, 1337-1343.

Colmenares, L. U.; Liu, R. S. H. (1992): 9,11-Dicis-12-Fluororhodopsin. Photochem. Photobiol., in press.

Colmenares, L. U.; Liu, R. S. H. 19F-NMR Evidence for Restricted Rotation of the Retinyl Chromophore in Doubly Labelled Visual Pigment Analogs. J. Am. Chem. Soc., submitted.

Gerig, J. T. (1989): Fluorine NMR of Fluorinated Ligands. Methods Enzymol. 177, 3-22.

Gribble, W. G.; Keavy, J. D.; Olsen, R. E.; Rae, I. D.; Staffa, A.; Herr, T. E.; Ferraro, M. B.; Contreras, R. H. (1991): Fluorine Deshielding in the Proximity of a Methyl Group. An Experimental and Theoretical Study. Mag. Res. Chem. 19, 422-432.

Hull, E.; Sykes, B. D. (1976): 19F-NMR of Fluorotyrosine Alkaline Phosphatase. Biochemistry 15, 1535-1560.

Liu, R. S. H.; Matsumoto, H.; Asato, A. E.; Denny, M.; Shichida, Y.; Yoshizawa, T.; Dahlquist, F. W. (1981): Synthesis and Properties of 12-Fluororetinal and 12-Fluororhodopsin. A Model System for 19F NMR Studies of Visual Pigments. J. Am. Chem. Soc. 103, 7195-7201.

Liu, R. S. H.; Asato, A. E.; Denny, M.; Mead, D. (1984): The Nature of Restrictions in the Binding Site of Rhodopsin. J. Am. Chem. Soc. 106, 8298-8300.

Liu, R. S. H.; Mirzadegan, T. (1988): The Shape of a Three-Dimensional Binding Site of Rhodopsin Based on Molecular Modeling Analyses of Isomeric and Other Pigment Analogues. J. Am. Chem. Soc. 110, 8617-8623.

Liu, R. S. H.; Asato, A. E. (1990): The Binding Site of Opsin Based on Analog Studies with Isomeric, Fluorinated, Alkylated and Other Modified Retinals. In "Chemistry and Biology of Synthetic Retinoids" eds., M. Dawson & W. H. Okamura, CRC Press, 51-75.

Liu, R. S. H.; Crescitelli, F.; Denny, M.; Matsumoto, H.; Asato, A. E. (1986): Photosensitivity of 10-Substituted Visual Pigment Analogues: Detection of a Specific Secondary Opsin-Retinal Interaction. Biochemistry 25, 7026-7030.

Mateescu, G. D.; Abrahamson, E. W.; Shriver, J. W.; Copan, W.; Muccio, D.; Igbal, M.; Waterhaus, V. (1983): Solution and Solid State C-13 and N-15 NMR Studies of Visual Pigments & Related Systems: Rhodopsin & Bacteriorhodopsin. In "Spectroscopy of Biological Molecules" NATO ASI Series, eds. C. Sandorfy & T. Thegshanides, D. Reidd Publisher, 257-290.

Mead, D.; Loh, R.; Asato, A. E.; Liu, R. S. H. (1985): Fluorinated Retinoids via Crossed Aldol Condensation of 1,1,1-Trifluoroacetone. Tetrahedron Lett. 26, 2873-2876.

Mirzadegan, T.; Liu, R, S. H. (1991): Probing the Visual Pigment Rhodopsin and Its Analogs by Molecular Modelling Analysis and Computer Graphics. Prog. Retinal Res. 11, 57-74.

Mollevanger, L. C. P. J.; Kentgens, A. P. M.; Pardoen, J. A.; Veeman, W. S.; Lugtenburg, J.; deGrip, W. J. (1987): High-resolution Solid-state 13C-NMR Study of C-5 and C-12 of the Chromophore of Bovine Rhodopsin. FEBS Lett.,163, 9-14.

Smith, S.; Palings, I.; Miley, M.; Courtin, J.; deGroot, H.; Lugtenburg, J.; Mathies, R.; Griffin, R. (1990): Solid-state NMR Studies of the Mechanism of the Opsin Shift in the Visual Pigment Rhodopsin. Biochemistry 29, 8158-8164.

The rhodopsin-retinochrome system in the squid visual cell

Tomiyuki Hara, Reiko Hara, Ikuko Hara-Nishimura[1], Mikio Nishimura[1], Akihisa Terakita[2] and Koichi Ozaki[3]

Department of Biology, Kinki University School of Medicine, Osaka-Sayama, Osaka 589, [1] Department of Cell Biology, National Institute for Basic Biology, Okazaki 444, [2] Institute of Biology, Faculty of Education, Oita University, Oita 870-11, and [3] Department of Biology, Faculty of Science, Osaka University, Toyonaka, Osaka 560, Japan

Since vision is triggered by the photoreception of visual pigment, photoreceptive membranes of the visual cell need to be steadily replenished with rhodopsin. Interestingly, in the squid, the visual cell possesses two types of photopigment systems, each containing *rhodopsin* in the rhabdomeres of the outer segments and *retinochrome* in the myeloid bodies of the inner segments; irradiation of one photopigment finally assists in the regeneration of the other in the dark (1). At this time, 11-*cis*-retinal required for rhodopsin formation is generated by photoreaction of retinochrome to metaretinochrome, while all-*trans*-retinal for retinochrome formation is produced by conversion of rhodopsin to metarhodopsin. Between these two systems, a retinal-binding protein (*RALBP*) not only serves as an intracellular shuttle of the retinals but also reacts with meta-pigments to interchange the retinals (2). Consequently, in the visual cell, a set of three retinal proteins, rhodopsin, retinochrome and RALBP (Fig. 1) forms a conjugate system (*rhodopsin-retinochrome system*) to continue the recycling of retinal for the efficient regeneration of photopigments (3). In the squid retina which lacks structures similar to the pigment epithelium found in vertebrate retinas, the visual cell provides itself with the biochemical mechanisms necessary for maintaining its photosensitivity.

In order to examine the translocation of retinal between RALBP and meta-pigments, 3-dehydroretinal (retinal$_2$) was used as a tracer for retinal (4). All-*trans*-retinal$_1$-bearing RALBP was mixed with metaretinochrome$_2$-carrying membranes (myeloid bodies). After incubation in the dark for a few hours, the mixture was centrifuged to obtain a RALBP-containing supernatant and a membrane precipitate, each of which was assayed by spectrophotometry and HPLC.

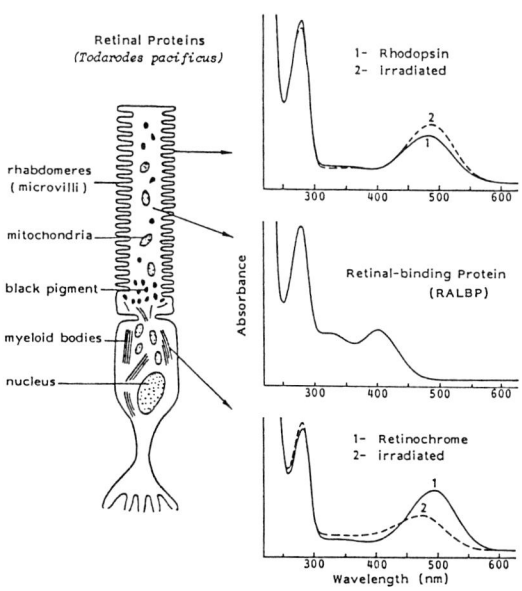

Fig. 1. Scheme of cellular constitution and absorption spectra of three retinal proteins.

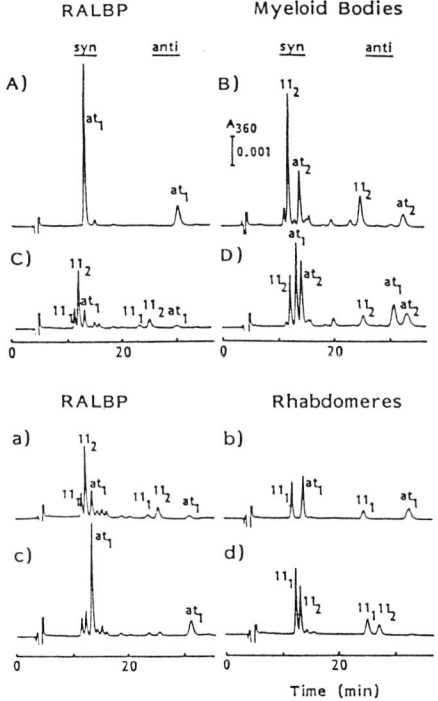

Fig. 2. The exchange of retinal during incubation of RALBP with metaretinochrome (upper) or metarhodopsin (lower) membranes. A and B (a and b), before incubation. C and D (c and d), after incubation.

The HPLC data are briefly presented in Fig. 2. During incubation, the all-*trans*-retinal$_1$ initially bound to RALBP (A) was moved to the membrane fraction (D), while the 11-*cis*-retinal$_2$ chromophore of metaretinochrome in the membranes (B) was transferred into the RALBP fraction (C). Since retinochrome$_1$ was found in the myeloid bodies after incubation, it was clear that, when mixed with all-*trans*-retinal$_1$-bearing RALBP, metaretinochrome$_2$ membranes took up all-*trans*-retinal$_1$ to form retinochrome$_1$, releasing 11-*cis*-retinal$_2$ to RALBP. In the following experiment, the 11-*cis*-retinal$_2$-bearing RALBP thus produced was incubated with metarhodopsin$_1$-containing membranes (rhabdomeres). The results by HPLC are also shown in Fig. 2. During incubation, the 11-*cis*-retinal$_2$ bound to RALBP (a) entered the membrane fraction (d), while the all-*trans*-retinal$_1$ chromophore of metarhodopsin in the membranes (b) moved into the RALBP fraction (c). After incubation, a newly formed rhodopsin$_2$ was detected by spectrophotometry in addition to rhodopsin$_1$ which had been present in the original membranes. It was thus clear that, when mixed with metarhodopsin$_1$ membranes, 11-*cis*-retinal$_2$-bearing RALBP took up all-*trans*-retinal$_1$ from metarhodopsin$_1$, releasing 11-*cis* ligand to form rhodopsin$_2$ in the rhabdomeres.

Based upon the above *in vitro* experiments, the actual movement of RALBP within the cell was further examined in connection with the regeneration of rhodopsin and retinochrome during the dark periods following irradiation of the eyes and eyecups. As the results, it was ascertained that the 11-*cis*-retinal of metaretinochrome is sent by RALBP to the outer segments to form rhodopsin in the rhabdomal microvilli, and that the all-*trans*-retinal of metarhodopsin is carried to the inner segments to form retinochrome in the myeloid bodies (3). The outline of the *rhodopsin-retinochrome system* is schematically presented in Fig. 3. In this system, photopigment regeneration is characterized by the exchange of *cis* and *trans* retinals which simultaneously proceeds between RALBP and meta-pigment; the retinal chromophore of meta-pigment is received by RALBP, while the retinal ligand of RALBP is received by pigment protein. In this place, the light and dark reactions developing in the retinochrome system play an important role in constantly providing the visual cell with only two forms of retinal, 11-*cis* and all-*trans*, eliminating the

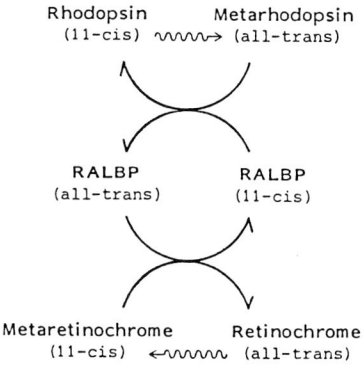

Fig. 3. Schematic presentation of the *rhodopsin-retinochrome system*.

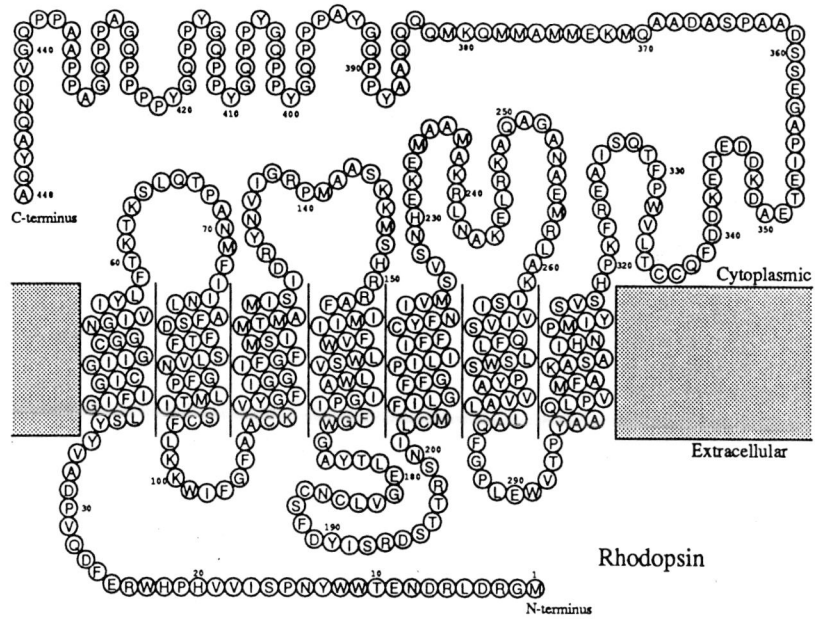

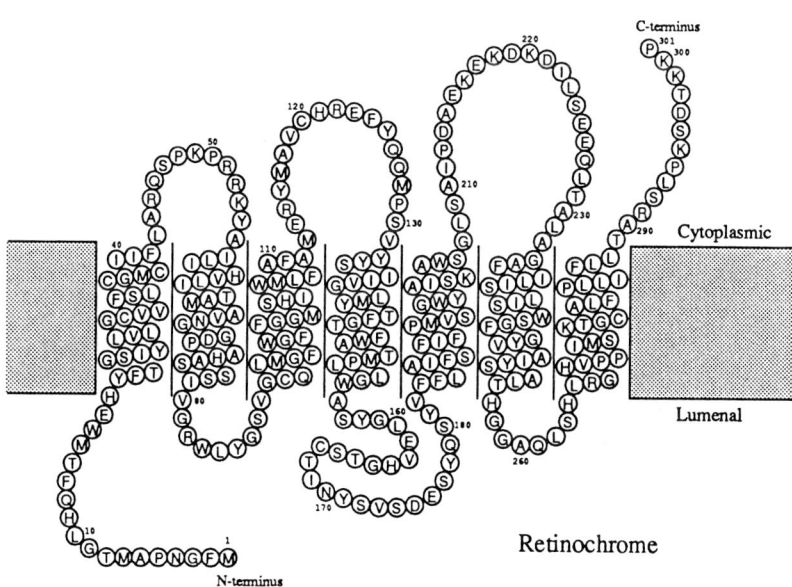

Fig. 4. Structural model of *Todarodes* rhodopsin and retinochrome.

other isomers of retinal (1). RALBP from the dark-adapted retina contains not only retinal but also retinol as endogenous ligands. When the RALBP is incubated with retinol dehydrogenase-containing membranes prepared from the retina, 11-*cis*-retinol ligand is specifically changed into the 11-*cis*-retinal, which may also be used for rhodopsin formation (3).

As described above, the two photoisomerization systems of rhodopsin and retinochrome are functionally combined with each other by the mediation of RALBP. However, details of the mechanism of the mutual exchange between RALBP and meta-pigments still remains to be examined physicochemically. Aiming at an understanding of this problem in the *rhodopsin-retinochrome system*, we set about the analyses of molecular structures of rhodopsin, retinochrome and RALBP in one species of the squid, *Todarodes pacificus*. They consisted of 448, 301 and 343 amino acids respectively, corresponding to M_r = 49,800, 33,500 and 39,200. Structural models of the two membrane proteins are shown in Fig. 4.

The molecular structures of cephalopod rhodopsins were previously analyzed in octopus (5) and *Loligo* (6). Comparison of them with *Todarodes* rhodopsin is separately described in this book. The analysis of hydropathicity demonstrated that *Todarodes* rhodopsin also possesses seven membrane spanning helices. The site of covalent binding of 11-*cis*-retinal chromophore was found at Lys-305 in the seventh helix. The size of the loop between helices 5 and 6 is far larger in squid rhodopsin than in vertebrate opsin. A characteristic feature of squid rhodopsin is the long C-terminal tail containing many blocks of repetitive sequences Pro-Pro-Gln-Gly-Tyr (PPQGY). Two possible N-linked glycosylation sites were observed at residues Asn-8 and Asn-185 in the extracellular space. Two cysteines, Cys-108 and Cys-186, may form an important disulfide bond to stabilize protein structure. Squid opsin also conserves the essential sequence Asp-Arg-Tyr near the cytoplasmic end of helix 3 for G-protein binding. Intramembranous proline residues which may serve to keep the polyene chain of retinal stable in the membrane were observed at positions 90, 170, 212, 276, 300 and 312 in helices 2, 4, 5, 6, 7 and 7, respectively. Unlike bovine rhodopsin, squid rhodopsin has the proline residue in helix 2 but not in helix 1. Close to helix 7 in the C-terminal region, two cysteines, Cys-336 and Cys-337 may be palmitoylated. In the C-terminal region of squid rhodopsin, there was no distinct serine- and threonine-rich part which has been indicated for light-dependent phosphorylation in vertebrate and insect rhodopsins. It is noticed that, even in the region except the PPQGY repetitive tail, squid rhodopsin is far more abundant not only in alanine but also in aspartic or glutamic acid than vertebrate visual pigments. The region from Asp-347 to Glu-358 appears to form a binding domain for calcium. We consider that this domain may play an important role for rhodopsin regeneration, associated with the retinal exchange reaction between RALBP and meta-pigment in the *rhodopsin-retinochrome system*.

In squid retinochrome (7), the C-terminal region is extremely short compared with rhodopsin. The retinal binding site of retinochrome (all-*trans*) was determined to be Lys-275 in the seventh helix, and the amino acid sequence around this site greatly differs from that of rhodopsin. Unlike rhodopsin, retinochrome lacks proline in helix 6. These findings are probably associated with the *trans*-to-*cis* photoisomerase activity characteristic of retinochrome. One possible glycosylation site is found at Asn-170, and the sequence Asp-Arg-Tyr near the cytoplasmic end of helix 3 in rhodopsin is altered to Glu-Arg-Tyr in retinochrome. The region from Asp-213 to Glu-225 may also be a significant domain for calcium binding.

REFERENCES

1) Hara, T. and Hara, R. (1987) In *Retinal Proteins*, ed. Yu.A. Ovchinnikov, pp. 457-466. Utrecht: VNU Science Press.
2) Hara, R., Hara, T., Ozaki, K., Terakita, A., Eguchi, G., Kodama, R. and Takeuchi, T. (1987) In *Retinal Proteins*, ed. Yu.A. Ovchinnikov, pp. 447-456. Utrecht: VNU Science Press.
3) Hara, T. and Hara, R. (1991) In *Progress in Retinal Research*, eds N. Osborne and J. Chader, pp. 179-206. Oxford: Pergamon Press.
4) Terakita, A., Hara, R. and Hara, T. (1989) *Vision Res.* 29, 639-652.
5) Ovchinnikov, Yu.A., Abdulaev, N.G., Zolotarev, A.S., Artamonov, I.D. Bespalov, I.A., Dergachev, A.E. and Tsuda, M. (1988) *FEBS Lett.* 232, 69-72.
6) Hall, M.D., Hoon, M.A., Ryba, N.J.P., Pottinger J.D.D., Keen, J.N., Saibil, H.R. and Findlay, J.B.C. (1991) *Biochem. J.* 274, 35-40.
7) Hara-Nishimura, I., Matsumoto, T., Mori, H., Nishimura, M., Hara, R. and Hara, T. (1990) *FEBS Lett.* 271, 106-110.

Structural comparison of retinal photopigments in cephalopods

Ikuko Hara-Nishimura, Maki Kondo, Mikio Nishimura, Reiko Hara* and Tomiyuki Hara*

*Department of Cell Biology, National Institute for Basic Biology, Okazaki 444 and *Department of Biology, Kinki University School of Medicine, Osaka-Sayama, Osaka 589, Japan*

In the molluscan retina, the visual cell possesses two types of photopigments, rhodopsin and retinochrome. Retinochrome is a photosensitive pigment that was originally extracted from the retina of the Japanese common squid, *Todarodes pacificus* (1). In cells, the visual pigment, rhodopsin, is located only in the rhabdomal microvilli of the outer segment, while retinochrome is stored in the myeloid bodies of the inner segments (2). The most marked difference between these photopigments lies in the stereoisomeric form of their chromophores, which is 11-*cis* in rhodopsin but all-*trans* in retinochrome. Irradiation of one photopigment assists in the regeneration of the other in the dark. At this time, 11-*cis*-retinal required for rhodopsin formation is generated by photoconversion of retinochrome to metaretinochrome. Particularly, retinochrome is capable of serving as an effective catalyst in the light to convert various isomers of retinal entirely into the 11-*cis* form (3).

We have been interested in the molecular structure of retinochrome, which severely differs from rhodopsin not only in its stereospecific photoisomerization of chromophoric retinal but also in intracellular location, photolytic behavior and other chemical properties. We previously reported the primary structure of *Todarodes* retinochrome deduced from the cDNA sequence (4), showing that retinochrome contains seven transmembrane spanning domains like rhodopsins. Recently we have identified the retinal-binding site in the retinochrome molecule, and compared the amino acid sequence around it with those of other retinal photopigments.

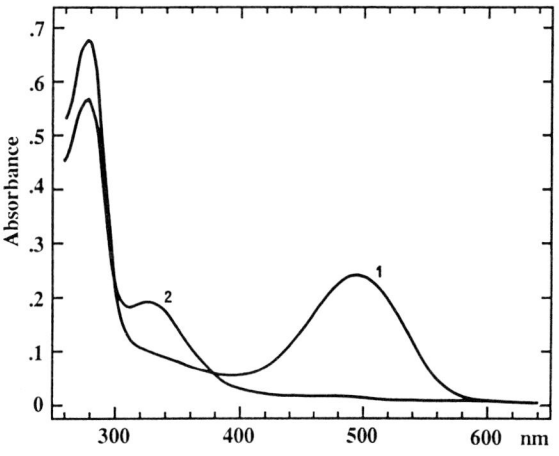

Fig. 1. Absorption spectra of retinochrome (1) and retinyl retinochrome (2).

Purified retinochrome was prepared from the retina of the squid (*Todarodes pacificus*) by our routine method (3) using 12 mM 3-[(3-cholamidopropyl) dimethylammonio]-1-propanesulfonate (CHAPS) in 67mM phosphate buffer, pH 6.5 (spectrum 1 in Fig. 1), acidified to

Table 1. Amino acid sequence of the retinyl peptide determined by Edman degradation.

Cycle No.	PTH-amino acid identified	Amount recovered (pmol)
1	Ser	13.3
2	Lys	3.9
3	Thr	8.6
4	Gly	47.7
5	-	-
6	Ala	8.0
7	Leu	6.9
8	Phe	5.7

pH 1.4 with HCl and reduced with borane dimethylamine. The absorption maximum was shifted from 495 to 330 nm, showing the formation of N-retinyl retinochrome (spectrum 2 in Fig. 1). The retinyl protein was then cleaved by cyanogen bromide (CNBr) under N_2 atmosphere. The resulted retinyl peptide was separated on HPLC while being monitored for absorbances at 215 and 330 nm. Finally, only one peak with absorbance at 330 nm was found through the chromatography. This fraction was subjected to amino acid sequence analysis by Edman degradation, using a gas phase automatic peptide sequencer. All these procedures were carried out under dim red light. The N-terminal amino acid sequence of the isolated retinyl peptide was determined to be Ser-Lys-Thr-Gly-X-Ala-Lue-Phe-Pro. This sequence was the same that we had found in the seventh membrane spanning domain of retinochrome (4). Table 1 shows the amount of PTH-amino acid from each cycle of the stepwise Edman degradation of the peptide. During Edman degradation, the yield of the PTH-lysine at the second cycle was lower than those of the other PTH-amino acids, indicating that the lysine of the peptide (Lys-275 of retinochrome) is involved in a retinal Schiff's base.

Sequence homology in retinal-binding regions of various photopigments is shown in Fig. 2. The amino acid sequence surrounding the retinal-binding lysine is very homologous among many rhodopsins and cone pigments. However, the sequence around the lysine in retinochrome is greatly different from the other visual pigments. It is only the proline at the seventh position following the retinal-binding lysine that is conserved throughout the photopigments. The phenylalanine at the second position preceding the lysine and the tyrosine at the tenth following the lysine are conserved in rhodopsins and cone pigments. In retinochrome, however, those phenylalanine and tyrosine are substituted by methionine and isoleucine, respectively. The above findings may explain the difference in the photoisomerization of chromophore between retinochrome and rhodopsin and the photoisomerase activity characteristic of retinochrome.

Squid (*Todarodes*) retinochrome	PPIMSKTGCALFPLLI
Squid (*Todarodes*) rhodopsin	PVMFAKASAIHNPMIY
Squid (*Loligo*) rhodopsin	PVMFAKASAIHNPMIY
Octopus rhodopsin	PVLFAKASAIHNPIVY
Drosophila rhodopsin	GACFAKSAACYNPIVY
Drosophila 8 cell pigment	GATFAKTSAVYNPVIY
Calliphora rhodopsin	GACFAKSAACYNPIVY
Bovine rhodopsin	PAFFAKTSAVYNPVIY
Ovine rhodopsin	PAFFAKSSSVYNPVIY
Porcine rhodopsin	PAFFAKSASIYNPVIY
Equine rhodopsin	PAFFAKSAAIYNPVIY
Chicken rhodopsin	PAFFAKSSAIYNPVIY
Chicken iodopsin	PAYFAKSATIYNPIIY
Human rhodopsin	PAFFAKSAAIYNPVIY
Human cone pigment (blue)	PSFFSKSACIYNPIIY
Human cone pigment (green)	PAFFAKSATIYNPVIY
Human cone pigment (red)	PAYFAKSATIYNPVIY

Fig. 2. Sequence homology in retinal-binding regions of various photopigments.

```
                                                            I
Todarodes  Rhodopsin   M--GRDLRDNETWWYNPSIVVHPHW-REFDQVPDAVYYSLGIFIGICGI
Loligo     Rhodopsin   .--...IP.........YMDI.....-KQ.....A...........A....
Octopus    Rhodopsin   .VESTT.V-.Q......TVDI....AK-...PI.......V......VV..


                                              II         100
IGCGGNGIVIYLFTKTKSLQTPANMFIINLAFSDFTFSLVNGFPLMTISCFLKKWIFGFAACKVYGFIGG
..V..V........................................M.Y.V..N.........L...
..IL...V...S..............M..LS..AI....K...A.M......KV..QL..LL..

  III                              IV
IFGFMSIMTMAMISIDRYNVIGRPMAASKKMSHRRAFIMIIFVWLWSVLWAIGPIFGWGAYTLEGVLCNC
...L....T................S........K......I..TI..................
......N.....................L.....M..IV.SV..V.N....VP..I.TS.

                  200             V
SFDYISRDSTTRSNILCMFILGFFGPILIIFFCYFNIVMSVSNHEKEMAAMAKRLNAKELRKAQAGANAE
.....T..T.......Y.FA.MC..VV.......................................
....L.T.PS...F....YFC..ML..I..A..................................S..

           VI                        300    VII
MRLAKISIVIVSQFLLSWSPYAVVALLAQFGPLEWVTPYAAQLPVMFAKASAIHNPMIYSVSHPKFREAI
.K.......T..................I...............................R.
.K....M..IT..M.......II......A.......E...L.........IV.........

SQT-FPWVLTCCQFDDKETEDDKDAETEIPAGESSDAAPSADAAQMKEMMAMMQKMQ---QQQAAY----
ASN-...I.....Y.E..I........A......Q.-GGET................AQQ...P..----
-..T...L........E..C..AN...E.VV.S.--RGGE.R..................---A.....QPPP

          400
PPQGYAPPPQGY---PPQG-Y-PPQGYPPQGYPPQGYPP---PPQGAPPQGAPP-AAPPQGVDNQAYQA*
.....--.....PPP....-.......................P....P..Q..............*
.....--.....---....A.P................QGY........VEA.QG............*
```

Todarodes	Rhodopsin:	Amino acid residues	448,	M_r	49,800
Loligo	Rhodopsin:		452,		50,600
Octopus	Rhodopsin:		455,		50,300

Fig. 3. Comparison of cephalopod rhodopsins. (.) represents identical amino acid. Dotted areas indicate transmembrane domains.

The squid visual cell contains rhodopsin, retinochrome and a retinal binding protein (RALBP) to form the *rhodopsin-retinochrome system*, which contributes to photopigment regeneration (5). To understand the mechanism of the flow of retinal within the cell, we also isolated a cDNA clone for *Todarodes* rhodopsin and deduced the protein structure. cDNA library constructed from poly(A)+RNA of the squid retina was screened. The resultant rhodopsin cDNA (3.1 kb) covered the whole coding regions of 1,334 bp. The hydropathicity profile suggested that the rhodopsin possesses seven membrane spanning domains. In Fig. 3, the amino acid sequence of *Todarodes* rhodopsin is presented together with those of *Loligo* (6) and octopus (7) rhodopsins to facilitate comparison. *Todarodes* rhodopsin fairly resembled *Loligo* and octopus rhodopsins in the primary structure, containing many blocks of Pro-Pro-Gln-Gly-Tyr (PPQGY) in the C-terminal region.

Unlike octopus rhodopsin, *Todarodes* and *Loligo* rhodopsins have the same amino acid sequence in helices 2 and 7, accompanied by only one different residue in helix 6. In these three rhodopsins, the sequences of the cytoplasmic loops 1-2, 3-4 and 5-6 are well conserved without marked difference of amino acid residues. They are also entirely the same in various points, i.e., the glycosylation site (Asn-8 in *Todarodes* rhodopsin), the formation of disulfide bond in the extracellular space (Cys-108, Cys-186), the conserved sequence of Asp-Arg-Tyr in cytoplasmic loop 3-4 for G-protein binding, the presence of some intramembraneous prolines and the acylation with fatty acids in the C-terminal region (Cys-336, Cys-337). A possible retinal-binding site is located at lysine-305 in the seventh helix. The histidine residue located at the fifth position following the retinal-binding lysine seems to be characteristic of cephalopod rhodopsins, since it is replaced by tyrosine in vertebrate visual pigments. As was suggested in *Loligo* rhodopsin (6), the tyrosine in helix 3 may play a role equivalent to the retinal Schiff's base counterion. This Tyr-111 is well conserved in all the cephalopod rhodopsins.

In the C-terminal region of cephalopod rhodopsins, serine or threonine residues are rather scattered, three in the accessory loop following helix 7, but five in *Todarodes* and only two each in *Loligo* and octopus in the subsequent stretch. It is also noticed that, even in the region except the PPQGY repetitive tail, cephalopod rhodopsins are far more abundant not only in alanine (over 7) but also aspartic or glutamic acid (over 13) than vertebrate visual pigments. The region from Asp-347 to Glu-358 appears to form an EF hand motif for calcium binding, which has also been found in squid retinochrome. We consider that this domain may play an important role for rhodopsin regeneration, probably associated with the retinal exchange reaction between RALBP and meta-pigment in the *rhodopsin-retinochrome system*.

Cephalopod rhodopsins possess the additional long region of repetitive sequences containing proline and glutamine, as distinguished from vertebrate rhodopsins (from position 383 to 436). In either of the three cephalopod rhodopsins, such a repeat of pentapeptide occurs about ten times, containing at least six complete sets of PPQGY sequence. Before and after the above-mentioned repetitive sequence, there are two well conserved regions composed of 16 (367-382) and 13 (437-448) amino acid residues, respectively. The former region is enchased with many methionines, and the sequence is in a symmetrical arrangement centering around Ala-376. In any case, the three cephalopod rhodopsins so far examined showed a fairly high similarity in opsin structure. However, *Todarodes* rhodopsin cDNA had a very long 3'-noncoding region which includes multiple polyadenylation signals (AATAAA), compared with those in *Loligo* and octopus. Such a long noncoding region was also observed in the cases of *Todarodes* retinochrome and RALBP cDNAs.

REFERENCES

1) Hara, T. and Hara, R. (1965) *Nature (London)* 206, 1331-1334.
2) Ozaki, K., Hara, R. and Hara, T. (1983) *Cell Tissue Res.* 233, 335-345.
3) Hara, T. and Hara, R. (1982) *Methods in Enzymology* 81, 827-833.
4) Hara-Nishimura, I., Matsumoto, T., Mori, H., Nishimura, M., Hara, R. and Hara, T. (1990) *FEBS Lett.* 271, 106-110.
5) Hara, T. and Hara, R. (1991) In *Progress in Retinal Research*, eds N. Osborne and J. Chader, pp. 179-206. Oxford: Pergamon Press.
6) Hall, M.D., Hoon, M.A., Ryba, N.J.P., Pottinger J.D.D., Keen, J.N., Saibil, H.R. and Findlay, J.B.C. (1991) *Biochem. J.* 274, 35-40.
7) Ovchinnikov, Yu.A., Abdulaev, N.G., Zolotarev, A.S., Artamonov, I.D. Bespalov, I.A., Dergachev, A.E. and Tsuda, M. (1988) *FEBS Lett.* 232, 69-72.

A structural study of retinochrome using fluorinated retinal analogues

Noriko Sekiya[1], Akio Kishigami[2], Fumio Tokunaga[2], Tetsuo Takahashi[1] and Kazuo Yoshihara[1]

[1] Suntory Institute for Bioorganic Research (Sunbor), 1-1, Wakayamadai, Shimamoto-cho, Osaka 618, Japan, and [2] Department of Biology, Faculty of Science, Osaka University, Machikaneyama, Toyonaka, Osaka 560, Japan

Retinochrome is a photosensitive protein in the visual cells of cephalopods. The protein has an all-trans-retinal chromophore bound to a lysine residue of the protein through a protonated Schiff base bond, and the chromophore photoisomerizes to the 11-cis-isomer (Hara & Hara, 1968; Hara & Hara, 1972). The other photosensitive protein, cephalopod rhodopsin (Hara & Hara, 1967;), has 11-cis-retinal Schiff base chromophore which photoisomerizes to the all-trans-isomer. The complementary isomerizations of the chromophore in rhodopsin and retinochrome indicate that the function of retinochrome is regeneration of 11-cis-retinal for rhodopsin synthesis (Terakita et al., 1989). So far, retinochrome appears to isomerize any of the known all-trans-retinal analogues to the 11-cis-isomers. Structural studies of retinochrome and rhodopsin are useful methods which provide data concerning the mechanism of selective photoisomerizations in retinal proteins. Photoreaction processes of retinochrome have been studied mainly by UV-visible absorption spectroscopy. On the photolytic process of retinochrome (λ_{max} 496 nm), intermediates, prelumiretinochrome (λ_{max} 465 nm) (Tokunaga et al., 1990; Kobayashi et al., 1986) lumiretinochrome (λ_{max} 475 nm), and metaretinochrome (λ_{max} 470 nm) (Hara et al., 1981), were detected. Chromophore structure of retinochrome intermediates has been characterized to be 11-cis by chromophore extraction experiment (Hara et al., 1981).

A variety of retinal analogues has been incorporated into the proteins to investigate chromophore-protein interactions in rhodopsin and bacteriorhodopsins, such as a series of alkylated and fluorinated analogues in rhodopsin (Crescitelli & Liu, 1988). The strong electron withdrawing property of fluorine atom is expected to affect both the electron distributions on the chromophore and the electrostatic interactions in the proteins without change in the steric environment (Shichida et al., 1987). This paper describes our preliminary study to detect such a chromophore-protein interaction in retinochrome analogues containing fluorinated (F-) retinals as the chromophore.

Regeneration of the analogue proteins (Fig.1 solid lines). Absorption spectra of the regeneration process of the pigments (pH 6.5, at 25°C) from incorporation of all-trans-10- and all-trans-12-F-retinals afforded each single absorption of λ_{max} at 495 and 500 nm, respectively, similar to the λ_{max} of native retinochrome (496 nm). This suggests that the fluorine atoms at C10 and C12 in the all-trans-chromophore do not significantly interact with the protein moiety.

Regeneration process of pigment from 14-F-retinal showed formation of two absorption bands, a weak red-shifted absorption with λ_{max} at 515 nm and another intense blue-shifted absorption with λ_{max} at 375 nm. In general non-protonated Schiff bases of retinals show blue-shifted absorption from the parent retinals. BSA washing caused no substantial decrease of the 375 nm absorption, indicating that the chromophore was covalently bonded with the protein. The major species with λ max at 375 nm and the minor species with λ_{max} at 515 nm are possibly retinochrome analogues with non-protonated and protonated Schiff base bonds, respectively. Probably the fluorine atom at C14, in the vicinity of the Schiff base, electrostatically interacts with the counterion and then partially interferes with Schiff base protonation from the counterion.

This kind of perturbation has been previously observed in 14-F-bacteriorhodopsin (Tierno et al., 1990). This analogue formed a relatively stable red-shifted species with all-trans chromophore, and an all-trans-15-syn structure was proposed for the chromophore. It was assumed that the fluorine atom interfered with Schiff base protonation from the native counterion by electrostatic interaction. It is likely that, in 14-F-retinochrome, the fluorine atom similarly interferes with the native counterion and permits only partial protonation of the Schiff base. This results in a larger population of non-protonated vs protonated species in the equilibrium. If correct, only 14-s-trans-15- anti conformation of the all-trans-chromophore (Fig. 2) allows the fluorine atom at C14 to interact with the counterion. It is not clear whether the native counterion or another carboxylate serves as the proton donor. Another possible explanation is that fluorination decreases the pKa value of the Schiff base proton (Sheves et al., 1986). However this decrease would be too small to cause Schiff base to be non-protonated in 14-F-retinochrome.

After acidification of the 14-F-retinochrome to pH 4.0, a new absorption of the λ_{max} at 496 nm was formed, while the absorption of λ_{max} at 375 nm disappeared. The 496 nm absorption could be due to Schiff base protonation by exogeneous ions. It is blue-shifted from 515 nm for the protonated species at pH 6.5. A similar observation was reported for E113Q-rhodopsin (Sakumar et al., 1989).

In a secondary structure for retinochrome composed of seven membrane spanning helices (Hara-Nishimura et al., 1990), lysine 275 in the seventh transmembrane domain was suggested to form the Schiff base with retinal. Similar to a squid rhodopsin (Hall et al., 1991), retinochrome has no aspartic acid or glutamic acid residue located in the extracellular site of the third helix, site for the counterion in bovine rhodopsin (Sakumar et al., 1989). Only three amino acid residues, Glu-17 in the first helix, Asp-71 in the second helix, and Glu-114 in the third helix, are candidates for the counterion of the Schiff base.

Photoreaction of the retinochrome analogues (Fig.1 dotted lines). Irradiation of retinochrome (550 ± 10 nm, 20°C) afforded metaretinochrome with a λ_{max} at 470 nm. Under the same conditions 10-F-retinochrome was converted to the meta-form with a λ_{max} at 480 nm. This suggests that no charged group in the protein significantly interacts with the fluorine atom at C10 in the 11-cis-chromophore.

Irradiation of 12-F-retinochrome and 14-F-retinochrome resulted in disappearance of the red-shifted absorptions and formation of new absorptions in the near UV region. Disappearance of these absorptions with BSA washing suggests that they were mainly due to the aldehydes with no chemical bonding with the protein. The chromophores in all the photoproducts were composed of more than 80 % of 11-cis isomers accompanying less than 10 % of all-trans-isomers.

The specific control of the isomerization suggests that the Schiff base bonds are cleaved after the isomerization. The bond cleavage is facilitated by the fluorine atoms at C12 and C14 in the 11-cis-chromophores. Although location of the C10 fluorine atom in both all-trans- and 11-cis- chromophores are the same, the location of the C12 and C14 fluorine atoms in the 11-cis-chromophores are changed to opposite side in the all-trans-isomers (Fig. 2). A possible explanation for the Schiff base cleavage is that the fluorine atoms causes conformational distortion of the chromophore by an electrostatic interaction with the charged groups of the protein. This results in destabilization of the Schiff base bonding. So far a few photochemical reactions of artificial retinochromes have been reported, and none have mentioned the formation of a meta-product.

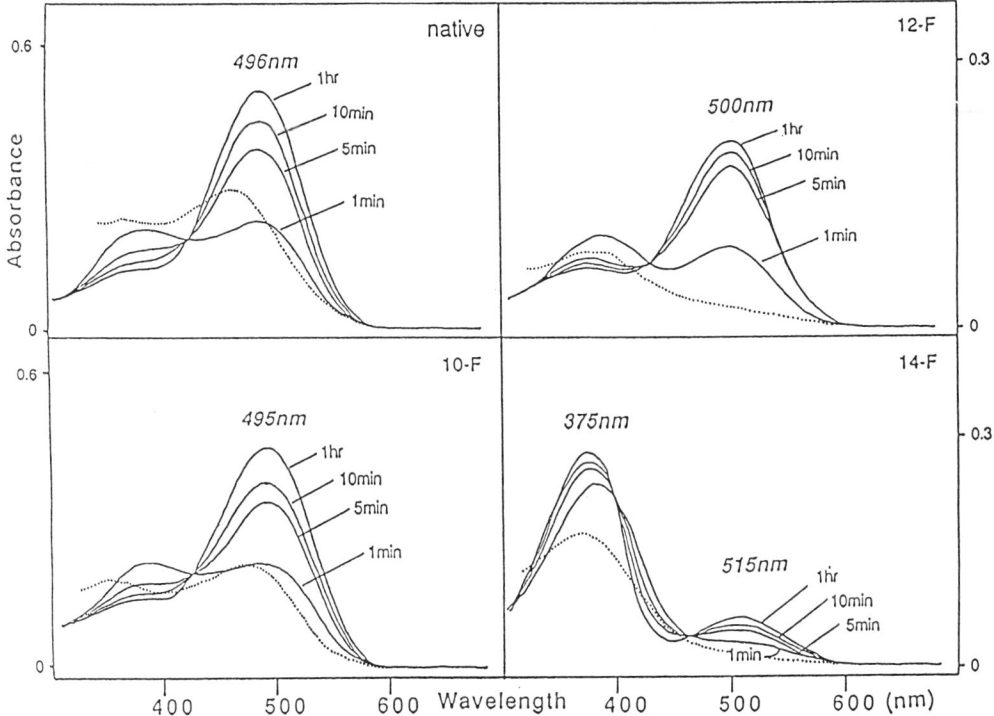

Fig. 1. Absorption spectra of pigment formation process with all-trans-retinal and the fluorinated analogues in solid lines and those photoproducts in dotted lines.

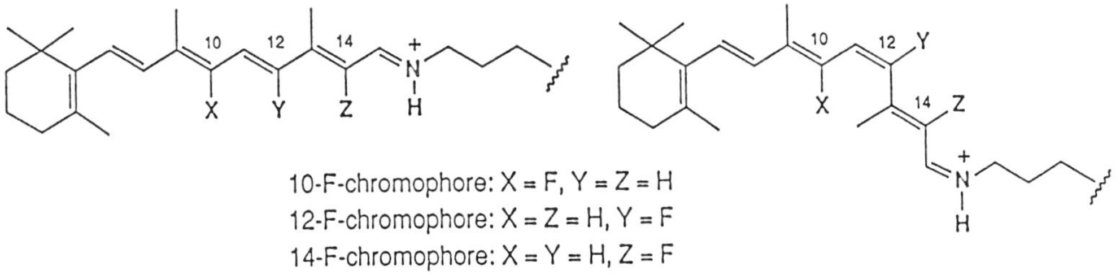

10-F-chromophore: X = F, Y = Z = H
12-F-chromophore: X = Z = H, Y = F
14-F-chromophore: X = Y = H, Z = F

Fig. 2. Structures of all-trans-15-anti-(left) and 11-cis-15-syn-chromophore (right).

Other 14-substituted Analogues. Incorporation of all-trans-14-chloro and 14-methylretinal analogues afforded red-shifted pigments with λ_{max} at 535 nm and 495 nm, respectively. It suggested that the protein environment near C14 hydrogen is not sterically restricted. The less negatively charged chlorine atom does not affect protonation of the Schiff base, however the substantial red-shift of the analogue from the native protein could be due to the longer distance between the counterion and the Schiff base caused by weak the chlorine-counterion interaction. Both pigments released free aldehydes after the photoreaction.

References

Crescitelli, F & Liu, R. S. H. (1988): The Spectral Properties and Photosensitivities of Analogue Photopigments Regenerated with 10 and 14-Substituted Retinal Analogues. In Proc. R. Soc. Lond. B Biol. Sci. 233 (1270), 55 - 76.

Kobayashi, T., Ogasawara, K., Koshihara, S., Ichimura, K. and Hara, R. (1986): Photochemistry of retinochrome studied by nanosecond and picosecond spectroscopy. In Primary Processes in Photobiology, ed. by Kobayashi, T. (Springer-verlag, Berlin), 125 - 133

Hall, M. D., Hoon, M. A., Ryba, N. J. P., Pottinger, J. D. D., Keen, J. N., Saibal, H. R., and Findlay, J. B. (1991): Molecular cloning and primary structure of squid (Loligo forbesi) rhodopsin, a phospholipase C-directed G-protein-linked receptor. Biochem. J. 274, 35 - 40.

Hara, T. and Hara, R. (1967): Vision in octopus and squid. Nature 214, 572 - 575.

Hara, T. and Hara, R. (1968): Regeneration of squid retinochrome. Nature 219, 450 - 454.

Hara, T. and Hara, R. (1972): Cephalopod retinochrome. In the Handbook of Sensory Physiology vol. VII/I ed. by Dartnel, H. J. A., (Springer, Berlin), 720 - 746.

Hara, T., Hara, R., Tokunaga, F., and Yoshizawa, T. (1981): Photochemistry of retinochrome. Photochem. Photobiol. 33, 883 - 891.

Hara-Nishimura, I., Matsumoto, T., Mori, H., Nishimura, M., Hara, R., and Hara, T. (1990): Cloning and nucleotide sequence of c-DNA for retinochrome, retinal isomerase from the squid retina. FEBS Lett. 271, 1 - 2.

Sakumar, T. P., Franke, R. R., and Khorana, H. G. (1989): Glutamic acid-113 serves as the retinylidene Schiff base counterion in bovine rhodopsin. Proc. Natl. Acad. Sci. USA 86, 8309 - 8313.

Sheves, M., Albeck, A., Friedman, N., and Ottolenghi, M. (1986): Controlling the pKa of bacteriorhodopsin Schiff base by use of artificial retinal analogues. Proc. Natl. Acad. Sci. USA 83, 3262 - 3266.

Shichida, Y., Ono, T., Yoshizawa, T., Matsumoto, H., Asato, A. E., Zingoni, J. P., and Liu, R. S. H. (1987): Electrostatic Interaction between Retinylidene Chromophore and Opsin in Rhodopsin Studied by Fluorinated Rhodopsin Analogues. Biochemistry 26, 4422 - 4428.

Terakita, T., Hara, R., and Hara, T. (1989): Retinal binding protein as a shuttle for retinal in the rhodopsin-retinochrome system of the squid visual cells. Vision Research 29, 639 - 652.

Tierno, M. E., Mead, D., Asato, A. E., Liu, R. S. H., Sekiya, N., Yoshihara, K., Chang, C. W., Nakanishi, K., Govindjee, R., and Ebrey, T. G. (1990): 14-Fluoro-bacteriorhodopsin and other fluorinated and 14-substituted analogues. An extra unusually red-shifted pigment formed during dark adaptation. Biochemistry 29, 5948 - 5953.

Tokunaga, F., Watanabe, T., Uematsu, J., Hara, R., and Hara, T. (1990): Photoreaction of retinochrome at very low temperatures. FEBS Lett. 262, 266 - 268.

Retinal photoisomerase from honeybee compound eye : physiological role

I.M. Pepe, C. Cugnoli, J.N. Keen and J.B.C. Findlay

Institute of Cybernetics and Biophysics of CNR, via Dodecaneso 33, 16146 Genova, Italy and Department of Biochemistry and Molecular Biology, University of Leeds, Leeds, UK

INTRODUCTION

In the visual cells of invertebrates, rhodopsin is converted by light to metarhodopsin, which is reconverted to rhodopsin by another photon absorption. In addition to this fast process of photoregeneration , rhodopsin is also regenerated by a slow process of membrane renewal, which takes days to complete and involves the biosynthesis of opsin.

Since 11-cis retinal is the only isomer which can combine with opsin to form rhodopsin, an isomerizing enzyme should exist, capable of transforming all-trans retinal, released from the degradation of metarhodopsin, into the 11-cis retinal isomer.

It is known that rhodopsin regeneration in cephalopod (Hara & Hara, 1984) and blowfly (Schwemer, 1984) visual cells, is based on a light-dependent process of isomerization involving a retinal-protein complex.

Two retinal isomerases have been isolated and well characterized: retinochrome, first isolated in photoreceptors of cephalopods (Hara & Hara, 1982) and a photoisomerase from the retina of honeybees (Pepe & Cugnoli, 1980; Schwemer et al., 1984). In both cases light in the visual range is required for the isomerization of all-trans retinal to 11-cis retinal.

Spectrophotometric characteristics
Retinal-photoisomerase has been extracted from homogenized honeybee heads in Tris-glycine buffer, incubated with tritiated retinol, isolated by preparative electrophoresis and further purified on a DEAE column (Pepe & Cugnoli, 1980; Schwemer et al., 1984).

The protein, first incubated with retinol and then purified, shows an absorbance maximum at 330 nm, which is typical of retinol free in solution, indicating that the binding of the protein with retinol is presumably not covalent. When the protein, prior to purification, is incubated with retinal instead, it absorbs maximally at 440 nm, indicating the formation of a protonated Schiff-base linkage between retinal and protein (Pepe et al., 1987).

Therefore the protein is able to bind retinol as well as retinal. The complete displacement of radioactive retinol bound to protein by the addition of an excess of retinal suggests that the binding site for retinol is presumably the same as that for retinal. (Pepe et al., 1987).

The pigment has the characteristic absorbance behaviour of protonated (440nm) and unprotonated (365nm) Schiff-bases which form between retinal and amino groups. In fact, when the pH is changed from 6.5 to 9.5 in the dark, the 440-pigment is converted into an alkaline form which absorbs maximally at 365 nm, with a pK=8.4 (Pepe et al.,1982).

The molecular weight of the protein was determined to be 50,000 by gel filtration, and 27,000 by SDS polyacrylamide gel electrophoresis, suggesting that the protein consists of a dimer (Pepe & Cugnoli, 1980; Pepe et al., 1987). The binding between all-trans retinal and protein was studied, and the stoichiometry resulted of one molecule of retinal bound to one molecule of protein (dimer) with a dissociation constant of $K=2 \times 10^{-6}$ M (Pepe et al., 1987).

When irradiated with green light, the pigment bleaches with a decrease in the absorbance at 440 nm accompanied by the formation of a photoproduct with absorbance maximum at 370 nm. A fall in absorbance at 440 nm also occurs in the dark in the presence of hydroxylamine, but does not occur in the presence of cyanoborohydride. In the presence of the latter reagent, the pigment only bleaches in the light, suggesting that the binding site is buried in the interior structure of the protein. A change in conformation triggered by light could cause the Schiff-base linkage to become accessible to the reducing agent and consequently to be deprotonated (Pepe et al., 1982).

Enzymic activity
Upon irradiation, the extinction coefficient decreases from about 47,000 M^{-1} cm^{-1} of the 440-pigment to about 24,000 M^{-1} cm^{-1} of the photoproduct (Pepe et al., 1982), suggesting a conversion of the chromophore from all-trans retinal to a cis-isomer. When analysed by high performance liquid chromatography, the photoproduct was in fact identified as 11-cis retinal (Schwemer et al., 1984).

When all-trans retinal was added in excess to a solution containing the protein, and the mixture was then continuously irradiated with green light of wavelength 528 nm, which is only absorbed by retinal bound to protein and not by free retinal, it was found that all-trans retinal was almost exclusively transformed into 11-cis retinal (Schwemer et al., 1984).

Considering that photoisomerization of free retinal produces a mixture of isomers, with all-trans isomer being the major component (Kropf & Hubbard, 1970), the result of retinal irradiation in the presence of bee photoisomerase is remarkable. The protein is in fact able to direct the isomerization of retinal made by light almost exclusively toward the 11-cis retinal formation.

When other starting isomers, different from all-trans retinal, were used, such as 13-cis or 9-cis retinal, irradiation in the presence of bee photoisomerase always led to the stereospecific formation of 11-cis retinal (Cugnoli et al., 1989).

Physiological role

Bee photoisomerase may have different functions. Being water-soluble, it may act as a carrier for the transport of retinol or retinal between the different compartments of the compound eye and, in addition to these mere functions of transport, it could be involved in the fundamental process of visual pigment regeneration.

All-trans retinal released by the degradation of metarhodopsin could bind to photoisomerase and be transformed into 11-cis retinal in light of the visible range. The 11-cis retinal so produced might then be directly transferred to opsin in order to regenerate rhodopsin. This latter function is suggested by the following experiment: when a solution of photoisomerase, previously loaded with all-trans retinal and subsequently irradiated, was mixed with a suspension of opsin membranes from bleached bovine rod outer segments, bovine rhodopsin was reconstituted (Pepe et al., 1990).

However, it is not very likely that photoisomerase could provide the 11-cis retinal for the fast regeneration of rhodopsin as this occurs via the photon absorption by metarhodopsin . Photoisomerase, instead, could be involved in the slow renewal of rhabdomal membranes which implies the "ex novo" synthesis of rhodopsin, and which seems to be a light-dependent process as it is in flies.

Partial protein sequence

The protein was further purified by electroeluting the corresponding band from SDS-PAGE. Since the N-terminal was found to be blocked,the protein was digested with Staphylococcus aureus V8 protease . Four peptides obtained from the digestion were electroeluted after separation on SDS-PAGE. One peptide gave the following sequence of 29 aminoacids:

GLDRAIFIAXDVSKNEQFQESFKXVXDTY

A good homology with human prostaglandin dehydrogenase was found after scanning the ISIS protein database with this sequence (Akrigg et al., 1988). This latter protein, structured as a dimer (MW of monomer: 29,000), belongs to the family of short - chain alcohol dehydrogenases, which were first discovered in insects.

This brings to mind the fact that bee photoisomerase is able to bind retinal as well as retinol, so it may have dehydrogenase activity. However, experiments performed on photoisomerase solutions in the presence of NADPH and retinal or NADP and retinol in different conditions failed to detect any dehydrogenase activity , and also when the lability commonly associated to such activity was taken into account by the addition of glycerol to the extraction and purification steps of the protein.

REFERENCES

Akrigg, D., Bleasby, A.J., Dix,N.I.M., Findlay, J.B.C., North, A.C.T., Parry-Smith, D., Wootton, J.C., Blundell, T.L., Gardner, S.P., Hayes, F., Islam, S., Sternberg, M.J.E., Thornton, J.M.,

Tickle, I.J. and Murray-Rust, P. (1988): A protein sequence/structure database. *Nature* 335, 745 - 746.

Cugnoli, C., Mantovani, R., Fioravanti, R and Pepe, I.M (1989): 11-cis retinal formation in the light catalyzed by a retinal-binding protein from the honeybee retina. *FEBS Lett.* 257, 63-67.

Hara, T. and Hara, R. (1982): Cephalopod Retinochrome. *Methods Enzymol.* 81, 827 - 833.

Hara, R. and Hara, T. (1984): Squid m-retinochrome. *Vision Res.* 24, 1629-1640.

Kropf, A. and Hubbard, R (1970): The photoisomerization of retinal. *Photochem. Photobiol.* 12, 249 - 260.

Pepe, I.M., Cugnoli, C. (1980): Isolation and characterization of a water - soluble photopigment from honeybee compound eye. *Vision Res.* 20, 97 - 102.

Pepe, I.M., Cugnoli, C., Peluso, M., Vergani, L. and Boero, A. (1987): Structure of a protein catalyzing the formation of 11-cis retinal in the visual cycle of invertebrate eyes, *Cell Biophys.* 10, 15-22.

Pepe, I.M., Schwemer, J. and Paulsen, R. (1982): Characteristics of retinal - binding proteins from the honeybee retina. *Vision Res.* 22, 775 - 781.

Pepe, I.M., C.Cugnoli and J. Schwemer, Rhodopsin reconstitution in bleached rod outer segment membranes in the presence of a retinal-binding protein from the honeybee, *FEBS Lett.,* 268 (1990) 177-179.

Schwemer, J. (1984): Renewal of visual pigment in photoreceptors of the blowfly. *J. Comp. Physiol. A.* 154, 535 - 547.

Schwemer, J., Pepe, I.M., Paulsen, R. and Cugnoli, C. (1984): Light - induced trans - cis isomerization of retinal by a protein from honeybee retina. *J. Comp. Physiol.* 154, 549 - 554.

Smith, W.C. and Goldsmith, T.H. (1991): The role of retinal photoisomerase in the visual cycle of the honeybee, *J. Gen. Physiol.* 97, 143-165.

Retinal photoisomerase from honeybee compound eye: enzymic mechanism

C. Cugnoli, R. Fioravanti, O. Golisano and I.M. Pepe

Institute of Cybernetics and Biophysics of CNR, via Dodecaneso 33, 16146 Genova, Italy

INTRODUCTION

Retinal photoisomerase, isolated and purified from honeybee retina, is an important enzyme in the visual cycle which converts all-trans retinal released by the degradation of metarhodopsin to 11-cis retinal, which is the isomer necessary for rhodopsin regeneration (Pepe and Cugnoli, 1992 ; Smith and Goldsmith, 1991).

It is well known that the photoisomerization of free retinals takes place after irradiation in the near UV region. A photoequilibrium is rapidly reached and the resulting isomer mixture - containing the all-trans isomer as the major component - has a constant composition, irrespective of which isomer is used. (Kropf and Hubbard 1970). Instead, when retinal is bound to a protein its absorbance maximum is shifted to the blue-violet and its photoisomerization could become highly stereospecific.

The water-soluble retinal photoisomerase of the honeybee binds all-trans retinal via a Schiff-base linkage with an absorbance maximum at 440 nm (Pepe and Cugnoli, 1980 ; Pepe et al., 1982). When all-trans retinal in excess is added to a solution of purified photoisomerase, continuous irradiation with green light leads mainly to the formation of 11-cis retinal (Schwemer et al. 1984).

RESULTS

In a typical experiment where two fold excess of all-trans retinal was added to a solution of the photoisomerase, the saturation of the binding between retinal and protein was followed in the dark by measuring the increase of the absorbance at 440 nm. Then the sample was irradiated with light of wavelength 531 nm, which is only absorbed by the retinal bound to protein and not by free retinal. Retinal isomers were extracted, as already described (Pepe and Schwemer, 1987) and analysed by High Performance Liquid Chromatography (H.P.L.C.).

Table 1 shows the results of retinal photoisomerization at different times of irradiation.

TABLE 1

Time course of all-trans retinal photoisomerization by continuous irradiation with wavelength of 531 nm.

Time (min)	All-trans	11-cis	13-cis	9-cis
0	95.4	0.0	4.6	0.0
15	72.1	21.2	6.7	0.0
30	58.6	30.7	5.9	4.8
60	43.4	40.7	8.5	7.4
120	32.8	44.8	11.2	11.2

All-trans retinal (4×10^{-6} M) was added to a solution of bee photoisomerase (2×10^{-6} M) in 0.1 M phosphate buffer at pH 7, in the dark at room temperature. Figures represent the percentage of retinal isomers extracted from aliquots (100 µl) collected from the sample (1 ml) at the time indicated and then analyzed by H.P.L.C.

TABLE 2

Time course of all-trans retinal photoisomerization by continuous irradiation with LASER light of 515 nm.

Time (sec)	All-trans	11-cis	13-cis	9-cis
0	95.0	0.0	3.8	1.2
30	76.6	17.9	3.4	2.1
60	56.2	33.2	5.4	5.2
90	49.4	42.4	6.1	2.1
120	34.4	54.0	7.6	4.0
180	19.1	64.8	7.7	8.4

The sample contained about 3×10^{-6} M photoisomerase and a six-fold excess of all-trans retinal in 0.1 M phosphate buffer at room temperature.
Figures represent the percentage of retinal isomers extracted as in Table 1.

When the light intensity of irradiation was increased by using a LASER light of wavelength 515 nm and about 3000 fold more intense than the actinic light used in the preceding experiments, the results did not show the expected increase in the kinetics, except during the first few minutes where the initial kinetics were intensity dependent. After about one hour of irradiation, the 11-cis isomer produced from all-trans retinal reached a value about only 15 % higher than the corresponding value in Table 1 (data not shown).

In order to speed up the production of 11-cis retinal with LASER light, irradiation was only performed on saturating concentration of retinal bound to protein by incubating the sample for 30 min in the dark, then irradiating with 30 sec pulse of LASER light and followed again by 30 min of darkness and so on. The results of the repeated pulses of light and dark intervals are shown in Table 2 where only the times of irradiation are reported so as to follow exclusively the reactions in the light.

The photoisomerization of all-trans retinal was still mainly directed toward the 11-cis retinal formation and the speed of photoisomerization was greatly enhanced: in fact after 3 min of LASER irradiation , about 60% of all-trans retinal was transformed to 11-cis retinal.

DISCUSSION

The "dark" association between the enzyme and all-trans retinal is the rate limiting reaction of the whole isomerization process. For this reason an increase in isomerization rate was not observed during irradiation with a higher intensity of light. The photoisomerization reaction was therefore isolated by irradiating the samples with pulses of LASER light, followed by 30 min darkness intervals. The time constant of the photoisomerization process in the particular experimental conditions above described is most likely to be less than 30 sec.

These results tend to suggest that the mechanism of retinal photoisomerization catalyzed by bee isomerase could be explained by different steps. The light-dependent process of isomerization which transforms all-trans retinal bound to the protein into 11-cis retinal is followed by dark reactions : the molecule of newly formed 11-cis retinal is displaced by a molecule of all-trans retinal which comes into contact with the enzyme while it is diffusing, and which is in turn transformed into 11-cis retinal by light.

The dark reactions could consist of two steps with different time constants: all-trans retinal diffusion and displacement of 11- cis retinal from the binding site. In order to investigate this model experiments are in progress which are using better time-resolving techniques and improved mathematical analysis.

Acknowledgements: This work was supported by P.F. Ingegneria genetica, CNR

REFERENCES

Kropf, A. & Hubbard, R. (1970): The photoisomerization of retinal. *Photochem. Photobiol.* 12, 249 - 260.
Pepe, I.M. & Cugnoli, C. (1980). Isolation and characterization of a water - soluble photopigment fromhoneybee compound eye. *Vision Res.* 20 , 97 - 102.
Pepe, I.M. et al., (1982). Characteristics of retinal - binding

proteins from the honeybee retina. *Vision Res.* 22, 775 - 781.

Pepe, I.M. & Schwemer, J. (1987). An improved high performanceliquid chromatography method for the separation of retinaldehyde isomers from visual pigments. *Photochem.Photobiol.* 45, 679 - 681.

Pepe, I.M. & Cugnoli, C. (1992). Retinal Photoisomerase ; role in invertebrate visual cells. *J. Photochem. Photobiol. B :Biol.,* 13, 5 - 17.

Schwemer, J. et al.,(1984). Light - induced trans - cis isomerization of retinal by a protein from honeybee retina. *J. Comp. Physiol.* 154, 549 - 554.

Smith, W.C. & Goldsmith, T.H. (1991). The role of retinal photoisomerase in the visual cycle of the honeybee. *J. Gen. Physiol.* 97, 143 - 165.

Photodamage to interphotoreceptor retinoid-binding protein

Irina B. Fedorovich, Karen M. Grant*, Antoinette M.J. Brennan*,
Michail A. Ostrovsky and Carolyn A. Converse*

Russian Academy of Sciences, Institute of Chemical Physics, Kosygin Street 4, 117334, Moscow, Russia.
* Department of Pharmaceutical Sciences, University of Strathclyde, Glasgow G1 1XW, UK

Retinoid-containing proteins are essential to the visual process. They included the visual pigments plus a large group of retinoid-binding proteins of which interphotoreceptor retinoid-binding protein (IRBP) is one.

This water-soluble protein is found in the interphotoreceptor matrix of the eye (Adler et al.,1985). IRBP's ligands in vivo may include different retinoids (Adler & Spenser,1991). The putative function of this protein is transport of retinoids between photoreceptor cells and pigment epithelium. This is means that IRBP probably plays a key role in the process of rhodopsin regeneration and any deterioration of this protein may impair vision. It is also known that deterioration of rhodopsin regeneration is a feature of some eye diseases, for instance retinitis pigmentoza. It is possible that deterioration of IRBP may play a role in this process.

Since ocular structures are designed to allow light entry, the eye is particularly liable to light-induced injury (Ostrovsky et al., 1987). Retinal-photosensitised damage to rhodopsin has been studied previously (Pogozeva et al.,1981). Photodamage to another retinoid-binding protein, IRBP, was the subject of the present investigation. The most possible native photosensitizer in this case may be retinal because the wave-lengths absorbed by retinol are screened out and normally don't reach the retina. In this study we examined the photo-oxidation of IRBP using *all-trans* retinal and *all-trans* retinol as photosensitizers.

IRBP was prepared from fresh bovine eyes as previously described (Al-Mahdawi et al., 1990). SDS polyacrylamide gel electrophoresis showed one protein band with an apparent molecular weight of ~ 140 KDa. The fluorescence test showed that this protein did not contain retinoids.The SH-group content was determined using Ellman's reagent (5,5´-dithio-his(2-nitrobenzoic acid)) in the presence of cethyltrimethyl ammonium bromide (Pogozeva et al.,1981). The oxidation of aromatic acids in the protein was monitoring by fluorescence. IRBP in Tris-HCl buffer (pH 8,0) at a concentration of about 10^{-6} M was illuminated by a high pressure mercury vapor lamp). During investigation of IRBP with *all trans* retinol we illuminated by full spectrum light at 0.033 W/cm^2, and with *all trans* retinal we illuminated in its absorption band using a blue filter, at 0.012 W/cm^2.

In both cases the illumination resulted in bleaching of retinoids (Fig.1-a,b). Comparison of the bleaching of the dyes with and without IRBP showed that the velocity of bleaching in the present of protein was less in both cases, i.e. IRBP had some protecting properties against retinoid oxidation.

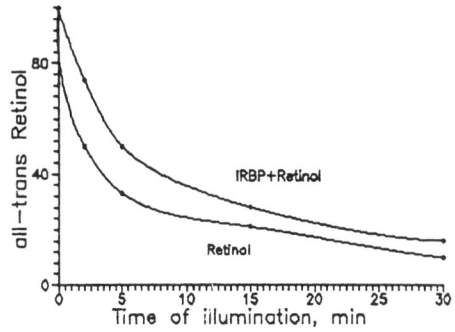
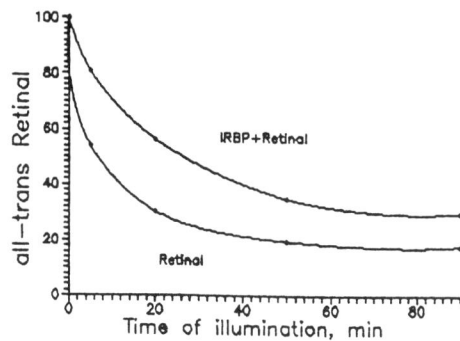

Fig.1. The velocity of all trans retinol (A) and all-trans retinal (B) bleaching during illumination with (1) and without (2) IRBP.

Also in both cases there was a dose-dependent decrease in the content of SH-groups (Fig.2). The adding of all trans retinal in the ratio 1.5:1 resulted in a 25% decrease in protein fluorescence. Illumination resulted in an additional dose-dependent decrease of protein fluorescence up to 36% after 60 min of illumination. This is thought to be the result of photo-damage to aromatic amino acids.

Thus the illumination of IRBP with retinal and retinol resulted in photosensitised oxidation of protein as was the case for rhodopsin. It is proposed that such photo-damage resulted in deterioration of IRBP function, that is, its ability to bind retinoids. This problem will be the object of further research.

Fig.2. Retinal sensitized photo-oxidation of IRBP SH-groups.

It is hoped that these studies will increase our knowledge of the function of this important retinoid-binding protein, and of the photo-induced deterioration of visual perception.

This research was supported by the Science and Engineering Research Council and the Royal Society - USSR Academy of Science Scientific Exchange Program.

References:

Adler A.J., Evans C.D. & Stafford,III W.R. (1985) Molecular properties of bovine interphotoreceptor retinol-binding protein. J.Biol.Chem.260: 4850-4855.

Adler A.J. & Spenser S.A. (1991) Effect of light on endogenous ligands carried by interphotoreceptor retinoid binding protein. Exp.Eye Res.53: 337-346

Al-Mahdawi S., McGettrick P.M., Lee W.R., Graham D.I., Shallal A. & Converse C.A. (1990) Experimental autoimmune uveoretinitis and pinealitis induced by interphotoreceptor retinoid-binding protein and S-antigen: induction of intraretinal and subretinal neovascularization. J.Clin.Lab.Immunol.32: 21-28

Ostrovsky M.A., Fedorovich I.B. & Dontsov A.E. (1987) The photooxidation processes in the eye structures. The protecting role of lens and screening pigments. Biofizika 32: 896-909

Pogozeva I.D., Fedorovich I.B., Ostrovsky M.A. & Emanuel N.M. (1981) Photodamage to rhodopsin molecule. SH-groups oxidation. Biofizika 26: 398-403

Opsin synthesis in blowfly photoreceptors is controlled by an 11-*cis* retinoid

Joachim Schwemer and Friederike Spengler*

*Department of Biophysics, University of Groningen, Westersingel 34, 9718 CM Groningen, The Netherlands and *Institute of Animal Physiology, Ruhr-University, Universitätsstrasse, 4630 Bochum 1, Germany*

INTRODUCTION

The visual pigment as well as the opsin content of blowfly photoreceptors R1-R6 depend upon the availability of vitamin A. When raised on a vitamin A-deficient diet, the adult blowflies lack visual pigment as well as opsin. Both, opsin and visual pigment increased concomitantly to the level of 'normal' flies after application of exogenous 11-*cis* retinal which was shown to be due to a *de novo* synthesis of opsin. All-*trans* retinal, however, did not trigger opsin synthesis, unless it was isomerized to the 11-*cis* congener, a reaction which is catalyzed by a photoisomerase (Schwemer, 1984, 1988; Paulsen and Schwemer, 1984). Moreover, the visual pigment of these photoreceptors undergoes a continuous renewal consisting of two processes, the degradation of the visual pigment in the metarhodopsin-state resulting in the release of the all-*trans* chromophore, and the biosynthesis of rhodopsin which requires the 11-*cis* chromophore. Both processes are linked by the light-dependent, highly stereospecific *trans* -> *cis* isomerization catalyzed by a soluble photoisomerase. Immediately after this isomerization, the aldehyde is reduced to the 11-*cis* alcohol. One of the two 11-*cis* retinoids seems to be involved in regulating opsin biosynthesis (Schwemer, 1988). The fact that the key reaction, the isomerization, depends upon (blue/violet) light allows the cycle to be dissected by either preventing or allowing the isomerization (Schwemer, 1984).

In the present study we analyzed the opsin mRNA levels of *Calliphora erythrocephala* and *Drosophila melanogaster* in relation to the availabilty of chromophore precursors and of specific isomers of the chromophore itself.

MATERIAL and METHODS

C. erythrocephala and *D. melanogaster* were used in this study. Chromophore and visual pigment content of the photoreceptor cells were modulated (1) by changing the vitamin A and carotenoid content of the larval diet (R^+-, R^--animals), (2) by application of exogenous retinal to R^- and (3) by experimental manipulation of the visual pigment cycle. The resulting visual pigment content was determined by microspectrophotometry *in situ* (*Calliphora*) and in digitonin extracts (*Drosophila*).

For Northern analysis, total RNA was isolated from eyes or heads according to a slightly modified method of Chirgwin et al. (1979). Poly(A)$^+$RNA was prepared according to Edmonds et al. (1971). In general, 2 µg of RNA per lane were electrophoresed on a 1% agarose-MOPS-HCHO-system. Fluorescence of the rRNA double bands was used to verify that the amounts of RNA applied to the gel were the same. RNA was then transferred to nylon membranes. The probes were hybridized with ^{32}P-labelled opsin cDNA from *Drosophila*. Dot blot analysis was carried out according to Thomas (1980). Relative opsin mRNA levels were estimated by densitometrically integrating and comparing the signals within single blots.

RESULTS and DISCUSSION

Since *Drosophila* and *Calliphora* are phylogenetically distant flies, the first essential of this study was to examine if hybridization of opsin cDNA from *Drosophila* to an eye-specific RNA from *Calliphora* could be observed in Northern blots. For this, total RNA was isolated from 50 whole heads, from 100 excised eyes as well as from the remaining 50 head capsules of R^+ which contained mainly the optic lobes and the brain. The three RNAs were electrophoresed, blotted and hybridized to ^{32}P-labelled opsin cDNA from *Drosophila*. The result of this Northern analysis (Fig. 1) clearly demonstrates a single band in the RNA preparation from heads, which is further shown to originate from the eyes, since no hybridization to RNA from head capsules was observed. These data indicate that *Drosophila* opsin cDNA indeed hybridizes to an eye-specific RNA which corresponds most likely to opsin mRNA. Densitometric evaluation of the signals yield a relative level of opsin mRNA of about 40 per cent in whole head RNA preparation compared to that from isolated eyes. This difference is due to the larger amount of total RNA obtained from heads, which finally results in a smaller opsin mRNA signal, because resuspension volumes as well as the amounts of the analyzed RNAs were identical in all three samples.

The following experiments were designed to analyze the opsin mRNA levels in *Calliphora* and *Drosophila* depending upon the availability of precursors of the visual pigment chromophore and, after that, more specifically upon the availability of specific isomers of the chromophore itself.

At first, opsin mRNA levels of R^+-flies which had access to chromophore precursors, were compared to those of R^--flies which were depleted of precursors during larval development. Precursor deficiency is known to cause not only a drastic reduction of the amount of visual pigment (Schwemer, 1984), but also of the visual pigment protein opsin (Paulsen and Schwemer, 1983). Poly(A)$^+$ RNA was isolated from heads of R^+- and R^--*Calliphora* and total RNA from heads of R^+- and R^--*Drosophila*. After electrophoresis of the RNAs, they were blotted and hybridized to ^{32}P-labelled opsin cDNA. Fig. 2 demonstrates that in both R^--fly species, the deficiency of chromophore precursors results not only in

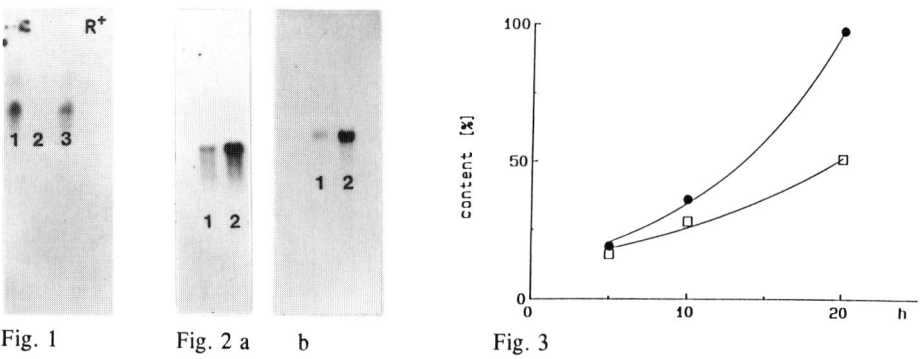

Fig. 1　　　　Fig. 2 a　　b　　　　Fig. 3

Fig. 1 Hybridization of ^{32}P-labelled opsin cDNA of *Drosophila* to total RNAs (2 µg per lane) isolated from excised eyes (1), from head capsules (2) and from intact heads (3) of R^+-*Calliphora*.

Fig. 2 Opsin mRNA levels depend upon the availability of the precursors of the visual pigment chromophore. The Northern analysis shows:
a) *Calliphora:* poly(A)$^+$ RNA was isolated from heads of vitamin A-deficient R^- (1) and 'normal' R^+ (2). Visual pigment content of R^- was about 4% of that of R^+.
b) *Drosophila:* total RNA was prepared from heads of carotenoid-deficient R^- (1) and 'normal' R^+ (2). Visual pigment content of R^- was about 2% of that of R^+.
2 µg RNA per lane was electrophoresed, blotted and hybridized with ^{32}P-labelled opsin cDNA.

Fig. 3 Opsin mRNA (●) and visual pigment levels (□) in R^--*Calliphora* increase following the application of exogeneous 11-*cis* retinal. Analyses were performed 5, 10 and 20 hrs after application. The data were obtained from densitometric evaluation of the dot blot analysis and from microspectrophotometry of intact eyes.

a severe reduction of the amount of visual pigment, but also in drastic reduction of the opsin mRNA levels to approximately 10 per cent of that of the corresponding R^+- controls. - The drastic decrease in opsin mRNA level was reversed in R^--blowflies when exogenous 11-*cis* retinal was applied to the eyes. Samples were taken 5, 10 and 20 hrs after application. Total RNA was prepared from heads and subjected to dot blot analysis. Fig. 3 shows the increase of opsin mRNA levels with time as deduced from densitometric evaluation of the blots. Microspecrophotometry was used to follow the increase in visual pigment *in situ*. In contrast, application of exogenous all-*trans* retinal did not lead to an increase in opsin mRNA level under these experimental conditions. Similarly, the opsin mRNA level and the amount of visual pigment returned to those of R^+, when the offspring of R^--*Drosophila* was raised on the carotenoid-depleted medium to which ß-carotene had been added.-

These results demonstrate that the drastically reduced opsin mRNA level of R^--flies is not only reversed by complementing chromophore precursor to the deficient medium (*Drosophila*), but even more specifically by the availability of 11-*cis* retinal (*Calliphora*). Since the all-*trans* congener is ineffective, the increase of opsin mRNA seems to require specifically the 11-*cis* isomer.

This requirement was further investigated by carefully directed changes of the level of 'free' (not covalently bound to opsin) endogenous 11-*cis* chromophore. As outlined above, the visual pigment cycle in photoreceptors R1-R6 of the blowfly is easily manipulated by e.g. preventing the blue/violet light-induced *trans* -> *cis* isomerization and, by that, depleting the photoreceptors of the 11-*cis* chromophore, the basic requirement for visual pigment synthesis. This is achieved by maintaining R^+-flies in e.g. green light (R^+g). Under these conditions, the M-state of the visual pigment is continuously degraded and the resulting all-*trans* chromophore is stored. On the other hand, biosynthesis of opsin and visual pigment is triggered, when these R^+g-flies after degradation of their visual pigment are returned to 'white' (room) light, which ensures the *trans* -> *cis* isomerization of the stored chromophore. This manipulation was used for studying the opsin mRNA levels after visual pigment degradation in R^+g, and, at different time intervals after the *trans* -> *cis* isomerization had

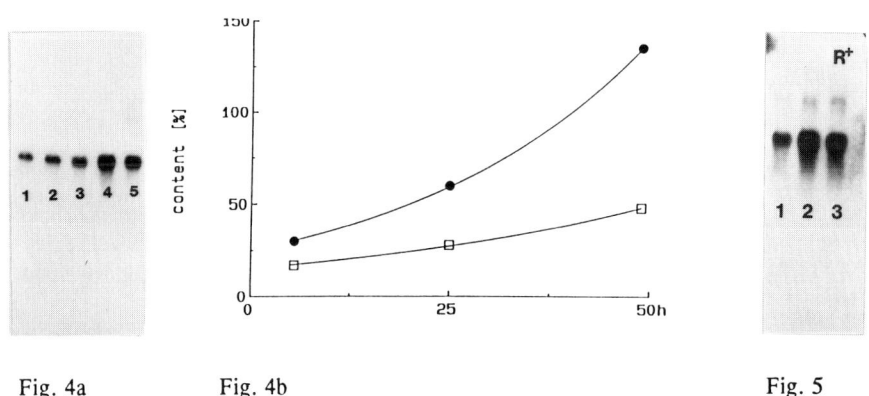

Fig. 4a   Fig. 4b   Fig. 5

Fig. 4 Opsin mRNA levels in R^+-flies depend upon the availability of 'free' endogenous 11-*cis* chromophore.
a: Northern analysis of total RNA preparations from heads of *Calliphora* (2 µg per lane); 1: R^+g after almost complete degradation of the total visual pigment; 2, 3 and 4: 5.5, 25 and 49 hrs after transfer of R^+g to room light; 5: R^+-control.
b: Estimation of opsin mRNA levels (●) by densitometric evaluation of the signals in (a) and microspectrophotometric measurements of visual pigment levels (□).

Fig. 5 Northern analysis of total RNA preparations from heads of R^+-*Calliphora* after adaptation (65 hrs) to red (1) and blue (2) light, respectively. R^+-control ('white' light; 12/12 hrs light/dark-cycle) is shown in lane 3. 2 µg RNA was applied per lane.

been permitted. Northern blot analysis of total RNA (Fig. 4) shows that depletion of the photoreceptors of the 11-*cis* chromophore by green light drastically reduces not only the visual pigment, but also the opsin mRNA level. Furthermore, when *trans* -> *cis* isomerization is allowed to occur, the opsin mRNA level increases again with time and may even reach a higher level than that of the R^+-controls (which have a smaller amount of 'free' 11-*cis* chromophore due to the experimental conditions). Although the visual pigment cycle of *Drosophila* is not yet known, we obtained comparable results by using the same light regimes as in *Calliphora*.

These data clearly indicate that opsin mRNA level changes with the level of 'free' endogenous 11-*cis* chromophore. Further support of this finding was obtained from experiments in which the rate of visual pigment turnover was manipulated by light in the following way: R^+-*Calliphora* was exposed to red and blue light, respectively. Red light is known to convert M570 quantitatively to R490, i.e. degradation is drastically reduced due to the lack of the M-state, and, at the same time, light induced isomerization of any residual all-*trans* chromophore is prevented. Blue light, on the other hand, leads to the highest fraction of M570 (R490 : M570 = 0.3 : 0.7) which can only be achieved under these experimental conditions. As a result of the high fraction of M, the rate of degradation is maximal. At the same time, blue light isomerizes the all-*trans* chromophore released from the M-state, i.e. visual pigment synthesis easily compensates for the breakdown of M (Schwemer, 1984). Fig. 5 shows that the opsin mRNA levels of R^+-*Calliphora* 65 hrs after exposure to red and blue light are quite different: red light caused a severe reduction of opsin mRNA, whereas blue light led to an increase when compared to that of R^+-controls. However, the visual pigment content of the photoreceptors of these three groups of flies is the same.

The various experiments presented here indicate that opsin mRNA levels change with the availability of a 'free' 11-*cis* retinoid (aldehyde or alcohol), whereas neither 'free' all-*trans* nor 11-*cis* chromophore bound to opsin affect opsin mRNA levels. The results suggest that an 11-*cis* retinoid or its binding protein is involved in regulating the expression of the opsin gene. In agreement with our data, Sun et al. (1990) reported a several-fold increase of CAT activity in opsin-promotor-CAT fusion *Drosophila*, after R^--animals were supplied with carotenoids. The specific regulatory mechanism is not yet known, but one of several possible modes of action is especially intriguing and would involve an 11-cis specific nuclear retinoid receptor which activates opsin gene transcription in a similar way as the recently identified nuclear retinoic acid receptors (for review: Wolf, 1990).

REFERENCES

Edmonds, M., Vaughan, M.H. & Nakazato, H. (1971): Polyadenylic acid sequences in the heterogenous nuclear RNA and rapidly labeled polyribosomal RNA of HeLa cells: possible evidence for a precursor relationship. *Proc. Natl. Acad. Sci.* USA 68, 1336 - 1340.
Paulsen, R. & Schwemer, J. (1983): Biogenesis of blowfly photoreceptor membranes is regulated by 11-*cis* retinal. *Eur. J. Biochem.* 137, 609 - 614.
Schwemer, J. (1984): Renewal of visual pigment in photoreceptors of the blowfly. *J. Comp. Physiol. A* 154, 535 - 547.
Schwemer, J. (1988): Cycle of 3-hydroxy retinoids in an insect eye. In *Molecular physiology of retinal proteins*, ed. T. Hara, pp. 299-304. Osaka: Yamada Science Foundation.
Sun, D., Harrelson, A. & Stark, W.S. (1990): Carotenoid replacement enhances chloramphenicol acetyltransferase (CAT) activity in opsin-promotor-CAT fusion *Drosophila*. Abstract, Molecular Neurobiology of *Drosophila*, Cold Sprin Harbor Lab, NY.
Thomas, R.S. (1980): Hybridization of denatured RNA and small DNA fragments transferred to nitrocellulose. *Proc. Natl. Acad. Sci.* USA 77, 5201 - 5205.
Wolf, G. (1990): Recent progress in vitamin A research: nuclear retinoic acid receptors and their interaction with gene elements. *J. Nutr. Biochem. 1*, 284 - 289.

ACKNOWLEDGEMENTS

The authors are indebted to M. Applebury (Chicago) for her supply of *Drosophila* opsin cDNA, to O. Pongs (Hamburg) for discussion and making available his resources, and to D.G. Stavenga (Groningen) for carefully reading the manuscript. This work was supported by the Deutsche Forschungsgemeinschaft.

VII. Rhodopsins : structure-function

VII. *Rhodopsine : structure-fonction*

On the molecular origins of thermal noise in vertebrate and invertebrate photoreceptors

Robert R. Birge[1], Robert B. Barlow Jr[2] and Jack R. Tallent[3]

[1] Department of Chemistry and [2] Institute for Sensory Research, Syracuse University, Syracuse, NY 13244, USA

INTRODUCTION

The weakest pulse of light that the human eye can reliably distinguish from background noise sends roughly 100 photons through the pupil and produces 10-20 activated rhodopsin molecules (Baylor et al. 1980; Aho et al. 1988; Barlow 1988). It has been proposed that single photon sensitivities are not possible due to dark noise associated with randomly occurring thermal isomerizations of the protein bound chromophore in rhodopsin (Aho et al. 1988; Barlow 1988). Indeed, Aho and coworkers have demonstrated that animals with low body temperature, which decreases the rate of thermally activated photoreceptor signals, have higher visual sensitivity (Aho et al. 1988). These investigators propose further that thermally activated isomerizations are the key source of dark noise in both vertebrate and invertebrate visual systems (Aho et al. 1988). In contrast, Barlow and coworkers have studied photoreceptor noise in *Limulus* and have concluded that thermal isomerization of the chromophore is unlikely to be the source of the dark noise (Barlow Jr. & Kaplan 1989; Barlow Jr. & Silbaugh 1989). The purpose of this paper is to provide a preliminary examination of the origin of photoreceptor noise from a molecular standpoint. We demonstrate that thermal isomerization of the protein bound protonated Schiff based chromophore cannot be responsible for the thermal noise. Three possible alternative mechanisms are examined.

ORIGIN OF THE LIGHT ACTIVATED PHOTORECEPTOR SIGNAL

Although there is no consensus regarding the origin of photoreceptor thermal noise, investigators agree that thermal and light activated photoreceptor signals are identical with respect to intensity and temporal profile (Aho et al. 1988; Barlow 1988; Barlow Jr. et al. 1989; Barlow Jr. et al. 1989). Thus, it is important to understand the nature of the light activation and amplification process. The thermal noise phenomenon must involve either a side reaction or corruption of this process prior to amplification or a side reaction that undergoes nearly identical amplification.

The light absorbing chromophore in the visual pigment rhodopsin is 11-cis retinal covalently bound to the opsin protein via a protonated Schiff based linkage to lysine 296 (for a recent review, see Birge 1990). The primary photochemical event is an 11-cis to 11-trans photoisomerization of the protonated Schiff base chromophore (Birge 1990; Schoenlein et al. 1991; Yan et al. 1991; Tallent et al.

1992). A series of dark reactions then occur which ultimately deprotonate the chromophore and activate the protein. The activated protein initiates a complex biochemical process that hyperpolarizes the plasma membrane of the rod cell in the retina (Stryer 1986; Liebman et al. 1987). A single molecule of photoactivated rhodopsin (R^*) catalyzes the activation of up to 1000 transducin molecules ($T_{\alpha\beta\gamma}$-GDP + R^* → R^*-$T_{\alpha\beta\gamma}$-GDP + GTP → R^*-$T_{\alpha\beta\gamma}$-GTP + GDP), and represents the initial stage in the amplification process (Stryer 1986; Liebman et al. 1987). The second stage of amplification involves a splitting off of the α subunit of transducin from the β and γ subunits, and activation of phosphodiesterase (PDE) by the α subunit (PDE + R^*-$T_{\alpha\beta\gamma}$-GTP → PDE + T_α-GTP + R^* + $T_{\beta\gamma}$ → PDE*-T_α-GTP + R^* + $T_{\beta\gamma}$). The binding of GTP to transducin releases activated rhodopsin (R^*) for continued catalytic activity via the initial stage. As noted below, this catalytic activity of the activated protein places important constraints on the mechanism of thermal noise. The activated phosphodiesterase complex (PDE*-T_α-GTP) hydrolyzes cyclic GMP (c-GMP) to 5'-GMP which closes the sodium ion channels {c-GMP + H_2O + open channel(s) → 5'-GMP + H^+ + closed channel(s)}. The transducin cycle returns to the starting point through deactivation of phosphodiesterase via hydrolysis of GTP bound to T_α and recombination of the β and γ subunits with T_α (PDE*-T_α-GTP + $T_{\beta\gamma}$ → PDE + $T_{\alpha\beta\gamma}$-GDP). Two mechanisms operate to close down the cascade and regenerate the resting state in preparation for reactivation by a subsequent photon absorption event. Activated rhodopsin (R^*) is removed through phosphorylation followed by the binding of arrestin (An) (R^* + ATP + An → R^*-P + ADP + An → An-R^*-P). Arrestin is an inhibitory protein than blocks the binding of transducin to photoactivated rhodopsin. The second mechanism involves restoration of the open channels via catalysis of GTP with guanylate cyclase (GC) followed by hydrolysis of pyrophosphate (PP) [closed channels + GTP + GC + H_2O → c-GMP + PP + H_2O → c-GMP + 2P + open channels] (Stryer 1986; Liebman et al. 1987).

THERMODYNAMICS OF THERMAL NOISE

Baylor and coworkers have carried out a detailed analysis of electrical dark noise in toad retinal rod outer segments, and assigned the thermodynamic properties of the thermally activated dark processes: (E_a = 21.9 ± 1.6 kcal mol^{-1}, $\Delta G^\ddagger$ = 31.9 ± 0.13 kcal mol^{-1}, $\Delta H^\ddagger$ = 21.6 ± 1.6 kcal mol^{-1}, $\Delta S^\ddagger$ = -35.3 ± 5.6 e.u.) (Baylor et al. 1980). The activation energies measured for *Limulus* are in agreement within experimental error (E_a = 26.3 ± 7.8 kcal mol^{-1} (day), 27.9 ± 6.5 kcal mol^{-1} (night), 26.5 ± 7.5 kcal mol^{-1} (*in vitro*)) (Barlow Jr. et al. 1989; Barlow Jr. et al. 1989). A comparison of these data with denaturation activation energies measured by Hubbard for cattle rhodopsin (E_a = ~100 kcal mol^{-1}), frog rhodopsin (E_a = ~45 kcal mol^{-1}) and squid rhodopsin (E_a = ~72 kcal mol^{-1}) indicates that protein denaturation is not the origin of the dark signal (Hubbard 1958). Measurements of thermal isomerization of 11-cis retinal, however, appear to offer a much more compatible set of thermodynamic properties (E_a = 22.4 kcal mol^{-1}, $\Delta G^\ddagger$ = 29.3 kcal mol^{-1}, $\Delta H^\ddagger$ = 21.7 kcal mol^{-1}, $\Delta S^\ddagger$ = -21.4 e.u.) (1-propanol solution) (Hubbard 1966). Comparison of the latter measurements on 11-cis retinal with those observed by Baylor on rod segments has prompted some investigators to propose that thermal isomerization of "the chromophore" is responsible for dark activation of rhodopsin (Baylor et al. 1980; Aho et al. 1988; Barlow 1988). Unfortunately, this hypothesis is not consistent with the energetics of ground state isomerization of the protein bound chromophore. As noted above, the protein bound chromophore is not 11-cis retinal, but the protonated Schiff base of 11-cis retinal. The ground state barrier to isomerization of the protein bound chromophore is estimated to be $\Delta H^\ddagger$ = 45 ± 3 kcal mol^{-1} (Birge 1990). We can establish a lower limit of $\Delta H^\ddagger$ ≥ 42 ± 3 kcal mol^{-1} based on the relative enthalpy of bathorhodopsin (ΔH_{RB} = 32.2 ± 0.9 kcal mol^{-1})(Schick et al. 1987) plus the activation enthalpy of the bathorhodopsin → lumirhodopsin dark reaction ($\Delta H^\ddagger$ = 10 ± 2 kcal mol^{-1}) (Grellmann et al. 1962) and assuming additive errors. Thus, thermal (ground state) isomerization of the native (protonated) chromophore cannot be responsible for thermal activation of the protein.

POTENTIAL MECHANISMS OF THERMAL NOISE

The above section demonstrates that neither isomerization of the native chromophore nor protein denaturation is responsible for photoreceptor thermal noise. Although denaturation has not to our knowledge been implicated as a potential source, thermal chromophore isomerization is the currently accepted model for the origin of photoreceptor noise. There are, in fact, three alternative mechanisms that are more attractive than the thermal isomerization proposal. We provide a brief examination of these mechanisms below. We cannot, at this stage, rule out any of the following three mechanisms, but we anticipate based on our preliminary studies that the third (two-step) mechanism is the most likely.

Spontaneous rhodopsin conformational activation. Rhodopsin may have sufficient conformational flexibility to undergo a conformational change ($R \xrightarrow{\Delta} R^{**}$) that is interpreted (incorrectly) by transducin ($T_{\alpha\beta\gamma}$-GDP) to represent photochemically activated rhodopsin (R^*). Thus the initial step in the amplification process takes place involving this thermally activated rhodopsin ($T_{\alpha\beta\gamma}$-GDP + $R^{**} \rightarrow R^{**}$-$T_{\alpha\beta\gamma}$-GDP + GTP $\rightarrow R^{**}$-$T_{\alpha\beta\gamma}$-GTP + GDP). The key problem with this mechanism, however, involves the observation that the thermal noise signals have intensities identical to the light activated signals. Thus, R^{**} (thermally activated rhodopsin) must have a lifetime nearly identical to R^* (photochemically activated rhodopsin). This would seem unlikely, and we conclude that this mechanism, while possible, is not very attractive.

Rhodopsin activation via chromophore deprotonation. The second possibility is that there is an equilibrium within the rhodopsin binding site coupling protonated versus unprotonated chromophores. We have carried out MNDO molecular orbital calculations that indicate a barrier to deprotonation in the range 22 - 35 kcal/mol. The calculated barrier is sensitive to the number of water molecules available within the binding site to stabilize the proton, and the location of the counterions within the binding site. The lowest value (22 kcal/mol) is based on an adiabatic surface with three water molecules stabilizing the proton. We will refer to the protein containing the deprotonated 11-cis chromophore as R_d. The experiments of Longstaff and Rando have demonstrated that deprotonation of the Schiff base of retinal is *obligate* for rhodopsin activation (Longstaff et al. 1986). This mechanism assumes that deprotonation of the Schiff base is *sufficient* for activation. More precisely, deprotonation generates a form of rhodopsin that is interpreted by transducin as activated, and the cascade is initiated. The problem with this mechanism is identical to the problem examined for the previous mechanism. The observation that the thermal noise signals have intensities identical to the light activated signals requires that R_d have a lifetime identical to R^*.

Rhodopsin activation via chromophore deprotonation followed by thermal isomerization. We believe the most likely mechanism for thermal activation of rhodopsin is a two step process. The first step is deprotonation of the 11-cis protonated Schiff base chromophore. The second step is thermal 11-cis to 11-trans isomerization of the chromophore. An hypothetical reaction path is shown below based on the molecular orbital calculations described above and the thermal activation studies of retinal ($\Delta E_a \approx 22$ kcal mol^{-1}) (Hubbard 1966):

$$R\ (RNH_{11\text{-cis}}) \underset{\Delta E_2 \sim 25 \text{ kcal/mol}}{\overset{\Delta E_1 \sim 30 \text{ kcal/mol}}{\rightleftarrows}} R_d\ (RN_{11\text{-cis}}) \xrightarrow{\Delta E_3 \sim 22 \text{ kcal/mol}} R'^*\ (RN_{\text{all-trans}})$$

An analysis of the above mechanism based on reaction rate theory (Forst 1973) yields an overall process that is experimentally indistinguishable from a one-step process with an "apparent activation energy", ΔE_{app}, of ~27 kcal/mol. The assigned values of ΔE_1 and ΔE_2 yield an equilibrium mixture containing ~0.01% R_d at ambient temperature. This fraction of deprotonated species is well below the 3% upper limit established spectroscopically, and more in keeping with experimental pK_a values measured for model compounds. The apparent rate, k_{app}, can be approximated by the formula

$k_{app} = (A_1 A_3 / A_2) \exp[(\Delta E_2 - \Delta E_3 - \Delta E_1)/RT]$. Thus, the apparent activation energy can be estimated by the formula: $\Delta E_{app} \sim \Delta E_3 + \Delta E_1 - \Delta E_2$. Accordingly, ΔE_{app} will invariably be larger than ΔE_3 by a few kcal/mol. This observation is consistent with the experimental data on the thermally activated dark processes (see above). We note that the **R'*** generated via the above mechanism would be virtually identical to the **R*** generated via the light induced photobleaching sequence, and that both species would decay to form all-trans retinal and opsin. We conclude that the above two step mechanism is viable with respect to observed energetics as well as observed thermal noise signal intensities.

ACKNOWLEDGEMENTS

This work was supported in part by a grant to R.R.B. from the National Institutes of Health (GM-34548). We thank A. Warshel and H.B. Barlow for interesting and helpful discussions.

REFERENCES

Aho, A. C., Donner, K., Hyden, C., Larsen, L. O. and Reuter, T. (1988): Low retinal noise in animals with low body temperature allows high visual sensitivity. *Nature* **334**, 348-350.

Barlow, H. B. (1988): The thermal limit to seeing. *Nature* **334**, 296-305.

Barlow Jr., R. B. and Kaplan, E. (1989): What is the origin of photoreceptor noise? *Biol. Bull.* **177**, 323.

Barlow Jr., R. B. and Silbaugh, T. H. (1989): Is photoreceptor noise caused by thermal isomerization of rhodopsin. *Invest. Ophthalmol. Vis. Sci. Suppl.* **30**, 61.

Baylor, D. A., Matthews, G. and Yau, K. W. (1980): Two components of electrical dark noise in toad retinal rod outer segments. *J. Physiol.* **309**, 591-621.

Birge, R. R. (1990): Nature of the primary photochemical events in rhodopsin and bacteriorhodopsin. *Biochim. Biophys. Acta* **1016**, 293-327.

Forst, W. (1973). *Theory of unimolecular reactions.* New York, Academic, 414.

Grellmann, K. H., Livingston, R. and Pratt, D. (1962): A Flash-photolytic investigation of rhodopsin at low temperatures. *Nature* **193**, 1258-1260.

Hubbard, R. (1958): The thermal stability of rhodopsin and opsin. *J. Gen. Physiol.* **42**, 259-280.

Hubbard, R. (1966): The stereoisomerization of 11-cis-retinal. *Journal of Biological Chemistry* **241**, 1814-1818.

Liebman, P. A., Parker, K. R. and Dratz, E. A. (1987): The molecular mechanism of visual excitation and its relation to the structure and composition of the rod outer segment. *Ann. Rev. Physiol.* **49**, 965-991.

Longstaff, C., Calhoon, R. D. and Rando, R. R. (1986): Deprotonation of the Schiff base of rhodopsin is obligate in the activation of the G protein. *Proc. Natl. Acad. Sci. USA* **83**, 4209-4213.

Schick, G. A., Cooper, T. M., Holloway, R. A., Murray, L. P. and Birge, R. R. (1987): Energy storage in the primary photochemical events of rhodopsin and isorhodopsin. *Biochemistry* **26**, 2556-2562.

Schoenlein, R. W., Peteanu, L. A., Mathies, R. A. and Shank, C. V. (1991): The first step in vision: femtosecond isomerization of rhodopsin. *Science* **254**, 412-415.

Stryer, L. (1986): Cycle GMP cascade in vision. *Ann. Rev. Neurosci.* **9**, 87-119.

Tallent, J. R., Hyde, E. Q., Findsen, L. A., Fox, G. C. and Birge, R. R. (1992): Molecular dynamics of the primary photochemical event in rhodopsin. *J. Am. Chem. Soc.* **114**, 1581-1592.

Yan, M., Manor, D., Weng, G., Chao, H., Rothberg, L., Jedju, T. M., Alfano, R. R. and Callender, R. H. (1991): Ultra-fast spectroscopy of the visual pigment rhodopsin. *Proc. Natl. Acad. Sci. USA* **88**, 9809-9812.

Light induced conformational changes of octopus rhodopsin

Motoyuki Tsuda, Masashi Nakagawa, Tatsuo Iwasa and Satoshi Kikkawa

Department of Life Science, Himeji Institute of Technology, Harima Science Garden City, Ako, Hyogo 678-12, Japan

Phototransduction of vertebrate photoreceptor cells are now established that G-protein (transducin) couples the photoexcited rhodopsin to cGMP hydrolysis, leading the closure of cGMP-dependent cation channels (Stryer et al., 1986). On the other hand, the signal transduction is terminated by phosphorylation of rhodopsin by specific kinase followed by binding with arrestin, which sterically hinders the G-protein activation (Sitaramaya & Liebman, 1983). In contrast to the vertebrate visual system, little is known concerning transduction and desensitization processes in invertebrate photoreceptors (Tsuda, 1987). We have characterized invertebrate photoreceptor G-proteins by pertussis toxin and cholera toxin (Tsuda et al., 1986, Tsuda & Tsuda, 1990). Phosphorylation of octopus rhodopsin was shown to be catalyzed by both receptor (opsin) kinase and cAMP dependent kinase (A-kinase) just like β-adrenergic receptor. Moreover, we found that activation of G-protein by GTP analogs strongly enhanced the phosphorylation of octopus rhodopsin, suggesting that kinases are involved in regulating the interaction between rhodopsin and G-protein in invertebrate photoreceptors. In this paper, we will show several evidences of conformational changes of octopus rhodopsin upon illumination which interact with G-protein and kinases.

In previous paper (Tsuda et al., 1989), we showed that both GTPγS and GppNHp, non-hydrolyzable analogs of GTP, enhanced phosphorylation of illuminated rhodopsin. These results suggest that the G-protein(s) activated by GTPγS would be dissociated from rhodopsin molecule, resulting in exposure of extra site(s) in the rhodopsin molecule available for further phosphorylation. Thus, determination of phosphorylation site(s) in rhodopsin with or without GTPγS would give important contribution to understand the role and mutual interactions of rhodopsin, G-protein and kinases in the photoactivation and deactivation process (Tsuda et al., 1992).

Washed microvillar membranes were incubated with [γ^{32}P]ATP either in the presence or absence of GTPγS, and resulting phosphorylated membranes were digested with lysyl endopeptidase. Digested preparations were subjected to an HPLC with a reverse phase column. The addition of GTPγS resulted in further stimulation of phosphorylation of each peptide with exception of appearance of a newly

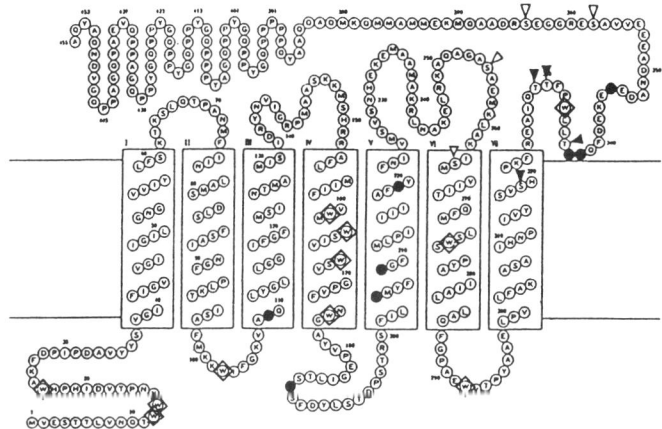

Fig. 1. Topological model of octopus rhodopsin.
▽; light-induced phospholylated amino acids without GTPγS.
▼; light-induced phospholylated amino acids with GTPγS.
●; Cysteine.
◇; Tryptophan.

phosphorylated peptide(s) recovered in peak C. Then the most heavily phosphorylated fractions including peaks A, B, without GTPγS and newly appeared peak C with GTPγS were collected and subjected to amino acid sequencing.

These results suggest that illuminated octopus rhodopsin is phosphorylated at S255 and S264 locating in the third cytoplasmic loop between helixes V and helix VI, and also at S358 and S364 in C-terminal peptide extended from helix VII. When G-protein is activated by GTPγS followed by dissociating from rhodopsin molecule, T329, T330 and/or T336 locating in the C-terminal region are phosphorylated, which in turn indicates that phosphorylation and dephosphorylation of this particular domain are regulatory for the association between rhodopsin and G-protein molecules.

Thus, it is very important to elucidate local conformational changes of this particular domain followed by rhodopsin-acid metarhodopsin transformation in order to understand interaction between rhodopsin, G-protein and kinases.

Spin-label and fluorescence label methods are excellent to investigate the local environment around the cysteine residues of the proteins. In previous paper (Kusumi et al., 1980), we showed that EPR spectra of MLS; (4-maleimide-2,2,6,-tetramethylpiperidinooxyl), -labeled octopus rhodopsin revealed at least two components, a broad peak due to a strongly immobilized signal and a sharp peak due to a weakly immobilized signal. When rhodopsin was photoconverted to acid metarhodopsin with blue light, the height of the broad peak in the spectra decreased and that of the sharp peak increased. The spectral changes of EPR indicated a change in local environment around the cysteine residues of rhodopsin, which causes less immobilized motion of part of the spin-label.

Compared with spin label method, fluorescence label method give distance information between fluorescent molecules as well as the local environment around cysteine residues. Moreover, time response of the fluorescence apparatus is fast enough to follow short lifetime intermediates in the photolysis of rhodopsin. Thus, further studies were done by fluorescence method.

ANM; (N-(1-anilinonaphthyl-4) maleimide) was used as an SH-directed fluorescence probe, because it has virtually no fluorescent in aqueous solution but shows fluorescence with a relatively high quantum yield and an accompanying shift of emission maximum upon binding to a hydrophobic region of a protein (Okamoto & Sekine., 1980). Octopus microvillar membranes in the saline were incubated with ANM solubilized with DMSO; (dimethyl sulfoxide). The resultant membranes were solubilized with detergent (SM1200) and ANM-labeled rhodopsin was isolated by Con-A affinity chromatography. Molar ration of rhodopsin and ANM of the preparation was estimated to be 1.4±0.3 from their absorbances.

In order to determine amino acid residues labeled by ANM, labeled rhodopsin was digested by lysylendopeptidase and resultant peptides were separated by a reverse phase HPLC monitored by UV absorbance and fluorescence. Fluorescent fractions from elution were collected and the amino acid sequence of the peptides in the fraction were analyzed by a gas phase peptide sequencer (ABI Model 470A). The sequence of fluorescent peptide corresponded to amino acid residues of 323-370 which contained two adjacent cysteines C337-C338 and a cysteine, C345. Since two adjacent cysteines were supposed to be palmitated and molar ratio of rhodopsin and ANM was estimated less than 2, C345 is the most probable candidate for being labeled by ANM. Peptide including C345 is the most interesting domain because present results suggest that G-protein and kinase are supposed to be associated with photolyzed rhodopsin around T326, T330 and T336.

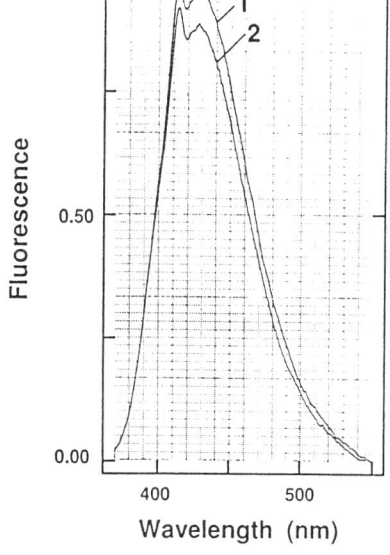

Fig. 2. Fluorescence emission spectra of ANM-labeled pigments excited at 360 nm.
1; ANM-labeled rhodopsin.
2; ANM-labeled acid metarhodopsin.

Next we studied fluorescence of ANM the changes in polarity around region C345 in accompany with rhodopsin-metarhodopsin transition upon illumination. Figure 2 shows that a decrease in intensity of fluorescence was observed when rhodopsin was converted to metarhodopsin with blue light illumination. When this metarhodopsin was photoregenerated to rhodopsin upon illumination of orange light, intensity of fluorescence was completely recovered to original level. These results suggest that domain included C345 exposed to rather hydrophilic environment when rhodopsin was photoconverted to metarhodopsin.

Intrinsic tryptophan fluorescence is another powerful method to monitor protein conformational changes, since tryptophan is highly fluorescence compared to the other aromatic amino acids and its fluorescence is very sensitive to a wide variety of environment conditions. Curves 1 and 2 in Fig. 3 shows the fluorescence spectra of rhodopsin and metarhodopsin excited at 280 nm. Intensity of fluorescence of rhodopsin at 330 nm increased when rhodopsin photoconverted to acid metarhodopsin, showing tryptophan in rhodopsin moved to hydrophobic environment when it converted to acid metarhodopsin.

Absorption spectrum of ANM has a high degree of overlap with the fluorescence emission spectrum of tryptophan. Because of the excitation minimum of ANM at 280 nm, energy transfer can be measured between tryptophan and ANM bound to C345. Curves 3 and 4 in Fig. 3 shows that fluorescence emission spectra of rhodopsin and metarhodopsin labeled with ANM excited at 280 nm. An abrupt decrease in intensity of fluorescence at 330 nm was observed with concomitant appearance of characteristic ANM fluorescent peak at 430 nm. These results strongly suggested that energy transfer from tryptophan residue (probably W333) to ANM at C345 took place.

In the present work, we showed that down stream of C-terminal peptides extended from helix VII in octopus rhodopsin takes part to form critical domain where G-protein and kinases can associate with. We successfully labeled fluorescence probe ANM to C345 which located in peptide extended from helix VII. Observation of a decrease in intensity from characteristic ANM fluorescence followed by transformation from rhodopsin to metarhodopsin showed that hydrophobicity around C345 were decreased upon this transformation. We observed energy transfer from tryptophan (W333) to ANM in C345. These results will give an important contribution to understand molecular dynamics of rhodopsin and signal coupling proteins such as G-protein, kinase and arrestin in invertebrate photoreceptors.

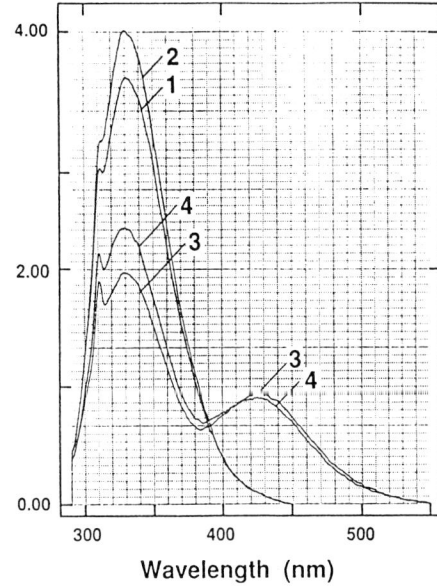

Fig. 3. Fluorescence emission spectra of pigments excited at 280 nm.
1; Rhodopsin. 2; Acid metarhodopsin.
3; ANM-labeled rhodopsin.
4; ANM-labeled acid metarhodopsin.

REFERENCES:

Stryer, L. (1986): Cyclic GMP cascade of vision. Ann. Rev. Neumann. 9, 87-119

Sitaramaya, A. and Liebman, P. A. (1983): Phosphorylation of rhodopsin and quenching of cGMP phosphodiesterase activity by ATP at weak bleaches. J. Biol. Chem. 258, 1206-1209.

Tsuda, M. (1987): Photoreception and Phototransduction in invertebrate photoreceptors. Photochem. Photobiol. 45, 915-931.

Tsuda, M., Tsuda, T., Terayama, Y., Fukada, T., Akino, T., Yamanaka, G., Stryer, L., Katada, T., Ui, M. and Ebrey, T.G. (1986): Kinship of cephalopod photoreceptor G-protein with vertebrate transducin. FEBS Lett. 198, 5-10.

Tsuda, M. and Tsuda, T. (1990): Two distinct light regulated G-protein in octopus photoreceptors. Biochim. Biophys. Acta 1052, 204-210.

Tsuda, M., Tsuda, T. and Hirara, H. (1989): Cyclic-nucleotides and GTP analogues stimulate light-induced phosphorylation of octopus rhodopsin. FEBS Lett. 257, 38-40.

Tsuda, M., Tsuda, T., Abdulaev, N. and Mitaku, S. (1991): Functional domains in octopus rhodopsin. in Photobiology (Edited by E. Riklis) pp13-22. Plenum Press, New York.

Tsuda, M., Hirata, H. and Tsuda, T. (1992): Interaction of rhodopsin, G-protein and kinase in octopus photoreceptors. Photochem. Photobiol. in press.

Changes in chromophore-opsin interaction in the photobleaching processes of visual pigments

Yoshinori Shichida

Department of Biophysics, Faculty of Science, Kyoto University, Kyoto 606-01, Japan

Many vertebrate visual pigments contain an 11-*cis*-retinal chromophore bound via a protonated Schiff base linkage to a specific lysine residue of the apoprotein opsin. Among the visual pigments, rhodopsin which is responsible for scotopic vision has been widely investigated using various spectroscopic and chemical techniques in order to elucidate the molecular mechanism of the visual transduction as well as the extraordinarily high photosensitivity. It is now well established that the primary photochemical event of rhodopsin is a cis-trans isomerization around the $C_{11}=C_{12}$ bond of the chromophore. For many years bathorhodopsin was considered to be the primary photoproduct. However, recently an earlier intermediate photorhodopsin has been detected (Shichida et al., 1984) which is formed from the excited state of rhodopsin within 200 fs (Schoenlein et al., 1991). Subsequent thermal reactions eventually led to the enzymatically active intermediate metarhodopsin II, which initiates the enzymatic cascade system in the photoreceptor cell. In the course of conversion from photorhodopsin to metarhodopsin II, a series of intermediates is formed whose absorption maxima are distinct from each other. However, little is known about changes in the chromophore-opsin interaction during and after the photoisomerization of chromophore. In this context, we have been trying to get clues to the conformational changes of the chromophore and subsequent changes in interaction between the chromophore and nearby protein using various spectroscopic techniques with the aids of synthetic retinal analogs.

PHOTOISOMERIZATION MECHANISM OF THE RHODOPSIN CHROMOPHORE

Evidence for isomerization of the chromophore in rhodopsin is based on the photoequilibrium that can be established at 77K between rhodopsin, bathorhodopsin and 9-*cis*-rhodopsin (which contains a 9-cis chromophore) (Yoshizawa and Wald, 1963). This was further confirmed by the experiments using rhodopsin analog containing the locked chromophore, 7-membered retinal or 5-membered retinal, in which the trimethylene or dimethylene bridge locks the $C_{11}=C_{12}$ bond in its cis configuration, respectively. Formation of bathorhodopsin was observed neither in 7-membered rhodopsin nor 5-membered rhodopsin (Mao et al., 1981; Fukada et al., 1984), indicating that the isomerization is essential for formation of bathorhodopsin. We have extended our investigation to elucidate the conformation of the chromophore in photorhodopsin, the precursor of bathorhodopsin, using three kinds of locked retinals having 5, 7 and 8-membered rings, respectively (Fig. 1) (Kandori et al., 1989; Mizukami et al., submitted). When the $C_{11}=C_{12}$ bond is rigidly fixed by

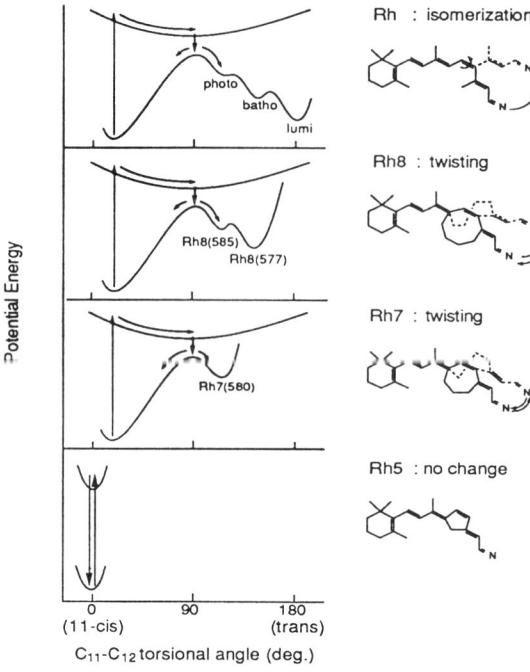

Fig. 1. Schematic diagrams of the ground and the excited state potential surfaces along the $C_{11}=C_{12}$ torsional cordinates of the chromophores of rhodopsin and its analogs.

5-membered ring (Rh5), excitation merely leads to an excited state that reverts to the original pigment. Fixation of the $C_{11}=C_{12}$ bond with a 7-membered ring (Rh7) which allows some flexibility around the double bond results in formation of a photorhodopsin-like intermediate through the excited state. When the flexibility of the $C_{11}=C_{12}$ bond is further increased by means of an 8-membered ring (Rh8), the formation of two intermediates corresponding to photorhodopsin and bathorhodopsin is observed. Now, we can safely speculate that photorhodopsin has a highly twisted all-trans chromophore which is formed by relaxation of the distorted chromophore in the excited state. Furthermore, above results clearly indicate that the appearance of the early intermediates is dependent on the flexibility of the moiety comprising the $C_{11}=C_{12}$ bond and the photoisomerization proceeds by sequential interaction between the chromophore and the protein, resulting in stepwise relaxation of the highly strained all-*trans*-retinal chromophore.

CHANGE IN CHROMOPHORE-OPSIN INTERACTION DURING THE BATHO-LUMI TRANSITION

Since the formation of bathorhodopsin after photon absorption of rhodopsin is extremely rapid (45 ps), it is easily believed that only minor rearrangement of amino acid residues constituting the chromophore-binding site could occur. On the other hand, a relatively larger change in chromophore-opsin interaction could take place during the batho-lumi transition. It has been suggested so far that a twisted all-trans chromophore of bathorhodopsin is relaxed concurrent with some rearrangement of the protein near the cyclohexenyl ring portion of the chromophore (reviewed in Shichida, 1986). To further clarify the nature of the changes in the chromophore-opsin interaction around the cyclohexenyl ring, we compared the kinetic properties of four kinds of 9-*cis*-rhodopsin analogs, each of which has a chromophore of acyclic retinal, α-retinal, acyclic α-retinal, or 5-isopropyl-α-retinal, respectively, with that of 9-*cis*-rhodopsin (Fig. 2) (Okada et al., 1991). The results clearly showed that the formation time constant of the lumi-intermediate at room temperature is dependent on the extent of modification of the ring portion of the

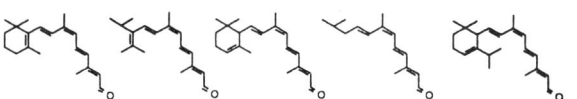

9-*cis*-retinal   acyclic retinal   α-retinal   acyclic-α-retinal   5-isopropyl-α-retinal

Formation Times of Lumi-intermediates

240 ns   200 ns   200 ns   <80 ns   2 μs

Fig. 2. Structures of 9-*cis*-retinal and 9-cis forms of acyclic retinal, α-retinal, acyclic-α-retinal and 5-isopropyl-α-retinal. Formation time constants of lumi-intermediates from 9-*cis*-rhodopsin and its analogs having these retinal analogs are shown below.

chromophore. Namely, the complete truncation of the cyclohexenyl ring [AcaRh(9)] resulted in decrease of the time constant, while the attachment of the isopropyl group [PaRh(9)] to the ring resulted in increase of the time constant. However, partial truncation of cyclohexenyl ring [AcRh(9)] or saturation of $C_5=C_6$ bond [aRh(9)] resulted in little change. These results are consistent with the notion that changes in chromophore-opsin interaction in the batho-lumi transition are primarily limited to the ring portion of the chromophore.

CHANGE IN CHROMOPHORE-OPSIN INTERACTION DURING THE LUMI-META I TRANSITION

In the course of investigation of the photobleaching process of 7-*cis*-rhodopsin, which has a 7-*cis*-retinal as its chromophore, we have obtained interesting results that could elucidate the change in chromophore-opsin interaction during the lumi-meta I transition. The experiments using picosecond and nanosecond laser phtolyses as well as low-temperature spectrophotometry (Shichida et al., 1991) showed that batho-intermediate from 7-*cis*-rhodopsin is different in spectral shape and decay time constant from those of rhodopsin and 9-*cis*-rhodopsin. These differences retained in lumi stage, but disappeared in meta I stage. The difference in batho stage between 7-*cis*-rhodopsin and the other two rhodopsins would be explained by unique relocation of the 9-methyl group of the 7-cis chromophore during the isomerization to all-trans form. In fact, the 9-methyl group of the chromophore of 7-*cis*-rhodopsin is oriented in the opposite direction relative to the protein from those of rhodopsin and 9-*cis*-rhodopsin. There are several evidences that indicate the unique chromophore-opsin interaction between the 9-methyl group and nearby protein in rhodopsin and 9-*cis*-rhodopsin (Blatz et al., 1969; Kropf et al., 1973; Ganter et al., 1989). The conversion of rhodopsin or 9-*cis*-rhodopsin to bathorhodopsin does not require any significant relocation of the methyl group, because the isomerization takes place by movement of half of the polyene chain near the Schiff base (Shichida et al., 1987). On the other hand, the conversion of 7-*cis*-rhodopsin to its batho-intermediate requires relocation of the 9-methyl group, possibly causing local protein perturbation which induces chromophore-opsin interaction different from that in bathorhodopsin (Fig. 3).

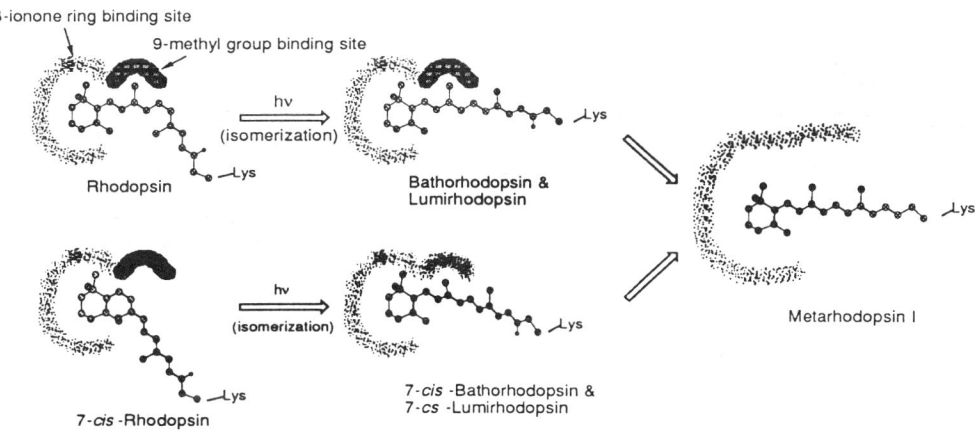

Fig. 3. Schematic representation of chromophore-opsin interactions in the photobleaching processes of rhodopsin and 7-*cis*-rhodopsin.

This may explain the difference in absorption spectrum and decay time constant between batho-intermediates from 7-*cis*-rhodopsin and the other two rhodopsins. That the difference disappears in meta I stage suggests the disappearance of the unique interaction between the 9-methyl group and the sorrounding protein during the lumi-meta I stage, probably owing to the conformational changes in the protein.

Now the changes in chromophore conformation and chromophore-opsin interaction during the process from rhodopsin to metarhodopsin I could be speculated as follows: The photon absorption by rhodopsin results in formation of highly twisted all-trans chromophore in photorhodopsin which relaxes in a stepwise manner to the still twisted all-trans chromophore in bathorhodopsin. These changes proceed by movement of the half of the polyene chain containing the Schiff base. The subsequent interaction change between cyclohexenyl ring portion of the chromophore and surrounding protein produces lumirhodopsin. A twisted conformation of the chromophore of bathorhodopsin may force to induce a conformational change of the protein near the cyclohexenyl ring of the chromophore. Lumirhodopsin is then converted to metarhodopsin I by disappearance of the interaction between 9-methyl group of the chromophore and surrounding protein.

REFERENCES

Blatz, P. E., Lin, M., Balasubramaniyan, P., Balasubramaniyan, V., and Dewhurst, P. B. (1969) A new series of synthetic visual pigments from cattle opsin and homologs of retinal. *J. Am. Chem. Soc. 91*: 5930-5931.

Fukada, Y., Shichida, Y., Yoshizawa, T., Ito, M., Kodama, A., and Tsukida, K. (1984): Studies on structure and function of rhodopsin by use of chclopentatrienylidene 11-cis-locked-rhodopsin. *Biochemistry 23*: 5826-5832.

Ganter, U. M., Schmid, E. D., Peres-Sala, D., Rando, R. R., and Siebert, F. (1989): Removal of the 9-methyl group of retinal inhibits signal transduction in the visual process. A fourier transform infrared and biochemical investigation. *Biochemistry 28*: 5954-5962.

Kandori, H., Matuoka, S., Shichida, Y., Yoshizawa, T., Ito, M., Tsukida, K., Balogh-Nair, V., and Nakanishi, K. (1989): Mechanism of isomerization of rhodopsin studied by use of 11-cis-locked rhodopsin analogues excited with a picosecond laser pulse. *Biochemistry 28*: 6460-6466.

Kropf, A., Wittenberger, B. P., Goff, S. P., and Waggoner, A. S. (1973): The spectral properties of some visual pigment analogs. *Exp. Eye Res. 17*: 591-606.

Mao, B., Tsuda, M., Ebrey, T. G., Akita, H., Balogh-Nair, V., and Nakanishi, K. (1981): Flash photolysis and low temperature photochemistry of bovine rhodopsin with a fixed 11-ene. *Biophys. J. 35*: 543-546.

Okada, T., Kandori, H., Shichida, Y., Yoshizawa, T., Denny, M., Zhang, B-W., Asato, A. E., and Liu, R. S. H. (1991): Spectroscopic study of the batho-to-lumi transition during the photobleaching of rhodopsin using ring-modified retinal analogues. *Biochemistry 30*: 4796-4802.

Shichida, Y., Matuoka, S., and Yoshizawa, T. (1984): Formation of photorhodopsin, a precursor of bathorhodopsin, detected by picosecond laser photolysis at room temperature. *Photobiochem. Photobiophys. 7*: 221-228.

Shichida, Y. (1986): Primary intermediate of photobleaching of rhodopsin. *Photobiochem. Photobiophys. 13*: 287-307.

Shichida, Y., Ono, T., Yoshizawa, T., Matsumoto, H., Asato, A. E., Zingoni, J. P., and Liu, R. S. H. (1987): Electrostatic interaction between retinylidene chromophore and opsin in rhodopsin studied by fluorinated rhodopsin analogues. *Biochemistry 26*: 4422-4428.

Shichida, Y., Kandori, H., Okada, T., Yoshizawa, T., Nakanishi, N., and Yoshihara, K. (1991): Difference in the photobleaching process between 7-cis- and 11-cis-rhodopsins: A unique interaction change between the chromophore and the protein during the lumi-meta I transition. *Biochemistry 30*: 5918-5926.

The role of the Cys110-Cys187 disulfide bond in rhodopsin investigated spectrophotometrically

D. Garcia-Quintana, P. Garriga, M. Duñach and J. Manyosa

Unitat de Biofísica, Departament de Bioquímica i de Biologia Molecular, Facultat de Medicina, Universitat Autònoma de Barcelona, 08193 Bellaterra, Catalonia, Spain

The series of conformational changes triggered by photoactivation of rhodopsin eventually results in the hydrolysis of the Schiff base bond between the retinal chromophore and the opsin moiety. *In vivo*, regeneration of rhodopsin with fresh 11-*cis*-retinal ensures the continuous functionality of the system. The kinetic parameters of the process suggest a close similarity between the conformation of rhodopsin and opsin (Ostroy, 1977).

Karnik *et al.* (1988) have demonstrated that the mutant opsins $Cys_{110}\rightarrow Ser$ and/or $Cys_{187}\rightarrow Ser$ are unable to generate rhodopsin, thus suggesting a role of $cystine_{110-187}$ in the preservation of the functional, regenerable conformation of opsin. Rhodopsin contains 10 cysteinyl amino acid residues, two of which -Cys_{110} and Cys_{187}-, are involved in a disulfide bond (DeGrip, 1988). These two cysteines are the only ones conserved throughout the whole rhodopsin-like receptors superfamily (Findlay & Pappin, 1986; Applebury & Hargrave, 1986), and presence of disulfide bond reducing agents has been reported to result in a decreased ligand binding in the ß-adrenergic receptor (Vauquelin *et al.*, 1979).

The effect of the reduction of the disulfide bond in rhodopsin in the presence of ditiothreitol (DTT) has been investigated spectrophotometrically.

EXPERIMENTAL PROCEDURES

The outer segments of rod cells were obtained from bovine retinas as described (Wilden & Kühn, 1982) except that no DTT was present in the buffers. Membranes were depleted of peripheral proteins by thorough washing in low (5 mM phosphate buffer, pH 6.9) and in middle

(100 mM) ionic strength buffer. Membranes were finally suspended in 50 mM Hepes (pH 7.0).

All procedures were carried out at physiological temperature. When indicated, rhodopsin, at a concentration around 5-6 μM, was incubated in buffer containing 5 mM DTT for 10 min before starting the experiment (DTT-treated rhodopsin). Metarhodopsin III (meta III) was obtained by photoactivation of the pigment by irradiation with light of wavelengths above 520 nm. At the temperature of work, opsin was obtained 1 h after photoactivation.

Absorption spectra at 0.1 nm resolution were scanned in a Perkin-Elmer 320 spectrophotometer, equipped with a head-on photomultiplier in order to minimize light scattering. Samples were thermostated and magnetically stirred. Spectra were directly transferred to a computer for subsequent mathematical treatment. Several spectra were coadded to achieve a good signal-to-noise ratio. Fourth-derivative was performed with a program assembled in our laboratory based on the algorithm of Savitzky & Golay (1964). Fourth-derivative analysis of the UV absorption spectrum of proteins is a well established technique developed in our laboratory (see Padrós *et al.*, 1984; and references therein).

RESULTS AND DISCUSSION

Incubation of rhodopsin with DTT does not result in a λmax shift in the visible region. It does neither alter significantly the 4th-derivative spectrum in the UV region. In addition, the Schiff

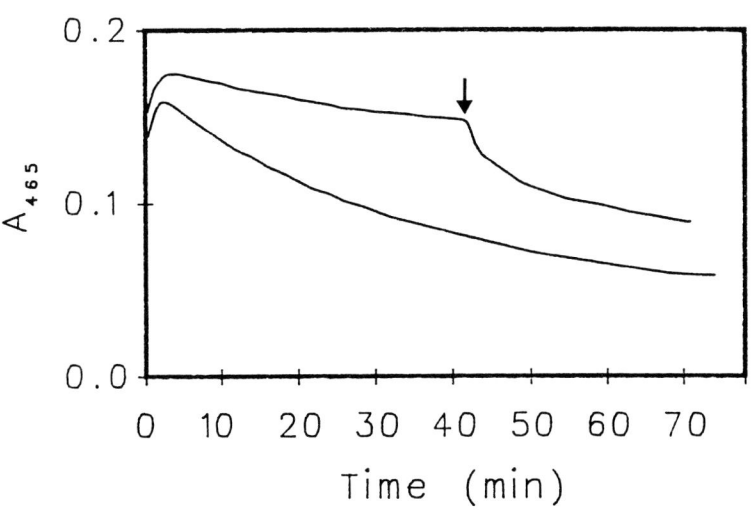

Fig. 1. The decay of meta III from rhodopsin (upper trace) and from DTT-treated rhodopsin (lower trace) after a 20 s flash ($\lambda > 520$ nm). Initial increase reflects ongoing formation of meta III. The arrow indicates addition of DTT to a concentration of 5 mM.

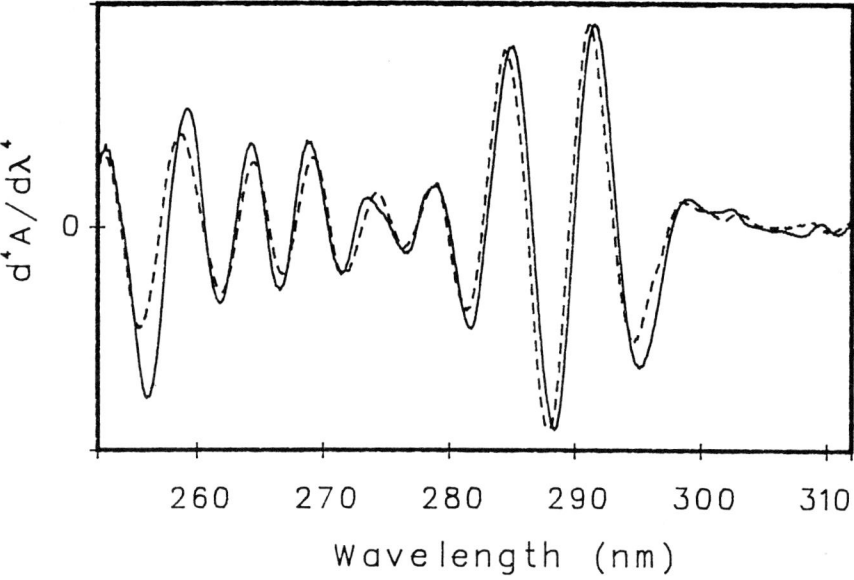

Fig 2. Fourth-derivative spectra in the ultraviolet region of rhodopsin (———) and DTT-treated opsin (- - - - -). Fourth-derivative greatly favors narrow transitions; this results in no peaks from oxidized DTT appearing in this region, so that the protein spectrum can be analyzed confidently.

base remains inaccessible to hydroxylamine in DTT-treated rhodopsin. Also, the thermal stability of rhodopsin, monitorized as A_{498} vs. temperature, is not modified by the presence of DTT, similar profiles being obtained. These findings suggest that the disulfide bond does not play a critical role in maintaining the overall structure of rhodopsin.

Although DTT treatment does not preclude the formation of meta III, the decay to opsin is affected (Fig. 1), suggesting a conformational destabilization of the former. In addition, no N-retinylidene opsin (NRO) is detected in the decay of DTT-treated meta III in the same pH conditions that NRO_{440} or NRO_{365} are definitely isolated from untreated meta III.

Differences are observed in the 4th-derivative UV spectrum of opsin resulting from the bleaching of DTT-treated rhodopsin (Fig. 2), in opposition to opsin from untreated rhodopsin, which remains unperturbed. A small although significant blue-shift (0.5-1.0 nm) is assessed in the former, in the 280-300 nm region, indicative of an increase in the hydrophylicity of the environment of some Trp and Tyr amino acid residues. Differences are also observed in the shorter wavelengths region, which reflects mainly the contribution from Phe residues. These results suggest a conformational alteration in opsin generated from DTT-treated rhodopsin compared to untreated opsin.

It is concluded that the Cys_{110}-Cys_{187} disulfide bond does not seem to play a structural role in rhodopsin, but rather it preserves the native structure of the protein after photoactivation, in accordance with the unability of $Cys_{110} \rightarrow Ser$ and/or $Cys_{187} \rightarrow Ser$ mutants to bind the chromophore described by Karnik *et al.* (1988). Differences in secondary structure are currently being investigated by means of FTIR spectroscopy and will be compared with the previously studied structure of rhodopsin (Garcia-Quintana *et al.*, 1991; Garcia-Quintana *et al.*, submitted for publication). Investigation of the photoreaction sequence from cystine110-187-cleaved rhodopsin, in particular the study of the ability of the resulting meta II to activate transducin, should yield further insight into the role of this evolutively-conserved disulfide bond.

This work was supported by grant DGICYT (PB89-0301)

REFERENCES

-Applebury, M.L. & Hargrave, P.A. (1986): Molecular biology of the visual pigments. *Vision Res.* 26, 1181-1895.
-DeGrip, W.J. (1988): Recent chemical studies related to vision. *Photochem. Photobiol.* 48, 799-810.
-Findlay, J.B.C. & Pappin, D.J.C. (1986): The opsin family of proteins. *Biochem. J.*, 238, 625-642.
-Garcia-Quintana, D., Garriga, P. & Manyosa, J. (1991): High-resolution FT-IR spectroscopy study of retinal rhodopsin. In *Spectroscopy of Biological Molecules*, ed. R.E. Hester & R.B. Girling, pp. 213-214. Cambridge: The Royal Society of Chemistry.
-Karnik, S.S., Sakmar, T.P., Chen, H.-B. & Khorana, H.G. (1988): Cysteine residues 110 and 187 are essential for the formation of correct structure in bovine rhodopsin. *Proc. Natl. Acad. Sci. USA* 85, 8459-8463.
-Padrós, E., Duñach, M., Morros, A., Sabés, M. & Manyosa, J. (1984): Fourth-drivative spectrophotometry of proteins. *Trends Biochem. Sci.* 9, 508-510.
-Savitzky, A. & Golay, M.J.E. (1964): Smoothing and differentiation of data by simplified least squares procedure. *Anal. Chem.* 36, 1627-1639.
-Vauquelin, G., Bottari, S., Kanavek, L. & Strosberg, A.D. (1979): Evidence for essential disulfide bonds in ß1-adrenergic receptors of turkey erythrocyte membranes. Inactivation by dithiothreitol. *J. Biol. Chem.* 254, 4462-4469.
-Wilden, U. & Kühn, H. (1982): Light-dependent phosphorylation of rhodopsin: number of phosphorylation sites. *Biochemistry* 2, 3014-3022.

Cysteine residues in rhodopsins

Tatsuo Iwasa, Nadic G. Abdulaev*, Masashi Nakagawa, Satoshi Kikkawa and Motoyuki Tsuda

*Department of Life Science, Himeji Institute of Technology, Harima Science Garden City, Ako, Hyogo 678-12, Japan, and *Shemyakin Institute of Bioorganic Chemistry, Russian Academy of Sciences, 117871 GSP Moscow V-437, Russia*

Rhodopsin is an integral membrane protein participating in the light transducing process that takes place in the photoreceptor cells of higher organisms. There is a striking similarity between the biochemical pathway of intracellular signaling triggered by light and those by hormones or neurotransmitters, and all the signal transduction systems appear to have a similar molecular base of action. It appears that the G-protein coupled receptors including visual pigments as well as hormone and neurotransmitter receptors show remarkable sequence similarities to each other and are believed to consist of seven transmembrane segments (Dixon *et al.*, 1986; Iwabe *et al.*, 1989). One outcome of the alignment and comparison of the deduced amino acid sequences of the various members of the G-protein coupled receptor family has been the identification of a number of highly conserved cystein residues which may play important structural or functional roles (Karnik *et al.*, 1988; Karnik & Khorana, 1990; Dohlman *et al.*, 1990).

In the present paper, the properties of cystein residues of bovine and octopus rhodopsins have been studied. Concerning cystein residues of bovine rhodopsin, the effect of disulfide and sulfhydryl reactive reagents on absorption spectrum and regenerability of rhodopsin were studied. And also the existence of a disulfide bond was examined by protein chemical methods. We found that an excess amount of dithiothreitol (DTT) had no effect on the both properties of several different preparations of bovine rhodopsin. Moreover the protein chemical studies on bovine rhodopsin suggest that there is no positive evidence that rhodopsin molecules on isolated disk membranes form a disulfide bond, which was also supported by the results from reverse phase HPLC analyses of lysyl endopeptidase-digested rhodopsin with or without DTT. The reaction of octopus rhodopsin with a fluorescent thiol probe, ANM (anilinonaphtylmaleimido), was studied using microvilli membrane suspension. We determined the number and position of ANM incorporated in octopus rhodopsin molecule.

Bovine rhodopsin contains ten cystein residues and the property of them has been studied for many years (DeGrip & Damen, 1982). Recent success in the expression of bovine rhodopsin gene or mutagenized gene enable to examine the role of cysteine residues. Karnik *et al.* (1988) found that replacement of Cys110 or Cys187 by serine residue resulted in proteins expressed at reduced levels, glycosylated abnormally and unable to bind 11-*cis* retinal. Using mutated rhodopsins expressed in Cos-1 cells, they concluded that these two cysteine residues form a disulfide bond essential for the formation of correct structure of rhodopsin (Karnik & Khorana, 1990). The structural motif of seven transmembrane helixes has been proposed for a large number of G-protein coupled receptor including bovine rhodopsin (Fig. 1). This model suggests that highly conserved two cysteins corresponding to Cys110 and Cys187 (in bovine rhodopsin) located in the cytoplasmic loop. Moreover, these systems form an intramolecular disulfide bond which may be critical in directing correct helix-helix interaction that forms the ligand binding pocket. If this is the case, disruption of the disulfide bond in rhodopsin are expected to change the absorption spectrum and/or regenerability from opsin and retinal.

The disk membranes were isolated at 4°C under dim red light. Unless otherwise stated, DTT was deleted from the buffers. The addition of DTT to cholate-solubilized disk membranes till 50 mM (2.8

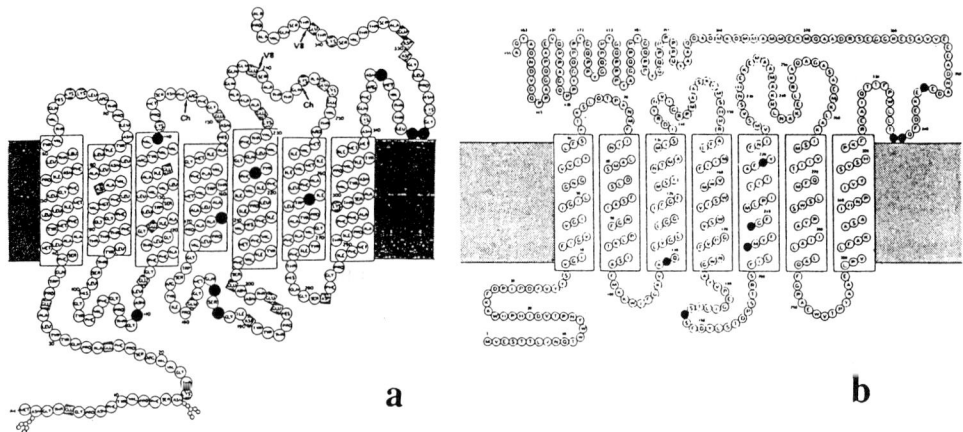

Fig. 1. The membrane structure model for bovine (**a**) and octopus (**b**) rhodopsins. The cysteine residues are represented with closed circles.

thousand times excess molar amount of cysteine residues in rhodopsin) has no effect in the visible absorption band (α-band) of rhodopsin. Further, we studied the effect of DTT on the absorption spectra of the other preparations (frozen-thawed disk, dodecyl maltoside-solubilized, and octyl glucoside-solubilized disk), but failed to observe any changes.

The regeneration of rhodopsin from opsin and 11-*cis* retinal is more sensitive to the degree of integrity of opsin conformation. If the disulfide bond between Cys110 and Cys187 is critical in maintaining the correct helix-helix interaction that form 11-*cis* retinal binding pocket, it is expected that disruption of the disulfide bond by DTT may affect regenerability of rhodopsin. In order to study the effect of DTT on regeneration of rhodopsin, the bleached disk membranes were incubated with DTT (final concentration; 100 mM) for 30 min in the dark, followed by the addition of 11-*cis* retinal. The longer incubation of bleached disk membranes with DTT for overnight gave the same results. The amount of regenerated rhodopsin was estimated from the absorbance change at 500 nm in the presence of hydroxylamine before and after the irradiation for 30 min. No differences were observed in the degree and the rate of the pigment regeneration depending on DTT (100 mM). The degree of regeneration in cholate-solubilized disk membranes was essentially same as that of the bleached disk membranes. The rate of regeneration of detergent-solubilized preparation became slower than that of membrane preparations. Spectroscopic results show that an excess amount of DTT did not affect the absorption spectra and regenerability of rhodopsin molecules in disk membranes or in detergent micells. A possible explanation for these results is that these cysteines are buried in the disk membrane and that DTT molecules cannot attack the disulfide bond. Kalinik and Khorana (1990) found that SH residues in the mutated rhodopsin were only attacked by SH-reagent at a denatured state. These results suggest that the secondary structure model of bovine rhodopsin as shown in Fig. 1a is still uncertain.

The existence of a disulfide bond in rhodopsin was studied by protein chemical methods. Intact disk membranes (mg/ml) were digested with α-chymotrypsin (1/50 or 1/100; rhodopsin/α-chymotrypsin in molar ratio) in a manganese supplemented Ringer solution (Adams *et al.*, 1982) at 37°C for 2 h in the dark. α-Chymotrypsin digested rhodopsin in intact disk membranes into three fragments; A(1-146), B(147-244) and C(245-348). The digested rhodopsin retained its absorption spectrum in detergent and was charged to Con A sepharose column. In order to analyse peptide fragments with Con A sepharose column, the α-chymotrypsin-digested disk membranes were solubilized with Ammonyx LO buffer and charged on Con A sepharose column (Fig. 2a). After the initial peak has eluted, rhodopsin charged on the column was bleached by irradiation from a projector lamp (300 W halogen lamp) through Toshiba O-55 filter (> 530 nm) for 30 min at room temperature. After rhodopsin was completely bleached by irradiation, fragments released by light were eluted by the Ammonyx LO buffer until the absorbance at 280 nm reached to steady level. The eluted fraction absorbed at 380 nm as well as 280 nm. The absorbance at 380 nm could be resulted from retinal, which suggests that this fraction contains a fragment with retinal-binding site, fragment C. The fragments still charged on the resin were eluted out by the same buffer containing sugar (200 mM methyl-α-D-mannopyranoside). The fractions eluted by sugar showed undigested rhodopsin (38 kDa) and 21 kDa peptide bands with minor bands at 29.5 and 18.5 kDa on SDS-PAGE (lanes 8 and 9 in Fig. 2b), indicating that the fragment banded at 21 kd contains the sugar binding site for Con A and thus corresponds to fragment A. The elute after irradiation gave bands at 25 and 23 kd (lanes 6 and 7 in Fig. 2b). Some degraded materials were observed below 20 kDa in lanes 6 - 9. Comparing the bands of the V8-protease-digested rhodopsin with those digested with α-chymotrypsin, the band at 23 kDa (lanes 6 and 7 in Fig. 2b) was assigned to fragment C and 25 kDa band to fragment B, respectively.

As a result, α-chymotrypsin cleaved rhodopsin in the intact disk membranes into three fragments, A, B and C, which form a complex in detergent solution. Isomerization of retinal by light induced conformational changes of protein and resulted in dissociation of this complex. After irradiation of rhodopsin, fragment B and C were eluted from the column. If Cys110 in fragment A and Cys187 in fragment B formed a disulfide bond, fragment B should be eluted concomitantly with fragment A. The results, however, clearly show that the fragment A was eluted alone, not with fragment B, strongly suggesting that there is no disulfide bond between Cys110 and Cys187.

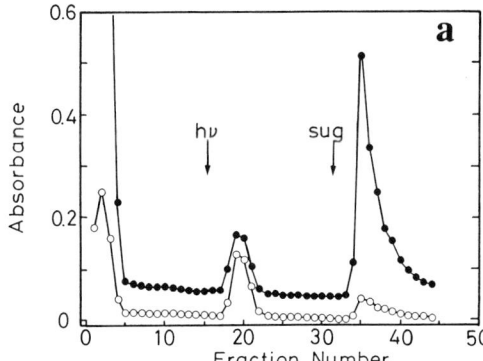
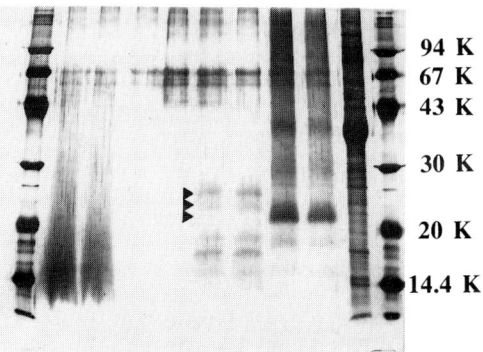

Fig. 2. The Con A sepharose column chromatography of α-chymotrypsin digested rhodopsin (a): Absorbances at 280 nm (●) and 380 nm (○) were monitored. The SDS-PAGE pattern of the column fractions (b): Lanes 1 and 11; molecular weight markers, Lanes 2 - 9; fractions 2, 3, 12, 15, 19, 20, 35 and 36, respectively. Lane 10; the disk membranes used in the experiments.

The results were also supported by reverse phase HPLC analyses of lysyl endopeptidase-digested rhodopsin with or without DTT (Fig. 3). Rhodopsin purified by Con A sepharose chromatography was denatured by SDS, mixed with N-ethylmaleimido and digested by lysyl endopeptidase (1/50-75, W/W) at 37°C overnight. The digested sample and the reference sample, which was prepared at the same time by the same procedure without rhodopsin, were incubated with or without DTT (100 mM) for at least 1 h at room temperature. They were analysed by an HPLC (Waters 510) with μBONDASPHERE C4 reverse phase column (Waters). After injection, the column was eluted with isocratically with solvent A (0.05 per cent TFA in water) for 10 min followed by a linear gradient elution with solvent B (0.05 per cent TFA in acetonitrile). As expected from the specificity of lysyl endopeptidase, Cys110 would belong to fragment(68-141) and Cys187, to fragment(142-231). This indicates that if Cys110 and Cys187 form a disulfide bond, fragment(68-141) and fragment(142-231) should be eluted as a large fragment. After the sample was incubated with DTT, the disulfide bond would be broken and resulted fragments would be eluted separately as smaller fragments.

Thus, a peak corresponds to the large fragment in Fig. 3a would disappear after the incubation with DTT and two new peaks correspond to the smaller fragments would be expected to appear on chromatogram in Fig. 3b. The obtained chromatogram, however, did not show any changes of peaks depend on DTT as expected (Fig. 3c). These results were consistent to those from Con A column chromatography. The results obtained from Con A column chromatography and HPLC analyses indicate that the rhodopsin molecules prepared from disk membranes without SH-reagents did not form a disulfide bond between Cys110 and Cys187. Although we used the buffer deleted of SH-reagent, it is not excluded that the disulfide bond was cleaved during the preparation.

Recently, Doi *et al.* (1990) formed several kinds of deletion mutants in intradiscal loops of rhodopsin and found that not only Cys110 and Cys187 but also other residues of the intradiscal loops are important for the membrane structure of rhodopsin. It is also noteworthy that bacteriorhodopsin contains no cystein residue, indicating that the membrane structure of seven transmembrane segments can be maintained without a disulfide bond.

Considering these results with the results by Karnik *et al.* (1988), it may be suggested that the disulfide bond is important for integration of a newly synthesized opsin molecule and for making a correct structure, but once a correct structure is formed that is maintained by the many parts of a intradiscal domains of rhodopsin.

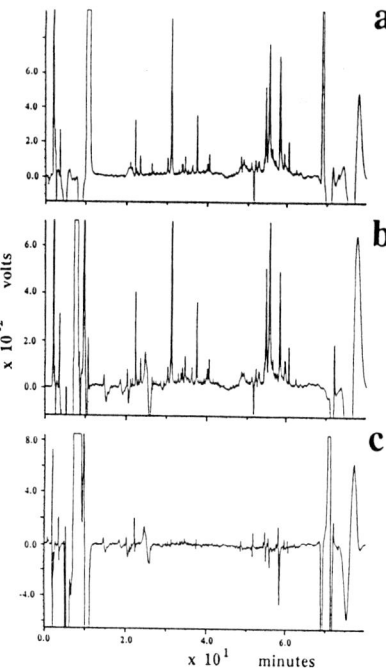

Fig. 3. The chromatograms of reverse phase HPLC of lysyl endopeptidase digested bovine rhodopsin without and with DTT (a and b, respectively). Chromatogram c was obtained as (a - b).

Octopus microvilli (MV) membranes were mixed with a thiol-reactive fluorescent probe, ANM, in the dark. The time course of the reaction was monitored with an increase in the intensity of the fluorescence, which was composed of two phases; very fast phase and slow phase. The fast phase was mainly due to an interaction of ANM with opsin molecule and the slow phase was that with lipids. The ANM-labelled preparations were solubilized with SDS solution and digested with lysyl endopeptidase. The digested fragments were analysed by a reverse phase HPLC and a fluorescent fragments, which included ANM-labelled peptides, were pooled and sequenced. The present results suggest that a candidate for ANM-labelled site of octopus rhodopsin in MV membranes was Cys 345.

This work was supported in part by Grants-in-Aid from the Ministry of Education, Science and Culture to M.T. (02304067, 03303056, 03305002) and to T.I. (01304061, 04680272).

REFERENCES

Adams, A.J., Tanaka, M. and Shichi, H. (1982) Isolation of intact disks by concanavaline A columns. *Methods Enzymol.* **81**, 61-64.

DeGrip, W.J. and Damen, F.J.M. (1982) Sulfhydryl chemistry of rhodopsins. *Methods Enzymol.* **81**, 223-236.

Dixon, R.A.F., Koblika, B.K., Strader, D.J., Benovic, J.L., Dohlman, H.G., Fruelle, T., Bolanowski, M.A., Bennett, M.A., Rands, E., Diehl, R.E., Mumford, R.A., Slator, E.E., Signal, I.S., Caron, M.G., Lefkowitz, R.J., and Strader, R.A. (1986) Cloning of the gene and cDNA for mammalian beta-adrenergic receptor and homology with rhodopsin. *Nature* **321**, 75-79.

Dohlman, H.G., Caron, M.G., DeBlasi, A., Frielle, T., and Lefkowitz, R.J. (1990) Role of extracellular disulfide-bonded cysteines in the ligand binding function of the β_2-adrenergic receptor. *Biochemistry* **29**, 2335-2342.

Doi, T., Molday, R.S., and Khorana, H.G. (1990) Role of the intradiscal domain in rhodopsin assembly and function. *Proc. Natl. Acad. Sci. USA* **87**, 4991-4995.

Iwabe, N., Kuma, K-I., Saitou, N., Tsuda, M., and Miyata, T. (1989) Evolution of rhodopsin supergene family. Independent divergence of visualpigments in vertebrate and insects and possibly in mollusks. *Proc. Japan Acad.* **65**, Ser. B, 195-198.

Karnik, S.S., and H.G. Khorana (1990) Assembly of functional rhodopsin requires a disulfide bond between cystein residues 110 and 187. *J. Biol. Chem.* **265**, 17520-17524.

Karnik, S.S., Sakmar, T.P., Chen, H-B., and Khorana, H.G. (1988) Cysteine residues 110 and 187 are essential for the formation of correct structure in bovine rhodopsin. *Proc. Natl. Acad. Sci. USA* **85**, 8459-8463.

Phospholipids associated with rhodopsin purified by concanavalin-A sepharose

N. Virmaux, L. Menguy*, N. Boukra, G. Nullans, A. van Dorsselaer* and P.F. Urban

INSERM U 338 and *CNRS URA 31, Centre de Neurochimie, 5 rue Blaise Pascal, 67000 Strasbourg, France

SUMMARY : After extraction from bovine retinal rod outer segments (ROS) by a non ionic detergent, the nonanoyl-N-methyl glucamide, rhodopsin was purified by Concanavalin-A Sepharose (Con-A) chromatography. By checking the lipid phosphorus content of the mannoside-eluted rhodopsin, we observed that about 3% of the total phospholipids (PL) remain associated with the protein. The major class of this protein-bound lipid fraction was found to be phosphatidylcholine (PC). Analyses of the PC molecular species were performed by positive Fast-Atom Bombardement-Mass Spectrometry (FAB-MS). The mass spectra of PC from PL of the Con-A purified rhodopsin and from PL not retained by Con-A reveal in addition to molecular species of classical masses between 734 and 856 other molecular species of unusual masses between 1018 and 1050. Since the dipolyunsaturated species of PC from ROS have been shown to contain very long chain (up to 34 carbons) polyunsaturated (4,5,6 double bonds) fatty acids (VLCPUFA), such unusual masses could be attributed to VLCPUFA-containing dipolyunsaturated species of PC. Furthermore the comparison of the two FAB-MS spectra shows that the PC of rhodopsin-bound PL are specially enriched in VLCPUFA-containing species. These results provide further evidence of the peculiarity of the lipid environment of rhodopsin.

Among the major phospholipids (PL) found in the retinal rod outer segments (ROS) about one third are dipolyunsaturated (Miljanich et al., 1979) and consists of 30% of phosphatidylcholine (PC), 20% of phosphatidylethanolamine (PE) and 50% of phosphatidylserine (PS) (Aveldaño & Bazán, 1983). In addition the dipolyunsaturated species of PC have been shown to contain unusual fatty acids identified as very long chain (24-34 carbons) polyunsaturated (4,5,6 double bonds) fatty acids (VLCPUFA) (Aveldaño, 1987; Aveldaño & Sprecher, 1987). They were found associated to rhodopsin after hexane treatment of ROS (Aveldaño, 1988). In an attempt to elucidate the role of these PL molecular species in the functional conformation of rhodopsin we first investigated if during classical purification of rhodopsin by Con-A Sepharose chromatography in the presence of dialyzable detergent some phospholipids with peculiar molecular species remained associated with the protein as it happens during hexane treatment of ROS.

METHODS - Rhodopsin was extracted from bovine retinal ROS by using a non-ionic detergent, the nonanoyl-N-methylglucamide (Mega-9) (Hildreth, 1982) in 10 mM Tris-HCl pH 7.4. After washing of the membranes with 0.25% Mega-9, the extraction was achieved with 0.8% Mega-9. The protein-lipid complex was then purified by chromatography on a Concanavalin-A Sepharose 4B column (Litman, 1982) equilibrated with 0.8% Mega-9. After deposit of 20 mg protein on the top of the Con-A column (7cm x 1.5cm), washing was done with 0.8% Mega-9 which eluted the phospholipidic material without protein. The protein fraction was detached from Con-A by alpha-methyl-D-mannoside 0.5M in presence of 1% Mega-9. During the washing and elution processes, the absorbance at 280nm and

500nm was followed to identify the presence of rhodopsin and the lipidic phosphorus of all the fractions was determined according to Ames (1966). After extensive dialysis of the lipidic and proteic fractions to eliminate the methylmannoside and the Mega-9, lipid extraction was carried out according to the classical method of Folch et al (1957). Separation of the PL classes was performed by thin layer chromatography (TLC) on silica gel plates according to the procedure of Leray et al (1987). The species of a given class of phospholipids have been separated and identified by mass spectrometry (MS) using an ionisation technique by fast atom bombardment (FAB). Positive ion mass spectra were obtained on a VG ZAB HF double focusing instrument with reverse geometry (B-E). The underivatized PC samples were embedded in a matrix of meta nitrobenzoyl alcohol (m-NBA) on the stainless steel FAB probe target. Desorption and ionisation of the sample was obtained with a 8 keV energy Xenon atom beam gun. Scanning was done by exponential magnetic field variation. The mass to charge range was 1500 amu at 8kV ion kinetic energy. Instrumental resolution was 1000 and total acquisition time was about 2 minutes for 8 scans. Monoisotopic pseudomolecular peaks (M+H)+ and accompanying isotopic profile were recorded on a VG11/250 data system (VG Analytical Ltd, Manchester, England).

RESULTS AND DISCUSSION- During the purification of rhodopsin by the affinity chromatography, based on glycans binding of the membrane protein to the Con-A Sepharose, the PL were separated in two fractions, free PL non retained by Con-A and bound PL asssociated with rhodopsin. The ratio of free PL to bound PL is about 97 to 3. Furthermore the classes of PL of the two fractions were determined by TLC on silicagel plates and the main classes of phospholipids were clearly identified in the protein-free fraction. These are PC, PS and PE. In the rhodopsin bound fraction PC is the main compound seen.

In order to know the molecular species present in the two fractions we analyzed the PC of both fractions by positive FAB-MS which realize simultaneous volatilization and ionisation of normally non volatile samples. The best matrix was found to be m-NBA. In positive ionisation mode, predominantly positive pseudomolecular ions (M+H)+ were formed and could be used for direct analysis of phospholipid species. Intensities of the ion (mass/charge) peaks were analyzed and recorded versus time covering the acquisition of 6 to 7 spectra in order to control the proportionality of the various species detected which should be nearly constant during the ejection process from the matrix.

The analysis of the molecular species of PC (Fig.1) showed the presence of classical species in the range of 700 to 850. We detected also dipolyunsaturated species with long chain fatty acids of masses ranging from 878 to 992. Beside these two ranges, we found also other molecular species of unusual masses between 1018 and 1050. For these m/z values corresponding to a certain number of carbon atoms (until 56) and double bonds (10 to 12) of the fatty acid part of PC we propose to attribute acyl chains corresponding to the VLCPUFAs reported by Aveldaño (1988) to the sn1 position.

m/z	C atoms and double bonds	sn1	sn2
1018	54 : 12	32 : 6 n-3	22 : 6 n-3
1020	54 : 11	32 : 5 n-3	22 : 6 n-3
1022	54 : 10	32 : 4 n-6	22 : 6 n-3
1046	56 : 12	34 : 6 n-3	22 : 6 n-3
1048	56 : 11	34 : 5 n-3	22 : 6 n-3
1050	56 : 10	34 : 4 n-6	22 : 6 n-3

In order to obtain some quantitative information on these various PC species, ions chromatograms of classical and VLCPUFA species were examined for several acquisitions. The ratio of the pseudomolecular ions of each species remained roughly constant.

Examination of mass spectra from lipid fraction (Fig.1, Part A) and protein fraction (Fig.1, Part B) showed that the protein fraction is enriched in species with long and very long chain fatty acid. Part C of Fig.1 (enlarged view of mass range 1018-1052) shows clearly the distribution of the VLCPUFAs in the protein-bound fraction.

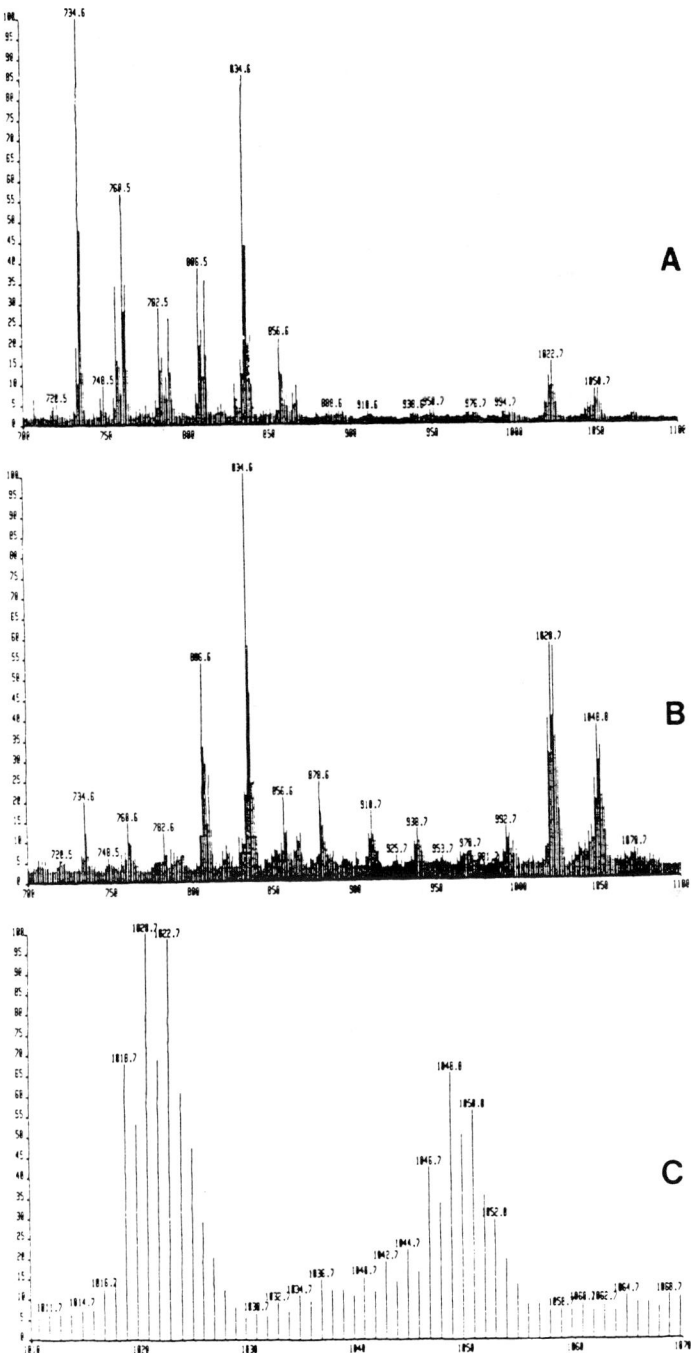

Fig.1. Positive FAB-MS spectra of underivatized PC isolated on Si-TLC from the Con-A Sepharose fractions. Part A: Lipid fraction free of rhodopsin. Part B: Lipid associated with rhodopsin. Part C: Enlarged view (from B) corresponding to the VLCPUFAs-containing molecular species.

The reported results provide further evidence of the peculiarity of the lipid environment of rhodopsin. This is in agreement with the hypothesis of an intimate annulus of lipids surrounding the rhodopsin which seemed very important for its functionnal activity.

REFERENCES

Ames, B. N. (1966): Assay of inorganic phosphate, total phosphate and phosphatases. Methods Enzymol. 8: 115-117.

Aveldaño, M. I. (1987): A novel group of very long chain polyenoic fatty acids in dipolyunsaturated phosphatidylcholines from vertebrate retina. J. Biol. Chem. 262: 1172-1179.

Aveldaño, M. I. (1988): Phospholipid species containing long and very long polyenoic fatty acids remain with rhodopsin after hexane extraction of photoreceptor membranes. Biochemistry 27: 1229-1239.

Aveldaño, M. I., and Bazán, N. G. (1983): Molecular species of phosphatidylcholine, -ethanolamine, -serine, and -inositol in microsomal and photoreceptor membranes of bovine retina. J. Lipid Res. 24: 620-627.

Aveldaño, M. I., and Sprecher, H. (1987): Very long chain (C_{24} to C_{36}) polyenoic fatty acids of the n-3 and n-6 series in dipolyunsaturated phosphatidylcholines from bovine retina. J. Biol. Chem. 262: 1180-1186.

Folch, J., Lees, M., Sloane-Stanley, G. H. (1957): A simple method for the isolation and purification of total lipids from animal tissues. J. Biol. Chem. 226: 497-509.

Hildreth, J. E. K. (1982): N-D-gluco-N-methylalkanamide compounds, a new class of non-ionic detergents for membrane biochemistry. Biochem. J. 207: 363-366.

Leray, C., Pelletier, X., Hemmendinger, S., and Cazenave, J.-P. (1987): Thin-Layer Chromatography of human platelet phospholipids with fatty acid analysis. J. Chromatogr. 420: 411-416.

Litman, B. J. (1982): Purification of rhodopsin by concanavalin A affinity chromatography. Methods Enzymol. 81: 150-153.

Miljanich, G. P., Sklar, L. A., White, D. L., and Dratz, E. A. (1979): Disaturated and dipolyunsaturated phospholipids in the bovine retinal rod outer segment disk membrane. Biochim. Biophys. Acta 552: 294-306.

VIII. Bacterial sensory rhodopsins : phototaxis

VIII. Rhodopsines sensorielles des bactéries : phototaxisme

Phototransduction by sensory rhodopsin I

John L. Spudich

Department of Microbiology and Molecular Genetics, The University of Texas Medical School at Houston, 6431 Fannin, Houston, Texas 77030, USA

Sensory rhodopsin I (SR-I) is a retinal-containing intrinsic membrane protein that functions as a phototaxis receptor in the archaebacterium *Halobacterium halobium* (Bogomolni & Spudich, 1982). SR-I controls swimming behavior of the cells by modulating the frequency of reorientation ("reversals") of their swimming direction (reviewed in Spudich & Bogomolni, 1988; Oesterhelt & Marwan, 1990). Orange light generates attractant signals which suppress reversals, whereas near-UV light generates repellent signals which induce reversals. The minimum time between photon absorption by the receptor and the flagellar motor response is 700 milliseconds (Sundberg et al, 1986). Two sequential processes occur in this interval: receptor activation in the first millisecond produces a receptor signaling conformation; at later times signal transduction by post-receptor components relays receptor signals to the flagellar motors.

RECEPTOR ACTIVATION

Photon absorption by SR-I (λ_{max} 587nm) causes isomerization around the C13-C14 double bond of the retinal chromophore (Tsuda et al., 1985) which is essential for receptor activation (Yan et al., 1990). The photoisomerization energy is transferred to the protein in a process requiring steric interaction between the retinal C13 methyl group and protein residues (Yan et al., 1991). A similar steric trigger occurs during the activation of the visual pigment rhodopsin (Ganter et al, 1989). These events produce in < 1 millisecond a spectrally distinct attractant signaling state (λ_{max} 373nm) of the SR-I protein which decays thermally within seconds to the original unstimulated receptor conformation (Bogomolni & Spudich, 1987; Yan & Spudich, 1991). During formation of the signaling state a proton is transferred from the chromophore attachment site to a proton acceptor, presumably on the protein (E.N. Spudich & B. Yan, unpublished). The repellent signal is generated by absorption of a second photon by the receptor during the lifetime of its attractant signaling state (Spudich & Bogomolni, 1984).

In recent experiments Bing Yan has used retinal analogs to further examine steric chromophore/protein interactions in SR-I. The analog chromophore binding rates, photocycles of SR-I analogs, and phototaxis of bacteria containing analog SR-I pigments are altered by the deletion of substituents from retinal and increasing the size of the retinal C13 substituent. Deletion of the β-ionone ring renders the analog pigment inactive, while deletion of the 9-methyl, ring methyls and part of the ring were previously found to not significantly reduce the photochemical activity and physiological function of the analog pigments. Substituting the 13-methyl with a bulkier ethyl group slows the binding 30-fold, decreases the photocycle rate 20-fold and reduces the relative photocycling activity of the analog pigment to only ~10% of that of SR-I. The slower photocycle is attributable to the slower decay of a photointermediate which corresponds to the S_{373} intermediate of native SR-I. Prolonged attractant light step-up responses occur as expected from retarded decay of the S_{373}-like intermediate. Compared to SR-I, bacteriorhodopsin (BR) is more tolerant of these retinal modifications; analog BR pigments lacking the 13-methyl or β-ionone ring are active and the 13-ethyl BR analog behaves like the native BR. These results and the previous finding that deletion of the 13-methyl group disables SR-I (Yan et al., 1991) indicate that the protein and the retinal are subject to tight steric constraints around the 13-methyl and the C4-C6 unit on the β-ionone ring. Loss of either of these steric constraints abolishes photochemical and physiological activities of SR-I analog pigments (Yan et al, in preparation).

THE SR-I TRANSDUCER (HtrI)

A fundamental question concerns the components responsible for transmission of these signals to the flagellar motor. The gene-derived primary structure of SR-I reveals a 7-transmembrane helix protein largely embedded in the membrane similar in its structure to the related proton pump BR (Blanck et al., 1989). Fumarate (Marwan et al., 1990), thiomethylation (Lebert et al, 1992), cyclic nucleotides and Ca^{++} (Schimz & Hildebrand, 1987) and a G-protein (Schimz et al., 1989) have been suggested to be involved in post-receptor signaling. However, no post-receptor component capable of sensing SR-I conformations and transducing this information to modulate the motor has been demonstrated in these studies.

Phototaxis mutant analysis identified a methylated membrane protein of M_r 97kDa, expression of which was tightly correlated with the SR-I protein of 25 kDa (Spudich et al., 1988). The methylation linkage (carboxylmethyl esterification; Spudich et al., 1988; Alam et al., 1989) is of the same type as that found in the chemotaxis signal generators ("transducers") of eubacteria (e.g., *Escherichia coli*), chemoreceptor proteins which transmit signals from the membrane to a cytoplasmic sensory pathway (Bourret et al., 1991). SR-I attractant and repellent signals modulate methyl group turnover *in vivo* as do chemotaxis transducers. Based on these findings the methyl-accepting protein (also called MPP-I) was postulated to be the transducer for SR-I signals (Spudich et al, 1989). Analysis of taxis mutants and revertants (Sundberg et al, 1990) and antigenic cross-reactivity of the 97kDa M_r protein with eubacterial transducers (Alam & Hazelbauer, 1991) further strengthened the analogy to the chemotaxis system.

Virginia Yao has isolated the proposed transducer protein, cloned its gene based on partial protein sequences, and sequenced the gene (Yao and Spudich, submitted). The 1611bp *htrI* (*h*alobacterial *t*ransducer for sensory *r*hodopsin *I*) gene ends at the beginning of the *sopI* gene which encodes the sensory rhodopsin I apoprotein. Putative promotor elements are located in an AT-rich region upstream of *htrI*. Comparison of the translated nucleotide sequence with N-terminal sequence of the purified protein shows the protein is synthesized without a leader peptide and the N-terminal methionine is removed in the mature HtrI. The predicted size of 57kDa for the protein indicates an aberrant electrophoretic migration on SDS gels as occurs with other highly acidic halophilic proteins. The sequence predicts two transmembrane helices at the N-terminal which would anchor the protein to the membrane. Beyond this hydrophilic region of 46 residues, the remainder of the protein (535 amino acid residues total) is hydrophilic. The C-terminal 270 residues contain a region homologous to the signaling domains of eubacterial transducers (e.g. *E. coli* Tsr protein), flanked by two regions homologous to the methylation domains of the transducer family. These results substantiate the proposal that HtrI functions as a signal transducing relay between SR-I and cytoplasmic sensory pathway components.

EXPRESSION OF A CASSETTE *SOPI* GENE

Elena Spudich and I, in collaboration with Mark Krebs and H. Gobind Khorana, designed a *sopI* gene optimized for site-specific mutagenesis and helix-swapping with other retinylidene proteins. The synthetic *sopI* gene was substituted for the bacterioopsin gene in an *H. halobium* expression vector (Krebs et al, 1991). This construct encodes a polypeptide with an extended N-terminus (enSR-I) of 21 amino acids from the bacterioopsin precursor that includes a 13 amino acid leader sequence that may be processed. The vector was introduced in the *H.*

halobium strain Pho81, which does not contain SR-I nor MPP-I. The enSR-I protein was expressed in transformants at 4-fold higher levels than was SR-I in wild type *H. halobium* and had a slightly higher M_r on SDS-PAGE than SR-I, consistent with an extended N-terminus. The absorption spectrum and flash-induced absorption difference spectrum of enSR-I in membranes from transformed Pho81 were indistinguishable from those of SR-I in wild-type membranes. There were, however, notable differences in the photochemical activity of these preparations. The thermal decay of the S_{373} intermediate was accelerated by both lower pH and azide under conditions in which S_{373} in wild type membranes remained unaffected. Either of two factors, the extended N-terminus of enSR-I or the absence of MPP-I in Pho81 may be responsible for these differences. The cause of the anamolous properties of enSR-I are being studied by making other constructs without the N-terminal extension and expression in both MPP-I$^-$ and MPP-I$^+$ strains. An interesting possibility raised by these results is that MPP-I may interact with SR-I in a way which influences the process of reprotonation of the Schiff base, an essential reaction in the transition of S_{373} to $SR-I_{587}$. Further expression studies will test this possibility.

REFERENCES

Alam, M. & Hazelbauer, G.L. (1991): Structural Features of Methyl-Accepting Taxis Proteins Conserved between Archaebacteria and Eubacteria Revealed by Antigenic Cross-Reaction. *J. Bacteriology* 173, 5837-5842.

Alam, M., Lebert, M., Oesterhelt D., & Hazelbauer, G.L. (1989): Methyl-accepting taxis proteins in *Halobacterium halobium*. *EMBO J.* 8, 631-639.

Blanck, A., Oesterhelt, D., Ferrando, E., Schegk, E.S., & Lottspeich, F. (1989): Primary structure of sensory rhodopsin I, a prokaryotic photoreceptor. *EMBO J.* 8, 3963-3971.

Bogomolni, R.A. & Spudich, J.L. (1982): Identification of a third rhodopsin-like pigment in phototactic *Halobacterium halobium*. *Proc. Natl. Acad. Sci. USA* 79, 6250-6254.

Bogomolni, R.A. & Spudich, J.L. (1987): The photochemical reactions of bacterial sensory rhodopsin-I: Flash photolysis study in the microsecond to eight second time window. *Biophysical J.* 52, 1071-1075.

Bourret, R.B., Borkovich, K.A., & Simon, M.I. (1991): Signal Transduction Pathways Involving Protein Phosphorylation in Prokaryotes. *Annu. Rev. Biochem.* 60, 401-441.

Ganter, U.M., Schmid, E.D., Peres-Sala, D., Rando, R.R., & Siebert, F. (1989): Removal of the 9-methyl group of retinal inhibits signal transduction in the visual process. A Fourier transform infrared and biochemical investigation. *Biochemistry* 28, 5954-5962.

Krebs, M.P., Hauss, T., Heyn, M.P., RajBhandary, U.L., & Khorana, H.G. (1991): Expression of the bacterioopsin gene in *Halobacterium halobium* using a multicopy plasmid. *Proc. Natl. Acad. Sci. USA* 88, 859-863.

Lebert, M.R., Oesterhelt, D., Nitz, S., Kollmannsberger, H. & Hazelbauer, G.L. Methylthiolation and methylation, alternative chemistries in an archaebacterial sensory system, submitted.

Marwan, W., Schafer, W., & Oesterhelt, D. (1990): Signal transduction in *Halobacterium* depends on fumarate. *EMBO J.* 9, 355-362.

Oesterhelt, D. & W. Marwan. (1990): Signal transduction in *Halobacterium halobium*. *Symp. Soc. Gen. Microbiol.* 46, 219-239.

Schimz, A. & Hildebrand, E. (1987): Effects of cGMP, calcium and reversible methylation on sensory signal processing in halobacteria. *Biochim Biophys Acta* 923, 222-232.

Schimz, A., Hinsch, K.-D. & Hildebrand, E. (1989): Enzymatic and immunological detection of a G-protein in *Halobacterium halobium*. *FEBS Lett.* 249, 59-61.

Spudich, E.N., Hasselbacher, C.A. & Spudich, J.L. (1988): A methyl-accepting protein associated with bacterial sensory rhodopsin I. *J. Bacteriology* 170, 4280-4285.

Spudich, E.N., Takahashi, T. & Spudich, J.L. (1989): Sensory rhodopsins I and II modulate a methylation/demethylation system in *Halobacterium halobium* phototaxis. *Proc. Natl. Acad. Sci. USA* 20, 7746-7750.

Spudich, J.L. & Bogomolni, R.A. (1984): The mechanism of colour discrimination by a

bacterial sensory rhodopsin. *Nature* 312, 509-513.
Spudich, J.L. & Bogomolni, R.A. (1988): Sensory rhodopsins of halobacteria. *Annu. Rev. Biophys. and Biophys. Chem.* 17, 193-215.
Sundberg, S.A., Alam, M. & Spudich, J.L. (1986): Excitation signal processing times in *Halobacterium halobium* phototaxis. *Biophysical J.* 50, 895-900.
Sundberg, S.A., Alam, M., Lebert, M., Spudich, J.L., Oesterhelt, D., & Hazelbauer, G.L. (1990): Characterization of mutants of *Halobacterium halobium* defective in taxis. *J. Bacteriology* 172, 2328-2335.
Tsuda M., Nelson B., Chang C.-H., Govindjee R., Ebrey T.G. (1985): Characterization of the chromophore of the third rhodopsin-like pigment of *Halobacterium halobium* and its photoproduct. *Biophys J.* 47, 721-724.
Yan, B. & Spudich, J.L. (1991): Evidence the repellent receptor form of sensory rhodopsin I is an attractant signaling state. *Photochem. Photobiol.* 54, 1023-1026.
Yan, B., Nakanishi, K. & Spudich, J.L. (1991): Mechanism of activation of sensory rhodopsin - I: Evidence for a steric trigger. *Proc. Natl. Acad. Sci. USA* 88, 9412-9416.
Yan, B., Takahashi, T., Johnson, R., Derguini, F., Nakanishi, K. & Spudich, J.L. (1990): All-*trans*/13-*cis* isomerization of retinal is required for phototaxis signaling by sensory rhodopsins in *Halobacterium halobium*. *Biophysical J.* 57, 807-814.

Signal transduction in *Halobacterium halobium* mediated by the switch factor fumarate

Wolfgang Marwan, Marco Montrone and Dieter Oesterhelt

Max-Planck Institut für Biochemie, 8033 Martinsried, Germany

Halobacterium halobium is a rod-shaped archaebacterium that swims with a polarly inserted flagellar bundle. The flagella are driven by rotary motors at their bases. The cells can swim back and forth simply by switching the rotational sense of the bundle. Switching occurs spontaneously causing random movement of the cells in their environment. Light and some chemical stimuli can induce or suppress motor switching thereby introducing a bias in the random walk which ultimately results in an accumulation in a preferred region of the biotope.

Although these bacteria are considerably small cells with an average volume of about 1 fl, they show all the features known from sensory systems in higher organisms: excitation by sensory receptors, amplification of the sensory input, integration of light and chemical stimuli, adaptation to the stimulus background and finally the behavioral response (for review see Oesterhelt & Marwan, 1990; Schimz & Hildebrand, 1988; Spudich, 1991).

Halobacteria up to now are the only bacteria known that use retinal proteins as photoreceptors. The primary structure of one of these so-called sensory rhodopsins, SR-I, has already been determined. It is also a seven-helix transmembrane protein and shares considerable sequence identity with the halobacterial light-driven ion pumps bacteriorhodopsin and halorhodopsin but not with the corresponding G-protein coupled receptors in eukaryotes (Blanck et al., 1989). We are interested in the mechanism of photoreception and the biochemical basis of signal processing. Here we briefly discuss the results obtained on the signal transduction mechanism.

PHOTORECEPTORS TALK TO THE FLAGELLA BY CHEMICAL SIGNALLING

The first question to be answered was whether excitation transmission from the photoreceptors to the flagellar motor is of electrical nature or not. We used artificially long grown cells for microbeam irradiation experiments and showed that many of these cells are bipolarly flagellated. The flagellar movement could be directly observed by dark field microscopy. Light stimulation did not cause a flagellar response when the middle part of the cell was irradiated, however the bundle responded when light excitation occurred at the pole of the cell. Interestingly, the flagellar bundle switched independently indicating that a signal cannot be transmitted through the cell. This result excludes electrical signalling and at the same time suggests a chemical messenger that can be degraded upon diffusion through the cell (Oesterhelt & Marwan, 1987).

COMPUTER-ASSISTED CELL TRACKING AND KINETIC ANALYSIS OF THE SIGNAL CHAIN

An important technical tool in the study of signal transduction is the exact measurement of the motor response from thousands of cells after application of a stimulus. To obtain this information, a computer-assisted image processing system was optimized. The motion of the cells was recorded in infrared observation light which was not absorbed by the cellular photoreceptors. The video frames recorded by a CCD camera were digitized and the outlines of the cells were used to reconstruct the swimming paths by a computer.

Two different parameters can be used to characterize the output of the signal chain: the probability for the flagellar bundle to switch and the time it takes until switching occurs. If these parameters are quantified as the response to defined stimuli, kinetic data on the photoreceptors or on the signal formation process can be extracted from behavioral experiments (Marwan & Oesterhelt, 1987; Marwan & Oesterhelt 1990). The switching signal acting on the motor is formed in the second range. Using this type of input-output analysis, it was shown that the rate of signal formation is proportional to the number of sensory rhodopsin molecules activated by a blue light pulse. The results in addition suggested that activated sensory rhodopsin I and II molecules catalyse the formation of the signal (see below).

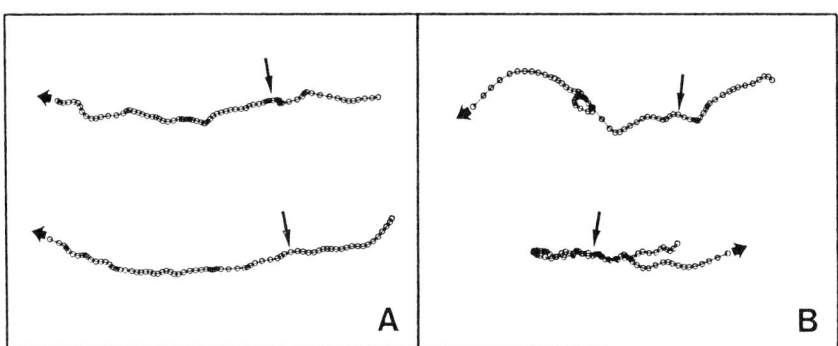

Fig. 1. Swimming paths of mutant cells (strain M415) defective in switching the rotational sense of the flagellar motor (A). The ability to switch can be restored by reconstitution with fumarate (B). The centers of the cells are displayed with a time resolution of 67 ms. The thin arrows mark the onset of blue stimulus light.

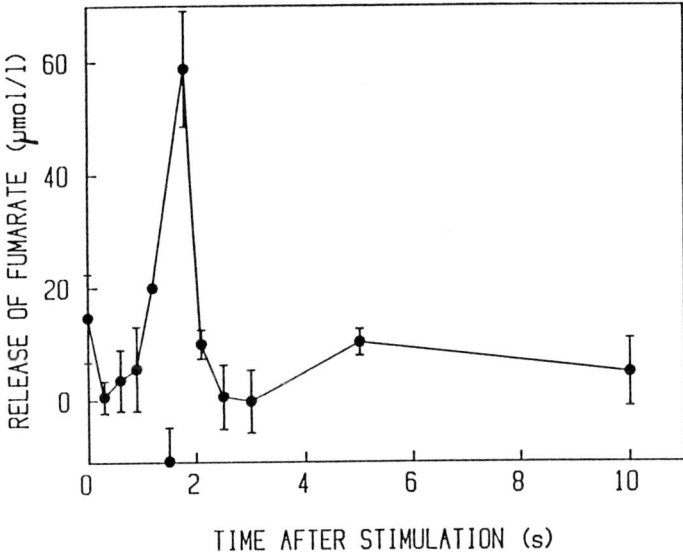

Fig. 2. Light-induced release of fumarate. Cells were stimulated with a blue light pulse of 50 ms and lysed after the delay indicated on the abscissa. The lysate was subjected to ultrafiltration and the fumarate concentration in the filtrate determined with an enzymatic assay. Error bars indicate the standard error of the mean.

IDENTIFICATION OF FUMARATE AS SWITCH FACTOR

In order to search for a substance which regulates the flagellar motor switch, we used a behavioral mutant which rotates its flagellar motor only in the clockwise direction. These cells do not switch the motor neither spontaneously nor stimulus-induced. Switching was re-introduced through permeabilizing the cells by mild ultrasonication in the presence of extracts prepared from wild-type cell membranes. Figure 1 shows swimming tracks recorded from mutant and reconstituted mutant cells. The reconstitution assay was used to purify the active substance which subsequently was identified as fumaric acid (Marwan et al., 1990). Fumaric acid reconstituted motor switching even if only a few molecules per cell were applied. It was therefore concluded that fumaric acid itself, or a close derivative, is the active compound rather than a biosynthetic intermediate derived from it.

LIGHT-INDUCED RELEASE OF FUMARATE

Identification of fumarate as a switch factor raises the question whether it acts as a cofactor required for the switching process or as a response regulator. The latter has to be assumed when the cellular concentration of fumarate is under receptor control.

A suspension of wild-type cells was exposed to blue light stimulation and after a defined period of time, the cells were lysed by rapid mixing with lysis buffer of low ionic strength. After ultrafiltration which removed the high molecular weight compounds, the switch factor activity in the filtrate was measured by its ability to reconstitute motor switching in mutant cells. Blue light, detected by SR-II, or a decrease in orange light, sensed by SR-I, caused a release of the switch factor (fumarate) from a protein-bound pool (Marwan & Oesterhelt, 1991).

By using an enzymatic assay for fumarate, which detects the compound in the lower pmol range, we measured the kinetics of fumarate release after activation of SR-II. The concentration of fumarate released from the protein-bound pool was drastically increased after the application of a blue light pulse as shown in Fig. 2. The kinetics of fumarate release correlate well with the probability of motor switching in intact cells as measured by computer-assisted motion analysis. Cells defective in retinal biosynthesis, and therefore lacking functional photoreceptors, showed only a light-induced release of fumarate when the sensory opsins were reconstituted in the cell by addition of retinal. From this experiment it was concluded that fumarate release is caused by receptor activation rather than by a light-induced change in the activity of metabolic enzymes. Under our experimental conditions at least 200 molecules of fumarate were released per activated SR-II molecule (Montrone et al., manuscript in preparation). This strongly suggests that amplification of the signal occurs and is consistent with the model of photocatalytic signal formation as derived from behavioral measurements.

REFERENCES

Marwan, W. & D. Oesterhelt (1987): Signal formation in the halobacterial photophobic response mediated by a fourth retinal protein (P 480). J. Mol. Biol. 195, 333-342.

Marwan, W. & D. Oesterhelt (1990): Quantitation of photochromism of sensory rhodopsin-I by computerized tracking of Halobacterium halobium cells. J. Mol. Biol. 215, 277-285.

Marwan, W., W. Schäfer & D. Oesterhelt (1990): Signal transduction in Halobacterium depends on fumarate. EMBO J. 9, 355-362.

Marwan, W. & D. Oesterhelt (1991): Light-induced release of the switch factor during photophobic responses of Halobacterium halobium. Naturwiss. 78, 127-129.

Oesterhelt, D. & W. Marwan (1987): Change of membrane potential is not a component of the photophobic transduction chain. J. Bacteriol. 169, 3515-3520.

Oesterhelt, D. & W. Marwan (1990): Signal transduction in Halobacterium halobium. In Biology of chemotactic response, ed. J. P. Armitage, J. M. Lackie, pp. 219-239. Symposium of Soc. Gen. Microbiol. Vol 46, Cambridge: Cambridge university press.

Schimz, A. & E. Hildebrand (1988): Photosensing and processing of sensory signals in Halobacterium halobium. Bot. Acta 101, 111-117.

Spudich, J. L. (1991): Color discriminating pigments in Halobacterium halobium. In Biophysics of photoreceptors and photomovements in microorganisms, ed. F. Lenci et al., pp. 243-248. NATO ASI Series Vol 211, New York : Plenum.

A carboxyl group is protonated during the photocycle of the photophobic receptor psR-II from *Natronobacterium pharaonis*

Birgit Scharf, Martin Engelhard and Fritz Siebert*

*Max-Planck Institut für Ernährungsphysiologie, Rheinlanddamm 201, 4300 Dortmund, Germany; *Institut für Biophysik und Strahlenbiologie, Albertstrasse 23, 7800 Freiburg, Germany*

The phototactic response of the archaeon *Halobacterium halobium* is mediated by two photoreceptors the sensory rhodopsins sR-I and sR-II (for a recent review see Oesterhelt and Marwan, 1990). The absorption of one photon by sR-I initiates the photoattractant response of the bacteria. In a two photon process sR-I triggers also the photophobic answer which has its maximum at 373 nm. A second maximum of the photophobic behavior of *Halobacteria* is found at 490 nm which is due to the second sensory rhodopsin sR-II (Tomioka et al., 1986; Wolff et al., 1986). Whereas quite a few data about the physiological role and the structure of sR-I have been accumulated (Blanck et al., 1989) little is known about sR-II. Reasons are found in its minute concentration in the cellular membrane and its instability towards low salt concentrations.

A different situation is encountered in the observation of an sR-II like pigment in *Natronobacterium pharaonis* (Bivin and Stoeckenius, 1986) and which was spectroscopically analyzed in a couple of publications (Imamoto et al., 1992). A comparison of the biochemical and photochemical properties of sR-II and psR-II revealed striking similarities between the two receptors (Scharf et al., 1992). One major difference was found in the tolerance of psR-II towards low salt concentrations which made it a possible target for a simpler purification scheme and biophysical studies. In this short communication the partial purification of sR-II is described and the results from difference FTIR-spectroscopy experiments are presented.

The membranes from 40 l cell culture were solubilized by dodecylmaltosid and subsequently fractionated on an anion exchange column. Two final purification steps, hydroxylapatit chromatography and gelfiltration resulted in a preparation which had an absorption spectrum characteristic for psR-II (Takahashi et al., 1990; Imammoto et al., 1992) with a maximum at 500 nm and a shoulder at lower wavelengths (Fig. 1). The band at 420 nm is indicative for a small impurity of a cytochrome. The overall enrichment factor was 124. The SDS-polyacrylamide gel showed a major band at 23 kDa, similar to that found by Imamoto et al. (1991).

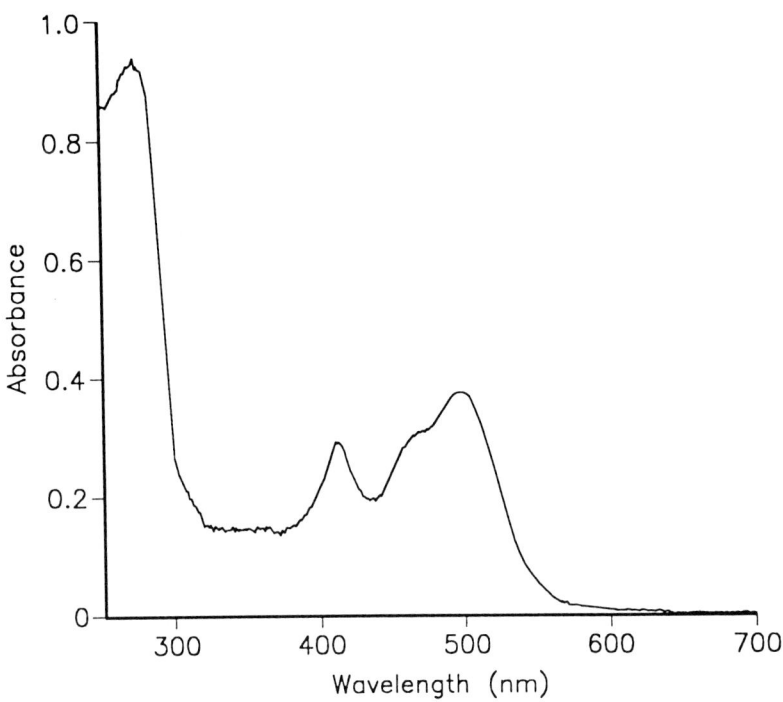

Fig. 1. Absorption spectrum of psR-II

For the FTIR-experiments a sample of psR-II was concentrated to about 30 nmol/ml and dried on a CaF_2 - window. The photolytic activity was not inhibited except for a small decrease in the turnover rate of the photocycle. For the FTIR-experiments the sample was measured in the dark and during irradiation with light above 500 nm. 20 light/dark cycles with 256 scans each were accumulated and the difference between the photostationary and the dark adapted state were determined.

The difference spectrum is shown in figure 2 which displays typical features of that of a retinylidene protein. The negative bands at 1156 cm^{-1}, 1201 cm^{-1}, and 1242 cm^{-1} belong to the C—C stretch modes. The intensive difference band with a minimum at 1547 cm^{-1} and the maximum at 1569 cm^{-1} is due to the C=C stretching vibration of psR-II and the main component of the photostationary state.

The ethylenic stretch mode at 1547 cm^{-1} corresponds to an absorption maximum of 496 nm which is quite close to the measured value of 498 nm. The positive band at 1569 cm^{-1} is found at a similar position as in the M-intermediate of bR or in sR-I_{373} (Bousché et al., ,1991) although contributions from amide II modes have to be taken into account. It can be concluded that the intermediate observed in this steady state experiment is characterized by a deprotonated Schiff base and is identical to psR-II_{390} which was first described by Imamoto et al. (1991).

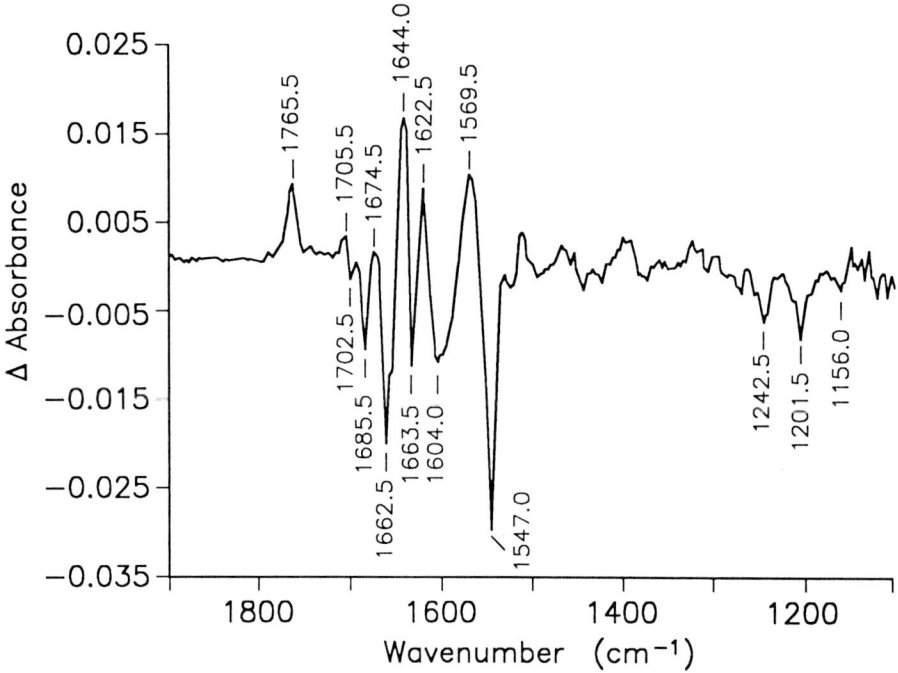

Fig. 2. FTIR difference spectrum of psR-II$_{390}$ - psR-II

The vibrational modes at higher wavenumbers are in the region of the amide bonds. The intensities are comparable to the ethylenic mode of retinal which indicates that gross conformational changes are occuring on the formation of the psR-II$_{390}$ intermediate. The extent of the underlying structural changes are similar to those observed with rhodopsin (Ganter et al., 1992). Due to these strong amide vibrations the C=N mode of the Schiff base cannot be identified from this experiment.

The positive band at 1765.5 cm^{-1} can be attributed to the protonation of a carboxylic group. A further assignment to Asp or Glu is at this stage not possible because neither the amino acid sequence is known nor isotope labelling has been undertaken. The protonation of a carboxyl group mirrors the properties of rhodopsin (Ganter et al., 1992) and bacteriorhodopsin (Engelhard et al, 1985). It is unlike sR-I where only a differential band is observed (Bousché et al., 1991). This would indicate that the corresponding amino acids in sR-I and psR-II have a different charge pattern in the ground state: Whereas the carboxyl group in sR-I is protonated in psR-II it is deprotonated. The reason for this observation might be connected to the colour regulation of the chromophore. One would expect that the protonation of a negative charge close to the chromophore would give rise to a bathochromic shift as it is observed in the acid form of bR.

A comparison of the vibrational difference spectra of sR-I, bR, and rhodopsin with sR-II reveals striking correspondencies of sR-II with rhodopsin and partially with bR. The similarities are observed mainly in the gross conformational changes of the protein back bone and the protonation of a carboxyl group concomittantly with the deprotonation of the Schiff base. Furthermore, the pattern in the finger print region around 1200 cm^{-1} are for bacterial rhodopsin and sR-II quite analogous.

References

Bivin DB, Stoeckenius W (1986) Photoactive Retinal Pigments in Haloalkaliphilic Bacteria. Jour Gen Microbiol 132:2167-2177

Blanck A, Oesterhelt D, Ferrando E, Schegk ES, Lottspeich F (1989) Primary structure of sensory rhodopsin I, a prokaryotic photoreceptor. EMBO Journal 8:3963-3971

Bousché O, Spudich EN, Spudich JL, Rothschild KJ (1991) Conformational changes in sensory rhodopsin I: Similarities and differences with bacteriorhodopsin, halorhodopsin, and rhodopsin. Biochem 30:5395-5400

Engelhard M, Gerwert K, Hess B, Kreutz W, Siebert F (1985) Light-driven protonation changes of internal aspartic acids of bacteriorhodopsin: An investigation by static and time-resolved infrared difference spectroscopy using [4-^{13}C] aspartic acid labeling purple membrane. Biochem 24:400-407

Ganter UM, Charitopoulos T, Virmaux N, Siebert F (1992) Conformational changes of cytosolic loops of bovine rhodopsin during the transition of Meta II. Photochem. Photobiol. in press

Imamoto Y, Shichida Y, Yoshizawa T, Tomioka H, Takahashi T, Fujikawa K, Kamo N, Kobatake Y (1991) Photoreaction cycle of phoborhodopsin studied by low-temperature spectrophotometry. Biochem 30:7416-7424

Imamoto Y, Shichida Y, Hirayama J, Tomioka H, Kamo N, Yoshizawa T (1992) Chromophore configuration of *pharaonis* phoborhodopsin and its isomerization on photon absorption. Biochem 31:2523-2528

Oesterhelt D, and Marwan W Signal transduction in Halobacterium halobium. In: *Biology of the Chemotactic Response Society for General Microbiology Symposium Vol. 46*, edited by Armitage, JP and Lackie, JM Cambridge: Cambridge University Press, 1990, p. 219-239

Scharf B, Pevec B. Hess B, Engelhard, M. (1992) Biochemical and photochemical properties of the photophobic receptors from *Halobacterium halobium* and *Natronobacterium pharaonis.* EurJ Biochem in press

Takahashi T, Yan B, Mazur P, Derguini F, Nakanishi K, Spudich JL (1990) Color regulation in the archaebacterial phototaxis receptor phoborhodopsin (sensory rhodopsin II). Biochem 29:8467-8474

Tomioka H, Takahashi T, Kamo N, Kobatake Y (1986) Flash spectrophotometric identification of a forth rhodopsin-like pigment in *Halobacterium halobium*. Biochem Biophys Res Comm 139:389-395

Wolff EK, Bogomolni RA, Scherrer P, Hess B, Stoeckenius W (1986) Color discrimination in halobacteria: Spectroscopic characterization of a second sensory receptor covering the blue-green region of the spectrum. P Natl Acad Sci 83:7272-7276

Chromophore configuration and photoreaction cycle of phoborhodopsin from *Natronobacterium pharaonis*

Yasushi Imamoto[1], Yoshinori Shichida[1], Junichi Hirayama[2], Hiroaki Tomioka[2], Naoki Kamo[2] and Tôru Yoshizawa[3]

[1] *Department of Biophysics, Faculty of Science, Kyoto University, Kyoto 606,01, Japan,* [2] *Department of Biophysical Chemistry, Faculty of Pharmaceutical Sciences, Hokkaido University, Sapporo 060, Japan, and* [3] *Department of Applied Physics and Chemistry, The University of Electro-Communications, Chofu, Tokyo 182, Japan*

pharaonis Phoborhodopsin (ppR; λ_{max}=498 nm) is a photoreceptor protein for the negative phototactic response of haloalkaliphilic bacterium, *Natronobacterium pharaonis* (Tomioka *et al.*, 1990). It has several unique properties distinguishable from the other retinal proteins. Among them, the following two properties should be noted. First, ppR displays an absorption spectrum having a unique vibrational fine structure, while the other retinal proteins show the bell-shaped absorption spectra [*i.e.* bacteriorhodopsin (bR), halorhodopsin (hR), sensory rhodopsin (sR), visual pigments, *etc.*]. Second, in analysis of the photocycle of ppR by low-temperature spectroscopy, its L-intermediate is not detected (Hirayama *et al.*, 1992). These spectroscopic properties are very similar to those of phoborhodopsin (pR) of *Halobacterium halobium* (Takahashi *et al.*, 1990; Imamoto *et al.*, 1991). Because of the unique properties of ppR, it is of interest to investigate its chromophore configuration and photocycle at room temperature. Recently, we have established the procedure for purification of ppR (Hirayama *et al.*, 1992), which enables us to study its spectroscopic properties in detail much more than those of pR. Using the purified ppR sample (buffer condition: 25 mM Tris-HCl, 4 M NaCl and 0.5 % octyl-glucoside, pH 7.2), the chromophore configurations of ppR and its intermediates were determined (Imamoto *et al.*, 1992a), and several intermediates of ppR were identified with the laser photolysis on the nano- to microsecond time scale (Imamoto *et al.*, 1992b).

Chromophore configuration of ppR and its intermediates.
To clarify the chromophore configurations of ppR and its intermediate(s), the purified ppR sample was first treated with the following three conditions. 1. It was incubated at 4 °C for 2 days in the dark (dark-adapted ppR). 2. Dark-adapted ppR was incubated at 20 °C in the presence of 100 mM hydroxylamine until ppR was completely bleached to retinal oxime and protein moiety (Fig. 1a, curve 31; dark-bleached ppR). 3. Dark-adapted ppR was irradiated with 501-nm light for 128 min in the presence of 10 mM hydroxylamine until ppR was completely bleached to retinal oxime and protein moiety (Fig. 1b, curve 8; photo-bleached ppR). The chromophores of these samples were then extracted according to the standard procedure (Shichida *et al.*, 1988). Namely, the equivalent volumes of methanol and dichloromethane were added to the samples for the denaturation of the protein moiety of ppR, followed by extraction of the chromophores with hexane. In the case of dark-adapted ppR, final 100 mM of hydroxylamine was added to the sample for the formation of retinal oxime prior to extraction. The configurations of extracted retinal oximes were analyzed by HPLC.
The retinal oximes extracted from dark-adapted ppR (Fig. 1c) and dark-bleached ppR (Fig, 1d) were

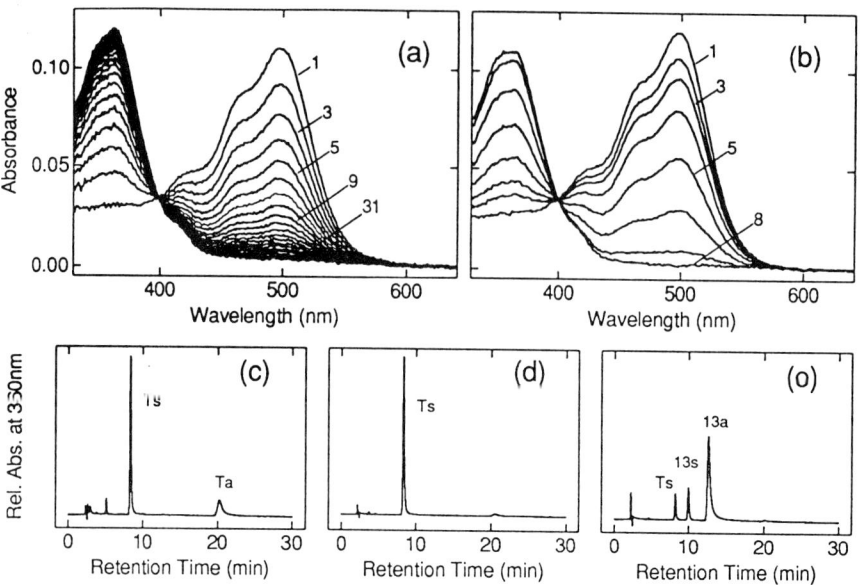

Fig. 1. The analyses of chromophore configurations of ppR and its photo-intermediates. (a) The bleaching process of ppR with hydroxylamine in the dark (dark-bleached ppR). Dark-adapted ppR sample was supplemented with final 100 mM of hydroxylamine (curve 1), and incubated at 20 °C until ppR was completely bleached (15 h). The spectra were recorded at the intervals of 30 min. (b) The photo-bleaching process of ppR in the presence of hydroxylamine (photo-bleached ppR). Dark-adapted ppR sample was supplemented with final 10 mM of hydroxylamine (curve 1), and then irradiated with a 501-nm light for a total of 2, 4, 8, 16, 32, 64 and 128 min at 20 °C (curves 2-8, respectively). (c), (d), (e) The HPLC patterns of the retinal oximes extracted from dark-adapted ppR (curve 1 in a), dark-bleached ppR (curve 31 in a), and photo-bleached ppR sample (curve 8 in b), respectively. The solvent was composed of 98.8 per cent of benzene, 1.0 per cent of diethylether and 0.2 per cent of 2-propanol. Ts, all-*trans*-15-*syn*-retinal oxime; Ta, all-*trans*-15-*anti*-retinal oxime; 13s, 13-*cis*-15-*syn*-retinal oxime; 13a, 13-*cis*-15-*anti*-retinal oxime.

in the all-trans form, indicating that ppR has an all-*trans*-retinal as its chromophore and exhibits no dark isomerization of the chromophore. It is also found in sR (Tsuda *et al*., 1985), but is in contrast with bR and hR. Therefore, absence of dark isomerization of the chromophore may be one of the distinguishable properties between photo-sensors and ion-pumps. On the other hand, from the photo-bleached ppR, 10 per cent of all-*trans*- and 90 per cent of 13-*cis*-retinal oximes were extracted (Fig. 1e). This fact indicates that the chromophore of ppR is isomerized from all-trans to 13-cis form on photon absorption like light-adapted bR, hR, and sR_{587} systems. Under the experimental conditions, ppR, ppR_M and ppR_O were present in the sample during the irradiation, among which ppR is stable against hydroxylamine. Therefore, retinal oximes formed during the irradiation would originate from ppR_M and ppR_O. The 13-*cis*- and all-*trans*-retinal oximes extracted from photo-bleached ppR would be the chromophores of ppR_M and ppR_O, respectively.

It should be noted that the all-*trans*-retinal oxime extracted from dark-bleached ppR was mainly in the 15-syn form (Fig. 1d) whereas the 13-*cis*-retinal oxime from photo-bleached ppR was mainly in the 15-anti form (Fig. 1e). Under these experimental conditions, hydroxylamine attacked the Schiff base linkage surrounded by the protein moiety in the native conformation. On the other hand, retinal oxime extracted from dark-adapted ppR, where the structure of the protein moiety should had been destroyed before the formation of oxime, was composed of 70 per cent of 15-syn and 30 per cent of 15-anti forms (Fig. 1c). Therefore, the amino acid residue(s) near the chromophore induced the stereoselective reactions on C_{15}-N structures of the retinal oximes.

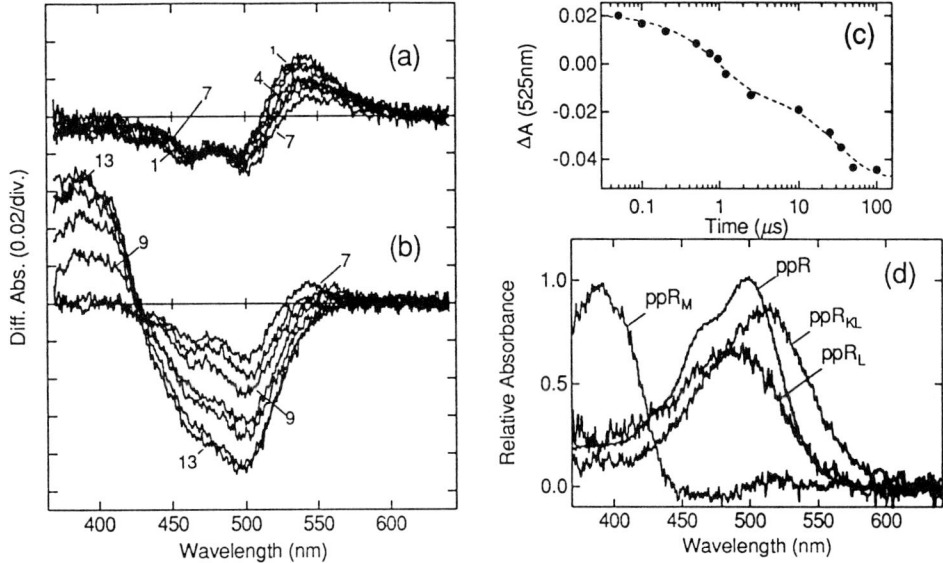

Fig. 2. The analysis of photocycle of ppR. (a) Difference spectra between original sample and its transients measured 50, 100, 200, 500, 750, 950 ns and 1.2 μs (curves 1-7, respectively) after excitation with a nanosecond blue pulse (wavelength, 460 nm; duration, 17 ns). These spectral shifts were mainly due to the conversion from ppR_{KL} to ppR_L. (b) Difference spectra between original sample and its transients measured 1.2, 2.5, 10, 25, 35, 50 and 100 μs after the excitation (curves 7-13, respectively). These spectral shifts were mainly due to the conversion from ppR_L to ppR_M. (c) The absorbance decreases at 525 nm in panels (a) and (b) were plotted against time. The curve approximated to the double-exponential curve whose $\tau_{1/e}$ were 0.99 and 32 μs. (d) The absolute absorption spectra of ppR, ppR_{KL}, ppR_L and ppR_M at 20 °C.

Identification of the primary intermediates of ppR.

The photoreaction cycle of ppR at room temperature was studied by means of nanosecond laser photolysis. The ppR sample, whose absorbance at 500 nm was 0.5, was excited by nanosecond blue pulse (wavelength, 460 nm; duration, 17 ns), and the transient difference spectra were recorded in the time scale from 50 ns to 100 μs. On this time scale, two conversion processes were observed, because two distinct isosbestic points appeared (Fig 2a and b). For further analysis of the conversion process, the absorbance changes at 525 nm were plotted, which approximated to the double-exponential curve whose $\tau_{1/e}$ were 990 ns and 32 μs (Fig. 2c). They are comparable to those of KL (2.2 μs) and L (55 μs) of bR, respectively. Then the absolute absorption spectra of these photoproducts were calculated, and the λ_{max} and ε_{max} of former and latter intermediates were 512 and 488 nm, and 0.85- and 0.68-times smaller than that of ppR, respectively (Fig. 2d). These absorption character-

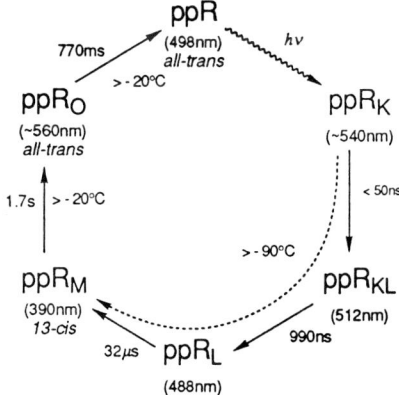

Fig. 3. The scheme of photoreaction cycle of ppR. ppR_K was detected by low-temperature spectroscopy and our recent picosecond laser photolysis (Mizukami et al., to be published). At low temperature, neither ppR_{KL} nor ppR_L was detected (dotted line). Indicated temperatures are the transition temperatures of corresponding intermediates detected by low-temperature spectrophotometry. The absorption maxima of ppR and its intermediates are shown in the blankets.

istics are also similar to KL and L of bR; λ_{max} and ε_{max} of KL and L are 596 and 543 nm, and 0.80- and 0.66-times smaller than that of bR, respectively (Shichida et al., 1983). Therefore, two intermediates observed in the present experiments could be assigned to be ppR_{KL} and ppR_L, respectively. These assignments are further supported by the fact that the precursor of ppR_{KL}, whose absorption characteristics were very similar to K of bR, was detected by picosecond laser photolysis (Mizukami et al., to be published). The photocycle of ppR is shown in Fig. 3.

As described, the present experiments showed that ppR_{KL} and ppR_L appeared in the photocycle of ppR at room temperature, whereas only ppR_K, ppR_M and ppR_O were detectable in our previous low-temperature spectroscopy. That is because ppR_K is thermally more stable (-90 °C; Hirayama et al., 1992) than K of bR (-140 °C; Iwasa et al., 1980). This fact indicates that the activation entropy of ppR_K is larger than that of K, considering from the fact that the $\tau_{1/e}$ of ppR_K at room temperature is comparable to that of K (about 10 ns; Shichida et al., 1983). It would imply that the state for ppR_K has a small entropy due to more rigid structure of the chromophore of ppR_K than that of K.

REFERENCES

Imamoto, Y., Shichida, Y., Yoshizawa, T., Tomioka, H., Takahashi, T., Fujikawa, K., Kamo, N. & Kobatake, Y. (1991): Photoreaction cycle of phoborhodopsin studied by low-temperature spectrophotometry. *Biochemistry* 30: 7416-7424.

Imamoto, Y., Shichida, Y., Hirayama, J., Tomioka, H., Kamo, N. & Yoshizawa, T. (1992a): Chromophore configuration of *pharaonis* phoborhodopsin and its isomerization on photon absorption. *Biochemistry* 31: 2523-2528.

Imamoto, Y., Shichida, Y., Hirayama, J., Tomioka, H., Kamo, N. & Yoshizawa, T. (1992b): Nanosecond laser photolysis of *pharaonis* phoborhodopsin: appearance of KL and L intermediates in the photocycle at room temperature. *Photochem. Photobiol. in press.*

Iwasa, T., Tokunaga, F. & Yoshizawa, T. (1980): A new pathway in the photocycle of *trans*-bacteriorhodopsin and the absorption spectra of its intermediates. *Biophys. Struct. Mech.* 6: 253-270.

Hirayama, J., Imamoto, Y., Shichida, Y., Tomioka, H., Kamo, N. & Yoshizawa, T. (1992): A photocycle of phoborhodopsin from haloalkaliphilic bacterium (*Natronobacterium pharaonis*) studied by low-temperature spectrophotometry. *Biochemistry* 31: 2093-2098.

Shichida, Y., Matuoka, S., Hidaka, Y. & Yoshizawa, T. (1983): Absorption spectra of intermediates of bacteriorhodopsin measured by laser photolysis at room temperatures. *Biochim. Biophys. Acta* 723: 240-246.

Shichida, Y., Nakamura, K., Yoshizawa, T., Trehan, A., Denny, M. & Liu, R. S. H. (1988): 9,13-*dicis*-Rhodopsin and its one-photon-one-double-bond isomerization. *Biochemistry* 27: 6495-6499.

Takahashi, T. Yan, B., Mazur, P., Nakanishi, K. & Spudich, J. L. (1990): Color regulation in the archaebacterial phototaxis receptor phoborhodopsin (sensory rhodopsin II). *Biochemistry* 29: 8467-8474.

Tomioka, H., Otomo, J., Hirayama, J., Kamo, N. & Sasabe, H. (1990): Isolation and characterization of a phoborhodopsin in phototactic haloalkalophile, *Natronobacterium pharaonis*. *IVth International Conference on Retinal Proteins* (Santa Cruz, Ca).

Tsuda, M., Nelson, B., Chang, C.-H., Govindjee, R. & Ebrey, T. G. (1985): Characterization of the chromophore of the third rhodopsin-like pigment of *Halobacterium halobium* and its photoproduct. *Biophys. J.* 47: 721-724.

A molecular genetic approach for studying the halobacterial photoreceptor sensory rhodopsin I

Elisa Ferrando and Dieter Oesterhelt

Max-Planck-Institut für Biochemie, Am Klopferspitz 18 a, D-8033 Martinsried, Germany

Introduction

Halobacteria, which belong to the phylogenetic branch Archea, provide a unique example of colour-discriminating photobehaviour mediated by retinal-containing sensory pigments. Two photoreceptors, sensory rhodopsin I (SR I) and sensory rhodopsin II (SR II) or P480 have been identified thus far in halobacterial membranes (Bogomolni & Spudich, 1982; Spudich et al., 1986). P480 mediates a repellent phototactical response to blue light (λ_{max}=480 nm), while sensory rhodopsin I is responsible for an attractant response to orange light (λ_{max}=587 nm) and a repellent response at 380 nm. SR I thus displays a dual light sensitivity which confers a simple colour discriminating ability to halobacterial cells, and which is a result of the two-photon cycle of this pigment: photoexcitation of the ground state absorbing at 587 nm delivers an attractant signal to the cell and generates a long-lived intermediate absorbing at 373 nm. This intermediate can relax either thermally to the ground state or convert rapidly by absorption of a second photon. This second photoreaction elicits a repellent phototactical stimulus whose occurrence is therefore dependent on orange background illumination (Spudich & Bogomolni 1988).

The halobacterial photoreceptors are closely related to the ion pumps bacteriorhodopsin (BR) and halorhodopsin (HR). All four proteins share the same secondary structure and consist of seven transmembrane helices with retinal as a chromophore. They occur, however, in very different amounts in halobacterial cell membranes: 1000/100/10/1 for BR/HR/SRI/P480.

Sensory rhodopsin I is the first and hitherto only photoreceptor to be isolated from membranes of *Halobacterium halobium* (Schegk & Oesterhelt 1988). Due to the low abundance of SR I and its poor stability after membrane solubilization, only 300 µg of purified protein can be isolated from 60 l of bacterial culture. The presence of exogenous retinal, which must be supplied to the purified SR I to prevent its inactivation, hampers the biophysical characterization of this pigment and, in particular, the determination of its retinal isomers.

The development of a transformation system for halobacteria by Cline *et al.* (1989) set the stage for studying halobacterial retinal proteins with the help of molecular genetic tools. The gene coding for bacteriorhodopsin has been already expressed in *H. halobium* using shuttle vectors bearing a mevinolin resistance marker (Ni **et al.** 1990; Ferrando **et al.** subm.). A possible approach to enhance the amount of retinal proteins other than bacteriorhodopsin in halobacterial cell membranes is to fuse their genes to the strong BR promoter. Overexpression of halorhodopsin in a HR⁻ strain by means of gene fusion is reported elsewhere in this book (Heymann **et al.**). Functional expression of sensory rhodopsin I requires a *H. halobium* strain, which is defective in the gene coding for SR I, *sop I*. Here we describe the construction of such a strain by targeted gene disruption and the first results of homologous expression of *sop I*.

The *sop I* gene, its targeted disruption, and its expression as a *bop-sop I* fusion

sop I was identified in a genomic digest of halobacterial DNA using a degenerated oligonucleotide probe, which was derived from the N-terminal sequence of the purified protein (Blanck **et al.** 1989). The structural *sop I* gene comprises 720 bp. The deduced amino acid sequence was confirmed by amino acid sequencing of tryptic peptides obtained from the purified protein. The SR I sequence has 26% identity to BR and 23% identity to HR.

For expression of *sop I* two *H. halobium* strains, which were phenotypically SR⁻ were tested and found to be inadequate because of secondary mutations (Ferrando **et al.** in prep.). Therefore, a *sop I*⁻ recipient strain was constructed in which the entire structural *sop*

I has been deleted. This deletion was achieved by means of site-specific homologous recombination, which has been shown to occur at high frequency in *H. halobium* (Ferrando **et al.** subm.). A suicide vector was constructed which carried a novobiocin resistance gene (Holmes **et al.** 1991) flanked by the 5´and 3´ nucleotide sequences of *sop I*. Transformation of halobacteria with this vector resulted in about 50% novobiocin resistant clones in which the novobiocin resistance marker had integrated into the genome at the desired location which is the replacement of *sop I*. These cells had lost the SR I specific phototactical response and no SR I specific photochemical activity could be detected in their membranes. This provided the first true correlation between the *sop I* DNA and the physiological function of the protein.

For expression of SR I in *H. halobium* a hybrid gene was constructed consisting of the upstream region and the first 39 bp of *bop* and the entire structural *sop I* gene. This hybrid gene encodes the SR I protein fused to the first 13 amino acids of BR which are probably required for efficient translation of BR mRNA. Transformation of halobacterial cells with a mevinolin resistance vector carrying the *bop-sop I* hybrid gene did not result in SR I expression at a level comparable to that of BR. The amount of SR I present in the transformants was increased by a factor of 2-3 compared to the wild-type as determined by Western blots and by the measurement of the photochemical activity. Analysis of the *sop*-specific transcript in these cells showed a 10-20 fold increase of the mRNA level after transformation. However, the amount of *sop*-specific mRNA was far lower than that of *bop* specific mRNA in cells transformed with a *bop* control plasmid. On the basis of these results, it was concluded that the copy number of SR I in *H. halobium* is regulated both at the protein and the mRNA level. Efforts to identify stability determinants on the SR I protein or on its mRNA are under way. The principal questions that may be now adressed with the help of the described deletion strain and of vectors suited for expression and mutagenesis of *sop I* are: first, understand the interactions of SR I with the phototactical signal chain, and second, gain higher amounts of protein in the membrane, which would bring much advantage to biophysical studies on this retinal protein.

Acknowledgement

We thank Judy Shiozawa for critical reading of the manuscript.

References

Blanck, A., Oesterhelt, D., Ferrando, E., Schegk, E.S., and Lottspeich, F.. (1989): Primary structure of sensory rhodopsin I, a prokaryotic photoreceptor. EMBO J. 8, 3963-3971.

Bogomolni, R.A. & Spudich J.L. (1982): Identification of a third rhodopsin-like pigment in phototactic *Halobacterium halobium*. Proc. Natl. Acad. Sci. USA 79, 6250-6254.

Cline, S.W., Lam, W.L., Charlebois, R.L., Schalkwyk, L.C., and Doolittle, W.F. (1989): Transformation methods for halophilic archaebacteria. Can. J. Microbiol. 35, 148-152.

Holmes, M.H., Nuttall, S.D., and Dyall-Smith, M.L. (1989): Construction and use of halobacterial shuttle vectors and further studies on *Haloferax* DNA Gyrase. J. Bac. 173, 3807-3813.

Ni, B., Chang, M., Duschl,A., Lanyi J., and Needleman,.R. (1990):. An efficient system for the synthesis of bacteriorhodopsin in *H. halobium*. Gene 90, 169-172.

Schegk, E. S., and Oesterhelt, D. (1988): Isolation of a prokaryotic photoreceptor: sensory rhodopsin from halobacteria. EMBO J. 7, 2925-2933.

Spudich, J.L., and Bogomolni, R.A. (1988): Sensory rhodopsins of halobacteria. Ann. Rev. Biophys. Biophys. Chem. 17, 193-215.

Spudich, E. N., Sundberg, S.A., Manor, D., and Spudich, J.L. (1986): Properties of a second sensory receptor protein in *Halobacterium halobium* phototaxis. Proteins 1, 239-246.

Overexpression of halorhodopsin: new perspectives for structure-function studies

Jürgen A.W. Heymann, W.A. Havelka and D. Oesterhelt

Max-Planck Institut für Biochemie, Am Klopferspitz 18A, 8033 Martinsried, Germany

Halobacteria, a member of the family of Archaea, have been used as a system to study proteins that are involved in photosynthesis, osmoregulation and phototaxis. A seven-helical structure has also been found in receptor proteins of eucaryotes and contains the common motif of these integral membrane proteins (Henderson & Schertler, 1990). Some of them take part in phototaxis and light-dependent transport processes and contain retinal as the chromophore. This is bound as a Schiff base to the protein via a lysine residue and is the absorber of visible light while interacting with the apoenzyme opsin.

Absorption of light drives the isomerisation of retinal leading to an activated species followed by thermal relaxation. In the case of rhodopsin, the chromoprotein will decay into free retinal and inactive opsin. Halobacterial chemistry of rhodopsins differs that perturbation by light is followed by relaxation and exhibits a catalytic photoycle. This leads to different responses depending on the protein environment of the energy transducing retinal; which includes vectorial ion translocation and induction of signal chains.

The best system known so far is bacteriorhodopsin (BR, *bop* gene) (Oesterhelt & Stoeckenius, 1971) which has been investigated during the last twenty years. Halorhodopsin (HR, *hop* gene) (Schobert & Lanyi, 1982) was the second retinal protein discovered in *H.halobium*. Both proteins are light-dependent electrogenic ion pumps, in which BR pumps protons outward and HR pumps chloride inward (Oesterhelt & Tittor, 1989). Anaerobiosis induces the expression creating a simple photosynthetic system that couples light energy with proton translocation omitting electron flow.

EXPRESSION OF HALOBACTERIAL RETINAL-PROTEINS

Wild-type BR could be isolated in large amounts from natural overproducers, like strain S9, however HR and the sensory proteins could not. Structure-function studies deal with wild-type or mutated protein species. The expression level and host genetics are often limiting and will therefore require time consuming purification and enrichment techniques. In early times, procaryotic *E.coli* has successfully been used to express BR (Dunn et al., 1987). Later, an eucaryotic *S. pombe* system, showed interesting features of heterologous expression by utilizing the protein translocation machinery of the cell (Hildebrandt et al., 1989). The use of heterologous systems often implies restrictions in terms of transcription and translation efficiency, postranslational modification and processing, folding and, the absence of cofactors.

Bearing this in mind, these systems have provided a wealth of information, however, ideally a homologous system would be best. There are requirements for homologous expression: vectors which can be stably maintained in different organisms and therefore contain multiple origins of replication, selection markers and unique cloning sites. Finally, a system has to be established that allows efficient DNA transfer into a host strain that contains an inactive gene. These requirements are a challenge without discussing the genetics of gene-overexpression. Recently, it has been shown that halobacteria could be transformed with plasmid DNA (Lam & Doolittle, 1990). In consequence, these vectors were used to express BR in *H.halobium* (Ni et al., 1990). We successfully followed the idea of homologous expression with HR and established shuttle vectors, vectors for site-specific mutagenesis (Heymann et al., manuscript in prep.) and a new expression system (Heymann et al., submitted).

FROM GENE TO PROTEIN

The *hop* gene was detected by hybridisation analysis using oligonucleotide probes from peptid-sequence information. The *hop* gene was therefore cloned from a cosmid library containing halobacterial chromosomal DNA (Blanck & Oesterhelt, 1987). Two different shuttle systems have been used to express HR in *H.halobium* (see Fig.). Both vectors contain origins that maintain replication in *E.coli* and *H.halobium*, antibiotic resistance genes for the selection of transformants in both organisms and unique cloning sites for the exchange of HR. By substitution of the *bop* gene for *hop* we could drive the expression of *hop* under the control of the *bop* promotor. A second shuttle vector was constructed by cloning a chromosomal DNA fragment containing the entire *hop* gene within a 4.7 kb DNA fragment. Recipient strain HN5, courtesy of K.Rumpel, is BR and HR minus and could be transformed with both shuttle vectors.

A system for the site-specific mutagenesis of HR was also established. This contains the *hop* gene, a phage origin of replication, to produce single stranded DNA, and a selection of cleaving sites which enables one to use gene cartridges for the construction of multiple mutated HR species.

The overexpression of HR is related to the strong *bop* promotor sequence and produces a protein with full activity. The differences between HR from the strain OD2w which is a natural overexpressor, and the HR from homologous system will be discussed (Heymann et al., submitted). The strain OD2w expresses HR 10 times lower than that of BR, and at this concentration the protein does not form crystalline arrays *in vivo*. The first purification of HR was achieved under denaturing conditions using solubilized membrane preparations for gel filtration and hydroxylapatite chromatography. Physical properties of HR show discrepancies which depend on the detergent system used in purification. This indicates the overall problem dealing with artificial systems, including reconstitution of protein function and thereby, neglecting intermolecular forces. Meanwhile, HR can be reconstituted with lipids from purple membrane (PM) trying to mimic the *in vivo* environment for analysis.

The *bop-hop*-fusion drives the expression of HR to a level of 50 to 60% of BR with following advantage: HR could be easily purified by isolating specific membrane fractions through equilibrium buoyant density centrifugation. This contrasts to HR from OD2w which requires detergent plus high salt concentrations which restricts further manipulation of the protein. Two fractions from the overexpressed HR could be distinguished according to their buoyant densities. The first, equals monomeric HR, as seen in OD2w, the second, resembles PM and shows arrays of a crystalline protein. Increasing amount of expressed protein forces *in vivo* crystallization and enables 2D structural analysis. Since protein quantities are not further limiting, 3D and spectroscopic analysis will help to study structure-function relationships in wild-type and mutated species which can also be applied to other 7-transmembrane helical proteins. Membrane fractions could be analysed directly without perturbing the native environment when applying detergents.

HR could also be expressed without the use of the *bop* promotor as described in the second shuttle construct. Overexpression was detected at higher levels than in OD2w, but reduced compared to the construct with the *bop* promotor. HR from this strain does not form PM like membrane fractions on a sucrose gradient. It will be interesting to see whether *in vivo* crystallization affects HR's properties.

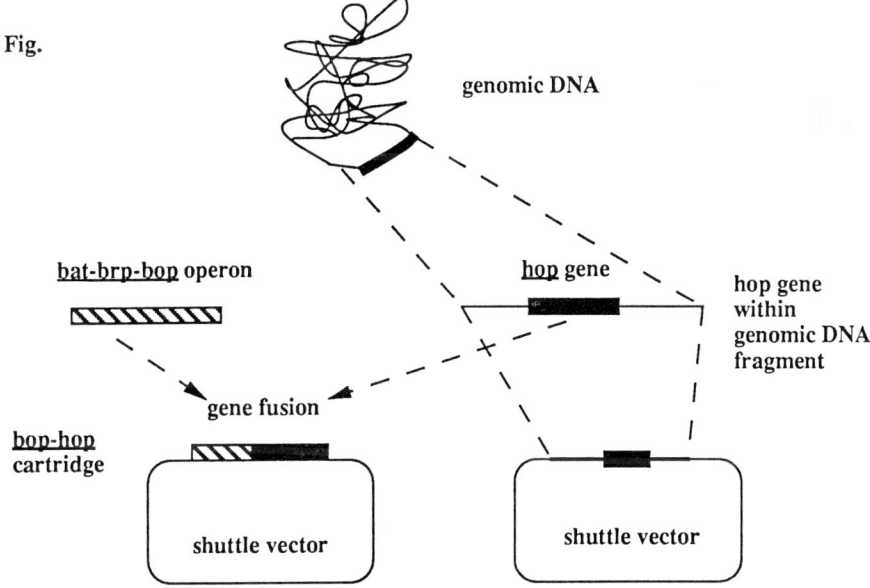

Fig.

SUMMARY

Overexpression of halorhodopsin in *H.halobium* was achieved by employing a gene-fusion approach. The *bop* promotor expressed HR to an extent where *in vivo* crystallization was observed. The homologous expression system established will allow for detailed structure-function analysis of HR in its natural environment and will be a powerful tool in combination with site-directed mutagenesis. This was the first successful overexpression of a 7-helical transmembrane protein and may be extended to other proteins of this family.

REFERENCES

Blanck, A. & Oesterhelt, D. (1987): The halo-opsin gene. II. Sequence, primary structure of halorhodopsin and comparison with bacteriorhodopsin. *EMBO J.* 6, 265-273.

Dunn, R.J. et al. (1987): Structure-function studies on bacteriorhodopsin. I. Expression of the bacterio-opsin gene in *Escherichia coli. J. Biol. Chem.* 262, 9246-9254.

Henderson, R. & Schertler, G.F.X. (1990): The structure of bacteriorhodopsin and its relevance to the visual opsins and other seven-helix G-protein coupled receptors. *Phil. Trans. R. Soc. Lond. B* 326, 379-389. (140)

Heymann, J.A.W. et al.: Overexpression of Halorhodopsin in *Halobacterium halobium*. *EMBO J.* submitted.

Hildebrandt. V. et al. (1989): Genetic transfer of the pigment bacteriorhodopsin into the eucaryote *Schizosaccharomyces pombe*. *FEBS Lett.* 243, 13.

Lam, W.L. & Doolittle, W.F. (1989): Shuttle vectors for the archaebacterium *Halobacterium volcanii*. *Proc. Natl. Acad. Sci. USA* 86, 5478-5482.

Ni, B. et al. (1990): An efficient system for the expression of bacteriorhodopsin in *Halobacterium halobium*. *Gene* 90, 169-172.

Oesterhelt, D. & Tittor, J. (1989): Two pumps, one principle: Light driven ion transport in halobacteria. *Trends Biochem. Sci.* 14, 57-61.

Oesterhelt, D. & Stoeckenius, W. (1971): Rhodopsin-like protein from the purple membrane of *Halobacterium halobium*. *Nature* 233, 149-152.

Schobert, B. & Lanyi, J.K. (1982): Halorhodopsin is a light driven chloride pump. *J. Biol. Chem.* 257, 10306-10313.

Direct evidence for the involvement of membrane potential changes in the photosensory transduction of halobacteria

Sergei I. Bibikov[1], Ruslan N. Grishanin[1], Andrey D. Kaulen[1], Wolfgang Marwan[2], Dieter Oesterhelt[2] and Vladimir P. Skulachev[1]

[1] A.N. Belozersky Institute of Physico-Chemical Biology Moscow State University, Moscow 119899, Russia;
[2] Max-Planck Institut für Biochemie, Martinsried, München, D-8033, Germany

PREFACE.

Halobacterial cells are rod-shaped and bipolarly flagellated. Their ability to sense both light and chemicals has prompted the intensive studies on the molecular mechanism responsible for this complex behavior. Two retinal proteins were shown to be directly involved in photosensory transduction: sensory rhodopsin 1 (sR1) and sensory rhodopsin 2 (sR2 or phoborhodopsin) (Spudich & Bogomolni, 1984; Takahashi et al, 1985). The photo-intermediates of the sR photocycle seem to activate the corresponding methyl-accepting phototaxis proteins (Spudich et al, 1988), which in turn initiate the series of cytoplasmic signalling components as has been shown earlier in other bacteria. The net result of the signal transduction is the reversal of the flagellar bundles and, hence, cells in response to a decrease in intensity of orange light or an increase in intensity of blue or near UV-light. Mutants which lack all retinal proteins in halobacteria are phototactically but not chemotactically deficient (Sundberg et al., 1990).

There are two other retinal proteins in halobacteria, namely bacteriorhodopsin (BR) and halorhodopsin (HR), which are involved in pumping protons out of the cytoplasm or chloride ions into the cells respectively (Oesterhelt & Stoeckenius, 1971; Schobert & Lanyi, 1982). The role of these energy converting proteins in sensory transduction has been a point of controversy. In spite of a remarkable coincidence of the absorption spectrum of BR and the action spectrum of the photoresponse to long-wavelength light, mutants defective in BR synthesis were capable of photosensing due to the presence of sR (Spudich & Bogomolni, 1984); and the transmission of this signal did not require changes in membrane potential (Oesterhelt & Marwan, 1987). However BR seemed to contribute to photosensory cycle in wild type strains particularly at high light intensities (Bibikov & Skulachev, 1989). In order to evaluate the role of bacteriorhodopsin and to test its possible dual role in halobacteria, we transformed the phototactically deficient strains of H.halobium with a set of plasmids carrying normal and site-specifically mutated bop-genes and analysed the behavior of the transformants.

RESULTS AND DISCUSSION.

Bacteriorhodopsin was introduced into the cells of the phototactically deficient mutant Pho81. Transformant cells acquired the capability to sense light if subjected to anaerobiosis or starvation (Bibikov et al.,1991). Cells accumulated in an illuminated area under the microscope and changed the direction of movement upon a decrease in light intensity (Fig.1)

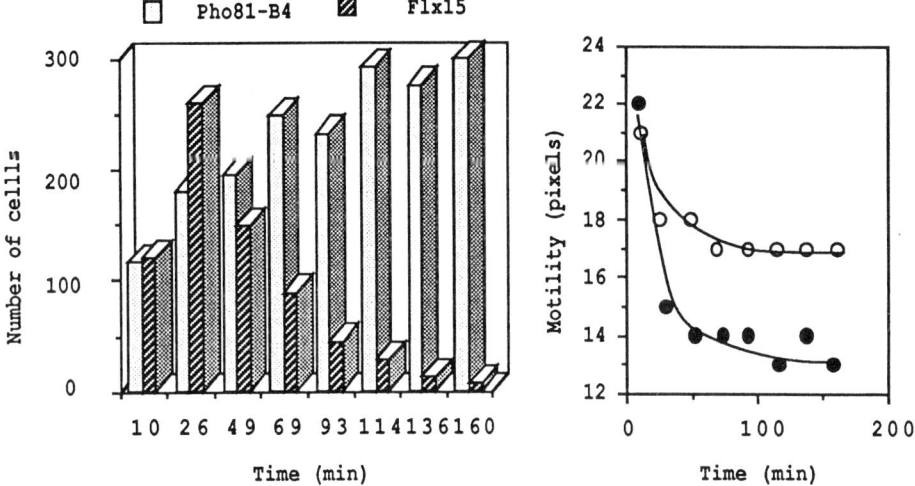

Fig.1. Accumulation of cells was analyzed using a motion analysis system. Only cells which were motile and responded to stimuli were registered. The number of motile cells in the light spot, average motility in pixels of the monitor screen and % of the cells which reversed upon the application of a light stimulus were analyzed synchronously. Cells were resuspended in a starvation medium and subjected to partial anaerobiosis by sealing the coverslip with petroleum jelly. Light source was a mercury lamp. Orange ($\lambda>540nm$) or blue ($\lambda<370nm$) light beams were applied through the condensor and objective of the Leitz microscope, respectively. Empty circles - Pho81-B4 transformant containing wild type BR as the only retinal protein; filled circles - Flx15 strain, from which strain Pho81 is derived, containing sR1 and sR2.

The action spectrum revealed a broad maximum so that not only orange light but also blue light elicited a reversal upon a decrease in intensity. Background intensity was shown to be an important factor in distinguishing bacteriorhodopsin and sensory rhodopsin mediated responses. Small changes in the intensity of orange light at high background levels were shown to be less effective for the sensory rhodopsin-dependent system than for the bacteriorhodopsin-dependent. Maximum response to a defined drop of intensity occured in the region of the maximum slope of the fluence response curve. At high light intensities both sR and BR dependent photosensory systems are saturated but addition of an uncoupler of oxidative phosphorylation increased the response of the transformant cells (not shown). At the same time uncoupler inhibited the response in these cells at low light intensities. The latter fact was in line with the idea that bacteriorhodopsin was signalling to flagella not directly and not in the same manner as sR but via a sensor of the proton-motive force - protometer (Bibikov & Skulachev, 1989) (Fig.2).

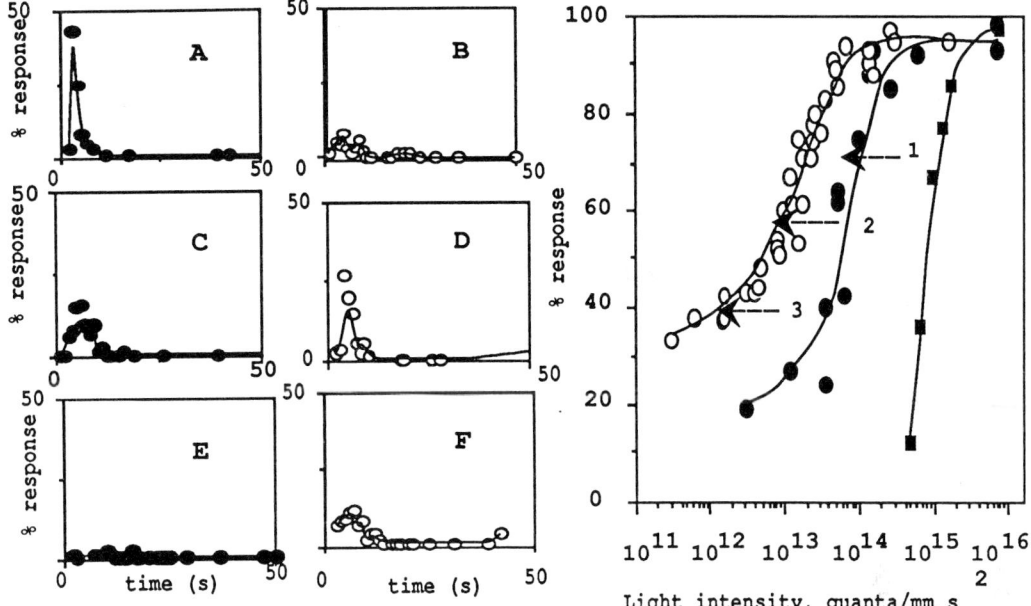

Fig.2. Reaction of the BR and sR containing cells to long-wavelength light of different intensities. Each point represents the data collected from at least 75 cells. Fluence response curves (right) are shown for Flx3 (sR) (empty circles), Pho81-B4 (BR) (filled circles) and Pho81-B4 in the presence of 2 μM CCCP (filled squares). Stimulus was turning off the light by electronic shutter. Step responses at different background intensities (left) are shown for Pho81-B4 (filled circles) and Flx3 (empty circles). Each pair corresponds to 85% decrease in light intensity shown by the arrow (direction of a step) on the right plot (A,B - 1; C,D - 2; E,F - 3).

Response to light in a transformant was not immediate. A delay of approximately 1.5s corresponded well with the time delay for the sensory rhodopsin signal. Direct influence of the membrane potential changes on flagella was rejected also by introducing bacteriorhodopsin into the switch mutant M17 which was not able to respond to light stimuli unless somatically supplied with fumarate.

Bacteriorhodopsin carrying the D96N point mutation is known to have a slow photocycle due to a relatively slow protonation of the Shiff base (Holz et al, 1989). However the photocycle may be promoted by addition of azide which works as an intramolecular donor of protons (Tittor et al, 1989). When introduced into Pho81, a phototactically deficient mutant, this protein was incapable to support the photoresponse to orange light unless azide was added. It was shown earlier that in the D96N bacteriorhodopsin mutant the M-intermediate of the photocycle with maximum at 412nm is accumulated to a relatively high level with orange background illumination (Butt et al, 1989). Intensive blue light stimulated the immediate return of all the molecules to the ground state together with the capture of protons from the periplasmic space. Massive movement of protons inside the cell causes depolarization. We checked the response of D96N mutant to blue light stimuli and observed immediate reversals in the cells (Fig.3). Azide inhibited the response to blue light.

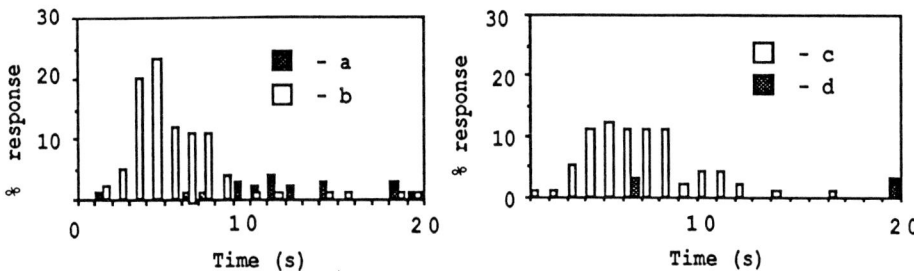

Fig.3. Photoreactions in the Pho81-D96N mutant which contains BR with a mutation Asp->Asn at position 96. The percentage of all the registered cells which reversed during a defined 1s time interval is presented as a bar at the corresponding time interval. a,b -responses to reduction in orange light intensity; c,d - responses to increase in blue light intensity. Curves b and d reflect the reaction of the cells when 5mM azide was applied to the medium.

This result indicates that the electrical component of the proton-motive force ($\Delta\psi$) is a possible link between bacteriorhodopsin and the signal-transduction chain.

REFERENCES

Bibikov,S.I. & Skulachev,V.P. (1989): Mechanisms of phototaxis and aerotaxis in Halobacterium halobium. FEBS Lett.243,303-306.
Bibikov,S.I.,Grishanin,R.N.,Marwan,W.,Oesterhelt,D. & V.P.Skulachev (1991): The proton pump bacteriorhodopsin is a photoreceptor for signal transduction in Halobacterium halobium. FEBS Lett.295,223-226.
Butt,H.J.,Fendler,K.,Bamberg,E.,Tittor,J. & Oesterhelt,D.(1989):Aspartic acids 96 and 85 play a central role in the function of bacteriorhodopsin as a proton pump. EMBO J.8,1657-1663.
Holz,M.,Drachev,L.A.,Mogi,T.,Otto,H.,Kaulen,A.D.,Heyn,M.P.,Skulachev,V.P. & Khorana,H.G. (1989): Replacement of aspartic acid - 96 by asparagine in bacteriorhodopsin slows both the decay of the M intermediate and the associated proton movement. Proc.Natl.Acad. Sci. USA 86, 2167-2171.
Oesterhelt,D. & Stoeckenius,W. (1971): Rhodopsin-like protein from the purple membrane of Halobacterium halobium. Nature New.Biol. 233, 149-152.
Oesterhelt,D. & W.Marwan. (1987): Change of membrane potential is not a component of the photophobic transduction chain in Halobacterium halobium. J.Bacteriol.169, 3515-3520.
Schobert,B. & Lanyi,J.K.(1982): Halorhodopsin is a light-driven chloride pump. J.Biol.Chem. 257,10306-10313.
Spudich,J.L. & Bogomolni.R.A.(1984): Mechanism of colour discrimination by a bacterial sensory rhodopsin. Nature (London).312, 509-513.
Spudich,E.N.,Hasselbacher,C.A. & Spudich,J.L. (1988): Methyl-accepting protein associated with bacterial sensory rhodopsin 1. J. Bacteriol. 170, 4280-4285.
Sundberg,S.A.,Alam,M.,Lebert,M.,Spudich,J.L.,Oesterhelt,D. & Hazelbauer,G.L. (1990): Characterization of Halobacterium halobium mutants defective in taxis. J. Bacteriol.172, 2328-2335.
Takahashi,T.,Tomioka,H.,Kamo,N. and Kobatake,Y.(1985): A photosystem other than PS-370 also mediates the negative phototaxis of Halobacterium halobium. FEMS Microbiol.Lett. 28, 161-164.
Tittor,J.,Soell,C.,Oesterhelt,D.,Butt,H.-J. & Bamberg,E. (1989): A defective proton pump, point-mutated bacteriorhodopsin Asp96-->Asn is fully reactivated by azide. EMBO J.,8,3477-3482.

IX. Rhodopsin-like pigments in *Chlamydomonas*

IX. Pigments de type rhodopsine chez Chlamydomonas

Light excited electric signals from *Chlamydomonas*

L. Keszthelyi

Institute of Biophysics, Biological Research Center of the Hungarian Academy of Sciences, H-6701 Szeged, Hungary

INTRODUCTION

It is well documented that the photoreceptor of the unicellular alga *Chlamydomonas reinhardtii* (which is responsible for its phototaxis) is a retinal containing protein (Foster et al., 1984). The photoreceptor is located in the eye spot on one side of the alga and the signal due to light absorption is guided to the two flagella of the cell (for a review see Nultsch and Häder, 1988). Recently, Derguini et al. (1991) have shown that all-trans retinal is the chromophore in the photoreceptor and Beckmann and Hegemann (1991) found the retinal bound to a protein of molecular weight of 32000. The study of the effect of the light absorption of this rhodopsin, the amplification and transmission of the signal to the flagella may open new insight into the different photosensory processes. The pioneering work of Litvin et al. (1978) on *Hematococcus pluvialis* - which behaves similarly to *Chlamydomonas* but its size is larger - by the so called suction pipette method and similar investigations by Sineshchekov et al. (1990) clearly demonstrated that a multicomponent electric potential appears in these cells after light absorption. The similar measurements on *Chlamydomonas* are more difficult because of their non-elastic cell walls (Harz and Hegemann, 1991) and in general the time resolution of this technique is rather poor. We developed a method for the easy registration of the receptor potential by the light gradient effect (for a recent review see Leibl and Trissl, 1990). It is based on the asymmetry of the location of the eye spot in the cells and the asymmetry of the generated potential propagating in the direction of the flagella (Sineshchekov et al. 1992). A fast, well measurable electric signal is obtained from cells in suspension if the direction of the illuminating light is properly related to the electrodes in the solution. The most important results are the good time resolution, the multicomponent nature and more importantly the strict retinal dependence of the light generated electric potential.

MATERIALS AND METHODS

Two different strains of *Chlamydomonas reinhardtii* : 495 - a green mutant with high phototactic sensitivity and 494/30 - a white mutant without chlorophylls and retinal (kindly provided by Dr. A. Chunaev from St. Petersburg) were used in the experiments. Vegetative cells were transformed into gamets and suspended in a solution of low ion concentration (0.1 mM K^+ in the form of phosphate buffer at pH 7, 0.05mM $CaCl_2$). The density of cells in a specially designed cuvette (Fig. 1) was ~ 5 x 10^6 cells/cm^3. A dye laser (coumarine 307 dye) pumped by

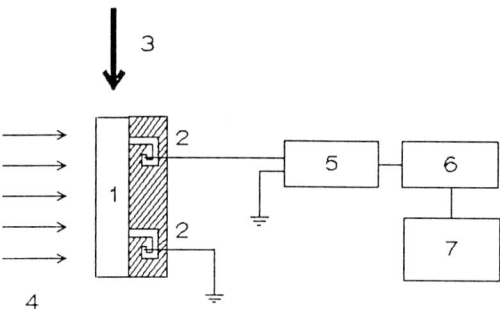

Fig. 1. The measuring system. 1 - cuvette, 2 - electrodes (thin Pt wire), 3 - laser light, 4 - red background illumination, 5 - amplifier, 6 - transient recorder, 7 - computer.

an excimer laser illuminated the cells at 500 nm. The energy was ~ 1 mJ, the length of the flash ~ 20 ns. The intensity of the red background light (Fig. 1) was ~ 2 W/m^2 in a wavelength range of > 630 nm. (The electrodes were well protected from the direct and scattered light of excitation.) The resistance of the solution was in the MΩ range. The signals were amplified 100x by a home made fast voltage amplifier (rise time ~ 0.3 µs) with an input resistance of 1 MΩ and recorded by a voltage digitizer (Thurlby DSA 524) and an IBM compatible computer.

RESULTS AND DISCUSSION

In Fig. 2 the time course of typical electric signals in case of green mutant is shown in logarithmic time base. The two signals measured with electrodes oppositely directed to the exciting beam are mirror images of each other. This shows that they are caused by the light gradient effect. The signal of ~ µs duration is due to the photosynthetic system of the cell, the next one which rises in ~ 200 µs and decays with a lifetime of ~ 3 ms is due to the photoreceptor potential and the next in the range of tens of ms is the regenerative response.

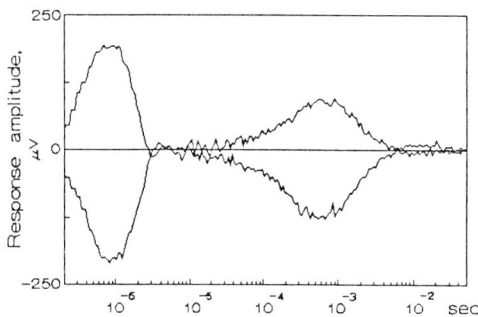

Fig. 2. Electric signals measured with alternated direction of exciting light related to the electrodes. Cell line 495, number of repetitions 20.

Fig. 3 shows that the red background light diminishes the amplitude of the fast signal because it drives the photosynthesis, i.e. it closes part of the reaction centers which in this case do not contribute to the signal. The amplitude of the photoreceptor potential increases in the presence of red background (Fig. 3). The increase can be explained by the hyperpolarisation of plasmalemma due to the active photosynthesis (Sineshchekov et al. 1992). The photopotential is generated by photons absorbed by the all-trans retinal bound to a protein which is part of the

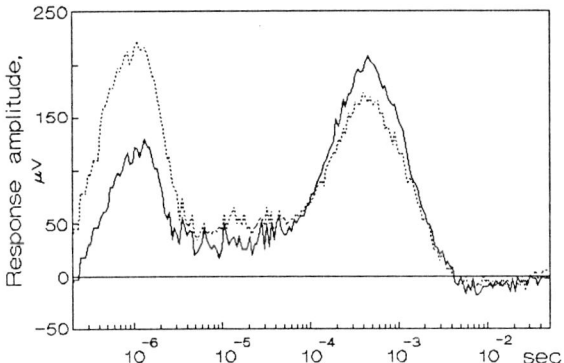

Fig. 3. Electric signals in absence (dotted line) and presence (solid line) of the red background illumination. Cell line 495, number of repetitions 20.

eye spot of the cell (Beckmann and Hegemann, 1991). In Fig. 4 we show that no signal is generated in the white cells. By adding, however, all-trans retinal to the suspension the photoreceptor potential appears. This figure is considered also as a clear demonstration that no signal occurs in the μs range if photosynthesis is absent.

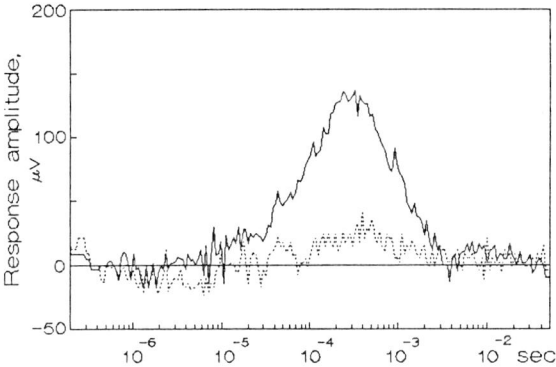

Fig. 4. Electric signals measured in case of white cells (494/30, dotted line) and with added all-trans retinal (1 μM) - solid line. Number of repetitions 20.

An important question is how does the retinal containing protein trigger the appearance of the photopotential. It is known from different studies that the first event after light absorption in rhodopsins is a sudden isomerization of the retinal. In such cases when electric response after laser excitation has been measured - like bacteriorhodopsin, halorhodopsin (Keszthelyi and Ormos, 1989) and the eye rhodopsin (Bolshakov et al., 1979) - a very fast signal was easily measurable. We expect that such a signal should be measurable with added retinal in the white mutants when the fast signal due to photosynthesis is not a disturbing factor. Such a signal is not seen in Fig. 4. An upper limit of $V_1 < 10$ μV is set by the noise, ~ 15 times smaller than the maximum of the photoreceptor potential V_2. We now estimate the lower limit of the amplification in the light sensing process using the following arguments (Keszthelyi and Ormos, 1989): the N_1 elementary charge Q set into motion by light absorption in the rhodopsin will be displaced by d_1 cm in the membrane with dielectric constant ε. The fast displacement current will charge the capacitance C to a voltage.

$$V_1 = \frac{N_1 Q d_1}{\varepsilon C D} \qquad (1)$$

where D is the distance of the electrodes. The photopotential is a slow signal, in this case the maximum is

$$V_2 = \frac{N_2 Q d_2}{\varepsilon D} Rk \qquad (2)$$

where R is the measuring resistance and k the decay rate. We assume $d_1 \approx 0.2$ nm (measured value in case of bacteriorhodopsin) $d_2 \approx 5$ nm (the thickness of the membrane). RC = 2×10^{-6} s and k = $3 \times 10^2 s^{-1}$ then $N_2/N_1 > 1000$. This estimation tells that even in the primitive light sensing system of an eukariotic alga huge amplification of the primary process exists.

This paper is based mainly on the work of O.A. Sineshchekov, E.G. Govorunova, A. Dér and W. Nultsch. Their contribution is greatly acknowledged.

REFERENCES

Beckmann, M. and Hegemann, P. (1991): In vitro identification of rhodopsin in the green alga *Chlamydomonas*. Bioelectrochemisty 30: 3962-3967.
Bolshakov, V.I., Drachev, A.L., Drachev, L.A., Kalamkarov, G.R., Kaulen, A.D., Ostrovskii, M.A. and Skulachev, V.P. (1979): Community of properties of bacterial and visual rhodopsins: light energy conversion to electric potential difference. Dokl. Ak. Nauk. 249:1462-1466.
Derguini, F., Mazur, P., Nakanishi K., Starace, D.M., Saranak, J. and Foster, K.W. (1991): All-trans retinal is the chromophore bound to the photoreceptor of the alga *Chlamydomonas reinhardtii*. Photochem. Photobiol. 54: 1017-1021.
Foster, K.W., Saranak, L., Patel, N., Zarilli, G., Olabe, M., Kline, T. and Nakanishi, K. (1984): A rhodopsin is the functional photoreceptor for phototaxis in the unicellular eucaryote *Chlamydomonas*. Nature. 311: 756-759.
Harz, H. and Hegemann, P. (1991): Rhodopsin regulated calcium currents in *Chlamydomonas*. Nature. 351: 489-491.
Keszthelyi, L. and Ormos, P. (1989): Protein electric response signals from dielectrically polarized systems. J. Memb. Biol. 109: 193-200.
Leibl, W. and Trissl, H.-W. (1990): Relationship between the fraction of closed photosynthetic reaction centers and the amplitude of the photovoltage from light-gradient experiments. Biochim. Biophys. Acta. 1015: 304-312.
Litvin, F.F., Sineshchekov, O.A. and Sineshchekov, V.A. (1978): Photoreceptor electric potential in the phototaxis of the alga *Hematococcus pluvialis*. Nature. 271: 476-478.
Nultsch, W. and Häder, D.P. (1988): Photomovement of motile microorganisms-II. Photochem. Photobiol. 47: 837-869.
Sineshchekov, O.A., Govorunova, E.G., Dér, A., Keszthelyi, L. and Nultsch, W. (1992): Photoelectric responses in phototactic flagellated algae measured in cell suspension. J. Photochem. Photobiol. 13, 119-134.
Sineshchekov, O.A., Litvin, F.F. and Keszthelyi, L. (1990): Two components of photorecetor potential in the phototaxis of the flagellated green alga *Hematococcus pluvialis*. Biophys. J. 57: 33-39.

Retinal-induced photoelectric responses in *Chlamydomonas reinhardtii* "blind" mutants

O.A. Sineshchekov[1,2], E.G. Govorunova[1,2], A. Dér[2] and L. Keszthelyi[2]

[1] Biology Department, Moscow State University, 119899 Moscow, Russia and [2] Biological Research Center, Hungarian Academy of Sciences, H-6701 Szeged, Hungary

Reconstitution of rhodopsin activity in "blind" (carotenoid-less) *Chlamydomonas* mutants by exogenous retinal was studied by means of photoelectric measurements in cell suspension. Dark-grown mutant cells show no photoelectric responses to the flashes of actinic light, but within a few minutes after addition of all-trans retinal in submicromolar concentrations to the cell suspension both photoreceptor potential and flagella regenerative response were registered. Photoreceptor potential in retinal-reconstituted mutant cells resembled that measured in wild type cells concerning the value of threshold fluence, action spectrum (maximum at 500 nm) and sensitivity to hydroxylamine bleaching, but had the opposite sign due to the "lens effect" of practically transparent cell bodies of the mutant cells.

INTRODUCTION

The first experimental evidence that normal phototactic behavior of "blind" (carotenoid-deficient) *Chlamydomonas* mutants can be easily restored by addition of exogenous retinoids had been obtained by observation of net long-term photomotile responses of cell populations (Foster et al., 1984). Later reconstitution of *Clamydomonas* rhodopsin available in mutant cells in its apo-form had been studied with more sophisticated methods of individual cell motion analysis (Lawson et al., 1991) and detection of flash-induced light scattering changes in cell suspension (Hegemann et al., 1991). Among various retinoid compounds testified, all-trans retinal had been found the most effective and therefore supposed to be the natural chromophore of the *Chlamydomonas* rhodopsin. However, these studies probed the later elements of the signal transduction chain in *Chlamydomonas* photoreception rather than primary events in the photoreceptor site.

By development of the special modification of extra-cell microelectrode measurements it had been shown that photoreception in green algae occurs via generation of the photoreceptor potential difference in a small portion of plasmalemma (Litvin et al., 1978; Sineshchekov et al., 1990). The method of photoelectric measurements in cell suspension (Sineshchekov, 1991; Sineshchekov et al., 1992) allowed us to study primary events in retinal-reconstituted phototaxis of *Chlamydomonas* "blind" mutants with submillisecond time resolution. *Chlamydomonas* mutants deficient in carotenoid biosynthesis ("white") were kindly provided by Dr. A. Chunaev (St. Petersburg, Russia). The description of the measuring set-up and experimental conditions can be found in Sineshchekov *et al.*, 1992.

RESULTS AND DISCUSSION

In suspensions of gamets, produced by dark-grown cultures of carotenoid-deficient *Chlamydomonas* mutants, no photoelectric responses to the flashes of actinic light were observed. After addition of exogenous retinal (in form of the methanol solution) to the suspension of "blind" gamets the signals appeared already in about 1 minute (showing saturation in amplitude after about 15 min of incubation), and could be induced by retinal concentration as low as 5nm without clear dependence on it within submicromolar range. Retinal-induced photoelectric responses measured in "white" mutant cell suspension were similar to those, typical for wild-type ("green") cells (Fig.1) and could be attributed to the generation of the photoreceptor potential (with the time course in millisecond range) and the slower flagella regenerative response (Sineshchekov, 1991).

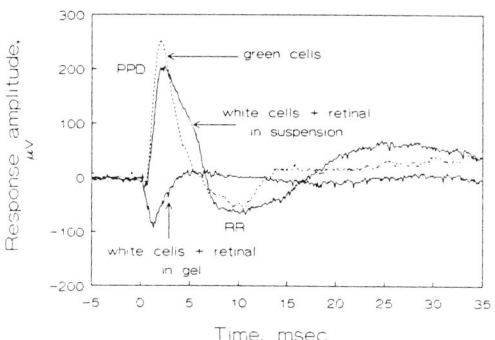

Fig.1. Photoelectric responses in the vertical cuvette in all-trans retinal reconstituted "white" cells and wild type cells. PPD - photoreceptor potential difference, RR - regenerative response.

In some experiments the same signal of much smaller amplitude could be registered after illumination of the cell suspension with repetitive flashes of actinic light in absence of exogenous retinal most probably due to photoinduction of carotenoid biosynthesis in "blind" mutants. The amplitude of the retinal-induced photoelectric response depended on fluence of the exciting flash. The value of threshold fluence (about 1-2 mJ/m^2) was very close to the value, measured in the green wild-type *Chlamydomonas* cells (Fig.2, curves 1,2). Hydroxylamine effected both the amplitude of the photoreceptor potential and the value of the threshold fluence (Fig.2, curves 2-5). 30 min of incubation with 2 mM hydroxylamine under continuous blue-green illumination (60 W/m^2) drastically decreased the response amplitude, and the threshold fluence for hydroxylamine-treated cells was about 2 orders of magnitude higher than in control. Under dark conditions no hydroxylamine effect was observed during at least 1 hour of incubation. These facts indicate specific hydroxylamine action on the photoreceptor molecules rather than on further components of the signal-transduction chain. The action spectrum of photoelectric potential in retinal-reconstituted "white" cells, based on reciprocal fluences eliciting responses of the same amplitude

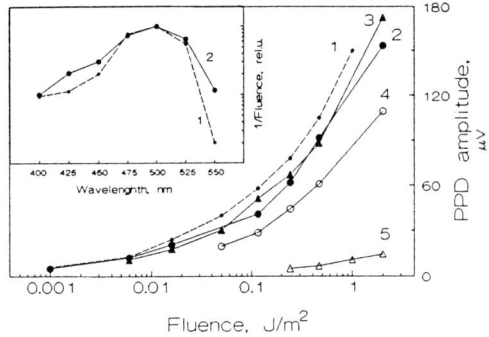

Fig.2. Characteristics of the retinal-induced photoreceptor potential. Fluence dependence: 1 - wild type cells; 2-5 - retinal-doped "white" cells (2 - dark incubation, control; 3 - dark incubation, hydroxylamine; 4 - light incubation, control; 5 - light incubation, hydroxylamine). Action spectra (insertion): 1 - wild type cells, 2 - retinal-doped "white" cells.

for exciting flashes of different wavelengths, is shown in the insertion in Fig.2.
The spectrum measured in "white" retinal-reconstituted cells resembled that measured in "green" wild-type cells (with broad maximum around 500 nm).

The sign of the signal in wild-type cells is determined by the sign of potential difference between the response in cells oriented with their photoreceptors facing the light source during the exciting flash and those having photoreceptors shielded by stigmata and cell bodies (thus generating the

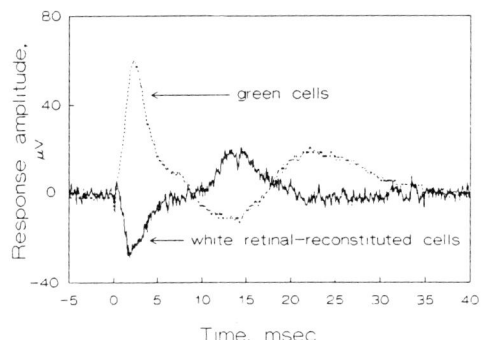

Fig.3. Photoelectric responses in the horizontal cuvette.

response of the smaller amplitude.) Since photoreceptor current is conveyed by the inward movement of cations across the photoreception portion of plasmalemma, the net photoelectric response of the cell suspension is positive with respect to the electrode remote from the light source (Sineshchekov et al., 1992). However, the retinal-induced signal measured in the "white" cell suspension in the horizontal experimental cuvette was negative (Fig.3). It means that the response of the bigger amplitude is generated by the cells with the photoreceptors oriented from the exciting flash. The higher intensity of the light impinging the photoreceptors of such cells is probably maintained by its focusing by the roughly spherical surfaces of the transparent (chlorophyll content less than 10 per cent of the wild-type) cell bodies (the "lens effect", similar to that observed in *Phycomyces* sporangiophores). In the vertical cuvette the sign of the response in the suspension of "white" cells was positive and did not depend on the direction of the exciting flash (Fig.1). These data reveal the vertical preorientation of the cells in suspension (most likely due to gravitaxis rather than to aerotaxis, for the effect was not sensitive to the air depletion of the medium).

The results obtained show that the photoreceptor potential generation involved in *Chlamydomonas* phototaxis is mediated by a rhodopsin-like photoreceptor pigment.

REFERENCES

Foster, K.W., Saranak, L., Patel, N., Zarilli, G., Okabe, M., Kline, T., and Nakanishi, K. (1984): A rhodopsin is the functional photoreceptor for phototaxis in the unicellular eukaryote *Chlamydomonas*. *Nature* 311, 756-759.

Hegemann, P., Gärtner, W., and Uhl, R. (1991): All-trans retinal constitutes the functional chromophore in *Chlamydomonas* rhodopsin. *Biophys. J.* 60, 1477-1489.

Lawson, M.A., Zacks, D.N., Derguini, F., Nakanishi, K., and Spudich, J.L. (1991): Retinal analog restoration of photophobic responses in a blind *Chlamydomonas reinhardtii* mutant. *Biophys. J.* 60, 1490-1498.

Litvin, F.F., Sineshchekov, O.A., and Sineshchekov, V.A. (1978): Photoreceptor electric potential in the phototaxis of the alga *Haematococcus pluvialis*. *Nature* 271, 476-478.

Sineshchekov, O.A. (1991): Electrophysiology of photomovements in flagellated algae. In *Biophysics of photoreceptors and photomovemwnts in microorganisms*, ed.F. Lenci et al., pp.191-202. New York: Plenum Press.

Sineshchekov, O.A., Litvin, F.F., and Keszthelyi, L. (1990): Two components of phohotoreceptor potential in phototaxis of the flagellated green algae *Haematococcus pluvialis*. *Biophys. J.* 57, 33-39.

Sineshchekov, O.A., Govorunova, E.G., Dér, A., Keszthelyi, L., and Nultsch, W. (1992): Photoelectric responses in phototactic flagellated algae measured in cell suspension. *J. Photochem. Photobiol.* 13, 119-134.

Rhodopsin-like pigment and G-proteins in the eyespot of *Chlamydomonas reinhardtii*

Inna Dumler, Sergei Korolkov and Maria Garnovskaya

Sechenov Institute of Evolutionary Physiology and Biochemistry, Torez pr., 44, 194223 St.-Petersburg, Russia

The photosensitive pigments of invertebrates which provide their movement responses are of great variability. In the same time still little is known about the biochemical nature and properties of these pigments. Under consideration of these data the question about the probable structural and functional similarity between different photosensitive pigments arises. This problem is of principal importance in the aspect of evolution of photosensitive molecules, their functional domains and the general mechanism of cellular signalling.

We tried to investigate some of these questions using as a model system the green algae *Chlamydomonas reinhardtii*. This unicellular microorganism responds to week flashes of light by changing it's swimming direction. This behaviour is controlled by a photosensitive pigment as the functional photoreceptor (Harz & Hegemann, 1991).

As it was shown by Foster et al. (1984), the chromophore part of this pigment corresponds, as well as in visual rhodopsin, to 11-cis- retinal. The structure of the protein part of this rhodopsin-like molecule is unclear, though it is evident that the protein domains must play the most important role in photosignalling.

Ch. reinhardtii wild type strain 494c(-) was used. The cultivation of the strain, isolation of cells and eyespot fractions were done as described previously (Dumler et al., 1988; Korolkov et el., 1989). The homogeneity of eyespot fraction was confirmed by electron microscopy.

In the immuniblotting analysis the eyespot proteins were recognized by anti-bovine-rhodopsin antibodies. It should be noticed that the immunological similarity between photosensitive pigments of some other invertebrates and visual rhodopsin also was shown (Nakaoka et al., 1991).

In an attempt to explore this similarity in more details we used the reconstituted systems, consisted of purified components of the chain of visual transduction from bovine retina - rhodopsin, transducin (T) and cyclic GMP phossphodiesterase (PDE). The photoinduced activation of PDE could be obtained in this system in the presence of GTP or Gpp(NH)p and usually serves as a functional test in photoreception study (Korolkov et al., 1989). In some probes *Ch.reinhardtii* photopigment was used instead of visual rhodopsin.

The results of these experiments are shown on Fig. 1. It is evident that the Ch.reinhardtii eyespot fraction could effectively work in the reconstituted systems, providing the photoinduced activation of PDE.

It might be concluded, that rhodopsin-like pigment of Ch.reinhardtii includes in the structure of it's molecule the domains which are capable to interact with T and participate in the visual transduction.

Taking into account these data, the question arises whether the unicellular algae posses their own G-protein(s) and what their possible functional role is.

In our effort to determine and identify the G-protein(s) in Ch.reinhardtii different experimental approaches were used. The details of these procedures are described elsewhere (Korolkov et al., 1990).

The measurement of GTP-γ-35S binding by all the fractions obtained during the eyespot purification revealed that the degree of specific binding was maximal in the fraction of eyespot. These data allow us to suppose the presence of GTP-binding proteins in Ch.reinhardtii and gave some reason to assume their possible localization in eyespot.

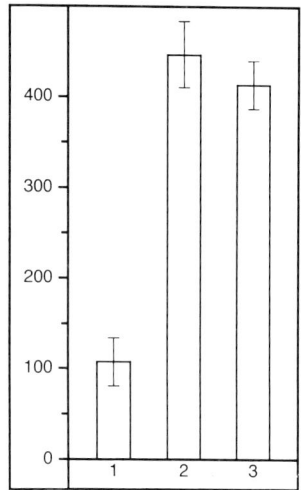

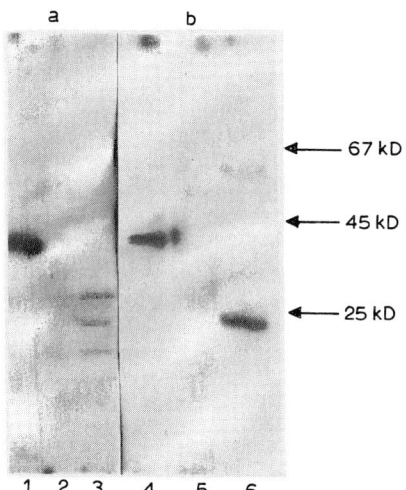

Fig. 1
Photoinduced activation of PDE in the reconstituted systems. Volues for PDE activity are expressed as nmol GMP/mg protein/min
1, Bovine photoreceptor membranes +PDE+T; 2, the same in the presence of $5 \cdot 10^{-5}$ M Gpp(NH)p; 3, eyespot fraction of Ch. reinhardtii + PDE + T in the presence of $5 \cdot 10^{-5}$ M Gpp(NH)p

Fig. 2
Immunoblott of Ch. reinhardtii proteins with βcommon (a) and αcommon (b) antiserum. Lanes: 1, 4, control; 2, 5, whole homogenate; 3, 6, eyespot fraction

The determination of high-affinity GTPase activity gave support to this supposition. Gpp(NH)p-inhibited GTPase, inherent in G-proteins, was also revealed in eyespot fractions (not shown).

The Western-blott analysis of Ch. reinhardtii eyespot proteins was done using two types of antibodies raised against highly conserved domains of vertebrate G-protein subunits, α*common* (AS8) and β—*common* (AS 11). The AS 8 antibodies showed a major band with Mr 24 kDa (Fig. 2). A weak band with Mr 57 kDa may be explained as a result of negligible tubulin contamination. In the case of AS 11 antibodies two additional bands with molecular mass of about 21 and 29 kDa were revealed (Fig. 2). It should be mentioned that both types of antibodies interacted on immunoblots with corresponding proteins only in the case of eyespot fraction and gave no reaction with separated whole cell homogenate and of intermediate fractions.

Our results allow to suggest, that Ch. reinhardtii eyespot contains at least three types of G-proteins with Mr of about 21, 24 and 29 kDa, which are homologous to vertebrate G-proteins α– and β– subunits. The homology between photosensitive G-proteins of vertebrates and some invertebrates and higher plants has been found by other authors (Pottinger et al., 1991; Warpeha et al., 1991). The fact of 24-kDa protein interaction with both As 8 and AS 11 antisera is of special interest.

We did not succeed in our attempt to perform the ADP-ribosylation of revealed G-proteins in the presence of pertussis and cholera toxins. This property of some low molecular weight G-proteins is known (Hall, 1990).

Taking into account an essential role of visual G-protein, T, in the photoexcitation process, possible functional significance of Ch. reinhardtii G-proteins is of considerable interest.

The important property of T, which allows it to be involved in cGMP cascade of reactions, is the light-dependence of it's interaction with photoreceptor membranes (Ting & Ho, 1991). It was shown that the extraction of Gpp(NH)p-inhibited high affinity GTPase from Ch.reinhardtii eyespot is also dependent on illumination (Fig. 3). In contrast to extractability of T increasing in darkness, the Ch. reinhardtii eyespot GTPase was extracted much better under illumination (Fig 3).

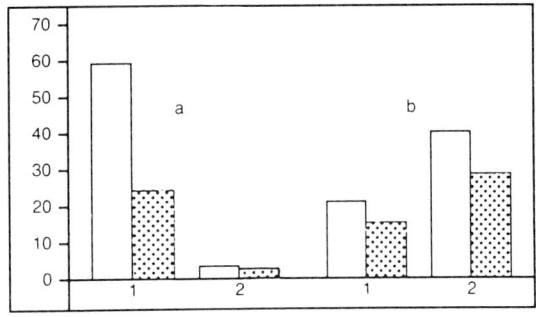

Fig. 3
GTPase activity of the eyespot fraction of Ch. reinhardtii. The ordinate axes: activity, pmol/mg protein/min. (a) Extractiion was performed under room light. (b) Extraction was performed in darkness, 1, extract; 2, pellet; control with Gpp(NH)p; 1.10^{-5} MGpp(NH)p was added in the incubation mixture for GTPase activity measurement.

The most intriguing question is about the possible mechanism of *Ch. reinhardtii* photoresponse to light stimulus. As it was shown by us previously, the systems of cyclic nucleotides does not play any functional role in this signalling (Dumler et al., 1988). In accordance to some data (Harz & Hegemann, 1991) the rhodopsin-regulated calcium currents could be involved in it. The cross-talking of different signalling pathways can not be also excluded. From this point of view the investigation of probable role of revealed G-proteins in the regulation of intracellular calcium via inositol polyphosphates metabolism seems to be very attractive.

References

Dumler, I.L., Basov, A.S., Chunaev, A.S. (1988): The enzymes of cyclic nucleotide metabolismm in unicellular algae *Chlamydomonas reinhardtii*. *J. Evol. Biochem. Physiol.* (in Russian). 24, 692 - 695.

Hall, A. (1990): The cellular functions of small GTP-binding proteins. *Science*. 249, 635 - 640.

Harz, H. & Hegemann, P. (1991): Rhodopsin-regulated calcium currents in *Chlamydomonas*. *Nature*. 351, 489 - 491.

Korolkov, S.N., Garnovskaya, M.N., Basov, A.S., Dumler, I.L. (1989): On photosensitive pigment from *Chlamydomonas reinhardtii*. *J. Evol. Biochem. Physiol.* (in Russian). 6, 777 - 780.

Nakaoka, Y., Tokioka, R., Shinosawa, T., Fujita, J., Usukara, J. (1991): Phororeception of Paramecium cilia: localization of photosensitivity and binding with anti-frog-rhodopsin JgG. *J. Cell Science*. 99, 67 - 72.

Pottinger, J.D.D., Ryba, N.J.P., Keen, J.N., Findlay, J.B.C. (1991): The identification and purification of the heterotrimeric GTP-binding protein from squid *(Loligo forbesi)* photoreceptors. *Biochem. J.*, 279, 323 - 326.

Ting, T.D. & Ho, Y.-K. (1991): Molecular mechanism of GTP hydrolysis by bovine transducin: pre-steady-state kinetic analysis. *Biochemistry*. 30, 8996 - 9007.

Warpeha, K.M.F., Hamm, H.E., Rasenick, M.M., Kaufman, L.S. (1991): A blue-light-activated GTP-binding protein in plasma membranes of etiolated peas. *Proc.Natl.Acad.Sci. USA*. 88, 8925 - 8929.

Choroplastic membrane-bound arrestin-like immunoreactive proteins in tobacco and *Chlamydomonas* cells

Ali Mirshahi[1], Aimé Nato*, Danièle Lavergne[1], Georges Ducreux[1], Jean-Pierre Faure[2] and Massoud Mirshahi[2]

[1] Laboratoire de Morphogenèse Végétale Expérimentale, URA 115 CNRS, Université Paris-Sud, Bât. 360, 91405 Orsay Cedex; [2] INSERM U 86, Centre de Recherches Biomédicales des Cordeliers, 15, rue de l'Ecole de Médecine, 75270 Paris Cedex 06, France

SUMMARY

In plants, the relatively few studies so far completed have raised important questions concerning the mechanisms of perception and transduction of signals. In a previous report (Mirshahi et al., 1991), we demonstrated the presence in tobacco and green *Chlamydomonas* cells of proteins immunoreactive with monoclonal antibodies against S-antigen, a protein of retinal photoreceptors, also named "arrestin "or "48 K protein". In the present study, using immunoblotting and ELISA with antibodies to S-antigen, we detected the presence of an S-antigen-like protein of 41 kDa in soluble extracts of cells from green wild type and albino mutant leaf of *Nicotiana tabacum*. Enriched chloroplast fraction was used to evidence the association of another S-antigen-crossreactive protein with grana lamellae. Its molecular weight was 29 kDa. A similar polypeptide was detected in thylakoid preparation of green wild type strain of *Chlamydomonas reinhardtii*. The relationship with chlorophyll was supported by the absence of this protein in yellow Y-1 strain of *Chlamydomonas* grown in the dark where the maturation of plastids is altered.

INTRODUCTION

Molecular recognition signalling mechanisms in plants are fundamental to a wide range of biological processes including cell division and growth, differenciation and development, fertilization, host-pathogen interactions, symbiosis and stress responses. The variety of transduction pathways for essential primary stimuli like light and hormones have still to be explored in plants and little is known about the mechanism of action of any of them (Guern et al., 1990). Light, an essential environmental factor, is used not only for photosynthesis, but also as a modulator of complex developmental and regulatory mechanisms. It is now well known that the plant responses to light involves massive changes in gene expression (Tobin and Silverthorne, 1985). Most of the work on light-controlled development has dealt with the etiolated or dark / greening growth system. During seed germination and leaf growth in darkness, the proplastids increase in size and develop to etioplasts. They contain large amount of internal membranes including a characteristic prolamellar body. Etioplasts lack chlorophyll and some other components required for photosynthesis. They undergo a transformation to photosynthetic competence upon illumination. The responses of etioplasts to light are complex and several photoreceptors seem to be involved. These include the chlorophyll precursor protochlorophyllide a and the photoconvertible pigment phytochrome (Mösinger et al., 1985). Following illumination, the most important changes are the dispersal of the prolamellar body and increase in the amount of thylakoid membrane. Photoactivation of phytochrome results in stimulation of the chlorophyll biosynthesis pathway and accumulation of chlorophyll-binding proteins. The synthesis of chloroplast soluble ribulose 1,5-biphosphate carboxylase/oxygenase (RuBisCo) and many other photosynthetic Calvin cycle enzymes are strongly stimulated by light. Light affects the expression of genes both in the nucleus and plastids (Mösinger et al., 1985). The plastome encoded photogene 32, a thylakoid protein of 32 kDa involved in photosystem II electron transport, that appears during illumination. Another important thylakoid protein synthesized in response to light and under nuclear genome control is the light-harvesting chlorophyll a/b protein complex (LHCP). The in vitro translation products of the mRNAs have molecular weights of 29 to 32 kDa. These mRNAs appear after illumination but the LHCP do not accumulate unless chlorophyll synthesis occurs (Schäfer et al., 1986).

In the animal kingdom, S-antigen (arrestin or 48 kDa-protein) regulates phototransduction in photoreceptor cells of the retina. The function of this protein is to arrest the enzymatic cascade of phototransduction in retinal rods through its binding to photoactivated and phosphorylated rhodopsin (Kühn et al., 1984; Pfister et al., 1984). Recently, homologous proteins have been detected in non-photosensitive cells of vertebrates and invertebrates (Mirshahi et al., 1989; see Razaghi et al., this symposium). In a previous report (Mirshahi et al., 1991), we demonstrated the presence of proteins immunoreactive with monoclonal antibodies against S-antigen in plant cells and unicellular green alga. We report here the presence of such a protein, with a molecular weight of 29 kDa, in the thylakoid fraction of green tobacco leaves and Chlamydomonas cells. This result suggests that in these organisms, the arrestin-like proteins can be involved in chloroplast phototransduction mechanism. This idea is strengthened by the absence of these proteins in yellow Y-1 strain of Chlamydomonas grown in the dark and in the albino mutant leaf of tobacco.

MATERIALS AND METHODS

The green wild type and the albino phenotype mutant of Nicotiana tabacum (cv Xanthi), and the characteristics of tobacco cell cultures were described by Nato et al. (1981) and Nguyen-Quoc et al. (1989). Two strains of Chlamydomonas reinhardtii, the green wild type (kindly offered by Dr. G. Garnier, C.N.R.S., Gif-sur-Yvette) and the Y-1 strain (gift from Dr. Ohad, Jerusalem) were cultivated according to the methods of Levine and Ebersold (1958). For the culture of Y-1 strain in the dark, we added glucose (10 g/liter) in the medium. Soluble extracts from tobacco leaves and cultured cells and alga cells were prepared according to Mirshahi et al. (1991). The microsomal fractions from cells were obtained according to Hoarau et al. (1991). The preparation of enriched chloroplast fraction and thylakoid membranes from tobacco leaves was conducted according to the methods of Schuler and Zielinski (1989). With the alga strains, the preparation of chloroplast fraction and thylakoid membranes was performed according to the procedure of Chua and Bennoun (1975). ELISA and Western blots were performed as described by Mirshahi et al. (1989).

RESULTS

Data obtained by ELISA show that the soluble 41 kDa S-antigen-like protein can be released by washing the microsomal fraction of both the green wild type and albino mutant tobacco cells. After several washings, treatment by Triton X100 (1%) released a new form of S-antigen-crossreactive protein, only from green cell microsomal fraction (Fig. 1). Western blot analysis of green and albino tobacco leaf soluble extracts revealed the 41 kDa polypeptide. In the chloroplast thykaloïd fraction, a 29 kDa protein band was detected only in the green leaf, and therefore seems to be related to the presence of chlorophyll. This protein was well recognized by three monoclonal antibodies directed at distinct epitopes located in the C-terminal region of S-antigen (M4H10, S9E2, S1A3). A mixture of these three antibodies was used in the experiments shown in figures 1-3. We extended this study to two strains of *Chlamydomonas reinhardtii*. Soluble extracts from both the green wild type and the Y-1 strains (grown in either light or dark) did not contain any soluble protein immunoreactive with antibodies to S-antigen, when they were not contaminated by membrane fragments. However, both strains contained a polypeptide of 29 kDa and some other bands of lower molecular weights revealed by the same antibodies in their chloroplast thylakoid membrane fraction. These proteins were present when *Chlamydomonas* of both strains were grown in the light, but they were absent from the yellow Y-1 strain grown in the dark (Fig. 3). This suggests that the chloroplast membrane-bound S-antigen-like proteins appear during the greening process.

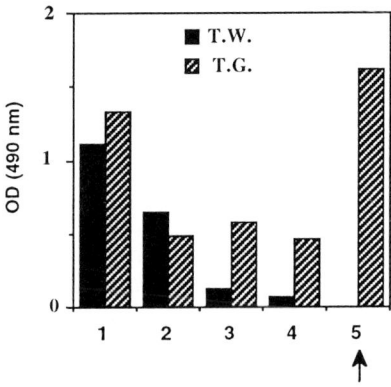
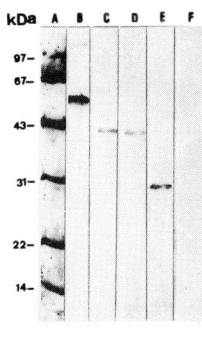

Fig. 1. ELISA analysis of extracts from green tobacco wild type (T.G.) and albino mutant (T.W.) tobacco cells cultured in vitro with the monoclonal antibodies to retinal S-antigen. 1-4: immunoreactivity of soluble extracts obtained from four successive washings of the materials in buffer; 5: immunoreactivity of the insoluble residual materials solubilized in Triton X100.

Fig. 2. Western blotting of soluble and insoluble fractions of green (T.G.) and albino (T.W.) tobacco leaves with the monoclonal antibodies to S-antigen. A: molecular weigth markers; B: bovine retinal S-antigen; C, D: soluble extracts of T.G. (C) and T.W. (D); E, F: Triton-solubilized fractions from T.G. and T.W.

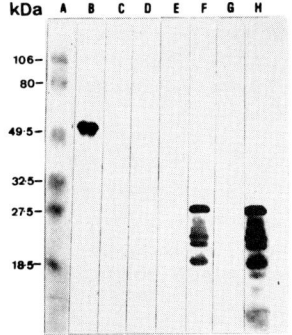

Fig. 3. Western blotting of soluble and insoluble fractions from *Chlamydomonas reinhardtii* with the monoclonal antibodies to S-antigen. A: molecular weigth markers; B: retinal S-antigen; C: soluble and D: Triton-solubilized fractions from Y-1 strain grown in the dark; E: soluble and F: Triton-solubilized fractions from Y-1 strain grown in the light; G: soluble and H: Triton-solubilized fractions from *Chlamydomonas* wild type grown in the light.

CONCLUSION

We previously described the presence of an S-antigen-like protein with apparent molecular weight 41 kDa in soluble extracts of tobacco cells. In the present study, we detected an S-antigen crossreactive protein in the chloroplast thylakoid fraction of tobacco leaves. Its molecular weight was 29 kDa. The albino mutant of tobacco, deficient in chlorophyll synthesis and chloroplast structuration, did not contain this membrane-bound protein. A similar protein was present in green *Chlamydomonas*, but absent from the yellow Y-1 strain of *Chlamydomonas* grown in the dark. However, both strains contained the thylakoid 29 kDa protein when cultivated in the light. These results suggest that a protein related to S-antigen is integrated in the chloroplast thylakoid, thus possibly playing a role in the phototransduction mechanism. The appearance of this protein during the greening process of the dark grown Y-1 strain suggests that its synthesis is under light control. A stimulating working hypothesis is to establish the sequential and functional analogies between the S-antigen--like protein and LHCP in green plants. Reversible modification of the LHCP is thought to control light energy distribution between photosystem I and II, thus maximizing the photosynthetic quantum yield.

REFERENCES

Chua, N.H., and Bennoun, P. (1975): Thylakoid membrane polypeptides of *Chlamydomonas reinhardtii* : wild type and mutant strains deficient in photosystem II reaction center. *Proc. Natl. Acad. Sci. USA* 72 ,2175-2179.

Guern, J. Ephritikhine, G., Imhoff, V., and Pradier J.M. (1990): Signal transduction at the membrane level of plant cells. In *Proceedings of the VIIth Intern. Cong. on Plant Tissue and Cell Culture*, ed. H.J.J. Nijkamp and L.H.W. Van der Aarthrijk, pp. 466-479, Dordrecht/Boston: Kluwer Acad.Pub.

Hoarau, J., Nato, A., Lavergne D., Flipo, V. and Hirel, B. (1991): Nitrate reductase activity changes during a culture cycle of tobacco cells : the participation of a membrane-bound form enzyme. *Plant Science* 79 ,193-204.

Kühn, H., Hall, S.W., and Wilden, U. (1984): Light-induced binding of 48 kDa protein to photoreceptor membranes is highly enhanced by phosphorylation of rhodopsin. *FEBS Lett.* 176 , 473-478.

Levine, R.P., and Ebersold ,W.T. (1958): The relation of calcium and magnesium to crossing over in *Chlamydomonas reinhardtii*. *Z. Vererbungslehere* 89,631-639.

Mirshahi, M., Borgese, F., Razaghi, A., Scheuringf U., Garcia-Romeu F., Faure, J.P., and Motais, R. (1989): Immunological detection of arrestiin, a phototransduction regulatory protein in the cytosol of nucleated erythrocytes. *FEBS Lett.* 258, 240-248.

Mirshahi, M., Nato, A., Razaghi, A., Mirshahi, A., and Faure, J.P. (1991): Présence de protéines apparentées à l'arrestine (antigène-S) dans des cellules végétales. *C.R. Acad. Sci., Paris*, 312, Série III , 441-448.

Mösinger, E., Batschauer, A., Schäfer, E., and Apel, K. (1985):Phytochrome control of in -vitro transcription of specific genes in isolated nuclei from barley (*Hordeum vulgare*). *Eur. J. Biochem.* 147 ,137-142.

Nato, A., Brangeon, and Dulieu, H. (1981): Nuclear gene mutation effects on the plastid structure and RUBPCase properties in *Nicotiana tabacum* (cv Xanthi). In *Proceeding of Photosynthesis V. Chloroplast Development* , ed. G. Akoyonoglou, pp. 821-830, Philadelphia: Balaban International Science Service.

Nguyen-Quoc, B., Prioul, J.L., Nato, A., and Dulieu, H. (1989): Light dependent expression of RUBISCO mRNAs in a chlorophyll-less mutant of tobacco. *Plant Physiol. Biochem.* 27 ,811-816.

Pfister, C., Dorey ,C., Vadot, E., Mirshahi, M., Deterre, P., Chabre M., and Faure, J.P. (1984): Identité de la protéine dite "48 K" qui interagit avec la rhodopsine illuminée dans les bâtonnets rétiniens et de l' "antigène-S rétinien", inducteur de l'uvéo-rétinite auto-immune expérimentale. *C.R. Acad. Sci. Paris* 299, série III, 261-265.

Schäfer,E., Apel, K., Batschauer, A., and Mösinger, E. (1986): Phytochrome: the molecular biology of action. In *Photomorphogenesis in Plants*, ed. R.E. Kendrick and G.H.M. Kronenberg, pp. 83-98, Dordrecht: Martinus Nijhoff/Dr W. Junle Publishers.

Schuler, M.A., and Zielenski, R.E. (1989): Preparation of intact chloroplasts from Pea. In *Methods in Plant Molecular Biology*. Academic Press, Inc. Harcourt Brace Joranovich Publishers.

Tobin, E.M., and Silverthorne, J. (1985): Light regulation of gene expression in higher plants. *Ann. Dev. Plant Physiol.* 36, 569-593.

Proteins related to retinal S-antigen (arrestin) in non photosensitive cells

A. Razaghi[1], F. Borgese[2], B. Fiévet[2], R. Motais[2], A. Nato[3], J. Oliver[4], A. Vandewalle[5], M. Mirshahi[1] and J.-P. Faure[1]

[1] INSERM U 86, Centre de Recherches Biomédicales des Cordeliers, 15, rue de l'Ecole de Médecine, 75270 Paris Cedex 06; [2] CEA, Villefranche-sur-Mer; [3] Université de Paris XI, Orsay; [4] Université de Montpellier II; [5] INSERM U 246, Paris, France

S-antigen (also named "48 K protein" or "arrestin") is a soluble protein with an apparent molecular weight of 48 kDa, abundant in photoreceptor cells of the retina, which displays autoantigenic properties responsible for induction of autoimmune disease in the eye (reviewed in Faure, 1980, 1992). This protein was demonstrated immunohistochemically in photoreceptors in all tested vertebrates and several invertebrates (Mirshahi et al., 1985) and in the photoreceptor-derived cells of the pineal organ of various vertebrates (Mirshahi et al., 1984; Collin et al., 1986; Van Veen et al., 1986). Genes encoding closely related proteins have been isolated from retinas of several mammals (Yamaki et al., 1987; Shinohara et al. 1989, 1990), from rat pineal gland (Abe et al., 1989) and from Drosophila (Smith et al., 1990; Hyde et al., 1990; Yamada et al., 1990). The presently known function of this protein in the retina is to arrest the enzymatic cascade of phototransduction in retinal rods, thus the name "arrestin". Arrestin binds to photoactivated and phosphorylated rhodopsin and thereby desensitizes rhodopsin, quenching the activation of cGMP phosphodiesterase (Wilden et al., 1986; Pfister et al., 1984, 1985).

Therefore S-antigen has been considered to be a specific component of photosensory cells and photoreceptor-derived cells. However, on the basis of the strong structural and functional homologies between phototransduction and other G-protein-related transduction systems, it seems likely that analogs of arrestin could exist, which could play such regulatory role in the transduction of various chemical signals. In this study, we present the evidence that proteins related to S-antigen / arrestin are widely distributed in many non photosensitive animal cells. These proteins were characterized by their immunoreactivity with a panel of antibodies to retinal S-antigen and their apparent molecular weight similar to that of retinal S-antigen. Crossreactive proteins were also detected in plant cells.

After fractionation of soluble extracts of different bovine organs by gel filtration and/or immunoaffinity chromatography, we detected an S-antigen-like protein in myocardium, kidney, liver, lung and cerebellum extracts using Western blotting with monoclonal or polyclonal antibodies to retinal S-antigen. The immunoreactive protein(s) was revealed as a single band with a molecular weight of about 48 kDa (Fig. 1). Several monoclonal antibodies that recognize different epitopes located in either the N-terminal or the C-terminal regions of retinal S-antigen reacted with these proteins. On sections of rat kidneys, these antibodies labeled epithelial cells of collecting and distal tubules and some cells in glomeruli (Fig. 2). These S-antigen-like proteins of non photosensitive mammalian organs are probably different, but analogous in their primary structure, to retinal S-antigen.

The gene of an S-antigen analog has been isolated from a bovine brain cDNA library by Lohse et al. (1990). The protein, named β-arrestin, expressed in transfected cells, inhibits the signalling function of phosphorylated β-adrenergic receptor (β-AR) in a reconstituted system. The amino acid sequence of β-arrestin displays 59% identity with bovine retinal S-antigen. Though retinal S-antigen can potentiate the desensitizing effect of β-AR phosphorylation (Benovic et al., 1987), β-arrestin is much more effective than S-antigen for β-AR desensitization while S-antigen is much more effective than β-arrestin for inhibiting rhodopsin function (Lohse et al., 1992).

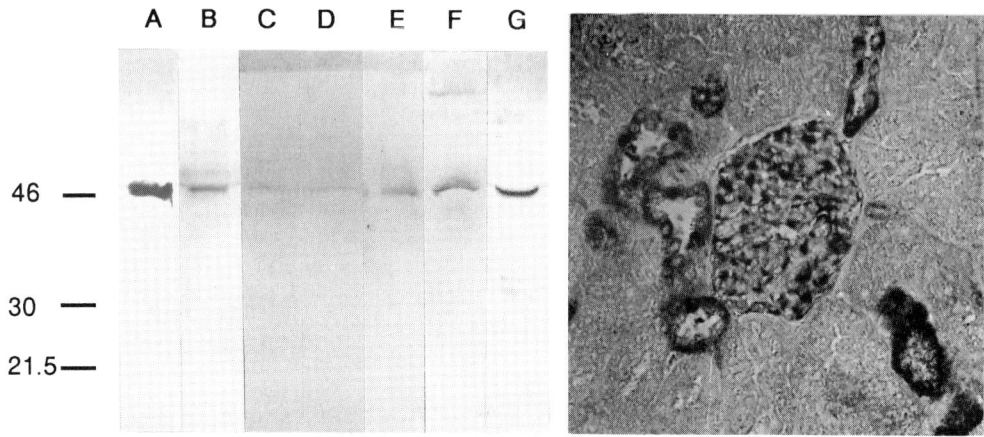

Fig. 1. Immunoblots of purified bovine retinal S-antigen (A) and fractions enriched by gel filtration or immunoaffinity chromatography from bovine myocardium (B), kidney (C), liver (D), lung (E), cerebellum (F) and pineal gland (G). A single 48 kDa band is revealed by a rabbit antiserum against purified retinal S-antigen.

Fig. 2. Immunoperoxidase staining of cells in distal tubules and glomerulus of the rat kidney using the monoclonal antibody S9E2 against retinal S-antigen.

In quails, immunohistochemical study of brain sections revealed strong immunostaining of choroid plexus epithelial cells using monoclonal antibodies S6H8 and S2D2 directed against a highly conserved epitope in the N-terminus of S-antigen (Oliver et al., 1987, 1992). On Western blots, these antibodies labelled a 45 kDa band in extracts of choroid plexuses or retina (Oliver et al., 1991). In trout and turkey nucleated erythrocytes, a protein was immunodetected with several monoclonal antibodies to S-antigen in cytosolic extracts enriched by immunoaffinity chromatography. The molecular weight was slightly lower than bovine S-antigen in trout and slightly higher in turkey erythrocytes (Mirshahi et al., 1989) (Fig. 3). Scheuring et al. (1990) then demonstrated that the trout red blood cell protein functionally behaves like retinal arrestin, i.e. binds to photoactivated and phosphorylated bovine rhodopsin. S-antigen-like proteins were also detected in the cytosol of human platelets (Mirshahi et al., 1991b) and the human leukocyte line Reh (unpublished). There is a growing evidence that proteins with structure and immunoreactivity analogous to S-antigen are present in many cells in vertebrates. All non ocular tissues or cells where these proteins were detected in various vertebrates are known to contain G-protein-coupled receptors, and arrestin-like regulatory functions of these proteins in various transduction systems is suspected.

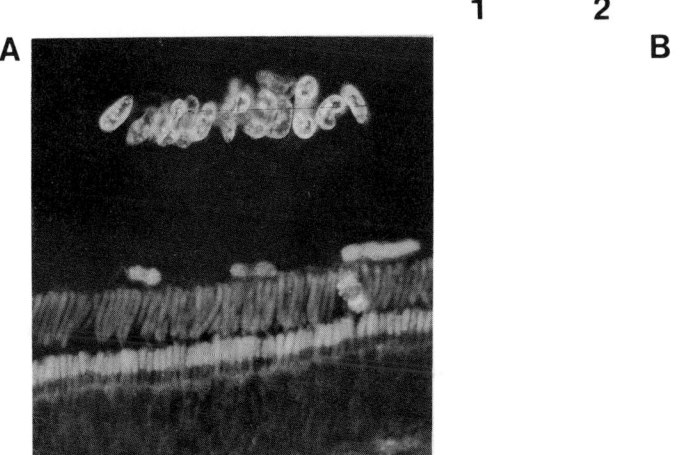

Fig. 3. S-antigen-like immunoreactivity in nucleated erythrocytes. A: Immunofluorescent labeling of a fish retina with the monoclonal antibody S7D6 against S-antigen. The immunoreactivity is observed in photoreceptors and in erythrocytes present in choroidal vessels. B: Immunoblot of bovine retinal S-antigen (lane 1) and a fraction from gel filtration of turkey erythrocyte cytosolic extract (lane 2). A single band is revealed by the monoclonal antibody S6H8 against S-antigen.

The high phylogenetic conservation of S-antigen is shown by its immunodetection in photoreceptors from a large range of animal species : all vertebrates, Amphioxus, Annelids, Nemerteans, Molluscs (Mirshahi et al., 1985), and by the isolation of two "S-antigen" genes from *Drosophila* (Smith et al., 1990; Hyde et al., 1990; Yamada et al., 1990). Proteins encoded by these genes have more than 40% sequence identity with human S-antigen, whereas the major *Drosophila* opsin is only 22% identical to human rhodopsin. We extended this study to the vegetal kingdom and isolated proteins immunoreactive with several antibodies to S-antigen from *Nicotiana tabacum* and *Chlamydomonas reinhardtii* (M. Mirshahi et al., 1991b; A. Mirshahi et al., this symposium). The function of these proteins in plants is at present unknown. Jeansonne et al. (1991) detected a 48 kDa protein antigenically and structurally similar to retinal S-antigen in the budding yeast *Saccharomyces cerevisiae*. This yeast protein is a component of the DNA-replicative complex : it is associated with DNA polymerase I - primase. Bovine S-antigen stimulates DNA polymerase activity in yeast.

It seems that the "S-antigen" family of genes derives from a very primitive ancestor gene. Products of these genes expressed in photoreceptors are associated with the fundamental function of photoreception in animals. Other products can be involved in other basic functions in living beings.

REFERENCES

Abe, T., Yamaki, K., Tsuda, M., Singh, V.K., Suzuki, S., McKinnon, R., Klein, D.C., Donoso, L.A., and Shinohara, T. (1989): Rat pineal S-antigen: sequence analysis reveals presence of alpha-transducin homologous sequence. *FEBS Lett.* 247, 307-311.

Benovic, J.L., Kühn, H., Weyand, I., Codina, J., Caron, M.G., and Lefkowitz, R.J. (1987): Functional desensitization of the isolated beta-adrenergic receptor by the beta-adrenergic receptor kinase : potential role of an analog of the retinal protein arrestin (48-kDa protein). *Proc. Natl. Acad. Sci. USA* 84, 8879-8882.

Collin, J.P., Mirshahi, M., Brisson, P., Falcon, J., Guerlotté, J., and Faure, J.P. (1986): Pineal-retinal molecular relationships : distribution of "S-antigen" in the pineal complex. *Neuroscience* 19, 657-666.

Faure J. P. (1980): Autoimmunity and the retina. In *Current Topics in Eye Research* vol.2, ed. J. A. Zadunaisky and H. Davson, pp.215-302, New York: Academic Press.

Faure J. P. (1992): Autoimmune diseases of the retina. In *Molecular Pathology of Autoimmune Diseases*, ed. C. Bona, K. Siminovitch, M. Zanetti and A. N. Theophilopoulos, Philadelphia: Gordon Breach Science Pub., sous presse.

Hyde, D.R., Mecklenburg, K.L., Pollock, J.A., Vihtelic, T.S., and Benzer, S. (1990): Twenty *Drosophila* visual system cDNA clones: one is homolog of human arrestin. *Proc. Natl. Acad. Sci. USA* 87, 1008-1012.

Jeansonne, N.E., Jazwinski, S.M., and Donoso, L.A. (1991): A 48-kDa, S-antigen-like phosphoprotein in yeast DNA-replicative complex preparations. *J. Biol. Chem.* 266, 14675-14680.

Lohse, M.J., Andexinger, S., Pitcher, J., Trukawinski, S., Codina, J., Faure, J.P., Caron, M.G., and Lefkowitz, R.J. (1992): Receptor-specific desensitization with purified proteins: stoichiometry, kinase-dependence and receptor specificity of β–arrestin and arrestin in the β2-adrenergic receptor and rhodopsin systems. *J. Biol. Chem.* 267, 8558-8564.

Lohse, M.J., Benovic, J.L., Codina, J., Caron, M.G., and Lefkowitz, J.R. (1990): β-arrestin : a protein that regulates β-adrenergic function. *Science* 248, 547-1550.

Mirshahi, M., Borgese, F., Razaghi, A., Scheuring, U., Garcia-Romeu, F., Faure, J.P., and Motais, R. (1989): Immunological detection of arrestin, a phototransduction regulatory protein, in the cytosol of nucleated erythrocytes. *FEBS Lett.* 258, 240-243.

Mirshahi, M., Boucheix, C., Collenot, G., Thillaye, B., and Faure, J.P. (1985): Retinal S-antigen epitopes in vertebrate and invertebrate photoreceptors. *Invest. Ophthalmol. Vis. Sci.* 26, 1016-1021.

Mirshahi, M., Faure, J.P., Brisson, P., Falcon, J., Guerlotté, J., and Collin, J.P. (1984): S-antigen immunoreactivity in retinal rods and cones and pineal photosensitive cells. *Biol. Cell* 52, 195-198.

Mirshahi, M., Nato, A., Razaghi, A., Mirshahi, A., and Faure, J.P. (1991a): Présence de protéines apparentées à l'arrestine (antigène-S) dans des cellules végétales. *C.R. Acad. Sci., Paris* 312, Série III, 441-448.

Mirshahi, M., Razaghi, A., Mirshahi, S.S., Tuyen, V.V., and Faure, J.P. (1991b): Immunopurification of an S-antigen-like protein from human platelets. *Thrombosis Res.* 64, 551-558.

Oliver, J., Herbuté, S., Mirshahi, M., Faure, J.P., Brisson, P., and Collin, J.P. (1987): Identification d'une protéine apparentée à l'antigène-S dans les plexus choroïdes de la caille. *C. R. Acad. Sci., Paris* 305, Série III, 1712-1716.

Oliver, J., Mirshahi, M., Herbuté, S., Péraldi-Roux, S., Trébuchon, L., Brisson, P., Collin, J.P., Faure, J.P., and Gabrion, J. (1991): Distribution of arrestin-like protein and beta subunit of GTP-binding proteins in quail choroid plexuses. *Cellular Signalling* 3, 461-472.

Pfister, C., Chabre, M., Plouët, J., Tuyen, V.V., de Kozak, Y., Faure, J.P., and Kühn, H. (1985): Retinal S-antigen identified as the 48 K protein regulating light-dependent phosphodiesterase in rods. *Science* 228, 891-893.

Pfister, C., Dorey, C., Vadot, E., Mirshahi, M., Deterre, P., Chabre, M., and Faure, J.P. (1984): Identité de la protéine dite "48 K" qui interagit avec la rhodopsine illuminée dans les bâtonnets rétiniens et de l' "antigène-S rétinien" inducteur de l'uvéo-rétinite expérimentale. *C.R. Acad. Sci. Paris* 299, Série III, 261-265.

Shinohara T., Donoso L., Tsuda M., Yamaki K. and Singh V. K. (1989): S-antigen: structure, function and experimental autoimmune uveitis (EAU). In *Progress in Retinal Research*, vol. 8, ed. N. N. Osborne and G. J. Chader, pp. 51-66. Oxford: Pergamon Press.

Shinohara, T., Singh, V.K., Tsuda, M., Yamaki, K., Abe, T., and Suzuki, S. (1990): S-antigen : from gene to autoimmune uveitis. *Exp. Eye Res.* 50, 751-757.

Smith, D.P., Shieh, B.H., and Zuker, C.S. (1990): Isolation and structure of an arrestin gene from *Drosophila*. *Proc. Natl. Acad. Sci. USA* 87, 1003-1007.

Van Veen, T., Elofsson, R., Hartwig, H.G., Gery, I., Mochizuki, M., Cena, V., and Klein, D.C. (1986): Retinal S-antigen : immunocytochemical and immunochemical studies on distribution in animal photoreceptors and pineal organs. *Exp. Biol.* 45, 15-25.

Wilden, U., Hall, S.W., and Kühn, H. (1986): Phosphodiesterase activation by photoexcited rhodopsin is quenched when rhodopsin is phosphorylated and binds the intrinsic 48-kDa protein of rod outer segments. *Proc. Natl. Acad. Sci. USA* 83, 1174-1178.

Yamada, T., Takeuchi, Y., Komori, N., Kobayashi, H., Sakai, Y., Hotta, Y., and Matsumoto, H. (1990): A 49-kilodalton phosphoprotein in the *Drosophila* photoreceptor is an arrestin homolog. *Science* 248, 483-486.

Yamaki, K., Takahashi, Y., Sakuraji, S., and Matsubara, K. (1987): Molecular cloning of the S-antigen cDNA from bovine retina. *Biochem. Biophys. Res. Commun.* 142, 904-910.

X. Signal transduction

X. *Transduction du signal lumineux*

Structures and Functions of Retinal Proteins. Ed. J.L. Rigaud. Colloque INSERM / John Libbey Eurotext Ltd.
© 1992, Vol. 221, pp. 361-364

Sites and mechanisms of interaction of rod G protein with rhodopsin and cGMP phosphodiesterase

Heidi E. Hamm[1], Nikolai O. Artemyev[1], John S. Mills[1], Nikolai P. Skiba[1], Helen M. Rarick[1], Christophe Lambert[2] and Edward A. Dratz[2]

[1] Department of Physiology and Biophysics, University of Illinois, Chicago, College of Medicine, Chicago IL 60612, USA. [2] Department of Chemistry and Biochemistry, Montana State University, Bozeman, MT 59717, USA

INTRODUCTION

The family of heterotrimeric GTP–binding proteins (G–proteins) is involved in cellular signal transduction by coupling cell surface receptors to effector molecules such as adenylyl cyclase (AC), cGMP phosphodiesterase (PDE), phospholipase C (PLC) and ion channels (K^+, Ca^{++}) (reviewed in Gilman, 1987). On the ROS disk membrane, the light–activated receptor, rhodopsin, stimulates the G–protein, transducin (G_t), which, in the GTP–bound form, activates the effector, cGMP phosphodiesterase (PDE). PDE cleaves cGMP to 5' GMP, resulting in plasma membrane hyperpolarization (reviewed in Liebman et al., 1987). Sites and mechanisms of interaction between rhodopsin, transducin and PDE have been studied using synthetic peptides to delineate regions on α_t of interaction with its neighbors.

RHODOPSIN–TRANSDUCIN INTERACTION

The background for this work is a series of studies of monoclonal antibodies that block the function of G_t. One of these antibodies, MAb 4A, was shown to block light activation of PDE (Hamm and Bownds, 1984), and the mechanism by which it worked was shown to be a blockade of interaction with rhodopsin (Hamm et al., 1987). In addition, its binding to α_t is blocked by light–dependent G_t binding to R* (Hamm, 1990). To determine functional sites of interaction with rhodopsin, epitope mapping was carried out, showing that MAb 4A binds to sites on the amino and carboxyl terminal regions of α_t (Hamm et al., 1987, 1988; Mazzoni et al., 1991). This evidence that the carboxyl terminal region of α_t is involved in interaction with rhodopsin was consistent with studies on pertussis toxin (PT) which ADP-ribosylates Cys^{347} and blocks interaction with rhodopsin (VanDop et al., 1985). When G_t is tightly bound to R*, PT is unable to ADP-ribosylate α (Hamm, 1990), suggesting that its access to the substrate site is blocked because it forms part of the interaction surface of α and R*.

Synthetic peptides from the regions of G_t that were predicted from the epitope mapping studies to form the interface with rhodopsin were tested for their ability to block rhodopsin-G_t interaction. Stabilization of Meta II by G_t was used as a direct assay of interaction. It was shown that these peptides can directly bind to Meta II and compete with G_t (Hamm et al., 1988). This method has been used to further map receptor interaction domains. Several peptides from the α subunit primary sequence blocked rhodopsin–G_t interaction. Two peptides that had been found to block MAb 4A binding to G_t also blocked interaction with rhodopsin. These sequences were at the carboxyl terminus, α_t–340–350, and just internal to the carboxyl terminal, α_t–311–329. The internal peptide was a more effective competitor after acetylation and amidation, while truncated versions were only partially effective. An analog peptide which substituted Cys^{321} with Ser was no longer a competitor, suggesting that this amino acid is important in promoting binding of G_t to rhodopsin. An amino terminal peptide, α_t–8–23, was also able to block interaction with rhodopsin, suggesting either that the amino terminus is directly involved in rhodopsin binding, or that the peptide blocked α–$\beta\gamma$ interaction and consequentially blocked G_t interaction with rhodopsin. As predicted by the competition studies, the two carboxyl terminal peptides directly bind to rhodopsin, and surprisingly are able to mimic G_t and directly stabilize Meta II (Hamm et al., 1988). The carboxyl terminal undecamer peptide was most potent in this effect.

Comparative studies with homologous carboxyl terminal peptides from α_s show that these peptides have similar effects on the β–adrenergic receptor adenylyl cyclase transmembrane signalling complex. First, α_s peptides corresponding to α_t–340–350 and 311–329 (α_s–384–394 and α_s–354–372) block β–adrenergic receptor–mediated stimulation of adenylyl cyclase (Lazarevich et al., 1991). Second, these peptides stabilize the active receptor conformation that has high affinity for agonists (Watanabe et al., 1991), similar to α_t peptides stabilization of Meta II. The effects are specific, since α_i or α_t peptides have no effect on β–adrenergic receptor–G_s interaction, while α_s peptides have no effect on rhodopsin–G_t interaction.

Thus there appears to be a great specificity of interaction with the appropriate receptor even in relatively short peptides, suggesting that peptides or peptide mimetics might be useful tools to directly modulate specific receptor–G protein interactions. To design more potent analogs of the most effective carboxyl–terminal peptide, it would be important to determine the active, receptor–bound peptide conformation. This has been accomplished by two–dimensional NMR (Dratz et al., 1991, 1992). The free peptide has very little structure, but the rhodopsin–bound peptide has a very well–defined structure. Interestingly, Cys^{347}, the target for pertussis toxin ADP–ribosylation, takes part in a β–turn which was shown to be critical for the activity of the peptide to stabilize Meta II (Hamm et al., 1991). These data provide structural insight into the mechanism of pertussis toxin action.

TRANSDUCIN ACTIVATION

The NMR studies also provided important structural information which suggest a mechanism of G_t activation by rhodopsin. Rhodopsin catalyzes a decrease in GDP affinity by causing a change in the conformation of the GDP binding site. This most likely occurs by changing the environment of the guanine ring, which can change GDP affinity by several orders of magnitude (mutagenesis studies). NMR studies (Dratz et al., 1992) showed that light–activation of rhodopsin causes a conformational change in the bound peptide around the Lys^{345} peptide bond, which is predicted to be part of the GDP binding site (Deretic and Hamm, 1987). If such a conformational change occurs in the native protein, this could lead to GDP release. A mutation in the α_t–Lys^{345} cognate residue in α_s, Arg^{389}, in the unc mutant of G_s, to proline, leads to uncoupling from the receptor (Sullivan et al., 1987). Recent studies in which the carboxyl terminal of α_o is truncated show that removal of 10–15 amino acids significantly decreases the affinity of GDP without affecting GTP affinity (Denker et al., 1992), providing evidence that the carboxyl terminus is important in determining GDP affinity.

Upon GDP release, GTP binding can occur, and the α subunit acquires a conformation that can interact with and activate the rod effector enzyme, cGMP phosphodiesterase (PDE). PDE consists of two catalytic subunits, α and β, and two inhibitory γ subunits. Activation by α_t-GTP is accomplished by releasing the inhibitory constraint imposed by the γ subunits. The mechanism of the interaction between α_t-GTP and PDE and PDE activation will be discussed in the following section.

TRANSDUCIN – PHOSPHODIESTERASE INTERACTION

In most systems, it is known that the GTP–bound α subunit is involved in effector activation, whereas the mechanism of that activation is unclear. We have investigated the site on α_t of interaction with and activation of cGMP phosphodiesterase. Activation by α_t-GTP is accomplished by releasing the inhibitory constraint imposed by the PDEγ subunits. This activation has recently been shown to be mediated by α_t-293-314 (Rarick et al., 1992). Peptides from along the α primary sequence were examined for their ability to either block α_t-GTP activation of PDE or to directly activate PDE. Peptide 293-314, a 22 amino acid peptide from near the carboxyl terminus of α_t, is fully competent to activate cGMP PDE, to the same extent and with a similar potency (Ka 8μM) as α_t-GTP (Ka 1-2μM). The peptide also mimicked the ability of α_t-GTP to dislodge the inhibitory γ subunit of PDE from the catalytic $\alpha\beta$ subunits, suggesting that this is the major site on α_t for effector activation. The fact that α_t-GDP has no ability to activate PDE suggests that this region must be buried in α-GDT and becomes accessible upon GTP binding. It appears that this region is important for effector activation by other G proteins as well, since a similar region of α_s is necessary for activation of adenylyl cyclase (Berlot and Bourne, 1992). Scanning mutagenesis of a chimeric α_s/α_i molecule containing a minimal α_s region which still activates adenylyl cyclase (amino acids 236-356) revealed three regions required for adenylyl cyclase activation, region I (α_s-236-241), region II (277-285), and region III (348-354). The first region corresponds to ras p21 Switch II region of GTP–induced conformational change, while the second region is close to the guanine binding consensus region NKXD, and was determined by Itoh and Gilman (1991) to impart binding affinity to adenylyl cyclase. The third region corresponds to residues 307-313 of α_t-, within the sequence of the activating peptide, 293-314. Thus it is attractive to hypothesize that this region is generally involved in effector activation by G proteins. It is of interest that this region is adjacent to the receptor activation domain (see below), thus the effector and receptor interaction domains of α subunits both map near the COOH terminus.

Other regions of α subunits clearly participate in G protein binding to effectors. In the same study of α_t peptide effects on PDE activity, two other peptides, α_t-53-65 and 201-215, inhibited activated PDE. Peptide 53-65 blocked both PDE activity activated by tryptic cleavage of the inhibitory γ subunit (tPDE) and α_t-GTPgS-activated PDE. Peptide 201-215, which inhibited only tPDE activity, corresponds to Switch II region of Ras p21. These regions may be involved in interaction with the PDE catalytic subunits.

How does α_t-GTP cause PDE activation? γ-PDE has a high affinity for $\alpha\beta$PDE provided by a two site interaction of residues 24-46 and 46-87 (Artemyev and Hamm, 1992). A recent study (Artemyev et al., 1992) on the interaction regions between α_t and PDE using a fluorescent assay of α_t-PDE binding showed that α_t-293-314 interacts with the carboxyl terminal region (46-87) of Pγ which is most important for PDE inhibition by Pγ (Artemyev and Hamm, 1992). Thus α_t (and the mimetic peptide) work by inducing a conformational change on Pγ that decreases its affinity for PDE $\alpha\beta$, leading to PDE activation.

The central polycationic region of Pγ also plays an important role in PDE activation. Normally, it lends affinity to the Pγ–$\alpha\beta$ interaction (Artemyev and Hamm, 1992). Upon binding of α_t–GTP, it loses its affinity for P$\alpha\beta$ and binds to α_t. Thus this may be the key region of α_t-induced conformational change involved in physical dissociation between the α_t–Pγ complex leading to fully active PDE. The site on α_t of Pγ–24–46 interaction is not yet known. This peptide was a good competitor of α_t–Pγ interaction, but had no effect on α_t–293–314–Pγ interaction, showing that it binds to a different, still unknown, region on α_t.

These studies provide a preliminary structural map of G$_t$–PDE interactions and their functional role in PDE activation. Future work will assess the structural basis for the conformational changes on PDEγ catalyzed by α_t–GTP.

REFERENCES

Artemyev, N. O. and H. E. Hamm. Biochem. J. 283, 273–279, 1992.

Artemyev, N. O, H. M. Rarick, J. S. Mills, N. P. Skiba and H. E. Hamm. J. Biol. Chem., in press, 1992

Berlot, C. H. and H. R. Bourne. Cell, 68, 911–922, 1992.

Denker, B. M., C. J. Schmidt and E. J. Neer. J. Biol. Chem. 267, 9998-10002, 1992.

Deretic, D., and Hamm, H. E. J. Biol. Chem. 262, 10831–10847, 1987.

Dratz, E. A., H. E. Hamm, R. Zwolinski, J. Furstenau and C. Lambert. Biophys. J. 59, 189a, 1991.

Dratz, E. A., C. Lambert, J. Furstenau, S. Paylian, T. Schepers and H. E. Hamm. Invest. Ophthalmol. Vis. Sci. (Suppl.), 33:873, 1992.

Deretic, D., and Hamm, H. E. J. Biol. Chem. 262, 10831–10847, 1987.

Gilman, A. G. Ann. Rev. Biochem. 56, 615–649, 1987.

Hamm, H. E. In A. Robison and P. Greengard (ed.) Advances in Second Messenger and Phosphoprotein Research. Vol. 24, pp. 76-82, 1990.

Hamm, H. E., Deretic, D., Hofmann, K. P., Schleicher, A., and Kohl, B. J. Biol. Chem. 262, 10831–10838, 1987.

Hamm, H. E., Deretic D., Arendt, A., Hargrave P. A., Koenig B., Hofmann K. P. Science 241, 832–835, 1988.

Hamm, H. E., R. Zwolinski, H. M. Rarick, J. Furstenau and E. A. Dratz. Biophys. J. 59, 347a, 1991.

Itoh, H. and A. G. Gilman. J. Biol. Chem. 266, 16226–16231, 1991.

Lazarevic, M. B., M. Watanabe, M. M. Rasenick and H. E. Hamm. Neurosci. 18, 1991.

Mazzoni M. R., Malinski, J. A. and Hamm, H. E. J. Biol. Chem., 266:14072–14081, 1991.

Rarick, H. M., N. Artemyev and H. E. Hamm. Science, 256, 1031-1033, 1992.

Sullivan, K. A., Miller, R. T., Masters, S. B., Beiderman, B., Heideman, W. and H. R. Bourne. (1987). Nature (Lond.) 330, 758–762.

Van Dop, C., G. Yamanaka, F. Steinberg, R. D. Sekura, C. R. Manclark, L. Stryer and H. R. Bourne. J. Biol. Chem. 259, 23-26, 1984.

Watanabe, M., M. B. Lazarevic, H. Hamm and M. M. Rasenick. Neurosci. 18, 1991.

Posttranslational modifications of retinal G proteins

Robert R. Rando

Department of Biological Chemistry and Molecular Pharmacology, Harvard Medical School, Boston, MA 02115, USA

INTRODUCTION

Signal transducing G proteins are post-translationally modified by isoprenylation (see below) (1). This hydrophobic modification is thought to direct the modified protein to a membrane (2). In the isoprenylation process, proteins containing a "CAAX" box (C=cysteine, A=aliphatic amino acid, and X=any amino acid) at their carboxyl termini are first S-alkylated at cysteine with either all-trans-farnesyl pyrophosphate (C15) or all-trans-geranylgeranyl pyrophosphate (C20), to produce an isoprenylated thioether (1). The products of the *ras* oncogenes are farnesylated (3) and the heterotrimeric G proteins are uniformly geranylgeranylated at their γ subunits, save for transducin, which is farnesylated (4-6). Indeed, geranylgeranylation occurs about 10-fold more frequently than farnesylation. After isoprenylation, proteolytic cleavage of the modified protein generates the mature protein containing the isoprenylated protein as the carboxyl terminal residue. Subsequent carboxymethylation of the modified cysteine by an S-adenosylmethionine-dependent methyltransferase produces the isoprenylated methyl ester (7,8). This latter reaction is the only potentially reversible one in the pathway and,

hence, the only one subject to regulation. We have recently uncovered a specific methylesterase which does in fact hydrolyze isoprenylated methyl ester substrates (9).

Little is known of the function of isoprenylation, although it is thought that this modification targets proteins to membranes. Recent studies suggest that isoprenylation and methylation together are important for G protein functioning (2,3). Association with membranes could be receptor-mediated or could simply arise by virtue of the increased hydrophobicity of the isoprenylated moiety. Furthermore, the enzymology of this pathway remains relatively unexplored. Only the isoprenyl transferases have thus far been purified (10). The methyltransferase has been studied to some extent, although it has yet to be purified (7,8). The protease and methylesterase have only very recently been identified (9).

Visual transduction mechanisms provide an ample laboratory to test hypotheses concerning the structure-function relationships underlying isoprenylation mechanisms. As mentioned above, transducin is farnesylated (4,5). In addition, the phosphodiesterase is farnesylated on its α subunit and geranylgeranylated on its γ subunit (10). Thus many of the elements of visual signal transduction are post-translationally modified by hydrophobic moieties. It is likely that these modifications are either involved in mediating protein-protein interactions between interacting elements or (and) are important for insuring that the proteins remain membrane-associated at the disk surface during transduction. This would insure efficient interaction between the signal transducing elements in two dimensions. It should also be noted that several "small" G proteins found in the rod outer segments are also modified by isoprenylation (8).

There are several important issues to be addressed in understanding protein isoprenylation and signal transduction. The molecular enzymology of isoprenylation will have to be clarified. The enzymes involved will have to be purified and their modes of regulation determined. Only the isoprenyl transferases have been purified thus far (10). The functional role(s) of isoprenylation in signal transduction will have to be determined. Since signal transduction mechanisms in vision are the best and most quantitatively understood of all transduction mechanisms, it is reasonable to begin to address functional issues in this system.

MOLECULAR ENZYMOLOGY OF ISOPRENYLATION

A. An Isoprenylated Protein Methyltransferase

The isoprenylation pathway requires a specific methyltransferase enzyme capable of methylating the isoprenylated cysteine carboxyl terminus. We have discovered such an enzymatic activity associated with retinal rod outer segment membranes.(8). For the methylase, the apparent K_M for SAM is 2 μM and the optimum pH is between 7.5 and 8.8 (12,13). Under physiological conditions, the enzyme is designed to recognize a prenylated G protein. Remarkably, we observed that N-acetyl-S-farnesyl-L-cysteine (AFC) is an excellent substrate for the ROS methyl transferase with a K_M of 23μM (12,13). The fact that a molecule as simple as AFC is a substrate for the enzyme suggests that the methyl transferase does not require the peptide moiety linked to the modified cysteine. Indeed, the K_M measured when AFC is the substrate is in the range of what has been determined for the methylation of synthetic peptides derived from *ras* proteins in other systems (7). It was also possible to inhibit G protein methylation by AFC in the latter's K_M range, demonstrating that both substrates are processed by the same ROS enzyme. The finding that a simple molecule like AFC is a substrate for the methyl transferase makes possible many interesting experiments on the biochemistry of the enzyme not previously possible. For example, inhibitor design and kinetic studies on the enzyme have proven to be straightforward (12,14). Although we have not done extensive cell biology on any of the methylase inhibitors, in collaboration with Professor Donald Kufe at the Dana Farber Cancer Center at Harvard, we have found that 25-50μM FTA strongly blocks the growth of a *ras* transformed cell line (HL-60) and that the inhibitor also blocks endogenous *ras* methylation in isolated membranes, suggesting that G protein methylase inhibitors may be interesting biologically.

B. Methylesterase

If methylation is to play a dynamic role in the isoprenylation pathway, then it would be expected that there would be a specific methylesterase present to reverse the methylation reaction. Methylesterase enzymes have been found in chemotactic bacteria and they are presumed to play an important role in the adaptive response. We have uncovered a specific methylesterase associated with rod outer segment membranes which may play an important role in the isoprenylation pathway. Specifically, we have found that AFC methyl ester is a substrate for a disk membrane-associated hydrolase (methylesterase) enzyme (9). These results further suggest a functional role for methylation in vertebrate signal transduction, perhaps similar to what is found in bacterial chemotaxis. Both L-AFC methyl ester and L-AGGC methyl ester are substrates for the enzyme but L-cysteine methyl ester is not, suggesting specificity for isoprenylated substrates. The enzyme is irreversibly inactivated by ebelactone B, suggesting that it is a serine esterase (9). Thus, at least at the substrate and inhibitor level, this methylesterase appears to have many of the attributes expected of an enzyme which could play an important signalling role. Further molecular studies on this enzyme will help to determine its mechanism of action and its functional roles.

C. A protease which specifically cleaves isoprenylated peptides.

As previously mentioned, proteolysis is a key event in the isoprenylation pathway. Using *in vitro* translation, Hancock et al. demonstrated that proteolysis is essential for efficient membrane association of the *ras* product (3). Proteolytic processing could occur by several different routes. The simplest route involves cleavage between the modified cysteine residue and the adjacent aliphatic amino acid to liberate the AAX tripeptide. Alternatively, proteolysis could begin anywhere within the AAX moiety and stop at the modified cysteine residue. Multiple proteases could be involved in this latter scenario. We have identified a microsomal proteolytic activity that produces a single cut between the modified cysteine residue and the AAX tripeptide, using a synthetic tetrapeptide substrate N-acetyl-S-farnesyl-L-cysteine(AFC)-val-ile-ser to probe the reaction (15). This latter sequence is based on the CAAX sequence of transducin γ. K_M and V_{max} values were measured to be 4.8 μM and 0.236 nmol/min/mg protein respectively. Proteolytic cleavage of the substrate is stereospecific, because the substitution of a farnesylated D-cysteine residue for the L amino acid leads to the abolition of substrate activity. A free terminal carboxyl group is also required for substrate activity, because methyl esterification renders the substrate inert. The tripeptide N-acetyl-S-farnesyl-L-cysteine-L-val-L-ile and the dipeptide N-acetyl-S-farnesyl-L-cysteine-L-val are also hydrolyzed by the protease. The measured K_M and V_{max} values for the tripeptide are 9.3μM and 58 pmol/min/mg respectively. The hydrolysis of the farnesylated tetrapeptide is not inhibited by a 5-fold excess of the non-farnesylated tetrapeptide and non-farnesylated peptides are not hydrolyzed, which taken together suggest that isoprenylation is important for substrate activity.

Possible Functional Consequences of the Isoprenylation Pathway in Visual Transduction

It has generally been assumed that the isoprenylation of proteins causes these proteins to become associated with membranes. It is of course true that virtually all isoprenylated proteins investigated thus far are membrane associated. This does not mean that isoprenylation per se leads to membrane association, but that isoprenylated proteins associate with other membrane-bound proteins. Indeed, the association of isoprenylated proteins with membranes is often quite weak. It is also difficult to conceive how the linking of a simple C.15 or C.20 isoprenoid to a protein would be sufficient in and of itself to cause association of the protein with a membrane. Model studies may bear on this point. Molecules like AFC methyl ester and AGGC methyl ester have very large n-octanol/water partition coefficients. However, when incorporated into phosphatidylcholine based liposomes, they rapidly undergo intermembranous transfer suggesting that membrane association is labile even when the isoprenoid is attached to a simple amino acid.

The possibility that isoprenylation is an essential component in protein-protein interaction is reasonable but little support has been been found for this possibility so far. Transducin farnesylation and methylation are important for efficient coupling of transducin to metarhodopsin 2 (16). A mechanism involving isoprene dependent interactions between the α and γ subunits of transducin has been suggested (16). This mechnism would suggest that molecules such as AFC methyl ester should specifically inhibit metarhodopsin 2-transducin coupling. Thus far, no specific effects have been observed.

ACKNOWLEDGEMENTS

The studies reported here was performed by Drs. J. Canada, B. Gilbert, R. Lai, Y. T. Ma, D. Perez-Sala, Y. Q. Shi, and E. W, Tan and supported by N. I. H. grant EY-03624.

REFERENCES

1. Maltese, W. A. (1990) *FASEB J.* **4**, 3319-3328.

2. Hancock, J. F., Magee, A. I., Childs, J. E., and Marshall, C. J. (1989) *Cell* **57**, 1167-1177.

3. Hancock, J. F., Cadwallader, K., and Marshall, C. (1991) *EMBO J.* **10**, 641-646.

4. Fukada, Y., Takao, T., Ohguro, H., Yoshizawa, T., Akino, T., and Shimonishi, Y. (1990) *Nature* **346**, 658-660.

5. Lai, R. K., Pérez-Sala, D., Cañada, F. J., and Rando, R. R. (1990) *Proc. Natl. Acad. Sci. USA* **87**, 7673-7677.

6. Mumby, S. M., Casey, P. J., Gilman, A. G., Gutowski, S., and Sternweis, P. C. (1990) *Proc. Natl. Acad. Sci. USA* **87**, 5873-5877.

7. Stephenson, R. C., and Clarke, S. (1990) *J. Biol. Chem.* **265**, 16248-16254.

8. Pérez-Sala, D., Tan, E. W., Cañada, F. J., and Rando, R. R. (1991) *Proc. Natl. Acad. Sci. USA* **88**, 3043-3046.

9. Tan, E. W., and Rando, R. R. (1992) *Biochemistry* (in press).

10. Reiss, Y., Goldstein, J. L., Seabra, M. C., Casey, P. J., and Brown, M. S. (1990) *Cell* **62**, 81-88.

11. Anant, J. S., Ong, O. C., Xie, H. Y., Clarke, S., O'Brien, P. J., and Fung, B. K. (1992) *J. Biol. Chem.* **267**, 687-690.

12. Tan, E. W., Perez-Sala, D., Canada, F. J., and Rando, R. R. (1991) *J. Biol. Chem.* **266**, 10719-10722.

13. Tan, E. W., Perez-Sala, D., and Rando, R. R. (1991) *J. Am. Chem. Soc.* **113**, 6299-6300.

14. Shi, Y., Q. and Rando, R. R. (1992) *J. Biol. Chem.* (in press).

15. Ma, Y. T., and Rando, R. R. *Proc. Nat. Acad. Sci.* (in press).

16. Ohguro, H., Fukada, Y., Shimonishi, Y., Yoshizawa, T., and Akino, T. (1991) *EMBO J.* **10**, 3669-3674.

Phosphorylation of rhodopsin in fly photoreceptor membranes is controlled by the light-dependent binding of an arrestin homolog

Anette Plangger and Reinhard Paulsen

Institut für Zoologie I, Universität Karlsruhe, Kornblumenstrasse 13, 7500 Karlsruhe 1, Germany

SUMMARY

Photoconversion of rhodopsin to metarhodopsin leads to the binding of the arrestin homolog Arr2, (49 KDa protein) to fly photoreceptor membranes. The amount of Arr2 bound is directly proportional to the amount of metarhodopsin formed. This suggests that Arr2 interacts with the metarhodopsin state itself. In isolated microvillar photoreceptor membranes the light-induced phosphorylation of rhodopsin by a protein kinase is efficiently activated by binding of purified Arr2. Contrary to what is known about the arrestin- rhodopsin- kinase system of vertebrate rod photoreceptors, phosphorylation of metarhodopsin appears not to be a prerequisite for arrestin binding. Phosphorylation experiments with urea-treated, isolated fly photoreceptor membranes indicate, that the rhodopsin kinase is an integral membrane protein or part of the rhabdomeral cytoskeleton. The primary function of arrestin might be to control the phosphorylation state of photoactivated rhodopsin.

INTRODUCTION

Arrestins have emerged as one family of proteins that appear to function as regulatory proteins in the termination of phototransduction by interacting with the photoactivated rhodopsin state. Vertebrate arrestin (48-kDa protein, S-antigen) binds tightly to photolyzed, phosphorylated rhodopsin and thereby efficiently quenches the activation of cyclic GMP-phosphodiesterase (Wilden et al., 1986). However, the situation with respect to arrestins in the invertebrate is not so clear. Molecular cloning of genes encoding a 41 kDa protein and a 45 kDa protein, respectively (Smith et al., 1990; Hyde et al., 1990; Yamada et al., 1990), showed that photoreceptors of the compound eye of *Drosophila* express two structurally distinct arrestin homologs, which have been designated arrestin 1 (Arr1) and arrestin 2 (Arr2) (LeVine et al., 1991). Arr2 of *Drosophila* has been characterized as a 49 kDa protein that undergoes Ca^{2+}-dependent, light-triggered phosphorylation (Yamada et al., 1990). The corresponding arrestin homolog of blowfly photoreceptors (a 49 kDa phosphoprotein) has been identified by cross-reactivity with monospecific antibodies directed against Arr2 of *Drosophila* (C. Zuker personal communication). This protein has been formerly isolated and partially purified on the basis of its reversible, light-dependent binding to isolated microvillar photoreceptor membranes. Arr2 of *Calliphora* binds specifically to photoreceptor membranes in which rhodopsin (P) of the major photoreceptors R1-6 has been photoconverted to its long-lived metarhodopsin state (M) (Bentrop and Paulsen, 1986).

It has been proposed that the light-dependent binding of Arr2 acts, in concert with the light-dependent phosphorylation of fly rhodopsin by a rhodopsin kinase, as a termination signal for the photoactivated state of fly rhodopsin (Paulsen et al., 1988). We here show that, opposite to what is known for vertebrate photoreceptors, the light-induced phosphorylation of metarhodopsin in the fly is a result of the interaction with Arr2 rather than a prerequisite for its binding.

METHODS

Isolation of fly photoreceptor membranes

Retinae were dissected from redlight-adapted male blowflies (*Calliphora erythrocephala*, chalky mutant) at 4 °C. This and subsequent procedures involving photoreceptor membranes were carried out under dim red light. Microvillar photoreceptor membranes (mPMs) were isolated from dissected retinae as described previously (Bentrop and Paulsen, 1986). Generally, isolated membranes were washed with a low Ca^{2+}-buffer (5 mM sodium phosphate buffer pH 6.2, containing 3 mM EGTA and 5 mM DTT). Endogenous protein kinase activity was removed by incubating mPMs for 5 min on ice in PBS (33 mM phosphate buffer, 0.1 M sodium chloride pH 6.5) which contains 4 M urea as denaturing agens. The extracted membranes were pelleted by centrifugation (10 min., 50000 x g). The rhodopsin content of these membranes was determined by difference spectrophotometry of digitonin extracted visual pigment (Bentrop and Paulsen, 1986).

Purification of Arr2

Fractions of soluble proteins enriched in Arr2 (approximately 320 µg from 2500 eyes) were obtained by light-induced reversible affinity binding to *Calliphora* eye membranes (Bentrop and Paulsen, 1986). Arr2 was precipitated from this fraction with 70 per cent ammonium sulphate, resuspended in 200 µl PBS and subjected to further purification by FPLC on a gelfiltration column (superdex 75, HR 10/30, Pharmacia). After gelfiltration, fractions containing Arr2 were loaded on heparin-agarose spin columns (Millipore Millex filter units with 10 µl heparin-agarose, Sigma). Loaded columns were washed three times with 70 µl PBS. Arr2 was eluted by centrifugation with 1.0 M sodium chloride in 20 mM Hepes, pH 7.5. For binding and phosphorylation experiments the eluted Arr2 fraction was dialyzed for two hours at 4 °C against PBS.

Rhodopsin phosphorylation

Light-induced phosphorylation of rhodopsin, using blue light to stimulate phosphorylation and $[\gamma-^{32}P]ATP$, followed the methods described previously (Bentrop and Paulsen, 1986). Phosphorylation of urea-extracted membranes was performed on Millex filter units. In this case urea-incubated mPMs were sedimented onto the membrane of the filter unit and washed three times with PBS. After addition of purified Arr2 in a phosphorylation medium, P to M conversion was performed by irradiating the samples with blue light for two minutes on ice. Control samples were either kept in the dark or PBS was added instead of Arr2. For phosphorylation, samples were incubated for 15 min at 25 °C. Phosphorylation was terminated by washing the membranes with PBS containing 2 mM EDTA. Membrane proteins were eluted by centrifugation after loading the filter unit with 4 per cent SDS in 65 mM Tris/HCL buffer pH 6.8

Proteins were separated by SDS-PAGE on 8-20 per cent acrylamide gradient gels (Laemmli, 1970). Phosphate incorporation into opsin was recorded by autoradiography. The relative amount of Arr2 in the samples was estimated by laser densitometry of silver stained gels with a laser densitometer (Pharmacia).

RESULTS

Light-dependent binding of Arr2 to photoreceptor membranes

Previous studies showed that Arr2 binds, upon photoconversion of P to M, to the microvillar membranes of blowfly photoreceptors. The binding affinity of membranes for Arr2 decreases if M is photoconverted to P (Bentrop and Paulsen, 1986). In order to test the relation between the M-content of membranes and the amount of Arr2 bound, we exposed retinae to blue light which establishes a photoequilibrium of 69 per cent M and 31 per cent P and leads to maximal binding of Arr2. Subsequently, retinae were homogenized and soluble proteins were removed by several washes with PBS. Eye membranes were then exposed to light of different wavelengths, which established new photoequilibria with a lower proportion of M, or to red light converting all M to P. The amount of Arr2 becoming extractable with a buffered salt solution (PBS) is directly proportional to the amount of M converted to P (Fig.1). This suggests that the binding site for Arr2 in photoreceptor membranes is located at the M-state of rhodopsin.

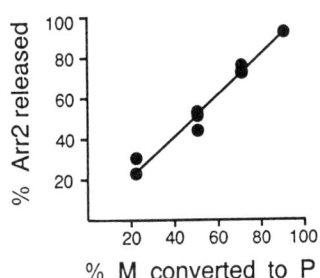

Fig. 1. Arr2 binding to irradiated photorecptor membranes is directly proportional to the metarhodpsin content of membranes. Values represent the relative amount of light-dependently bound Arr2 which becomes extractable with PBS from membranes after a portion of M was photoconverted to P.

Purification of Arr2

Previous studies of invertebrate arrestins used Arr2 fractions obtained by repeated light-dependent affinity binding to photoreceptor membranes (Paulsen et al., 1988). These fractions still contain minor amounts of other soluble proteins. In particular, it could not be excluded that a soluble rhodopsin kinase also underwent a light-dependent binding to photoreceptor membranes and was, therefore, copurified with Arr2. In an attempt to separate Arr2 from a putative soluble rhodopsin kinase, Arr2 was further purified by ammonium sulphate precipitation, gelfiltration and affinity binding to heparin-agarose. As shown in Fig. 2, Arr2 can be highly purified by light-dependent affinity binding to eye membranes (lane B), followed by FPLC gelfiltration on superdex 75 (lane C) and reversible binding to heparin-agarose (lane D). The gelfiltration was employed to remove proteins with M_r higher than that of Arr2, particularly a rhodopsin kinase which in vertebrate rod photoreceptors has a M_r of about 67 000 - 70 000 (Palczewski et al., 1988). The electrophoretic mobility of purified Arr2 on 8-20 per cent acrylamide gradient gels corresponds to a M_r of 45 000 (Fig. 2).

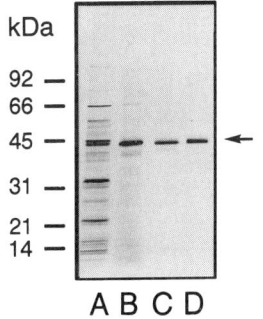

Fig. 2. Purification of Arr2. Silver-stained protein pattern obtained by SDS-PAGE of Arr2 containing fractions. Lane A shows the total PBS soluble proteins present in blowfly retinae. Purification of Arr2 by light-induced affinity binding to eye membranes (lane B) was followed by FPLC on superdex 75 (lane C) and affinity binding to heparin-agarose (lane D). Arr2 is indicated by an arrow.

Recently it was shown that heparin and other polyanions bind vertebrate arrestin. The proposed site for this interaction is also present in Drosophila Arr1 and Arr2 (Palczewski et al., 1991). Therefore we have included a further purification step using heparin-agarose placed on a Millex filter. This unit can be used as a spin column. Since Arr2 binds to heparin with a high affinity, buffered salt solutions containing 1.0 M sodium chloride are required to elute Arr2 from heparin-agarose (recovery 90 per cent). This step particularly removes proteins of M_r lower than that of Arr2. Silver staining of SDS-polyarylamide gels reveals that, after the heparin-agarose step, the Arr2 fraction is virtually free of other proteins.

Effect of Arr2 binding on opsin phosphorylation

Fractions enriched in Arr2 which have been obtained by light-dependent affinity binding to microvillar membranes stimulate light dependent phosphorylation of rhodopsin (Paulsen et al., 1988). This effect is also observed in the experiment summarized in Fig. 3. Here we recombined mPMs with different fractions of soluble proteins obtained by gelfiltration. Only Arr2-containing fractions stimulate light-dependent phoshorylation of rhodopsin efficiently (lanes C-E). Fractions containing soluble proteins with M_r higher (lane A and B) or lower than 49 000 (lane F) and reduced levels of Arr2 itself do not show this effect. Accordingly, light-induced phosphorylation of rhodopsin is strongly dependent on the presence of Arr2.

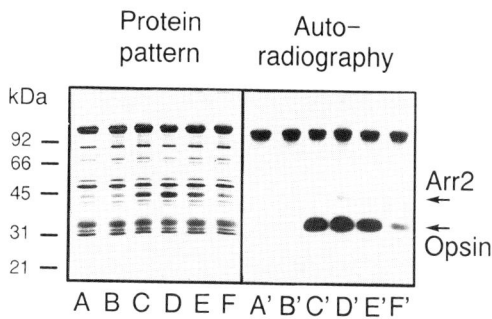

Fig. 3. Stimulation of light-induced opsin phosphorylation by Arr2. Rhabdomeric membranes were combined with different protein fractions obtained by separating partially purified arr2 by FPLC on a gelfiltration column. After after the light-induced phosphorylation reaction the membranes were subjected to SDS-PAGE. The coomassie-stained protein pattern (lanes A-F) shows the binding of Arr2 to membranes that were supplemented with fractions that contain relatively high amounts of Arr2 (lanes C-E). The corresponding autoradiograph (lanes A'-F') reveals that only these samples show light-activated phosphorylation of rhodopsin (lanes C'-E').

The following experiments were designed to investigate whether or not Arr2 itself acts as a protein kinase or, alternatively, controls M phosphorylation catalyzed by a separate membrane bound protein kinase.

In order to test the localization of rhodopsin kinase, we prepared protein kinase-free membranes by extraction with urea (Miller and Paulsen, 1975). Treatment with 5 M urea in PBS does not affect photoconversion of P to M (Fig. 4a) and urea-extracted mPMs still bind reversibly Arr2 in a light-dependent manner (Fig. 4b).

Phosphorylation experiments performed with urea-extracted membranes are summarized in Fig. 5. Light-dependent phosphorylation in isolated untreated mPMs is negligible in the absence of Arr2 (lane D). Addition of purified Arr2 leads to an more than eight-fold increase of of rhodopsin phosphorylation (lane F). This light-dependent phosphorylation of rhodopsin is abolished completely in urea-extracted mPMs, although urea treatment does not alter the binding capacity of M for Arr2 (see Fig. 4b). The results show that light-induced phosphorylation requires the presence of Arr2 and a membrane bound protein, most likely a protein kinase. Taken together the results of the experiments depicted in Fig. 4b and Fig. 5 also demonstrate that binding of Arr2 by M does not require that M is in a phosphorylated state, because the amount of Arr2 bound to M-state membranes does not depend on the phosphorylation status of M.

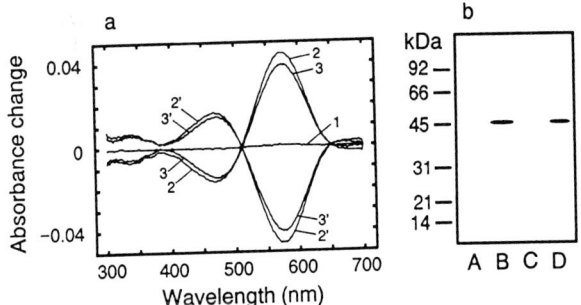

Fig. 4. Comparison of untreated and urea-extracted rhabdomeric membranes. (a) Spectral properties: Difference spectra of digitonin extracts recorded after photoconversion of P to M and M to P show no difference between untreated and urea-extracted membranes (compare 2 and 3, and 2'and 3', respectively; 1, baseline). (b) Binding and release of Arr2: Silver-stained protein pattern after SDS-PAGE showing no difference between untreated and urea-extracted membranes with respect to the amount of Arr2 bound to M-state membranes and released after photoconversion of M to P (lanes B and D, respectively). No Arr2 is extractable from M-state membranes before photoconversion of M to P (lanes A and C).

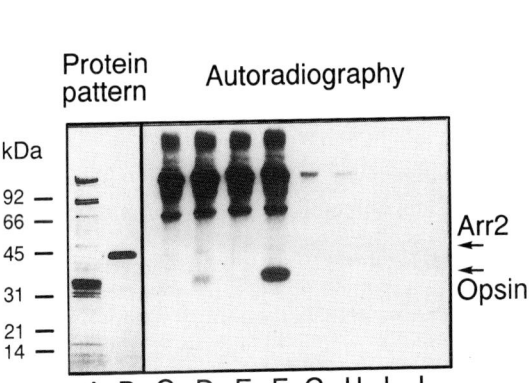

Fig. 5 Effect of Arr2 on the light-dependent rhodopsin phosphorylation in untreated and urea-treated photoreceptor membranes. Lanes A and B, Protein pattern obtained by SDS-PAGE of digitonin-extracted membrane proteins from isolated photoreceptor membranes (A) and purified Arr2 (B, see also Fig. 2, lane D). Autoradiography of proteins separated by SDS-PAGE (lane C-J) indicating opsin phosphorylation in untreated (C to F) and urea extracted (G to J) photoreceptor membranes in the absence of Arr2 (lane C and G dark controls; lane D and H, irradiated samples) and in the presence of chromatographically purified Arr2 (lane E and I, dark control; lane F and J, irradiated samples). Light-dependent rhodopsin phosphorylation is only observed in untreated membranes reconstituted with Arr2.

DISCUSSION

These results strongly suggest that the protein kinase responsible for phosphorylation of M in isolated mPMs is a membrane bound protein which is denatured or extracted by treating membranes with urea. Arr2 would then be a regulatory protein which, by binding to the photoactivated state of rhodopsin, allows the metarhodopsin to become multiply phosphorylated. This could happen by affecting the interaction of M with a rhodopsin kinase or, alternatively, by preventing the dephosphorylation of opsin by a protein phosphatase. Such a regulatory function of arrestin has been demonstrated earlier for the interaction of arrestin with the phosphorylated bovine rhodopsin (Palczewski et al., 1989). So far we have no evidence that the isolated mPMs contain a membrane bound protein phosphatase. Therefore, it might be primarily the interaction of M and rhodopsin kinase which is affected by the binding of Arr2. Candidate for a membrane bound protein kinase is for example the *ninaC* gene product (Porter et al., 1992). The findings presented here do not exclude, that Arr2 itself is a protein kinase and urea-extraction removes another regulatory component from the membrane. Urea-treatment could also alter the properties of metarhodopsin as a kinase substrate. Arr2 being a protein kinase, however, is less likely in view of the absence of apparent protein kinase consensus sequences in the Arr2 gene (Yamada et al., 1990).

In conclusion, our data suggest that in the microvillar photoreceptor membranes of flies phosphorylation is a consequence rather than a prerequisite for the binding of Arr2. The binding of

Arr2 to M might be the primary termination signal for the active rhodopsin state (M*). The interaction of M* with Arr2 might account for the rapid inactivation of phototransduction (< 150 ms) (Richard et al., 1992). Desensitization would then strongly depend on the ratio of free Arr2 to M*. A limited availability of Arr2 within the microvillar lumen could explain why experimentally induced non-physiological large shifts of P-M, e.g. 70 per cent of P into the M-state, lead to a prolonged depolarizing afterpotential (PDA), particularly in receptors which contain high levels of visual pigment (Hamdorf and Razmjoo, 1979). Phosphorylation of the M*-Arr2 complex might then produce a more permanently inactivated M-species. In this context it is important to note that gene manipulations which prevent expression of phosphorylation sites on rhodopsin do not particularly affect photoreceptor desensitization (C. Zuker, personal communication). This would be expected if phosphorylation is the primary event in the termination of the active state of fly rhodopsin.

Acknowledgments - This work was supported by the Deutsche Forschungsgemeinschaft (Pa 274/3-3). We thank T. P. Williams and J. Bentrop for helpful comments.

REFERENCES

Bentrop, J. and Paulsen, R. (1986): Light-modulated ADP-ribosylation, protein phosphorylation and protein binding in isolated fly photoreceptor membranes. *Eur. J. Biochem.* 161, 61-67.

Hamdorf, K. and Razmjoo, S. (1979): Photoconvertible pigment states and excitation in *Calliphora*; the induction and properties of the prolonged depolarizing afterpotential. *Biophys. Struct. Mech.* 5, 137-161.

Hyde, D.R., Mecklenburg, K.L., Pollock, J.A., Vithelic, T.S. and Benzer, S. (1990): Twenty *Drosophila* visual system cDNA clones: one is a homolog of human arrestin. *Proc. Natl. Acad. Sci. USA* 87, 1008-1012.

Laemmli, U.K. (1970): Cleavage of structural proteins during the assembly of the head of bacteriophage T4. *Nature* 227, 682-685.

LeVine III, H., Smith, D.P., Whitney, M., Malicki, D.M., Dolph, P.J., Smith, G.F.H., Burkhart, W. and Zuker, C.S. (1991): Isolation of a novel visual-system-specific arrestin: an *in vivo* substrate for light-dependent phosphorylation. *Mech. Develop.* 33, 19-26.

Miller, J.A. and Paulsen, R. (1975): Phosphorylation and dephosphorylation of frog rod outer segment membranes as part of the visual process. *J. Biol. Chem.* 250, 4427-4432.

Palczewski, K., McDowell, J.H. and Hargrave, P.A. (1988): Purification and characterization of rhodopsin kinase. *J. Biol. Chem.* 263, 14067-14073.

Palczewski, K., McDowell, J.H., Jakes, S., Ingebritsen,T.S. and Hargrave,P.A. (1989): Regulation of rhodopsin phosphorylation by arrestin. *J. Biol. Chem.* 264,15770-15773.

Palczewski, K., Pulvermüller, A., Buczylko, J. and Hofmann, K.P. (1991): Phosphorylated rhodopsin and heparin induce similar conformational changes in arrestin. *J. Biol. Chem.* 266, 18649-18654.

Paulsen, R., Bentrop, J., Hinsch, K.D. and Schultz, G. (1988): Invertebrate phototransduction: G-proteins and the function of a 49 kDa protein. In: *Proc. Yam. Conf. XXI*, ed. T. Hara, pp. 227-232, Osaka: Yamada Science Foundation

Porter, J.A., Hicks, J.L., Williams, D.S. and Montell, C. (1992): Differential localizations of and requirements for the two *Drosophila ninaC* kinase/myosins in photoreceptor cells. *J. Cell. Biol.*, 116, 683-693.

Richard, E.A. and Lisman, J.E. (1992): Rhodopsin inactivation is a modulated process in *Limulus* photoreceptor. *Nature*, 356, 336-338.

Smith, D.P., Shieh, B.-H. and Zuker, C.S. (1990): Isolation and structure of an arrestin gene from *Drosophila*. *Proc. Natl. Acad. Sci. USA* 87, 1003-1007.

Wilden, U., Hall, S.W. and Kühn, H. (1986): Phosphodiesterase activation by photoexcited rhodopsin is quenched when rhodopsin is phosphorylated and binds to the intrinsic 48-kDa protein of rod outer segments. *Proc. Natl. Acad. Sci. USA* 83, 1174-1178.

Yamada, T., Takeuchi, Y., Komori, N., Kobayashi, T., Sakai, Y., Hotta, N. and Matsumoto, H. (1990): A 49-kilodalton phosphoprotein in the *Drosophila* photoreceptor is an arrestin homolog. *Science* 248, 483-486.

The effect of rhodopsin phosphorylation on the formation and decay of metarhodopsin and rhodopsin-G_t interactions

Julia Kibelbek, Drake C. Mitchell and Burton J. Litman

Department of Biochemistry, University of Virginia Health Sciences Center, Charlottesville, VA, 22908

Visual pigments represent a subclass of a superfamily of receptors, whose signal transduction is initiated by their activation of a G-protein. Several receptors in this superfamily have been shown to be phosphorylated by receptor-specific kinases (for review see Palczewski & Benovic, 1991). A requirement of such phosphorylation is that the receptor be in the agonist-bound state. Phosphorylation has been shown to be associated with receptor desensitization in these systems (Miller et al., 1986; Kwatra & Hosey, 1986; Benovic et al., 1986), suggesting that down regulation of receptor function via phosphorylation of an agonist-stimulated receptor by a receptor-specific kinase may be a general mechanistic motif in receptor-mediated signal transduction pathways.

Among the best characterized of the G-protein mediated receptor systems is the visual transduction pathway, which is triggered by the photolysis of bovine rhodopsin (for review see Stryer, 1991). This system is an ideal one in which to study the mechanism of receptor down regulation by phosphorylation at the level of receptor conformation. Shifts in the spectrum of the retinal moiety of rhodopsin allow one to monitor both the kinetic and equilibrium properties related to the formation of metarhodopsin II (meta II) (cf. Straume et al., 1990). A variety of studies provide convincing evidence that meta II corresponds to the functionally active form of photolyzed rhodopsin, R^* (Emeis et al., 1982; Bennett et al., 1982; Kibelbek et al., 1991).

Phosphorylation may desensitize receptors through one of several mechanisms. These include a change in the inherent stability of the active agonist bound receptor conformation, thereby reducing the concentration of functional receptor, and a weakening the receptor G-protein interaction, resulting in less efficient G-protein activation due to the lower concentration of receptor-G-protein complex formed. The following the experiments were undertaken to discriminate between these mechanisms. The effect of phosphorylation on the steady state concentration of meta II was determined by comparing the K_{eq} for the meta I ↔ meta II equilibrium and the decay time, τ, for the conversion of meta II to meta III for two samples containing unphosphorylated and phosphorylated rhodopsin respectively. The extent of meta II-G_t complex formation in these two samples was also compared.

MATERIALS AND METHODS

Preparation, Characterization and Purification of Phosphorylated Rhodopsin and G_t   Rhodopsin in ROS suspensions was phosphorylated by bleaching in the presence of added ATP to produce rhodopsins with variable phosphorylation levels. Samples were regenerated with 11-cis retinal, purified by Concanavalin A-Sepharose chromatography, and applied to a chromatofocusing column to separate differentially phosphorylated species of rhodopsin. An additional Concanavalin A-Sepharose column removed ampholytes introduced in the chromatofocusing elution buffer (Aton et al., 1984). Unphosphorylated and phosphorylated samples consisted of column fractions containing 0 phosphates per rhodopsin and a pool of ≥ 4 phosphates per rhodopsin, respectively. These samples were individually reconstituted into large unilamellar POPC vesicles by a dilution reconstitution procedure (Jackson & Litman, 1985). Hypotonic extract containing G_t was prepared according to Miller et al. (1986). Overloaded SDS PAGE gels (30-60 µg per lane) showed only a trace contamination of 48K protein (arrestin) in the extract. Concentration of functional G_t was obtained using a filter binding assay (Fung & Stryer, 1980).

Spectrophotometric Measurements Spectral data were collected on a Hewlett-Packard 8452A diode array UV/vis spectrophotometer. Equilibrium meta I - meta II spectra were derived by subtracting the spectrum of the initial sample from that acquired ~3 sec following ca. 25% bleach by a green-filtered flash (500 nm. ± 20 nm bandpass filter) from a camera strobe and adding the spectrum of the bleached rhodopsin. The spectrum of the bleached rhodopsin was derived by simultaneously analyzing the 3 possible difference spectra formed from spectra acquired of the starting sample, following complete reaction with 30 mM hydroxylamine, and after complete bleaching. K_{eq} for the meta I ↔ meta II equilibrium was calculated following deconvolution of corrected difference spectra into meta I and meta II (Straume et al., 1990).

The decay time of meta II to meta III at 20 °C in pH 8.0 isotonic buffer was determined according to the detailed procedures outlined in Kibelbek et al. (1991). A series of 15-20 scans was collected at 45 sec to 5 min intervals beginning ~3 sec after the bleaching flash. The ~3 postflash spectrum was modeled as an equilibrium mixture of meta I and meta II and the last (> 30 min postflash) spectrum was modeled as a mixture of N-retinylidene oxime (NRO) and meta III, which are the decay products of the meta I-meta II equilibrium at 20 °C, pH 8.0 (Blazinsky & Ostroy, 1984). Corrected difference spectra from intermediate times were fit with linear combinations of the derived meta I-meta II and meta III-NRO absorbance profiles. The resulting peak heights for meta II and meta III were used to determine their rates of decay and formation, respectively, which were both modeled as single exponential processes.

Enhanced meta II formation, in both unphosphorylated and phosphorylated rhodopsin samples, was monitored for varying G_t to R^* ratio. Rhodopsin-containing vesicles and G_t (in the form of hypotonic extract) were reassociated by incubation on ice overnight. Equilibrium concentrations of meta I and meta II were determined from corrected difference spectra as described above. The analysis of meta II enhancement by G_t assumed that half of the rhodopsin is oriented with its cytoplasmic surface towards the inside of the vesicle, making it unavailable for interaction with G_t, and that the values of K_{eq} for both orientations are identical. The dissociation constants (K_D) of the meta II•G_t complex for phosphorylated and unphosphorylated rhodopsin were derived by analyzing the extra meta II formed at different R^*/G_t ratios following Schleicher et al. (1989).

RESULTS and DISCUSSION

Earlier work in this laboratory showed that rhodopsin phosphorylation resulted in lower levels of activated phosphodiesterase (PDE), where increased phosphorylation levels yielded correspondingly lower PDE activities (Miller et al., 1986). The current experiments were carried out to test several mechanisms whereby receptor phosphorylation might result in desensitization of the visual signal transduction pathway. The equilibrium between meta II, which activates G_t, and its precursor meta I establishes itself in the order of several ms. Any alteration of this equilibrium, which would result in a reduction in the steady state concentration of meta II, would be reflected in lower levels of G_t activation and a desensitization of the pathway. In these studies, the effect of phosphorylation on the extent of formation of meta II from photolyzed rhodopsin was determined by comparing the K_{eq} for the meta I - meta II equilibrium in phosphorylated and unphosphorylated samples, Table I. Measurements were carried out at pH 7 and 8 at 20 ° and 37 °C. K_{eq}'s measured at 20 °C show that phosphorylation <u>increased</u> the concentration of meta II ~6%, whereas at 37 °C, phosphorylation <u>decreased</u> the meta II concentration by ~5%. This small reduction in meta II concentration in the phosphorylated rhodopsin samples at 37 °C is not considered an effective mode of desensitization.

TABLE I Values of meta I ↔ meta II K_{eq} for phosphorylated and unphosphorylated rhodopsin[a]

Conditions	Unphosphorylated rho	Phosphorylated rho
pH 8.0, 20 °C	0.33 ± 0.03	0.36 ± 0.03
pH 7.0, 20 °C	0.70 ± 0.04	0.78 ± 0.05
pH 8.0, 37 °C	1.32 ± 0.07	1.25 ± 0.08
pH 7.0, 37 °C	1.65 ± 0.13	1.55 ± 0.12

[a] Uncertainties represent 1 standard deviation. Standard deviations of individual data sets were propagated for determination of all final values of K_{eq} and standard deviation. pH 8.0, 20 °C, n=7; pH 7.0, 20 °C, n=3; pH 8.0, 37 °C, n=3; pH 7.0, 37 °C, n=2.

In order to determine if the concentration of meta II is reduced by a change in its lifetime, the rate of conversion of meta II to meta III was measured by determining both the decay time of meta II and rise time of meta III, Table II. The measured time constants all show excellent agreement, demonstrating no differences between decay times and rise times for either phosphorylated or unphosphorylated rhodopsin, nor between the two samples themselves. Thus, phosphorylation does not change steady state concentrations of meta II by influencing the rate of its conversion to meta III.

TABLE II Time constants of meta II decay and meta III rise.[a]

unphosphorylated rhodopsin	4.7 ± 0.4	4.6 ± 0.3
phosphorylated rhodopsin	4.7 ± 0.6	4.5 ± 0.5

[a] Time constants in minutes, errors represent 1 standard deviation, n=3 for all samples.

Since the steady state concentrations of meta II are not affected by phosphorylation, measurements were carried out to determine if desensitization could be explained at the level of meta II-G_t interactions. Meta II·G_t formation was monitored spectrally by the extra meta II produced in a series of rhodopsin samples in which the bleach level of rhodopsin was kept constant and increasing amounts of G_t were added. Analysis of these experiments yielded K_D's for meta II·G_t dissociation of 35 ± 18 and 440 ± 110 nm for unphosphorylated and phosphorylated rhodopsin samples respectively. Hence, there is an order of magnitude reduction in the level of complex formation due to phosphorylation. This lower level of complex formation would lead to a lower level of efficiency in activating G_t at any given level of meta II, in agreement with the reduced PDE activity, which results from the bleaching of phosphorylated rhodopsin (Miller et al., 1986).

In summary, the data reported herein supports a mechanism whereby receptor phosphorylation induces desensitization of a G-protein mediated signalling pathway at the level of receptor-G-protein interacts, rather than by affecting the conformational equilibrium of the receptor itself. K_{eq} and τ measurements are both reflective of conformational changes associated with the retinal or ligand binding pocket, formed from the seven transmembrane helical segments. The correspondence of K_{eq} and τ for the phosphorylated and unphosphorylated rhodopsin samples suggests that phosphorylation does not change the helical orientation in the retinal binding pocket. This likely means that the conformation of the helical connecting loops is also unchanged by phosphorylation, since they are expected to be conformationally restricted due to their connection with the rigid helices. The primary site of phosphorylation is in the carboxyl terminus of rhodopsin, which is also the most flexible region of the protein on the cytosolic surface of the disk. We suggest that phosphorylation weakens the meta II-G_t interaction by either creating a highly negative surface charge, which results is an electrostatic repulsion between meta II and G_t, or that the negatively charged carboxyl terminus interacts with positively charged groups on the cytosolic loops of rhodopsin to sterically obscure the G_t binding site. Both of these mechanisms would reduce the probability of G_t binding and result in an increase in the K_D for the meta II$\cdot G_t$.

ACKNOWLEDGMENT: supported by NIH grant no. EY00548.

REFERENCES

Aton, B.R., Litman, B.J. & Jackson, M.L. (1984): Isolation and purification of the phosphorylated species of rhodopsin. *Biochemistry* 23, 1737-1741.

Bennett, N., Michel-Villaz, M. & Kuhn, H. (1982): Light-induced interaction between rhodopsin and the GTP-binding protein. Metarhodopsin II is the major photoproduct involved. *E. J. Biochem.* 127, 97-103.

Benovic, J.L., Strasser, R.H., Caron, M.G. & Lefkowitz, R.J. (1986): β-Adrenergic receptor kinase: identification of a novel protein kinase that phosphorylates the agonist-occupied form of the receptor. *Proc. Nat. Acad. Sci. U. S. A.* 83, 2797-2801.

Blazinsky, C. & Ostroy, S.E. (1984): Pathways in the hydrolysis of vertebrate rhodopsin. *Vis. Res.* 5, 459-470.

Emeis, D., Kuhn, H., Reichert, J. & Hofmann, K.P. (1982): Complex formation between metarhodopsin II and GTP-binding protein in bovine photoreceptor membranes leads to a shift of the photoproduct equilibrium. *FEBS Lett.* 143, 29-34.

Fung, B.K.K. & Stryer, L. (1980): Photolyzed rhodopsin catalyzes the exchange of GTP for bound GDP in retinal rod outer segments. *Proc. Nat. Acad. Sci., U.S.A.* 77, 2500-2504.

Jackson, M.L. & Litman, B.J. (1985): Rhodopsin-egg phosphatidylcholine reconstitution by an octylglucoside dilution procedure. *Biochim. Biophys. Acta* 812, 369-376.

Kibelbek, J., Mitchell, D.C., Beach, J.M. & Litman, B.J. (1991): Functional equivalence of metarhodopsin II and the G_t-activating form of photolyzed bovine rhodopsin. *Biochemistry* 30, 6761-6768.

Kwatra, M.M. & Hosey, M.M. (1986): Phosphorylation of the cardiac muscarinic receptor in intact chick heart and its regulation by a muscarinic agonist. *J. Biol. Chem.* 261, 12429-12432.

Miller, J. L., Fox, D. A. & Litman, B. J. (1986): Amplification of phosphodiesterase activation is greatly reduced by rhodopsin phosphorylation. *Biochemistry* 25, 4983-4988.

Palczewski, K. & Benovic, J.L. (1991): G-Protein coupled receptor kinases. *Tr. Biochem. Sci.* 16, 387-391.

Schleicher, A., Kuhn, H. & Hofmann, K.P. (1989): Kinetics, binding constant, and activation energy of the 48-kDa protein-rhodopsin complex by extra-metarhodopsin II. *Biochemistry* 28, 1770-1775.

Straume, M., Mitchell, D.C., Miller, J.L. & Litman, B.J. (1990): Interconversion of metarhodopsins I and II: a branched photointermediate decay model. *Biochemistry* 29, 9135-9142.

Stryer, L. (1991): Visual excitation and recovery. *J. Biol. Chem.* 266, 10711-10714.

Binding of Ca^{2+}, Mg^{2+} and Tb^{3+} to bovine retinal arrestin

Franz Bruckert and Claude Pfister

Laboratoire de Biophysique Moléculaire et Cellulaire. Département de Biologie Moléculaire et Structurale. Centre d'Etudes Nucléaires de Grenoble, BP 85X, 38041 Grenoble Cedex, France

Introduction

Arrestin (also known as S-antigen), is a 50 kD soluble protein which binds specifically to photoactivated and phosphorylated rhodopsin, and thus prevents photoactivated rhodopsin from activating continously transducin (Wilden et al. 1986). Analogues of arrestin seem to desensitize other G-protein coupled receptor systems (Lohse et al. 1992). In addition of its role in signal transduction, arrestin is the major antigen recognized by the antibodies secreted during autoimmune uveitis (Wracker et al. 1977), moreover, injection of arrestin induces all of the features of the disease (Pfister et al. 1985).

On the basis of titration of Ca^{2+} by arsenazo III and binding of radioactive Ca^{2+} to arrestin blotted on nitrocellulose, Huppertz et al. (1990) proposed that arrestin binds Ca^{2+}, one ion per protein molecule, with a affinity of 4 µM. This result was contradicted by Palczewski and Hargrave (1991), who found no Ca^{2+} binding to arrestin at physiological ionic strength, as determined by equilibrium dialysis, gel filtration and intrinsic fluorescence spectroscopy. Beside trivial explanations like contaminating membranes or proteins, the discrepancy may come from a very slow exchange rate of Ca^{2+} to arrestin, or from binding to another conformational state of arrestin, which is not present in every preparation.

Arrestin preparation, purification and intrinsic fluorescence: evidence of heterogeneity of the protein.

Crude arrestin was prepared from frozen bovine retinae as in Dorey et al. (1982), and further purified by two anion exchange chromatographies (Pharmacia Mono Q column: Weyand and Kühn 1990). Arrestin was then exchanged for the final solution buffer (20 mM Acetate, Mes, Hepes or Tris, depending on the pH desired) and salt (100 mM KCl) by gel filtration (Pharmacia PD-10 column) and concentrated by ultrafiltration (Millipore Millisep 30kD). Arrestin solutions were used within one week and stored at 4°C with 1 mM NaN_3. On isoelectric focusing gel, arrestin exhibited the microheterogeneity described by Weyand and Kühn (1990).

Bovine arrestin contains 14 tyrosines and 1 tryptophan (Shinohara et al. 1987). Figure 1 shows typical fluorescence spectra of arrestin in 0.1 M KCl at pH 7.0, after *one* Mono Q chromatographis step. Upon 275 nm excitation, the fluorescence spectrum is dominated by the emission of tyrosines (304 nm), but extends significantly beyond 360 nm, which indicates a contribution of the tryptophan to the fluorescence. Ecitation at 295 nm gives rise to a small but measurable peak centered at 334 nm, which confirms the existence of a tryptophan emission, 15 times weaker than that of tyrosines. This confirms that tryptophan fluorescence is quenched in the native protein (Kotake et al. 1991).

At the *second* Mono Q chromatographic step, the major peak (containing arrestin) displayed the same spectral characteristics as above. Moeover, the shape of the spectra were identical for the fractions corresponding to the leading edge or to the trailing edge of the peak. This excluded contamination by low amounts of residual proteins as origin of the tryptophan fluorescence. Yet, beside the major arrrestin peak, this second chromatography separated another peak, which contained a protein which was of same molecular weight as arrestin and was recognized by monoclonal antibodies prepared against arrestin. Sugar analysis in the major and the minor peaks revealed however that the minor peak contained at least 10 times more sugars (mannose and glucose) per mg of protein than the main arrestin peak (3 mannoses, 2 glucoses and 1 xylose per arrestin, and traces of galactosamine). Moreover, the fluoresence spectra of this species differed from that of arrestin. We suggest that this peak contains a more glycosylated subpopulation of arrestin, with a different conformation.

Kotake et al. (1991) showed that thermal denaturation of arrestin increased 5 fold the fluorescence of tryptophan and decreased by 30 % the fluorescence of tyrosines. Moreover, we observed that decreasing the pH of the solution to pH 4 induces the same fluorescence changes. In contrast, the fluorescence spectra were independant of the pH between 6.0 and 8.6, indicating that no major conformational change of arrestin occurs in this pH range.

Neither Ca^+, nor EGTA induce fluorescence changes on native arrestin:

Fluorescence emitted upon excitation of the sample at 280 nm was emission-filtered in order to measure mainly tryptophan emission (low-pass 335 nm filter; relative sensitivity 5.10^{-3}) or mainly tyrosine emission (low-pass 395 nm filter; relative sensitivity 10^{-3}). We were unable to detect any fluorescence change induced by addition of EGTA or Ca^{2+}, for free Ca^{2+} concentrations ranging from 10 nM to 1 mM, and EGTA up to 1 mM. Calcium binding, if any, would thus probably neither occur in the vicinity of the tyrosines and tryptophan chromophores, nor induce a large conformational change in these regions.

Terbium strongly influences the intrinsic fluorescence of arrestin:

Addition of 200 μM Tb^{3+} to a pH 7.0 arrestin solution increased the tryptophan fluorescence 2.4 fold and decreased tyrosines fluorescence by 20 % (figure 1). This effect saturated with Tb^{3+} and was completely reversed by EGTA. No precipitation of arrestin occured at this Tb^{3+} concentration; on the contrary, precipitation of arrestin accompanied the fluorescence changes at pH 7.5 (Palzewski and Hargrave 1991).

Exciting the arrestin solution at 280 nm, we recordered simultaneously the tryptophan fluorescence emitted (low-pass 335 nm filter) and the Rayleigh light scattered (inferference 280 nm filter). The increases in fluorescence and light scattering induced by 200 μM Tb^{3+} in four arrestin aliquots (pH 6.8, 7.0, 7.2 and 7.5) are shown on Fig.2. Precipitation of arrestin occured only at pH above 7.2 (small increases in light scattering at pH 7.0 were due to precipitation of Tb^{3+} with OH^- and N_3^-), whereas the fluorescence increase could be detected even at pH 6.8. The fluorescence changes induced by Tb^{3+} are thus distinct from the precipitation of arrestin. The amplitude of the fluorescence increase (Fig.2) as well as the kinetics of the fluorescence changes (not shown) depended on pH, but whatever the pH, the maximum fluorescence increase was 50 % of initial fluorescence level.

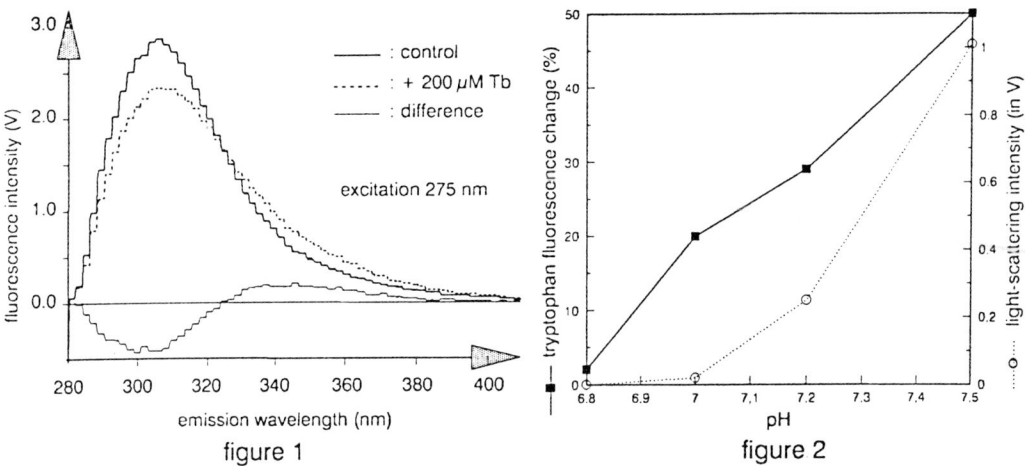

figure 1   figure 2

The fluorescence changes reveal that Tb^{3+} binds to arrestin and induces a reversible conformational change. The pH-dependence of the binding suggests two mechanisms of interaction between Tb^{3+} and arrestin: (i) Tb^{3+} binds to histidine or cysteine residues, whose side-chain pK are 6.5 and 8.5 respectively, or (ii) a $Tb(OH)_x$ complex, rather than Tb^{3+} alone, binds to arrestin. Although Tb^{3+} is known to bind at Ca^{2+} and Mg^{2+} binding sites on proteins, binding could occur to various other sites (Walters and Johnson 1990), thus competition experiments are needed to unravel the complexity of terbium-protein interactions.

Competition between Tb^{3+} and Ca^{2+} or Mg^{2+}:

We studied Ca^{2+} or Mg^{2+} competition with Tb^{2+} at pH 7.0 or 7.5. Addition of Ca^{2+} or Mg^{2+} *after* Tb^{3+} addition never reversed the fluorescence change induced by Tb^{3+}. However, we observed that preincubation with 80 µM Ca^{2+} or Mg^{2+} reduced by 50 % the amplitude of the fluorescence changes and of the precipitation induced by Tb^{3+} at pH 7.5, but we could not reproduce this effect at pH 7.0. This is currently under investigation, but it seems as yet unlikely that Tb^{3+} binds to an high-affinity Ca^{2+} site on arrestin.

References:

Dorey C. Cozette J. and Faure JP. 1982: A simple and rapid method for isolation of retinal S-antigen. *Ophtalmic Res. 14,249-255.*

Pfister C. Chabre M. Plouet J. Tuyen V.V. De Kozak Y. Faure J.P. and Kühn H. 1985: Retinal S-antigen identified as the 48K protein regulating light dependent phosphodiesterase in rods. *Science 228, 891-893.*

Huppertz B. Weyand I. and Bauer P.J. 1990:*J.B.C. 265,9470-9475.*

Kotake S. Hey P. Mirmira R.G. and Copeland R.A. 1991: Physicochemical characterization of bovine retinal arrestin. *Arch. Biochem. Biophys.: 285, 126-133.*

Lohse M.J., Andexinger S. Pitcher J. Trukawinski S. Codina J. Faure J.P. Caron M. and Lefkowitz R.J. 1992: Receptor specific desensitization with purified proteins. *J.B.C. 267, 8558-8564.*

Palczewski K.and Hargrave P.A. 1991: Studies of ligand binding to arrestin. *J.B.C.: 266, 4201-4206.*

Shinohara T., Dietzschold B., Craft C.M. Wistow G. Early J.J. Donoso L.A. Horwitz J. and Tao R. 1987: Primary and secondary structure of bovine retinal S antigen (48 kD protein). *Proc. Natl. Acad. Sci USA: 84, 6975-6979.*

J.D.Walters and J.D.Johnson 1990: Terbium as a luminescent probe of metal-binding sites in protein kinase C. *J.B.C. 265: 4223-4226*

Weyand I. and Kühn H. 1990: Subspecies of arrestin from bovine retina. *Eur. J. Biochem. 193, 459-467.*

Wilden U. Hall S.W. and Kühn H. 1986: Phosphodiesterase activation by photoexcited rhodopsin is quenched when rhodopsin is phosphorylated and binds the intrinsic 48K protein of rods outer segments. *Proc. Natl. Acad. Sci. USA 83: 1174-1178.*

Wracker W.B., Donoso L.A., Kalsow C.M., Yankeelov J.A. and Organisciak D.T. 1977: Experimental allergic uveitis. Isolation,characterization and localization of a soluble uveitopathogenic antigen from bovine retina. *J. Immunology. 119, 1949-1958.*

The fast GTP hydrolysis by transducin bound to phosphodiesterase

Frédérique Pages, Philippe Deterre and Claude Pfister

Laboratoire de Biophysique Moléculaire et Cellulaire, Département de Biologie Moléculaire et Structurale, Centre d'Etudes Nucléaires, Grenoble, France

ABSTRACT

The physiological response of a vertebrate rod to an incident photon is created in some 100-200 msec and recovers in some additional 200-400 msec (Baylor et al, 84). While the activation of the cGMP cascade is fast enough to account for the rise time of the electrical response (Vuong et al, 84), it still exists a discrepancy between the kinetics of recovery and most of the reported measurements on GTPase activity of transducin, which is one of the turn-off events. We demonstrate here that activated transducin (loaded with GTP) when binding to phosphodiesterase to activate it, hydrolyses its GTP with a faster rate than transducin not bound to phosphodiesterase.

INTRODUCTION

The physiological response of a vertebrate rod to an incident photon corresponds to an hyperpolarisation of the plasma membrane due to the closure of the cGMP-gated channels after cGMP concentration decreases in the cell. This involves a cascade of events, in wich transducin plays a centrale role: the α-subunits of transducin molecules (Tα) are induced by photoexcited rhodopsin (R*) to exchange GDP for GTP, dissociate from Tβγ and diffuse away to activate the cGMP phosphodiesterase (PDE) by interacting with PDE inhibitory subunits (PDEγ)(Chabre and Deterre, 89), and forming a membrane bound Tα-PDE complex (Clerc and Bennett, 92). The return to basal membrane conductance requires the replenishment of cGMP in the cell i.e. desactivation of PDE and activation of a cGMP-specific cyclase. Desactivation of PDE involves the deactivation of all the protagonists of the cascade: phosphorylation and blocking of R* by a specific kinase and arrestin (Wilden et al, 88), hydrolysis of the γ phosphate of the GTP bound to Tα: the affinity of TαGDP for PDEγ is then lowered while that for Tβγ is recovered. Finally, cGMP is synthesized by a cGMP cyclase activated by a Ca^{++} binding protein, recoverin.
The kinetic relevance of the deactivation of Tα by hydrolysis of its GTP has for long been a matter of debate: in vitro biochemical determination of GTP consumption by T in bleached rod outer segments (ROS) suspensions, or in reconstituted systems (with disc membranes and purified T), or by T in the absence of membranes, have given varying GTP turn-over numbers all of them

electrophysiological signal. Moreover they are difficult to compare because of the large differences in the conditions used, particularly in GTP and membranes concentration.

In the process of measuring GTPase activity by steady-state Pi formation, T undergoes several cycles of activation-deactivation, and it has been hypothesized that the usually slow observed GTPase rate would reflect a rate-limiting step in the cycle, as diffusion-limited collisions between $T\alpha.GDP-T\beta\Gamma$ and R* or between $T\alpha.GDP$ and $T\beta\Gamma$, or as membrane binding of $T\alpha.GDP-T\beta\Gamma$ (Chabre et al, 90). Moreover, GTP hydrolysis stricto sensu, i.e. formation of GDP.Pi inside the nucleotide site could be more rapid than the subsequent release of Pi in the cytosol (Ting and Ho, 91). Time-resolved microcalorimetric experiments, measuring the heat related to hydrolysis of GTP or cGMP in ROS suspensions, report GTPase times of less than 1 sec at 22°C (Vuong and Chabre, 90; Vuong and Chabre, 91).

We bring here a direct experimental proof that the GTPase activity of $T\alpha$ is higher when it is in the $T\alpha$-PDE complex. This provides the major argument to explain the descrepancies between the classical steady-state assays on ROS suspensions or in reconstituted systems and physiological or microcalorimetric experiments: actually, as T is at about a ten-fold excess over PDE, only part of $T\alpha.GTP$ gets involved in the activation of PDE (formation of a $T\alpha$-PDE complex) and this subpool of T is indeed the selectively observed species in physiological experiments; the main part of $T\alpha.GTP$ remains as an isolated species and essentially contributes to values observed by steady-state assays.

RESULTS

We have performed GTPase measurements on concentrated ROS membranes. These preparations had a significant dark GTPase activity, probably mainly reflecting the existence of GTPases other than transducin, or of ATPases of low specificity, or of guanylate cyclase, which are essentially light-independent processes, the transducin-related (light-dependent) GTPase activity (T-GTPase) has been calculated from the difference between light and dark activity. A typical result of a GTPase assay is presented in fig. 1A. The dark, light and T-GTPase activities increased with increasing GTP concentration (fig. 1B). We have after kept the standard GTP concentration in our assays at 1 mM, for which the observed activities reached a plateau and which is moreover the physiological range of nucleotide concentration in ROS.

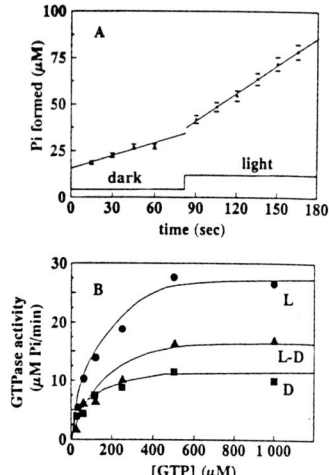

Fig. 1: The light-triggered GTPase activity of transducin in a ROS suspension and its dependance on GTP concentration

A: A dark-kept suspension of ROS membranes purified from fresh bovine retinas (5mg rhodopsin/ml in Hepes 20 mM, NaCl 120 mM, $MgCl_2$ 2 mM, DTT 2 mM, total volume 0.2 ml), was supplemented at time zero with an equal volume of GTP medium (2 mM in the same buffer). Temperature: 25°C. Every 15 sec, 30µl aliquots were taken and the reaction stopped by quenching in PCA medium for further Pi analysis by molybdate precipitation. After 75 sec a bright orange light was turned on (bleaching extent ca. 30% within 5 sec) and aliquots were again taken and treated as above. The results are expressed as the amounts of Pi formed as a function of time. Solid lines: linear regressions on the experimental points. Dark activity: 14 µM Pi/min, light activity: 29 µM Pi/min, T-GTPase activity (light activity - dark activity): 15 µM Pi/min

B: Dark (D), light (L), and transducin activity (L-D) measured in the same conditions as in A, as a function of GTP concentration from 30 µM to 1 mM. Solid lines: hand fitting of experimental data

Addition to a ROS suspension of exogenous purified PDEαβΓ$_2$ (PDE$_2$nat) (Deterre et al, 86) increased the amount of Tα involved in Tα-PDE complexes, as monitored in the experiment of fig. 2B, by an increasing in membrane attachment of Tα. In the GTPase assay (fig. 2A), such addition resulted in increased T-GTPase, with little or no influence on the dark activity. This effect was not reproduced by addition of BSA and most interestingly a slighty trypsinized PDE (PDE$_2$tr) (Catty and Deterre, 91) which still possesses its two inhibitory subunits, but is unable to interact with the membrane does not induce an enhancement of T-GTPase activity

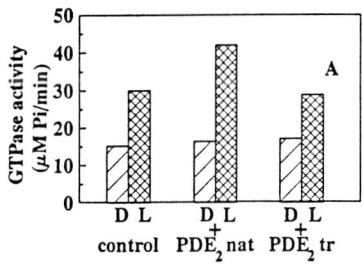
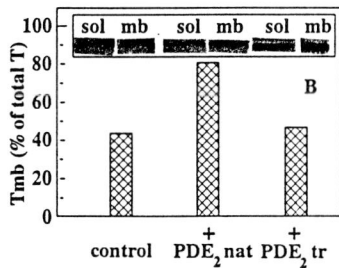

Fig. 2: Effect of PDEαβΓ$_2$ as increasing the GTPase activity of Tα: formation of a Tα-PDE membrane-bound complex.
Aliquots of ROS suspension were supplemented either with buffer (control), or with native inactive PDEαβΓ$_2$ (PDE$_2$nat), or with slightly trypsinized PDE (PDE$_2$tr).
A: GTPase activities measured as in fig. 1A, in the dark (D) or after illumination (L).
B: SDS-PAGE analysis of the soluble and membrane-bound fractions in the same suspension after illumination (sol. and mb. respectively) and the proportion of membrane-bound Tα deduced by scanning the T bands.
The given experiment corresponds to the following final concentrations: membranes 2.5 mg rhodopsin/ml; PDE$_2$nat added: 1.6 µM; PDE$_2$tr added: 1.6 µM

In a series of experiments, the amount of added PDEαβΓ$_2$ was varied, and the correlation was studied between the T-GTPase specific activity and the amount of Tα-PDE complexes formed (fig. 3). The observed linear correlation is consistent with the existence of two populations of Tα: those Tα not involved in Tα-PDE complexes (0% Tα in Tα-PDE on fig. 3) with slow turn-over rate (8.3 sec/GTP/T), and those involved in complexes (100% Tα in Tα-PDE on fig. 3) with fast turn-over rate (1.4 sec/GTP/T). Binding of Tα to PDE results in a ca. 4-fold increase in its specific activity.

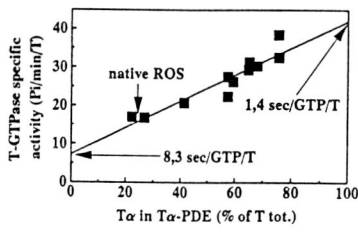

Fig. 3: Variation of T-GTPase specific activity as a function of the proportion of T involved in Tα-PDE complexes.
Activities were calculated as (L-D)/(T) from the difference between light and dark activities and from the total amount of T in the suspension (T). Protein concentration was determined after extraction on an aliquot of the ROS suspension by Coomassie blue staining, using BSA as standard; experimental values are corrected for the staining properties of the individual amino-acid in each protein. The proportion of T involved in Tα-PDE complexe was determined as in fig. 2 corrected for non specific attachment of Tα to the membrane which amounted to 20 % of soluble Tα (determined by an experiment on ROS membranes without any PDE, data not shown). Linear regression for the whole set of data is extrapolated for 0 and 100% of T involved in Tα-PDE complexes.

DISCUSSION

The retroaction on T of its binding to its effector PDE, which accelerates GTP hydrolysis, reminds of the effect of GAP protein (GTPase activating protein) on the "little" G-protein p21-ras. It would be interesting to know how PDE operates on T. It is now well documented that Tα interacts with the Γ subunits of PDE to activate it, but it has also been demonstrated that direct interactions between Tα and the αβ subunits of PDE also exist (Catty et al, 92; Clerc et al, 92). Preliminary results suggest that the enhancement of the T-GTPase activity we reported here is the consequence of the interaction between Tα and the catalytic subunits αβ of PDE and not a consequence of the interaction of Tα with PDEΓ.

REFERENCES

Baylor, D.A. Nunn, B.J. Schnapf, J.L. (1984): The photocurrent, noise and spectral sensitivity of rods of the monkey *Macaca fascicularis*. *J. Physiol.(London)* 357,575-607.

Catty, P. Deterre, P. (1991): Activation and solubilization of the retinal cGMP-specific phosphodiesterase by limited proteolysis. *Eur. J. Biochem.* 199, 263-269.

Catty, P. Pfister, C. Bruckert, F. Deterre, P. (1992): The cGMP phosphodiesterase-transducin complex of retinal rods. membrane binding and subunits interactions. *Submitted*.

Chabre, M. and Deterre, P. (1989): Molecular mechanism of visual transduction. *Eur. J. Biochem.* 179, 255-266.

Chabre, M. Deterre, P. Catty, P. Vuong, T.M. (1990): Regulation and rapid inactivation of the light induced cGMP phosphodiesterase activity in vertebrate retinal rods. In *Activation and desensitization of transducing pathways*, ed.T.M. Konijn, M.D. Houslay, P.J.M. Van Haastert (Springer-Verlag Berlin Heidelberg, 1990), vol H 44, pp.215-228.

Clerc, A. Bennett, N. (1992):Activated cGMP phosphodiesterase of retinal rods. *J. Biol. Chem.* 267, 6620-6627.

Clerc, A. Catty, P. Bennett, N. (1992): Interaction between cGMP phosphodiesterase and transducin α-subunit in retinal rods: a cross-linking study. *Submitted*.

Deterre, P. Bigay, J. Robert, M. Pfister, C. Kühn, H. Chabre, M. (1986): Activation of retinal rod cyclic GMP-phosphodiesterase by transducin: characterization of the complex formed by phosphodiesterase inhibitor and transducin α-subunit. *Proteines: structure, function, and genetics*. 1, 188-193.

Kühn, H. (1981): In *Current topics in membrane and transport*, ed W.H. Miller (Academic Press), pp.171-201

Ting, T.D. HO, Y.K. (1991): Molecular mechanism of GTP hydrolysis by bovine transducin: pre-steady-state kinetic analyses. *Biochemistry* 30, 8996-9006.

Vuong, T.M. Chabre, M. Stryer, L. (1984): Millisecond activation of transducin in the cyclic nucleotide cascade of vision. *Nature* 311, 659-661.

Vuong, T.M. Chabre, M. (1990): Subsecond deactivation of transducin by endogenous GTP hydrolysis. *Nature* 346, 71-74

Vuong, T.M. Chabre, M. (1991): Deactivation kinetics of the transduction cascade of vision. *Proc. Natl. Acad. Sci.* 88, 9813-9817.

Wilden, U. Hall, S.W. Kühn, H. (1986): Phosphodiesterase activation by photoexcited rhodopsin is quenched when rhodopsin is phosphorylated and binds the intrinsic 48-kDa protein of rod outer segments. *Proc. Natl. Acad. Sci.* 83, 1174-1178.

The retinal phosphodiesterase-transducin complex : membrane binding, subunits interactions and activity

Patrice Catty, Claude Pfister, Franz Bruckert and Philippe Deterre

Laboratoire de Biophysique Moléculaire et Cellulaire, Département de Biologie Moléculaire et Structurale, Centre d'Etudes Nucléaires, Grenoble, France

Abstract:

In vertebrate rod outer segments (ROS), the cGMP-specific phosphodiesterase (PDE$\alpha\beta\gamma_2$) is peripherally bound to the ROS discs membranes. In the presence of its activator, the α-subunit of transducin loaded with GTP (Tα^*), PDE$\alpha\beta\gamma_2$ displays a greatly enhanced membrane binding, due to the formation of an indissociated complex between Tα^* and PDE$\alpha\beta\gamma_2$. This was observed as well with purified PDE$\alpha\beta\gamma$ and with PDE$\alpha\beta$ species, which evidenced a direct link between Tα^* and the catalytic subunits PDE$\alpha\beta$.

The sequential interaction of two Tα^* per PDE$\alpha\beta\gamma_2$ as well as the membrane attachment of the formed complexes allow to reach maximal activation of PDE. We found that the first Tα^* interacts with PDE$\alpha\beta\gamma_2$ with a high affinity, and activates it to about 50-60%; the second Tα^* shows a much lower affinity, and allows complete maximal activation of PDE.

Introduction:

In retinal rod outer segments (ROS), light triggers a cascade of events, inducing a rapid hydrolysis of cGMP by a specific phosphodiesterase (PDE) which is activated by the α-subunit of Transducin loaded with GTP, Tα^* (Chabre et Deterre, 1989). In its inactive state, PDE is an heterotetramer, containing a catalytic core (PDE$\alpha\beta$) and two (identical) inhibitory subunits PDEγ (Deterre et al., 1988). Activation by Tα^* involves a direct interaction between Tα^* and PDEγ (Deterre et al. 1988). Maximal activation of PDE requires 2 Tα^* per PDE, and highly depends upon the existence and concentration of membranes (Bennett and Clerc, 1989; Clerc and Bennett, 1992). Tα^* is a soluble protein, but PDE is a peripheral protein, and we study here the mode of interaction of those proteins between themselves and with the membranes, and the relationship of those interactions with the obtained PDE activation.

Results:

Membrane attachment of PDE and of the Tα^*-PDE complexes: PDE$\alpha\beta\gamma_2$ is a peripherally bound protein, and can be detached from ROS membranes by dilution of the suspension in isotonic buffer (Fig. 1, open circles): half of the PDE content is released in the soluble fraction in a suspension at 0.1 mg rhodopsin (Rh) per ml. PDE$\alpha\beta\gamma$ and PDE$\alpha\beta$ behaved strictly identically (Fig. 1). This argues to the fact that PDEγ

has no role in the anchoring of PDE to the disc membranes, which is mainly related to the geranyl-geranylated and carboxymethylated C-terminal domain of the PDEβ subunit (Catty et Deterre 1992, Fung et al 1992).

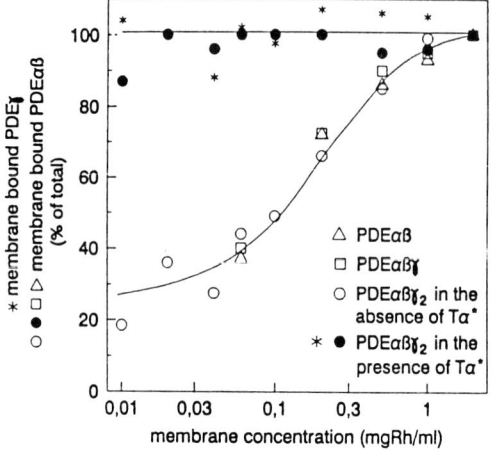

Fig. 1: Binding of PDE to ROS membranes, and effect of Tα*. A suspension of washed ROS membranes at 2 mgRh/ml in isotonic buffer (20 mM Tris pH 7.5, 120 mM KCl, 1 mM $MgCl_2$, 1 mM DTT, 0.1 mM PMSF) was supplemented with 500 nM of either PDEαβγ2 or PDEαβγ, or PDEαβ (obtained after activation of PDEαβγ2 by Tα*, here Tα loaded with GTPγS, and then FPLC-purified, as described in Deterre et al, 1986). The mixture was then diluted with the same buffer to the given membrane concentrations, in the absence (open symbols) or presence (closed circles and asterisks) of 3 mM Tα* (Tα loaded with GTPγS). The suspension was then centrifuged (the supernatant contained the soluble fraction of the proteins) and the pellet was resuspended in hypotonic buffer (5 mM Tris, pH 7.5) and centrifuged (the supernatant contained the membrane-bound proteins of the original mixture). The supernatants were analyzed by SDS-PAGE, proteins stained by Coomassie Blue and quantitized by scanning the gel. The amount of membrane-bound PDE in each condition was expressed as a percentage of the amount of membrane-bound PDE in the presence of 2 mgRh/ml of membranes.

The binding of PDEαβγ2 to membranes was noticeably enhanced in the presence of Tα*: with excess Tα* and at all the membrane concentrations tested, PDEαβγ2 was totally recovered in the membrane fraction (Fig. 1, compare closed and open circles) without release of its PDEγ subunits in the soluble fraction (Fig.1, asterisks). This argues to the fact that the membrane-bound complex formed between Tα* (at excess concentration) and PDEαβγ2, which contains 2 Tα* per PDE (Clerc and Bennett 1992), can be written as (Tα*)$_2$-PDEαβγ2.

In the presence of increasing amounts of Tα*, two transducin-PDE complex form sequentially: a first Tα* binds, and forms a (Tα*)-PDEαβγ2; then a second Tα* forms the final (Tα*)$_2$-PDEαβγ2 complex. An enhanced membrane binding of PDEαβγ2 is observed for both complexes (Fig. 2, circles).

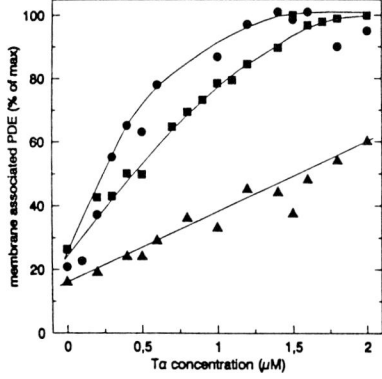

Fig. 2: Tα*-induced binding of PDEαβγ2, PDEαβγ and PDEαβ to ROS membranes as a function of Tα* concentration. The experimental procedures were as described in Fig.1, except for the concentration of Tα* which was varied as indicated. Circles: PDEαβγ2; squares: PDEαβγ ; triangles: PDEαβ

Figure 2 shows that a similar Tα*-induced enhancement in membrane binding of PDE is observed for PDEαβγ (squares), i.e. for the (Tα*)-PDEαβγ complex. Moreover and interestingly, Tα* was able to induce an enhanced binding of PDEαβ as well (Fig. 2, triangles), with however a reduced efficiency, and this gives experimental evidence that a direct interaction exists between Tα* and the catalytic subunits of PDE (PDEαβ).

Activation of PDE by Tα* in the presence of membranes:

The maximal activity of PDE (typically 2000 cGMP hydrolyzed/sec/PDE) can be obtained either with trypsin-activated PDE or with native PDE in the presence of excess Tα* (2-10 μM) and membranes (0.1- 2 mg Rh/ml). Bennett and Clerc (1989) reported, and we also observed, that activation of PDE by Tα* in the absence of membranes never exceeded 10-20% of this maximum.

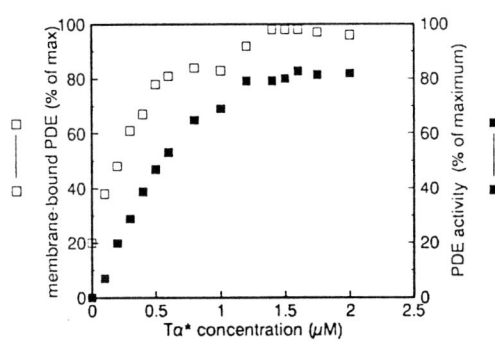

Fig. 3: Activation of PDEαβγ2 by Tα* in the presence of membranes at low concentration.

Activation curve (closed squares). Aliquots of a suspension at 0.1 mg Rh/ml and 25 nM PDEαβγ2 were supplemented with Tα* and with 2μM cGMP, incubated at 25°C for 30 sec, then quenched by addition of 20 volumes of cold PCA 10% and 500 μg/ml BSA and centrifuged; the supernatants were neutralized, centrifuged, and the supernatants were analyzed for nucleotide content by anion exchange chromatography onPI-SAX column. 100% activity refers to the values obtained above 2 μM Tα*.

Binding curve (open squares):aliquots of the same suspension (without cGMP) were analyzed as explained in Fig. 2.

At low membrane concentration (ca. 0.1 mg Rh/ml), the activation curve (Fig.3, filled squares) parallel the binding curve (Fig. 3, open squares), this merely reflecting the fact that PDE activity in the soluble fraction is negligible to that of the membrane-bound fraction (see above). Maximal binding is however attained before maximal activation (for example, at 1.5 μM Tα*, binding is over 90%, whereas activity is only 70%), meaning that an intermediate complex, logically (Tα*)-PDEαβγ2, is fully bound but not fully active. In the given experimental conditions, the concentration of Tα* always exceeds that of PDE, and the activation of PDE by Tα* is governed by the binding of the T-PDE complexes to the membranes, and its shape does not depend on the concentration of PDE (not shown).

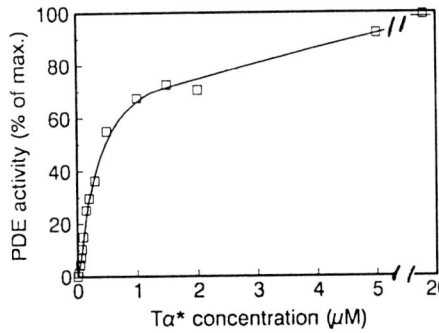

Fig. 4: Activation of PDEαβγ2 by Tα* in the presence of membranes at high concentration.

Aliquots of a ROS suspension at 2 mg Rh/ml and 350 nM PDEαβγ2 were supplemented with Tα* as indicated, and with 4 mM cGMP. cGMP hydrolysis was followed by the pH-metric determination of H^+ release (Liebman 1982)

In the presence of high membrane concentrations, for which isolated PDE is totally membrane-bound (ca. 2 mg Rh/ml, see Fig. 1), the activation curve as a function of the concentration of Tα* displays a biphasic shape (Fig.4): it first rises steeply up to ca. 50-60% activity, and then bends to a slow rising phase which extents above 5 μM Tα*. The concentration of Tα* at the bending region decrease when PDE concentration is lowered by depleting the membranes of PDE (by washes in isotonic buffer), and is roughly of the order of PDE concentration in the suspension (not shown). This behaviour can be interpreted as i) binding of one Tα* to PDE$\alpha\beta\gamma_2$ with high affinity, giving a (Tα*)-PDE$\alpha\beta\gamma_2$ complex with 50-60% activity (refered to maximum); ii) binding of a second Tα* to (Tα*)-PDE$\alpha\beta\gamma_2$, with a low affinity, and giving the final (Tα*)$_2$-PDE$\alpha\beta\gamma_2$ complex in which PDE reaches its maximal activity.

Conclusion

The attachement to the membrane of the complexes formed between Tα* and PDE is essential for the complete activation of PDE. In this attachement, the PDEγ subunits play no direct role. The interactions of the two PDEγ subunits with two Tα* appear to have clearly different characteristics: the binding of the first Tα* would be the most efficient, as it has a high affinity and elicits at least 50% of PDE activity. This would ensure *in vivo* an efficient activation of PDE at low levels of illumination. The binding of the second Tα* would complete the maximal activation of PDE only at very high levels of Tα* formation.

References

Bennett, N. and Clerc, A. (1989). Activation of cGMP phosphodiesterase in retinal rods: mechanism of interaction with the GTP-dependent protein (transducin). *Biochem.*

Catty, P. and Deterre, P. (1991). Activation and solubilisation of the retinal cGMP phosphodiesterase by limited proteolysis. Role of the C-terminal domain of the β-subunit. *Eur. J. Biochem.* 199, 263-269.

Chabre, M., and Deterre, P.(1989). Molecular mechanisms of visual transduction. *Eur. J. Biochem.* 179, 255-266.

Clerc, A. and Bennett, N. (1992). Activated cGMP phosphodiesterase of retinal rods. *J. Biol. Chem.* 267, 6620-6627

Deterre, P., Bigay, J., Robert, M., Pfister, C., Kühn, H., and Chabre, M. (1986). Activation of retinal cGMP phosphodiesterase by transducin: characterization of the complex formed by phosphodiesterase inhibitor and transducin α-subunit. *Proteins, structure, function and genetics* 1, 188-193.

Deterre, P., Bigay, J., Forquet, F., Robert, M. and Chabre, M. (1988). cGMP phosphodiesterase of retinal rods is regulated by two inhibitorysubunits. *P.N.A.S.* 85, 2424-2428.

Anant, J.S., Ong, O.C., Xie, H., Clarke, S., O'Brien, P.J., and Fung B.K-K., (1992). *In vivo* differential prenylation of retina cyclic GMP phosphodiesterase catalytic subunits. *J. Biol. Chem.* 267, 687-690.

Liebman, P. and Evanczuck (1982). Real time assay of rod disk membrane phosphodiesterase and its controllers enzymes.

Recovery of the photoresponse in vertebrate photoreceptors : role of Ca^{2+}-dependent regulation of guanylate cyclase

Karl-Wilhelm Koch

Institut für Biologische Informationsverarbeitung, Forschungszentrum Jülich, Postfach 1913, W-5170 Jülich, Germany

Vertebrate photoreceptors respond to illumination with the suppression of the dark current that flows into a dark adapted cell through the cGMP-gated cation channels located in the plasma membrane. Light triggers the hydrolysis of cGMP via an enzyme cascade involving Rhodopsin, a G-protein (transducin) and a phosphodiesterase (PDE). The light-induced decrease of cGMP causes the closure of the cGMP-gated channels (reviews e.g. Pugh & Lamb 1990; Stryer 1986). The light response of a vertebrate photoreceptor is composed of a *rising phase* and a *falling phase* (recovery). The biochemical reactions leading to the closure of the channels account for the *rising phase* of the photoresponse. The *falling phase* or the *recovery* to the dark state mirrors the reopening of cGMP-gated channels. Reactions that lead to the reopening include those that terminate the enzyme reactions of the cGMP cascade and those that catalyze the resynthesis of cGMP. Concomitant with the decrease of cGMP after illumination is a decrease of the cytoplasmic calcium concentration. This change in the cytoplasmic calcium concentration is a necessary step in restoring the cGMP level after illumination and in regulating the light sensitivity (light adaptation) of a photoreceptor cell (reviews e.g. Fain & Matthews 1990; Kaupp & Koch 1992). This paper discusses some biochemical aspects concerning the recovery of the photoresponse particularly the role of photoreceptor guanylate cyclase.

TERMINATION OF THE cGMP ENZYME CASCADE

The termination of the cGMP hydrolysis pathway requires the shutoff of all involved activated enzymes. Light-activated rhodopsin is inactivated by multiple phosphorylation at serine and threonine residues by a rhodopsin kinase. Binding of arrestin (48 K protein) to phosphorylated rhodopsin occurs in competition to the binding of transducin. This mechanism prevents further activation of transducin molecules (Palczewski & Benovic 1991). Another critical step in turning off the cGMP cascade is the deactivation of transducin and PDE. Studies done on isolated physiologically intact salamander rods provide evidence that PDE activity declines with a time constant of 1.5-2 sec when activated with moderate light flashes (Hodgkin & Nunn 1988). In a similar approach on isolated salamander rods Cobbs (1991) determined the apparent lifetime of light-activated PDE to be 0.9 sec. GTP hydrolysis by transducin is needed for the deactivation of the PDE (review Pugh & Lamb 1990). Thus, the observed lifetime of activated PDE sets a time frame for the GTPase rate of transducin. Recent determinations of a GTPase rate around 1 sec^{-1} seemed to fulfill this requirement (Vuong & Chabre 1991; Arshavsky et al. 1991).

SYNTHESIS OF cGMP BY RETINAL ROD GUANYLATE CYCLASE

In addition to the terminating reactions, the reopening of cGMP-gated channels depends on the replenishment of the cGMP pool by the activity of guanylate cyclase (GC). Photoreceptor GC is a membrane bound enzyme with a molecular weight of 112 kDa in bovine (Koch 1991) and of a 110/115 kDa double band in amphibian rod outer segments (ROS) (Hayashi & Yamazaki 1991). The basal activity of GC supplies a bovine photoreceptor cell with cGMP at a continous rate of 2-10 μM/sec. This rate is accelerated about 10-fold when the intracellular calcium level decreases below the dark value of 300 nM (Koch & Stryer 1988). The decrease of calcium is sensed by a small Ca^{2+}-binding protein, called recoverin or p26 (Dizhoor et al. 1991; Lambrecht & Koch 1991a). Recoverin is necessary for the activation of GC at decreased calcium levels. Halfmaximal activation occurs between 100 and 240 nM free calcium and is highly cooperative. The amino acid sequence exhibits three calcium binding motifs. Ca^{2+}-binding was directly monitored by $^{45}Ca^{2+}$-autoradiography and tryptophan fluorescence (Lambrecht & Koch 1991a; Dizhoor et al. 1991).

Recoverin does not act like an inhibitory subunit as for example the γ-subunits of the PDE. Removing recoverin from a suspension of ROS leads to a preparation of ROS membranes containing only the basal GC activity. Recoverin has to be reconstituted with these washed ROS membranes in order to increase the GC activity at low calcium (Lambrecht & Koch 1991a; Dizhoor et al. 1991).

Recoverin has little sequence homology to calmodulin. It also shows a remarkable difference to calmodulin when activating its target enzyme (GC): *It increases the activity of its target enzyme when the calcium concentration is low* (< 100 nM free Ca^{2+}).

An additional regulatory mechanism for the cGMP synthesis seemed to be the phosphorylation of recoverin. Phosphate is incorporated into recoverin when the free calcium concentration is below 200 nM (Lambrecht & Koch 1991b). The phosphorylation by an unknown kinase does not depend on illumination of the ROS membranes.

INTERACTION OF GUANYLATE CYCLASE WITH RECOVERIN AND OTHER PROTEINS

A quantitative estimation of the recoverin and GC content revealed that both are present in ROS at nearly equal amounts at a ratio to rhodopsin of ~ 1:100 (Lambrecht & Koch 1991a; Koch 1991). Assuming a 1:1 complex the high cooperativity would mainly result from the binding/unbinding of several Ca^{2+}-ions, but *in vivo* the interaction may be more complex. For example, polymorphism is described for vertebrate rod GC and it was suggested that the different variants may be regulated by multiple mechanisms (Hayashi & Yamazaki 1991). Furthermore, GC could be composed of several homooligomers. An indication that GC is either an oligomer or associated with large complexes came from gelfiltration experiments. Hakki and Sitaramayya (1991) reported that GC activities eluted shortly after the void volume when solubilized fractions of bovine ROS membrane proteins (rhodopsin, GC etc.) were chromatographed on a Biogel A column. Purified GC from toad ROS exhibits a similar elution on a Biogel A column as the bovine form (Hayashi & Yamazaki 1991). It was also suggested in earlier reports that GC is associated with cytoskeletal elements (Fleischman et al. 1980). Although GC has been solubilized and purified, it still remains unresolved, whether the enzyme exists as a homooligomer or whether it is tightly associated with other proteins. These interactions could cause its resistance to solubilization. A first experimental attempt at this problem is shown in Fig. 1. A fraction of ROS membrane proteins (including rhodopsin) is easily solubilized in detergents and can be removed from the GC by a selective extraction with low salt and 5 % Triton X-100 (Koch 1991). Under these conditions GC and other proteins remain insoluble. Sequential solubilization of GC is achieved with a buffer containing n-dodecyl-ß-D-maltoside and 1 M KCl. These extracts are devoid of rhodopsin but contain several proteins beside guanylate cyclase (112 kDa) as determined by a polyacrylamide gel electrophoresis. Application of this extract onto a gel filtration column (Fig. 1) revealed that the 112 kDa protein coelutes with several high molecular weight proteins (>170 kDa) shortly after the void volume. These bands have not been identified. However, they do not correlate with GC activities during further purification steps like anion exchange chromatography and GTP-agarose affinity chromatography and they are absent in purified GC preparations (Koch 1991). It remains an interesting possibility whether they are

part of a cytoskeletal system that interacts with cyclase similar to the spectrin like protein that is associated with the cGMP-gated channel protein (Molday et al. 1990).
The possible association of GC with other proteins points also to another open question, whether GC is uniformly distributed throughout the outer segment or restricted to either the plasma membrane or the disk membrane.

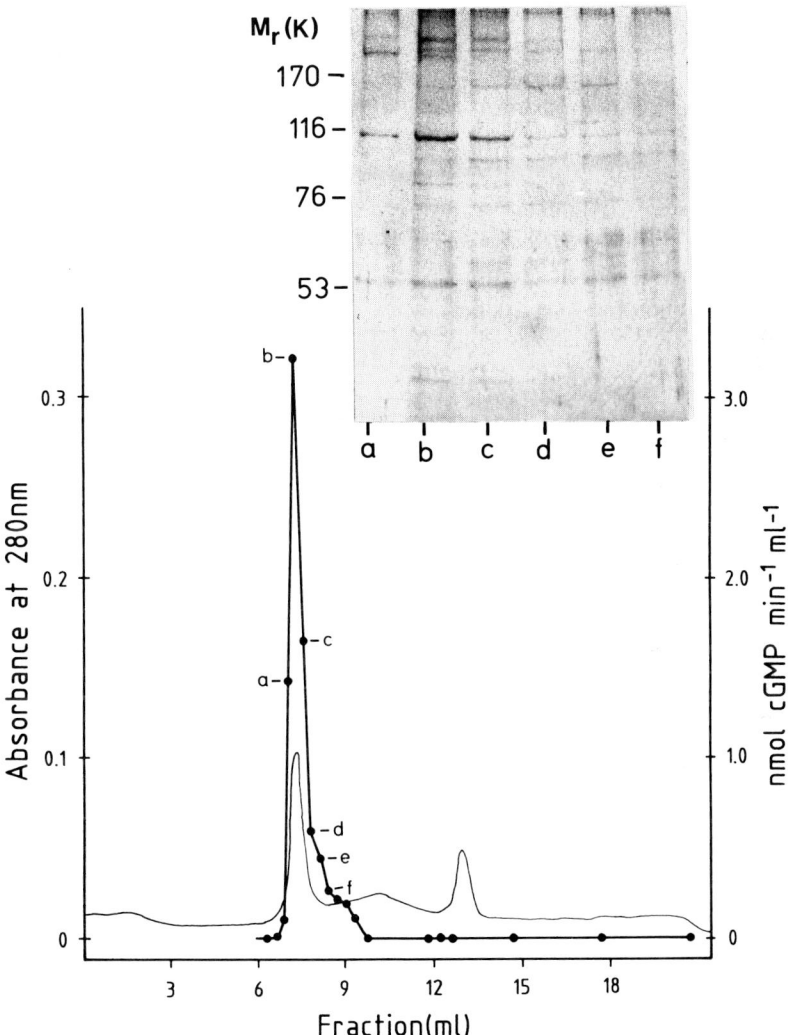

FIG.1: Gel filtration chromatography on a Superose 12 HR 10/30 column. ROS membrane proteins were solubilized and selectively extracted with buffers containing 5 % (v/v) Triton X-100 and 20 mM n-dodecyl-ß-D-maltoside. An extract containing GC in n-dodecyl-ß-D-maltoside buffer (Koch 1991) was applied to the column and fractionation was performed at a flow rate of 0.2 ml/min. *Lower part*: (-) absorbance at 280 nm; (-●-) guanylate cyclase activity expressed in nmol/min/ml. GC activity eluted shortly after the void volume (separation range of the column was 1-300 kDa). *Upper part*: Polyacrylamide gel electrophoresis of fractions containing the highest GC activities (a-f) obtained in the gel filtration chromatography.

IS THE SYNTHESIS OF cGMP DIRECTLY INFLUENCED BY LIGHT?

Pepe et al. (1986) reported that GC activity in disrupted ROS of the toad retina is increased at low calcium concentrations only after illumination. In contrast to this report, we observed that GC activity in bovine ROS strongly depends on the free calcium concentration but is independent of illumination (Koch & Stryer 1988). Rispoli and Detwiler (1992) described electrophysiological experiments in which the response recovery is accelerated even when the cytoplasmic calcium concentration is high. One possible explanation is that light directly affects GC activity. Bleaching of ROS membranes seems to influence at least the solubilization of GC (Horio & Murad 1991), although the interaction with other proteins appears to be stronger than with rhodopsin (s. above and Fig.1). Further studies on a reconstituted system with the purified components (GC, rhodopsin, recoverin etc.) is necessary to address these questions.

REFERENCES

Arshavski, V.Y., Gray-Keller, M.P. & Bownds, M.D. (1991): cGMP suppresses GTPase activity of a portion of transducin equimolar to phosphodiesterase in frog rod outer segments. *J. Biol. Chem.* 266, 18530-18537

Cobbs, W.H. (1991): Light and dark active phosphodiesterase regulation in Salamander rods. J. Gen. Physiol. 98:575-614

Dizhoor, A.M., Ray, S., Kumar, S., Niemi, G., Spencer, M., Brolley, D., Walsh, K.A., Philipov, P.P., Hurley, J.B. & Stryer, L. (1991): Recoverin: A calcium sensitive activator of retinal rod guanylate cyclase. *Science* 251, 915-918

Fain, G. & Matthews, H.R. (1990): Calcium and the mechanism of light adaptation in vertebrate photoreceptors. *Trends Neurosci.* 13, 378-384

Fleischman, D., Denisevich, M., Raveed, D. & Pannbacker, R.G. (1980). Association of guanylate cyclase with the axoneme of retinal rods. *Biochm. Biophys. Acta* 630, 176-186

Hakki, S. & Sitaramayya, A. (1990): Guanylate cyclase from bovine rod outer segments: solubilization, partial purification and regulation by inorganic pyrophosphate. *Biochemistry* 29, 1088-1094

Hayashi, F. & Yamazaki, A. (1991): Polymorphism in purified guanylate cyclase from vertebrate rod photoreceptors. *Proc. Natl. Acad. Sci. USA* 85, 94-98

Hodgkin, A. L. & Nunn, B.J. (1988): Control of light-sensitive current in salamander rods. *J.Physiol.* 403, 439-471

Horio, Y. & Murad, F., (1991). Solubilization of guanylyl cyclase from bovine rod outer segments and effects of lowering Ca^{2+} and nitro compounds. *J. Biol. Chem.* 266, 3411-3415

Kaupp, U.B. & Koch, K.-W. (1992): Role of cGMP and Ca^{2+} in vertebrate photoreceptor excitation and adaptation. *Ann. Rev. Physiol.* 54, 153-175

Koch, K.-W. (1991): Purification and identification of photoreceptor guanylate cyclase. *J. Biol. Chem.* 266, 8634-8637

Koch, K.-W. & Stryer, L. (1988): Highly cooperative feedback control of retinal rod guanylate cyclase by calcium ions. *Nature* 334, 64-66

Lambrecht, H.-G. & Koch, K.-W., (1991a): A 26 kd calcium binding protein from bovine rod outer segments as modulator of photoreceptor guanylate cyclase. *EMBO J.* 10, 793-798

Lambrecht, H.-G. & Koch, K.-W., (1991b): Phosphorylation of recoverin, the calcium sensitive activator of photoreceptor guanylate cyclase. *FEBS Lett.* 294, 207-209

Molday, L.L., Cook, N.J., Kaupp, U.B. & Molday, R.S. (1990): The cGMP-gated cation channel of bovine rod photoreceptor cells is associated with a 240-kDa protein exhibiting immunochemical cross-reactivity with spectrin. *J. Biol. Chem.* 265, 18690-18695

Palczewski, K. & Benovic, J.L. (1991): G-protein-coupled receptor kinases. *Trends Biochem. Sci.* 16, 387-391

Pepe, I.M., Panfoli, I. & Cugnoli, C. (1986): Guanylate cyclase in rod outer segments of the toad retina. *FEBS Lett.* 203, 73-76

Pugh, E.N. Jr. & Lamb, T.D. (1990): Cyclic GMP and calcium: The internal messengers of excitation and adaptation in vertebrate photoreceptors. *Vision Res.* 30, 1923-1948

Stryer, L. (1986): Cyclic GMP cascade of vision. *Ann. Rev. Neurosci.* 9, 87-119

Vuong, T.M. & Chabre, M. (1991): Deactivation kinetics of the transduction cascade of vision. *Proc. Natl. Acad. Sci. USA* 88, 9813-9817

Nitric oxide synthase activity in bovine retina

O. Goureau, M. Lepoivre*, F. Mascarelli and Y. Courtois

INSERM U 118, Unité de Recherches Gérontologiques affiliée au CNRS, 29 rue de Wilhem, 75016 Paris.
*URA 1116 CNRS, Bât. 432, Université de Paris-Sud, 91405 Orsay Cedex, France

Nitric oxide (NO) synthase is responsible for the synthesis of NO, a free radical, and L-citrulline from L-arginine (Moncada,1991). The existence of at least two different isoforms of NO synthase has been demonstrated by molecular cloning (Lyons, 1992; Bredt 1991). One form, is constitutively expressed and requires the presence of calcium and calmodulin in addition to the classical cofactors (NADPH, FAD and tetrahydrobiopterin) for its activation (Moncada, 1991). The second form, which is calcium/calmodulin independent, is induced by endotoxins and/or cytokines in several cell types including macrophages, neutrophils and some tumor cell lines of murine origin (Moncada, 1991).

The constitutive isoform of NO synthase has been previously identified in vascular endothelial cells (Palmer, 1987), where NO accounts for the biological properties of EDRF ("endothelium-derived-relaxing-factor"). A similar isoform has been described in some cell populations of the central nervous system (Bredt, 1991; Dawson, 1991). In these cells NO was described as the mediator of some amino acid neurotransmitters and can activate the soluble form of guanylate cyclase, however, its physiological role remains unknown. We have extended this system to a specific neural tissue, the retina, to assess the presence of the NO synthase. Three different types of bovine retinal cells, isolated photoreceptors, neural retina and pigmented epithelial cells, were used for this study.

RESULTS AND DISCUSSION

Retina fractionation (in order to separate neural retina and intact rod outer segments, and to collect retinal pigmented epithelial cells) was performed as previously described (Plouet,1987). Each preparation, rod outer segments (ROS), neural retinal cells without ROS and retinal pigmented epithelial cells (RPE), were sonicated and centrifuged at 100 000g. The supernatants (cytosolic preparations) were passed over column of Dowex AG 50W-X8 (Na^+ form) to remove endogenous arginine, and were then used to determine the activity of NO synthase by following the conversion of

L-[^{3}H]-arginine to L-[^{3}H]-citrulline, a reaction that stoichiometrically produces NO.

Incubation of ROS cytosols with 1.2μM L-[^{3}H]-arginine and 0.15mM NADPH for 10 minutes resulted in the synthesis of L-[^{3}H]-citrulline (Fig.1). In the presence of principle exogenous cofactors, namely calcium, calmodulin and tetrahydrobiopterin (BH4), in addition to NADPH, L-[^{3}H]-citrulline synthesis was increased two fold (9 pmol / min / mg of protein). The formation on radioactive citrulline was inhibited by the NO synthase inhibitor N-^{G}monomethyl-L-arginine (L-NMMA), demonstrating the presence of an NO synthase activity in the bovine ROS. The dependence of free calcium concentrations was also studied. When the incubations were performed in the presence of a potent calcium chelating agent, EGTA at 1mM, no radioactive citrulline could be detected (Fig.1), demonstrating that retinal NO synthase activity is calcium dependent.

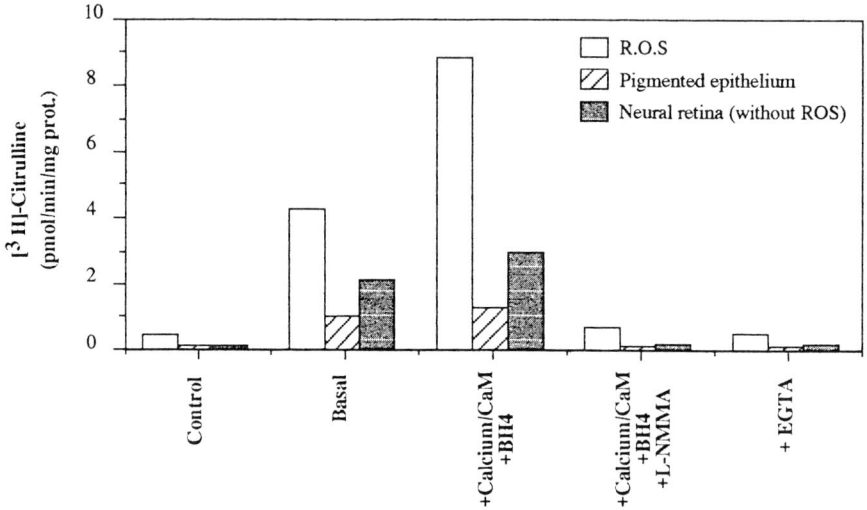

Figure 1: Constitutive NO synthase in the bovine retina
Calcium (1μM) ; CaM, calmodulin (0.2μM) ; BH4, tetrahydrobiopterin (0.1mM) ; L-NMMA (0.1mM) ; EGTA (0.1mM).

The same approach was used to examine neural retina homogenate, after ROS removal. This preparation contains different types of cells: rod inner segments, Müller cells, bipolar, horizontal, amacrine and ganglion cells, vascular cells (endothelium, pericytes) and some ROS, which are not removed. In cytosolic preparations of this homogenate, a small production of L-[^{3}H]-citrulline was observed (Fig.1), reflecting a weaker activity of NO synthase. This activity was also modulated by calcium (activation with calmodulin and inhibition with EGTA), and prevented by co-incubation with L-NMMA. The localization of NO synthase activity is difficult in view of the heterogeneity of the preparation : specific neurons, like amacrine cells (Dawson, 1991), could be involved, but the vascular system can not be excluded. Indeed, a model involving EDRF (NO), may play a role in the regulation of retinal arterial tone (Benedito, 1991). Furthermore, it is possible that the preparation is contaminated with small amount of ROS. In this case, a slight contamination by ROS NO synthase, could be responsible of citrulline production in neural retina preparation.

Concerning the third type of retinal cells, pigmented epithelial cells (RPE), a very weak NO synthase activity was detected (Fig.1), only 1-2 pmol per min per mg of protein. The same result was obtained in RPE cells in culture, confirming that NO synthase activity was associated with RPE cells, and not with ROS debris, which can be present in fresh RPE cells.

In view of these results, we have attempted to better characterize the NO synthase activity present in the bovine ROS, where the enzyme seems to be more active (or more expressed). Fig.2A shows the time course of L-[^{3}H]-citrulline formation in ROS cytosolic prepartions, in the presence of all cofactors. The apparition of radioactive citrulline, which reflects NO synthase activity, increased with time and reached a plateau after 10 at 15 minutes stimulation. Incubations of bovine ROS cytosolic preparations with increased L-arginine concentrations allowed us to determine the kinetic properties (Vm and Km values) of bovine ROS NO synthase for L-arginine (Fig. 2B). Vm = 33 pmol of citrulline / min / mg of protein and Km = 1.25 µM, which agree with values previously reported for the brain enzyme (Dawson, 1991).

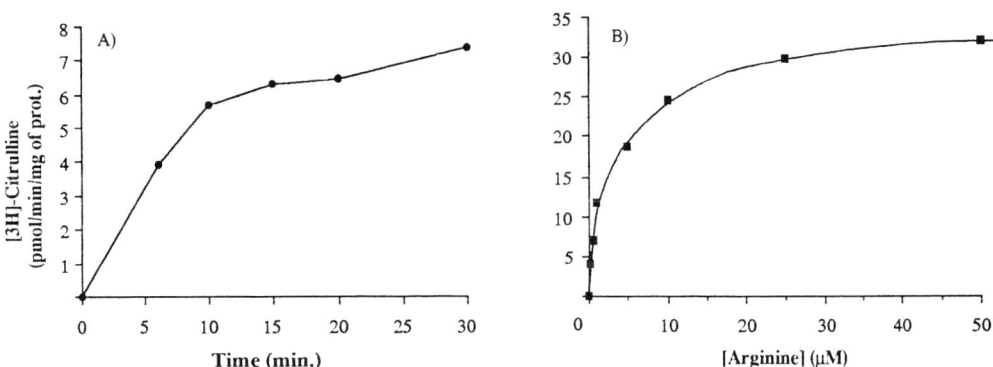

Figure 2 : A) Time course of [3H]-citrulline formation in ROS cytosolic extracts (L-arginine = 1.2 µM; proteins = 80 µg) B) Determination of Vm and Km values of bovine ROS - NO synthase for L-arginine (proteins = 80 µg ; 10 minutes). All cofactors (NADPH plus calcium, calmodulin and BH4) are present in each experiment.

Our studies demonstrate that the L-arginine / NO pathway is present in bovine retina, mainly in the photoreceptor outer segments (ROS). Furthermore, we have demonstrated that bovine ROS NO synthase is similar to the brain and endothelial enzymes in cofactor requirement and kinetic properties. However, the localization of the constituve NO synthase isoform in the retina is controversial; Snyder's group (Dawson, 1991) demonstrated by immunolocalization that the enzyme was present in the rat retina only in the amacrine cells and in some ganglions cells, whereas Moncada and his group (Venturini, 1992) found NO synthase activity in whole bovine retina and in ROS by direct measurement of NO formation. These results can be explained by the existence of distinct isoforms of the constitutive NO synthase, as suggested by Förstermann (1991).

Concerning the possible role of NO in the retina, we can postulate the involvement of the NO synthase in visual transduction or in the communication between photoreceptors and neighbouring cells, namely pigmented epithelial cells or Müller cells. In order to verify these hypotheses, the presence of potential activators of the NO synthase (excitatory aminoacids, like glutamate) and of potential targets of NO (differential proteins involve in visual transduction) is currently under investigation in the retina.

REFERENCES

Benedito, S. et al. (1991) : Role of the endothelium in acetylcholine-induced relaxation and spontaneous tone of bovine isolated retinal small arteries. Exp. Eye Res. 52, 575-579.

Bredt, D.S. et al. (1991) : Cloned and expressed nitric oxide synthase structurally resembles cytochrome P-450 reductase. Nature 351, 714-718.

Dawson, T.D. et al. (1991) : Nitric oxide synthase and neuronal NADPH diaphorase are identical in brain and peripheral tissues. Proc. Natl. Acad. Sci. USA 88, 7797-7801.

Förstermann et al. (1991) : Isoforms of Nitric oxide synthase : characterization and purification from different cell types. Biochem. Pharmacol. 42, 1849-1857.

Lyons, C.R. et al. (1992) : Molecular cloning and functional expression of an inducible nitric oxide synthase from a murine macrophage cell line. J. Biol. Chem. 207, 6370-6374.

Moncada, S. et al. (1991) : Nitric oxide: physiology, pathophysiology and pharmacology. Pharmacol. Rev. 43, 109-142.

Palmer, R.M.J. et al. (1987) : Nitric oxide release accounts for the biological activity of endothelium derived relaxing factor. Nature 327, 524-526.

Plouet, J. et al. (1988) : Regulation of eye derived growth factor binding to membranes by light, ATP or GTP in photoreceptor outer segments. EMBO J. 7, 373-376.

Venturini, C.M. et al. (1991) : Synthesis of nitric oxide in the bovine retina. Biochem. Biophys. Res. Comm. 180, 920-925.

A 23 kDa Ca^{2+}-binding putative cysteine protease in arthropod rhabdomeres

J.A. Clausen, L. Kelly*, M. Brown, E. O'Gara, A. Delaney and A.D. Blest

*Developmental Neurobiology Group, Research School of Biological Sciences, Australian National University, GPO Box 475, Canberra ACT 2601; *Department of Genetics, The University of Melbourne, Parkville Victoria 3052, Australia*

Rhabdomeral microvilli of some arthropods contain a labile cytoskeleton of one or more actin microfilaments linked to their plasmalemmae by side-arms (Blest *et al* 1982a,b; Arikawa *et al* 1990). Preservation of such microvillar cytoskeletons for electron microscopy requires pretreatments that either chelate Ca^{2+} or inactivate cysteine proteases by E-64 analogues (Blest *et al* 1982a,b). Although such pharmacological results imply that Ca^{2+}-dependent cysteine proteases may be implicated in the regulation of rhabdomeral intramicrovillar cytoskeletons, no direct evidence for their presence has so far been available.

A 23kDa Ca^{2+}-binding protein with cysteine protease activity has been isolated from whole *Drosophila* homogenates (Kelly, 1990 and unpublished results). Polyclonal antibodies show that it is present in gut, ovarian and neural tissues, but not in muscle.

Here we show that the *Drosophila* cysteine protease is present in photoreceptors of the Diptera *Calliphora* and *Musca* and a crab, *Leptograpsus*, using a library of monoclonal antibodies raised against the *Drosophila* cysteine protease.

METHODS

Monoclonal antibodies were raised to the Ca^{2+}-binding protein using standard techniques (Bundesen *et al.*, 1985). A library of eight monoclonal antibodies was used in subsequent work. Heat stable soluble protein extracts were isolated using whole fly homogenates of *Drosophila melanogaster* (Oregon R or Canton S wild type strains), *Calliphora auga* and *Musca domestica* and extracts of retinas dissected from compound eyes of the crayfish *Cherax destructor* and shore crabs *Leptograpsus variegatus*, following the method of Kelly (1990). Approximately 110 *Drosophila*, 13 *Calliphora*, 50 *Musca*, 48 crab retinas and 28 crayfish retinas were used to isolate the protein.

The proteins were subjected to SDS-PAGE (Laemmli, 1970) on either 12% or 15% gels. Western blots were performed on nitrocellulose membranes following the methods of Towbin *et al.* (1979). Visualisation of the antibodies was achieved using horse-radish peroxidase (HRP) labelled second antibody followed by enhanced chemiluminescence (Amersham) as the detection system. Variable amounts of total protein were loaded per sample, generally between 3 and 8 µg/track.

Thick frozen sections (0.5-1 µm) were cut from tissue lightly fixed in 4% paraformaldehyde, 0.25% gluteraldehyde, 0.01% saturated picric acid in 0.1M phosphate buffer, pH 7.4. Sections were incubated with antibodies to the Ca^{2+}-binding protein and visualised either using fluorescence (FITC-conjugated second antibody) or 5nm gold conjugated to the second antibody followed by silver intensification techniques.

RESULTS

One single band on Western blots at 18 kDa was labelled by the antibodies in homogenates from *Drosophila*, *Calliphora* and *Musca* as the source of antigens (Fig. 1a), although the *Drosophila* band ran slightly faster on the 15% gel, (not apparent on either 12% or gradient gels). The molecular weight of the Ca^{2+}-binding protein differed from that reported by Kelly (1990) and was most probably due to the use of dithiothreitol (DTT) as the reducing agent instead of 2-mercaptoethanol. Purified Ca^{2+}-binding protein (used as the antigen for preparation of the antibodies) also ran at 18 kDa under these conditions. However, from crab retinas a single band at 39 kDa bound the antibodies (Fig. 1b). Heat stable soluble protein extracts from crayfish retina antibodies bound to a single band at 42 kDa (Fig. 1b). The labelling of antibodies for both crab and crayfish protein extracts appeared weak on the western blots, most probably due to the level of protein present in the retinas. The Dipteran extracts gave a stronger label on western blots probably because whole flies were used as the source of homogenates. No protein bands at lower molecular weights bound to the antibodies in either crab or crayfish retina extracts.

The antibodies bound to both crab photoreceptor cells and the rhabdomeres (Fig. 2a) in the thick frozen sections. The degree of binding was not strong, in agreement with the weak band observed in the western blotting experiments. Retinal sections of both *Calliphora* and *Musca* bound antibodies to the photoreceptor cells, but it was difficult to determine if binding occurred in rhabdomeres (Fig. 2b, Calliphora results not shown). Further investigations at EM magnifications are required. *Drosophila* and crayfish retinal sections did not appear to bind the antibodies, but this was probably due to fixation problems.

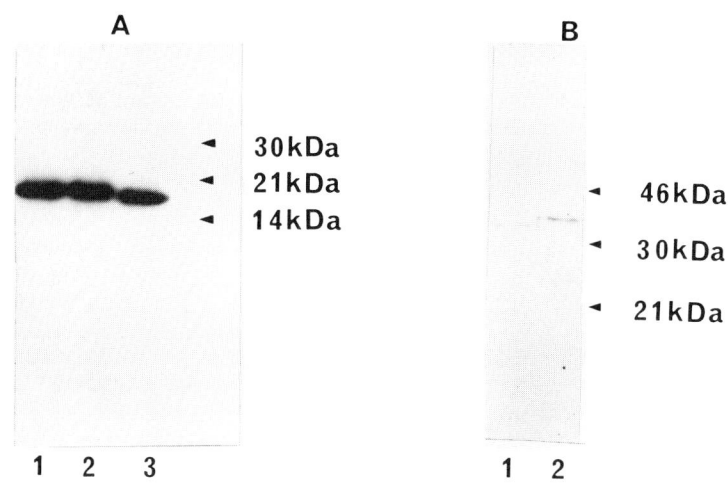

Fig. 1. Detection of the *Drosophila* Ca^{2+} binding protein using SDS-PAGE followed by western blotting. A. 15% SDS-PAGE, lane A1, *Calliphora* homogenates. Lane A2, *Musca* homogenates and lane A3, *Drosophila* homogenates. B. 12% SDS-PAGE, lane B1, crab retinal homogenates, lane B2, crayfish retinal homogenates. Total protein loaded per lane was; A1, 7μg; A2, 4μg; A3, 8μg; B1, 3μg and B2, 6μg.

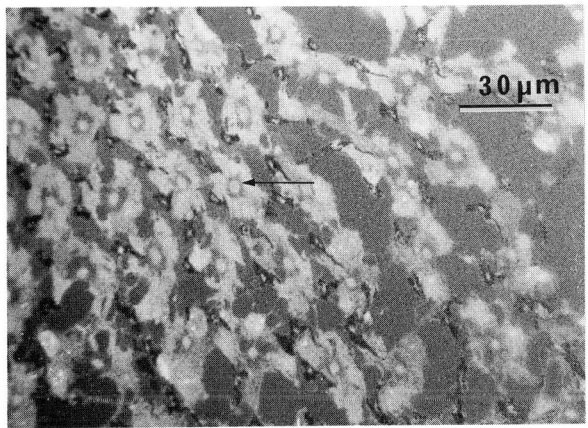

Fig. 2A. Immunofluorescence marking of crab retina by anti-*Drosophila* Ca^{2+}-binding protein antibodies. Antibodies labelled the rhabdomere (arrow) and associated photoreceptor cell cytosol.

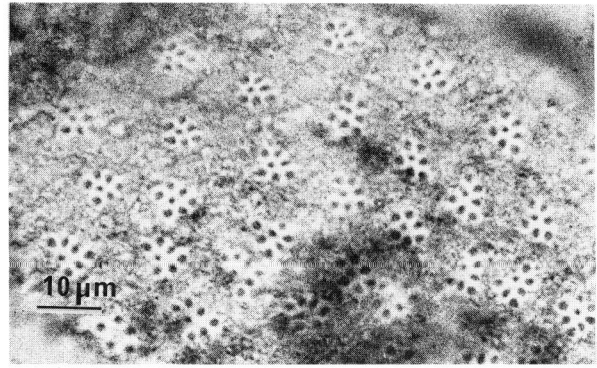

Fig. 2B. Immunogold followed by silver intensification of *Musca* retina by anti-*Drosophila* Ca^{2+}-binding protein antibodies. The cytosol and rhabdomeres of the photoreceptor cells were labelled.

CONCLUSIONS

Ca^{2+}-dependent cysteine proteases in the rhabdoms of arthropods were inferred from pharmacological evidence either by using specific cysteine protease inhibitors or Ca^{2+} chelation prior to fixation (Blest *et al*, 1982a,b), but there has been no direct evidence for their localisation.

Recently, a Ca^{2+}-activated protease was isolated from squid retinal tissues with the endogenous substrate shown to be squid rhodopsin (Oldenburg & Hubbell, 1990). The protease may allow rhodopsin to be freed from an underlying cytoskeleton to facilitate membrane turnover.

Drosophila rhabdomeres when fixed without prior incubation with specific cysteine protease inhibitors, or chelation of Ca^{2+}, show partial or complete degradation of the axial cytoskeleton and its numerous side-arms. It is tempting in light of the above results to suggest the role of the Ca^{2+}-dependent protease to be associated with axial microvillar filament breakdown, possibly acting through an actin binding protein rather than actin itself, as in human platelets (Markey et *al*, 1982). The role of the protease in the cytosol of the photoreceptors is far from clear at this stage.

REFERENCES

Arikawa, K., Hicks, J. L. & Williams, D. S. (1990): Identification of actin filaments in the rhabdomeral microvilli of *Drosophila* photoreceptors. *J. Cell Biol.* 110, 1993-1998.

Blest, A. D., Stowe, S. & Eddey, W. (1982a): A labile, Ca^{2+}-dependent cytoskeleton in rhabdomeral microvilli of blowflies. *Cell Tissue Res.* 223, 553-573.

Blest, A. D., Stowe, S., Eddey, W. & Williams, D. S. (1982b): The local deletion of a microvillar cytoskeleton from photoreceptors of tipulid flies during membrane turnover. *Proc. R. Soc. Lond. B.* 215, 469-479.

Bundesen, P. G., Wyatt, D. M., Cottis, L. E., Blake, A. S., Massingham, D. A., Fletcher, W. A., Street, G., Welch, J. S. & Rylatt, D. B. (1985): Monoclonal antibodies directed against *Brucella abortus* cell surface antigens. *Vet. Immunol. Immunopathol.* 8, 245-260.

Kelly, L. E. (1990): Purification and properties of a 23kDa Ca^{2+}-binding protein from *Drosophila melanogaster*. *Biochem. J.* 271, 661-666.

Laemmli, U. K. (1970): Cleavage of structural proteins during the assembly of the head of bacteriophage T_4. *Nature.* 227, 680-685.

Markey, F., Persson, T. & Lindberg, U. (1982): A 90 000-dalton actin-binding protein from platelets. Comparison with villin and plasma brevin. *Biochim. biophys. Acta* 709, 122-133.

Oldenburg, K. R. & Hubbell, W. L. (1990): Invertebrate rhodopsin cleavage by an endogenous calcium activated protease. *Exp. Eye Res.* 51, 463-472.

Towbin, H., Stehelin, T. & Gorden, J. (1979): Electrophoretic transfer of proteins from polyacrylamide gels to nitrocellulose sheets: procedure and some applications. *Proc. Natl. Acad. Sci. USA* 76, 4350-4354.

Light-activated channels in *Drosophila* are coded by the *trp* gene

Roger Hardie[1] and Baruch Minke[2]

[1] Department of Zoology, Cambridge University, Downing Street, Cambridge, UK, and [2] Department of Physiology, Hebrew University-Hadassah Medical School, Jerusalem, Israel

Excitation in invertebrate rhabdomeric photoreceptors is most likely mediated by the widely used inositol-lipid signalling cascade (reviews Ranganathan et al 1991; Pak, 1991; Nagy, 1991; Minke and Selinger 1992a,b). The trp (transient receptor potential) mutant of Drosophila (Cosens & Manning, 1969; Minke et al., 1975), and its counterpart, nss (no steady state) in Lucilia (Howard, 1984) are so-called because the light response decays to baseline during prolonged bright illumination. In addition, many manifestations of light adaptation, also appear to be blocked by this mutation (Minke et al., 1975; Howard, 1984; Minke & Selinger, 1992a). Since Ca^{2+} is known to be an intracellular messenger of adaptation (Lisman & Brown, 1972) and probably also excitation (Bolsover & Brown, 1985; Payne et al., 1986; Hardie, 1991a) in arthropod photoreceptors, it has been suggested that the mutation interferes with the normal rise in intracellular Ca^{2+} during light (Suss-Toby et al., 1991; Minke & Selinger, 1992a).

Recently it has proved possible to make voltage clamp measurements of the light-induced current (LIC) in Drosophila photoreceptors using the whole-cell recording technique on dissociated ommatidia (Hardie, 1991a,b; Ranganathan et al., 1991). To test if the trp gene product might in fact represent the light sensitive channels, the LIC has now been investigated in the trp mutant, and also in wild type flies in the presence of lanthanum which mimics many of the effects of the trp mutation (Hochstrate, 1989; Suss-Toby et al., 1991) and has been hypothesized to interfere with the trp gene product directly.

The trp phenotype can be mimicked by Ca^{2+} deprivation
Figure 1 shows responses to prolonged lights of varying intensities in cells clamped near resting potential (-50 mV). In wild type, dim light produces a maintained inward current. In the trp mutant the response to dim light looks nearly normal, however, at higher intensities the response decays completely to baseline, only recovering after ca. 1 minute in the dark. Figure 1c shows that prolonged exposure to nominally zero Ca_o (2.5 µM) combined with 5mM EGTA buffering of the internal pipette solution (free $[Ca_i]$ = 50 nM) can mimic the trp phenotype.

Evidence for two classes of light sensitive channels
If the trp gene codes for a subset of the light sensitive channels which are blocked by La^{3+} and if the opening of these channels is required for the steady-state of the receptor potential this may explain the trp phenotype. This in turn implies the existence of a second class of light sensitive channel which give the transient response in the trp mutant. Figure 2 shows that under certain conditions the LIC exhibits a biphasic reversal potential indicative of channels with different ionic selectivities as has also been reported in Limulus (Nagy, 1991). Under low extracellular Ca concentrations (0.5 mM), at a holding potential of around +10 mV, the LIC is initially outward, before becoming inward during the later phase of the response. This indicates that the early phase of the response is carried by channels with relatively low Ca^{2+} permeability (see Fig. 3), whilst highly Ca^{2+} permeable channels dominate the later phase - and also the plateau. At higher Ca^{2+} concentrations, the reversal potential appears to be unique as previously reported (Hardie, 1991a) and indicative of high Ca^{2+} permeability (Fig.

2). Apparently, the kinetics of the two classes of channel are now identical. In trp photoreceptors, the reversal potential appears to be unique with both low and high Ca_o (Fig. 2). With very bright flashes, there were indications of yet a third component to the responses (inset Fig. 2).

Ionic selectivity of the light sensitive channels

In wild type Drosophila photoreceptors the reversal potential (E_{rev}) of the LIC shows a strong dependence on extracellular Ca^{2+} (Ca_o) (Hardie, 1991; 1991a). Assuming constant field theory the Ca^{2+} dependent shift in reversal potential indicates that the channels are ca. 40x more permeable to Ca^{2+} than monovalent cations such as Na and Cs (Hardie, 1991a). Measurements performed in trp mutants (Fig. 3a, b) reveal a striking difference: E_{rev} is significantly more negative and its dependence upon Ca_o is much reduced. This clearly demonstrates that the ionic selectivity of the channels is altered in the mutant, and specifically that their permeability to Ca^{2+} ions is greatly reduced. Identical measurements and results were also obtained in wild type photoreceptors in the presence of 10 µM La^{3+} (Fig. 3a, b).

The role of the trp gene in receptor mediated Ca^{2+} entry.

Minke & Selinger (1992a) explained the decay of the light response in trp by assuming that a rise in $[Ca_i]$ is necessary for excitation, and that the limited amount of Ca^{2+} in the intracellular pools can only support excitation for a short time during intense light. Hardie and Minke (1992) provide direct evidence for an excitatory role of Ca^{2+} in Drosophila. Minke & Selinger's (1992a) hypothesis, which explains the trp phenotype, is based on more general models of receptor-mediated calcium entry in a variety of cells where the release of Ca^{2+} from the $InsP_3$-sensitive Ca^{2+} pools is followed by Ca^{2+} influx across the plasma membrane (Takemura et al., 1989; Berridge, 1990; Irvine, 1991). The mechanism underlying this co-ordinated process, which is important for efficient refilling of the pools, remain controversial (Berridge, 1990; Irvine, 1991) and represents one of the major unanswered questions in cellular Ca^{2+} homeostasis. The models postulate that a surface membrane Ca^{2+} channel (of unknown molecular identity) is linked to the endoplasmic reticulum via a direct interaction with the $InsP_3$ receptor, which has extensive sequence homologies with the ryanodine receptor (Furichi et al., 1989). How this hypothetical membrane channel is activated is a matter of intense speculation with both depletion of the luminal Ca^{2+} pool (Takemura et al., 1989; Berridge, 1990) binding of IP_4 or both having been suggested. Minke & Selinger (1992a) suggested that the trp gene product represented such a membrane channel or transporter. This suggestion is strongly supported by the recent study of Hardie and Minke (1992) and by a recent re-evaluation of the trp protein sequence (Phillips et al., 1992). Original reports of the primary structure (Montell & Rubin, 1989); revealed no homologies with known proteins. Phillips et al. (1992) however, have recently sequenced a trp homologue gene (trpl) from the Drosophila eye, and now report that both trp and trpl have significant homologies with vertebrate brain voltage sensitive calcium channels (di-hydropyridine receptors). A strong prediction of the model, namely that the trp protein is located close to the Ca^{2+} pools, thus allowing direct interaction with the $InsP_3$ receptor, has recently been confirmed by immunolocalization of the trp protein to the base of the microvilli (J.A. Pollock personal communication).

Together with these considerations, the results of Figs. 1-3 provide compelling evidence that the trp gene is a structural gene for a light-activated channel with high calcium permeability. Presumably the pathway leading to the activation of this channel must be the light-activated inositide cascade since, with the exception of the early receptor potential, all responsiveness to light in Drosophila is eliminated by null mutations of the norpA gene (Minke & Selinger,

1992b) which codes for a light-activated phospholipase C (Bloomquist et al., 1988).

References

Berridge, M.J. (1990). J. Biol. Chem. 265, 9583-9586.
Bloomquist B.T., Shortridge, R.D., Schneuwly, S., Pedrew, N., Montell, C., Steller, H., Rubin, G., Pak, W.L (1988). Cell 54, 723-733.
Bolsover, S.R., Brown, J.E. (1985). J. Physiol. Lond 364, 381-393.
Cosens, D.J. and Manning A. (1969). Nature 224, 285-287.
Furichi, T., Yoshikawa, S., Miyawaki, A., Wada, K., Madea, N. and Mokoshiba, K. (1989). Nature 342, 32-38.
Hardie, R.C. and Minke, B. (1992). Neuron 8, 1-20.
Hardie, R.C. (1991a). Proc. R. Soc. Lond. B 245, 203-210.
Hardie, R.C. (1991b). J. Neurosci. 11, 3079-3095.
Hochstrate, P. (1989). J. Comp Physiol. A 166, 179-188.
Howard, J. (1984). J. Exp. Biol. 113, 471-475
Irvine, R.F. (1990). FEBS Lett. 263, 5-9.
Lisman, J.E. and Brown, J.E. (1972). J. Gen. Physiol. 59, 701-719.
Minke, B. and Selinger, Z. (1992a). In, Progress in Retinal Research vol 11. Osborne N.N. and Chader, G.J. eds. Pergamon Press Oxford, pp. 99-124.
Minke, B. and Selinger, Z. (1992b). In, Sensory transduction. D. Corey and S.D. Roper (eds.) Rockefeller University Press.
Minke, B., Wu, C-F. and Pak, W.L. (1975). Nature 258, 84-87.
Montell, C. and Rubin, G.M. (1989). Neuron 2, 1313-1323.
Nagy, K. (1991). Q. Rev. Biophys. 24, 165-226.
Pak, W.L. (1991). Prog. Clin. Biol. Res. 362, 1-32.
Payne, R., Corson, D.W. and Fein, A. (1986). J. Gen. Physiol. 88, 107-126.
Phillips A.M., Bull, A. and Kelly, L. (1992). Neuron 8, 21-29.
Ranganathan, R., Harris, W.A., and Zuker, C.S. (1991). Trends Neurosci. 14, 486-493.
Suss-Toby, E. Selinger, Z. and Minke, B. (1991). J. Gen. Physiol. 98, 848-868.
Takemura, H., Hughes, A.R., Thastrup, O., and Putney, J.W. Jr. (1989). J. Biol. Chem. 264, 12266-12271.

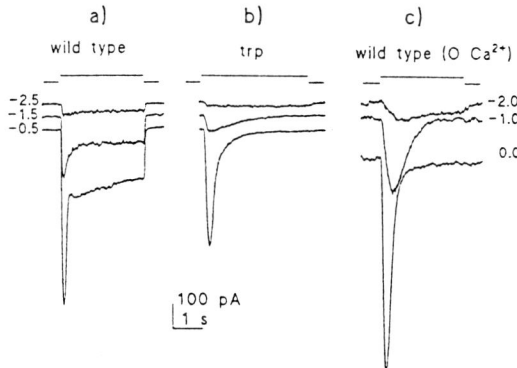

Figure 1. Whole-cell voltage-clamped inward currents in response to prolonged (3s) stimuli (relative log intensity indicated), at a holding potential of -40 mV. a) wild type and b) trp both recorded in normal Ringer with 2 mM Ca_o. c) the trp phenotype is closely mimicked in wild type photoreceptors following prolonged exposure (> 1 hr.) to nominally zero Ca^{2+} Ringer (2.5 μM Ca_o), in this case with Ca_i buffered to 50 nM with 5 mM EGTA.

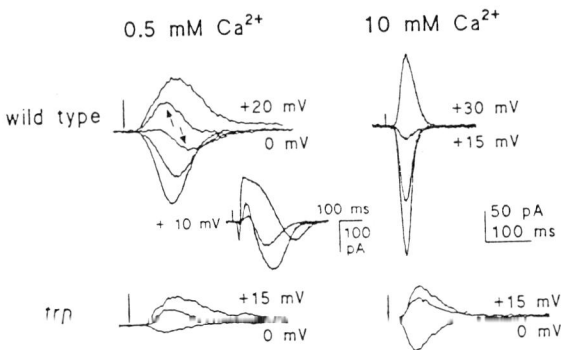

Figure 2. Biphasic reversal potential. Flash responses (10 ms flashes log -1.5, at time indicated by artefact), recorded at 5 mV intervals within the ranges indicated. Left: In 0.5 mM Ca_o, wild type photoreceptors show a biphasic reversal potential (arrows) between +10 and +15 mV. This biphasic behaviour was not observed at 10 mM Ca_o or in trp photoreceptors at either Ca^{2+} concentration (right). The inset shows three flash responses (wild type) to increasing intensity (-2.0, -1.0 & 0.0) all at +10 mV.

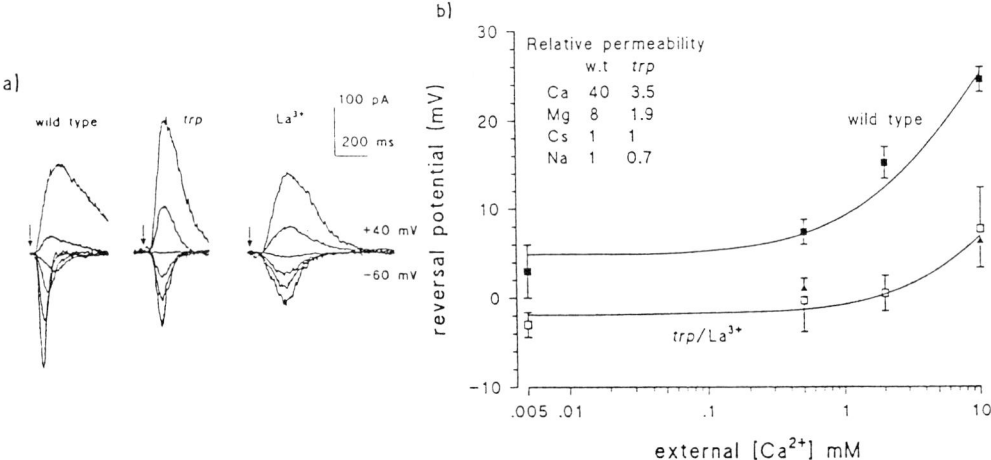

Figure 3a. Flash responses at different holding potentials (0.5 mM Ca_o). Voltage-clamped responses to identical 20 ms flashes (log -2.0) delivered at times indicated by arrows at holding potentials between -60 and +40 mV in 20 mV steps. Cells bathed in normal Ringer with 0.5 mM Ca_o. The wild type response kinetics show a strong voltage dependence, which is greatly reduced in trp flies or wild type flies in the presence of external La^{3+} (10 μM).

Figure 3b. Reversal potential (E_{rev}) as a function of [Ca_o] in wild type (solid squares), trp (open squares) and wild type photoreceptors in the presence of 10 μM external La^{3+} (solid triangles). Data points are means (± SD) from flash responses in at least 8 cells for each point, corrected for junction potential (-3 mV). The theoretical curves were calculated assuming constant field theory using the permeability ratios indicated.

Purification and identification of the βΓ-transducin complex of cone photoreceptor cells

Rehwa H. Lee[1,2], Bernice Lieberman[1], Harvey Yamane[3], Dean Bok[3] and Bernard D.K. Fung[3]

Molecular Neurology Laboratory[1], Veterans Administration Medical Center, Sepulveda, CA and Department of Anatomy and Cell Biology[2] and Jules Stein Eye Institute[3], University of California at Los Angeles, School of Medicine, Los Angeles, CA, USA

In the vertebrate retina, rod photoreceptors are responsible for night vision and cone photoreceptors for day/color vision. Rods are known to use a specific G protein (rod transducin) to transduce the light signal detected by rhodopsin into an increase in the enzymatic activity of a cGMP phosphodiesterase. The decrease in the intracellular level of cGMP leads to the closure of many cGMP-gated cation channels in the plasma membrane, resulting in a decrease in Na^+ conductance and the graded hyperpolarization of the cell (reviews, Stryer, 1986, Lolley and Lee, 1990). Although the molecular mechanism for visual excitation in cones is not as well understood (review, Pugh Jr. and Cobb, 1986), compelling electrophysiological evidence indicates that a cone-specific G protein (cone transducin)-mediated pathway very similar to that in rods is also involved in lowering the cGMP in cones following light stimulation (Haynes and Yau, 1985). Both rod and cone transducin belong to the family of G proteins which are heterotrimers composed of a Gα subunit and tightly complexed Gβ and GΓ subunits (GβΓ complex) (reviews, Fung, 1986, Gilman, 1987; Birnbaumer, 1990). Multiple isoforms of the three subunits exist which are encoded by the Gα, Gβ and GΓ families of homologous genes, respectively (review, Simon et al., 1991). The three polypeptides which constitute rod transducin have been identified and designated of as $Gα_r$, $Gβ_1$, and $GΓ_1$. Cone transducin has yet to be purified and characterized. In an earlier study, Lerea et al., (1986) reported the immunocytochemical localization of a cone-specific $Gα_c$ subunit to the outer segment of bovine cones, but its associated Gβ and GΓ subunits are unknown. Recently, Levine et al., (1990) detected in the human retina high levels of mRNA for a new isoform of Gβ ($Gβ_3$), suggesting the existence of a novel G protein which may have a specific function in retinal function.

As a first step in characterizing the $Gβ_3$ protein and determining its distribution in the retina, we developed three anti-peptide antisera directed against the most variable regions (residues 26 to 39) of the amino acid sequences of the three known Gβ isoforms ($Gβ_1$, $Gβ_2$, and $Gβ_3$, respectively) (Levine et al., 1990) and determined the specificity of the resultant antisera by immunoprecipitation using $Gβ_1$, $Gβ_2$, and $Gβ_3$ polypeptides which were synthesized by in vitro translation of the $Gβ_1$, $Gβ_2$ and $Gβ_3$ mRNAs in the presence of [^{35}S]-methionine (Fig. 1). It was found that incubation of all three polypeptides with the antiserum against $Gβ_1$ (β-636) resulted in selective precipitation of the $Gβ_1$ polypeptide. A similar degree of specificity was also observed for antisera β-637 and β-638 towards the $Gβ_2$ and $Gβ_3$ polypeptides against which these antisera are directed.

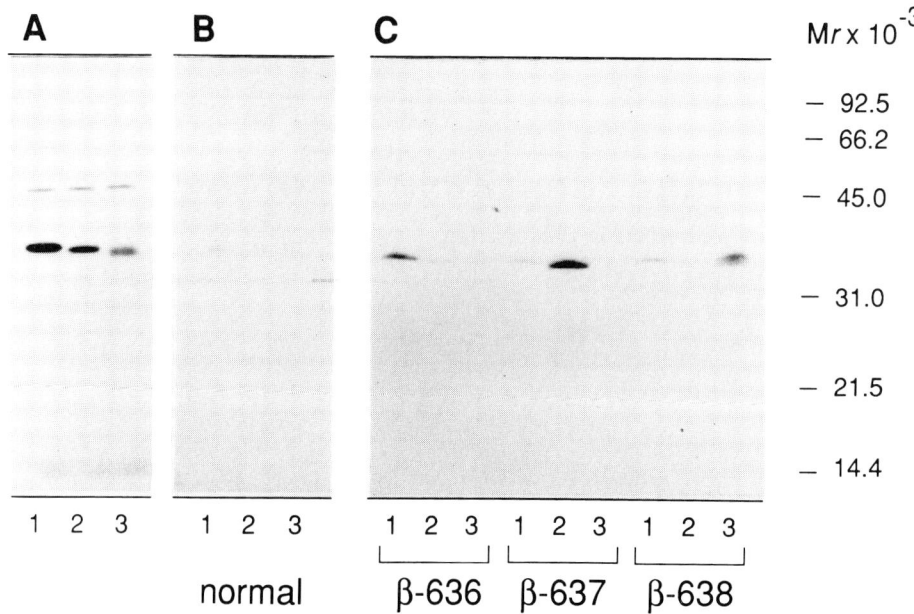

Fig. 1 Specificity of antisera. Human Gβ$_1$ (lane 1), Gβ$_2$ (lane 2) and Gβ$_3$ (lane 3) polypeptides synthesized by in vitro translation in the presence of [^{35}S]-methionine were incubated with antisera ß-636, ß-637 and ß-638 and precipitated with Protein A-sepharose. The immunoprecipitates were analyzed by SDS-PAGE and visualized by autoradiography. Panel A, autoradiogram of 0.2 μl of each of the three translation mixtures. Panel B, autoradiogram of immunoprecipitates obtained with normal rabbit serum. Panel C, autoradiogram of the immunoprecipitates obtained by using antisera ß-636, ß-637 and ß-638.

We used these antisera to examine the ß subunit of three purified protein preparations (Fig. 2): bovine retinal phosducin/ßΓ-transducin complex (Lee et al., 1987), transducin (Fung, 1983), and the GßΓ subunit from brain (Fung et al., 1990). As shown in the right panel, high levels of Gβ$_1$ immunoreactive to ß-636 were detected in all three preparations. In contrast, Gβ$_2$ was detected only in brain GßΓ but not in the transducin or in the phosducin/ßΓ-transducin complex, confirming the earlier observation that the phosducin/ßΓ-transducin complex, like rod transducin, contains only Gβ$_1$ and no Gβ$_2$ (Roof et al., 1985; Lee et al., 1987). Incubation with antiserum ß-638 revealed that Gβ$_3$ was present in all three samples, with the highest amount associated with the phosducin/ßΓ-transducin complex. These results indicate that the ß subunits of transducin and the phosducin/ßΓ-transducin complex, which were previously thought to be homogeneous, are actually mixtures of Gβ$_1$ and Gβ$_3$.

We compared the cellular distribution of Gβ$_1$ and Gβ$_3$ isoforms in bovine retinas by immunostaining with antisera ß-636 and ß-638 (Fig 3). Formaldehyde-fixed frozen sections incubated with ß-638 (panel C) showed staining throughout the entire cone photoreceptor, with the outer segments showing the highest intensity; yet all components of rod photoreceptors were unstained. In contrast, sections incubated with ß-636 showed fairly intense staining in the outer and inner segments, and synaptic terminals of rod photoreceptors; the cone inner segments and nuclei were unstained (result not shown). The selective localization of Gβ$_3$ in cones strongly implies it being the Gβ subunit of cone transducin.

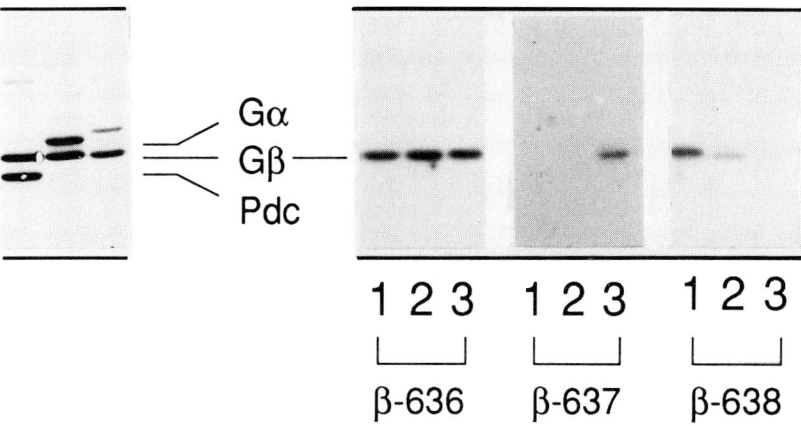

Fig. 2 Composition of Gß isoforms in purified protein preparations. The phosducin/ßΓ-transducin complex (lane 1), transducin (lane 2) and the brain GßΓ complex (lane 3), each containing 2 μg of ß subunits, were subjected to SDS-PAGE, followed by either Coomassie blue staining (left panel) or Western blot analysis using antisera ß-636, ß-637 and ß-638 (right panel). The bound antibodies were detected by [^{125}I]-protein A binding, followed by autoradiography. Pdc indicates the phosducin band.

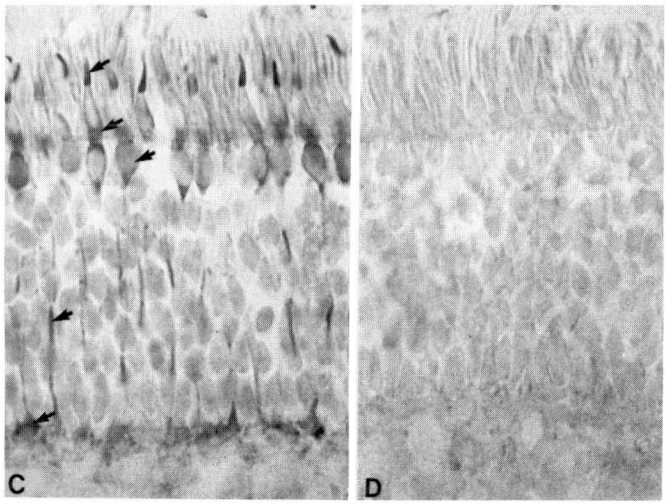

Fig. 3 Immunocytochemical localization of Gß$_3$ in bovine retinas. Formaldehyde-fixed frozen section of bovine retinas were incubated with 0.01 O.D. unit ($A_{280\ nm}$) of either immune or preimmune IgG. The sections were lightly counterstained with hematoxylin. Panel C, section incubated with immune IgG ß-638 and panel D, control for C with pre-immune IgG.

We have purified from bovine retinal extract the GβΓ complex which contains the Gβ$_3$ subunit. The total purification of Gβ$_3$Γ from Gβ$_1$Γ was achieved by subtractive affinity chromatography using an immobilized anti-GΓ$_1$ monoclonal antibody as the affinity matrix to remove the Gβ$_1$Γ complex. Western blot analysis with antisera ß-636 and ß-638 confirmed that the resultant Gβ$_3$Γ contains only Gβ$_3$, but not Gβ$_1$. We deduced a partial amino acid sequence from a tryptic fragment of the purified Gβ$_3$ isoform. This sequence, EGNVKVSRELSAHTGYLSCC-RFLDDNNI, aligned perfectly with residues 130 to 157 of that predicted by Gβ$_3$ cDNA (Levine et al., 1990) and it differed from the Gβ$_1$ sequence by four residues. This result conclusively establishes that the purified Gβ$_3$ is indeed the translation product of the Gβ$_3$ gene. Additionally, preliminary results showed that the associated GΓ subunit contains an internal sequence distinct from those of the seven known GΓ isoforms (results not shown). In summary, we have purified and identified from bovine retinas a novel GβΓ complex that is composed of Gβ$_3$ and a novel GΓ subunits. The specific localization of this Gβ$_3$Γ complex in the cone photoreceptors strongly implies that it is the GβΓ subunit of cone transducin, although its functional role in cone visual excitation needs to be confirmed biochemically. Our results provide evidence suggesting that G proteins of distinct cell types are formed from distinct subunits.

REFERENCES

Birnbaumer, L. (1990): G proteins in signal transduction. Ann. Rev. Pharmacol. Toxicol. 30, 675-705.

Fung, B. (1983): Characterization of transducin from bovine retinal rod outer segments. J. Biol. Chem. 258, 10495-10502

Fung, B. K.-K. (1986): Transducin: structure, function and role in phototransduction. Prog. Retinal Res. 6, 151-177

Fung, B. K.-K., Yamane, H. K., Ota, I. M. and Clarke, S. (1990): The Γ subunit of brain G -protein is methyl esterified at a C-terminal cysteine. FEBS Lett. 260, 313-317.

Gilman, A. G. (1987): G proteins: transducer of receptor-generated signals. Ann. Rev. Biochem. 56, 615-649.

Haynes, L., and Yau, K.-W. (1985): Cyclic GMP-sensitive conductance in outer segment membrane of catfish cones. Nature (London). 317, 61-66.

Lee, R., Lieberman, B. and Lolley, R. (1987): A novel complex from bovine visual cells of a 33,000-dalton phosphoprotein with beta, gamma-Transducin: purification and subunit structure. Biochemistry 26, 3983-3989.

Lerea, C. L., Somers, D. E., Hurley, J. B., Klock, I. B. and Bunt-Milam, A. H. (1986): Identification of specific transducin α subunits in retinal rod and cone photoreceptors. Science 234, 77-80.

Levine, M. A., Smallwood, P. M., Moen, P. T., Helman, L.J., and Ahn, T. G. (1990): Molecular cloning of ß3 subunit, a third form of the G protein ß subunit polypeptide. Proc. Natl. Acad. Sci. U.S.A. 87, 2329-2333.

Pugh, E. N. Jr., and Cobbs, W. H. (1986): Visual transduction in vertebrate rods and cones: a tale of two transmitters, calcium and cyclic GMP. Vis. Res. 26, 1613-1643.

Roof, D. J., Applebury, M. L. and Sternweis, P. C. (1985): Relationships within the family of GTP-binding proteins isolated from bovine central nervous system. J. Biol. Chem. 260, 16242-16249

Simon, M. I., Strathman, M. P., Gautham, N. (1991): Diversity of G protein in signal transduction. Science 2552, 8002-8008.

Stryer, L (1986): Cyclic GMP cascade of vision. Ann. rev. Neusci. 9, 87-119

A blue sensitive visual pigment based on 4-hydroxyretinal is found widely in mesopelagic cephalopods

Yuji Kito[1], Kinya Narita[1], Masatsugu Seidou[1], Masanao Michinomae[2], Kazuo Yoshihara[3], Julian C. Partridge[4] and Peter J. Herring[5]

[1] Department of Biology, Osaka University, Toyonaka 560 Japan, [2] Department of Biology, Konan University, Kobe 658, Japan, [3] Sunbor, Mishima-gun 618, Japan, [4] Department of Zoology, University of Bristol, Bristol, BS8 1UG, UK, [5] Institute of Oceanographic Sciences, Deacon Laboratory, Surrey GU8 5UB, UK

The firefly squid, Watasenia scintillans, has three visual pigments in its retina, each pigment being segregated in different parts of the retina and being based on a different chromophore. A retinal-based (A1) visual pigment (λmax: 484nm) is largely confined to the dorsal retina and a 3-dehydroretinal-based (A2) pigment (λmax: 500nm) is found in the proximal part of the rhabdoms of the ventral retina. The new 4-hydroxyretinal-based (A4) pigment (λmax: 470 nm) is found in the distal part of the rhabdoms of the ventral retina (Matsui et al., 1988 a; b; Seidou et al., 1991). This squid lives in the photic environment of monochromatic blue light in the deep-sea. Nevertheless, the squid bears three visual pigments with different spectral sensitivities, suggesting a spectral discrimination. This squid is well known for the intense bioluminescence from the large photophores on the tips of the fourth arms (Tsuji, 1986) and the animal has numerous small photophores on the skin of whole body for the purpose of counterillumination to the downwelling light. The counter-illumination of the squids close to the firefly squid was well demonstrated by Young and Roper (1980). Photographic observations on the lights emitted from the small skin photophores demonstrated that this squid could generate not only a blue light but also a greenish yellow light (Kito et al., 1988). The close relationship is considered between their bioluminescence and the vision. The lights emitted from the photophores were measured by a photon-counting spectrophotometer. From the ventral side of the live squid, the light with λmax at 470nm was recorded below 7°C and the light with λmax at 540nm above 15°C as shown in Fig.1. At the intermediary temperatures, the 470nm and 540nm lights were recorded. The mechanism of bioluminescence of this squid was suggested by T. Goto (1980): Luciferine and Luciferase reaction might emit light near UV, which could excite some fluorescent substances involved in the photophores. When individual small photophore was irradiated with a 360nm light, the fluorescent lights were very similar to lights from the live squid in Fig.1. A greenish yellow fluorescent substance and a blue fluorescent one were extracted in phosphate buffer solution from the photophores and partially purified mainly by the DEAE-cellulose column chromatography in aqueous solution.

Among three visual pigments, the main A1 pigment amounts to about 70 per cent of the total pigments in the retina and it is common in other mesopelagic cephalopods with the monochromatic vision (Seidou et al.,1991). The reddish A2 visual pigment is 10 per cent and in the rhabdom layer proximal to the black pigment granule layer in the ventral retina and is covererd by the yellowish rhabdom layer of the A4 pigment which amounts to 20 per cent. The absorbance of a 400 μm thick layer of the A4 pigment was measured with the microspectrophotometer to be about 3.0 at 470nm. With the electron microscopic observations it is shown that the rhabdoms bearing different pigments belong to each different visual cells (Masuda et al., 1988). The A4 pigment layer can act as a yellow filter on the A2 pigment layer as suggested in the deep-sea fish retina by Denton and Locket (1989) so that the spectral sensitivity curve of the A2 pigment is modified to shift towards the red end of the spectrum. After correction for the effect of the yellow filter, the A2 pigment becomes the most sensitive at 550nm. Thus the A2 pigment is maximally sensitive to the greenish yellow bioluminescence and the A4 pigment is also matched to the blue bioluminescence.

The A4 visual pigment has been known for a while only in the firefly squid retina in the animal kingdom. At last, we could find that some other squids and octopus, collected by the RRS Discovery in the mesopelagic zone of the Atlantic, had the new class of visual pigment. Squids, <u>Pyroteuthis</u> and <u>Pterigioteuthis</u>, and octopus, <u>Japetella</u> have two visual pigments based on the A1 and A4 chromophore. The squid, <u>Bathyteuthis</u> has three visual pigments based on the A1, A2 and A4 chromophores but in the different ratios from that in <u>Watasenia</u>. The absorbance maxima of their A1 pigment and A4 pigment were 484 nm and 470nm, respectively like the firefly squid. The fact that different visual pigments may generate from the same opsin indicates that there exists another way of the molecular adaptation of visual pigment in the animals living in the deep-sea blue light, which is different from the color vision by the vertebrate cone cells. Many vertebrates and insects developed various visual pigments by the evolutional modification in the primary structure of the opsin molecules.

The mechanism of the hypsochromic shift of the absorbance spectrum of visual pigment by use of the 4-hydroxyretinal may be mostly due to the steric perturbation on the retinal-protein interaction. The naturally occurring 4-hydroxyretinal in the firefly squid retina was the (S) configuration (Kito et al.,1992). In order to estimate the steric perturbation, when both enantiomers of the 11-cis 4-hydroxyretinal were added to the bovine opsin, both combined to generate the pigment with a λmax at 486nm in the solution of a detergent L-1695. The molar extinction coefficient and the shape of absorbance spectrum of the pigment were very similar to the pigment with 11-cis retinal. In the condition of the chromophore excess to the opsin, the (S) enantiomer could combine three times faster than the (R) enantiomer. The difference in photosensitivity and thermal stability between the synthetic pigments from the (S) enantiomer and from the enantiomeric mixture was not found clearly. So both enantiomers of 4-hydroxyretinal are accessible to the bovine opsin to form a stable pigment, and from the above combination experiments it could not find an adequate reason to explain the hypsochromic shift in the spectrum.

The spectral shift in the retinal Schiff's base compound is mainly determined with the positive charge introduced by the protonated nitrogen atom and with the solute-solvent interaction. The absorbance maxima of retinal and 4-hydroxyretinal and their Schiff base compounds with butylamine were examined in the different solvents using Trichloroacetic acid as the proton donor (Table 1).

In these solvent system, the spectral shifts of retinal was examined and characterized by Suzuki and Kito (1972). The degree of spectral shift in the protonated Schiff base compound of 4-hydroxyretinal is remarkably different in the nonpolar solvents from that of retinal. The environment surrounding retinal inside the visual pigment molecule may be highly hydrophobic and nonpolar as suggested by the known primary structure, and the polar 4-hydroxy group of the retinal can affect the spectrum, and that may be the reason why the 4-hydroxyretinal pigment is more blue.

Figure 1. Bioluminescence of small skin photophores at 2°C and 16°C. (See Text)

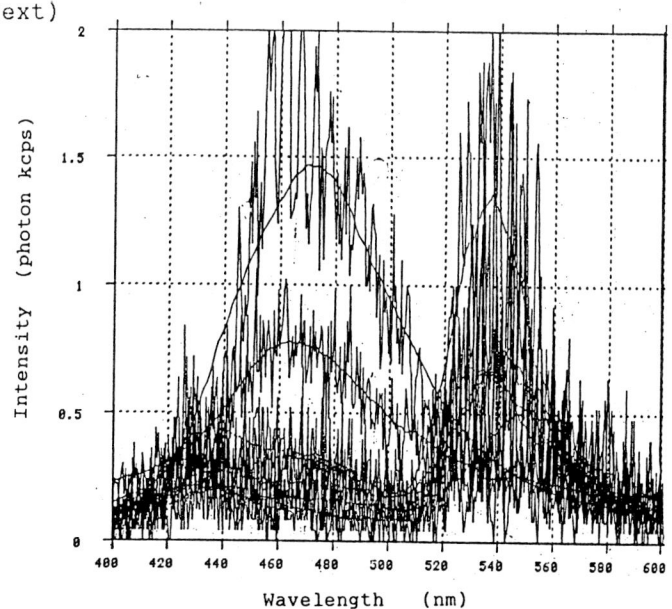

Table 1. Absorbance maxima of retinal and 4-hydroxyretinal and of N-retinylidenebutylamine in the absence (-) or presence (+) of 2M Trichloroacetic acid at 10℃

solvent	retinal	4-OH retinal	N-retinyl-butylamine	
n-hexane	368		357-	466+
		365	358-	442+
chloroform	390		368-	478+
		389	362-	450+
methanol	382		364-	448+
		378	370-	445+

References

Denton, E.J., and Locket, N.A. (1989): Possible wavelength discrimination by multibank retinae in deep-sea fishes. J. Mar. Biol. Assoc. UK. 69, 409-435.

Goto, T. (1980): Bioluminescences of marine organisms. In Marine natural products, ed P.J. Scheuer, p179-222, Academic Press. New York.

Kito, Y., Seidou, M., Matsui., Michinomae, M., Tokuyama, A., Sekiya, N., and K. Yoshihara. (1988): Vision and bioluminescence of a deep-sea cephalopod. Watasenia scintillans. In Proceeding of the Yamada Conference XXI, p285-290

Kito, Y., Partridge, J.C., Seidou, M., Narita, K., Hamanaka, T., Sekiya, N., and Yoshihara, K. (1992): The absorbance spectrum and photosensitivity of a new synthetic visual pigment based on 4-hydroxyretinal. Vision Res. 32, 3-10.

Masuda, H., Michinomae, M., Seidou, M., and Kito, Y. (1988): Light and electron microscopic investigation of cellular architecture in the reina of the firefly squid. Zool. Sci. 5, 1201.

Matsui, S., Seidou, M., Horiuchi, S., Uchiyama, I., and Kito, Y. (1988): Adaptation of a deep-sea cephalopod to the photic environment. J. Gen. Physiol. 92, 55-66.

Matsui, S., Seidou, M., Uchiyama, I.,Sekiya, N., Yoshihara, K., and Kito, Y. (1988): 4-Hydroxyretinal, a new chromophore found in the bioluminescent squid, Watasenia scintillans. Biochim. Biophys. Acta. 966, 370-374.

Seidou, M., Sugahara, M., Uchiyama, H., Hamanaka, T., Michinomae, M., Yoshihara, K., and Kito, Y. (1990): On the three visual pigment in the retina of the firefly squid, Watasenia scintillans. J. Comp. Physiol. 166, 769-773.

Suzuki, T., and Kito, Y. (1972): Absorption spectra of TCA-denatured rhodopsin and of a Schiff base compound of retinal. Photochem. Photobiol. 15,275-288.

Tsuji, F.I. (1985): ATP-dependent bioluminescence in the firefly squid, Watasenia scintillans. Proc. Natl. Acad. Sci. USA. 82, 4629-4632.

Young, R.E., and Roper, C.F.E. (1980): Bioluminescence in mesopelagic squid: Diel color change during counterillumination. Science, 208, 1286-1288.

Immunolocalisation of protein kinase A in bovine retina

Gregor Wolbring[1,2], Winfried Haase[3] and Neil J. Cook[1]

[1] Max-Planck Institut für Biophysik, Abteilung für Molekulare Membranbiologie, Heinrich-Hoffmann-Strasse 7, D-6000 Frankfurt am Main 71, Germany. [2] Max-Planck Institute für Biophysik, Kennedyallee 70, D-6000 Frankfurt am Main 70, Germany. Present address: [3] University of Calgary, Faculty of Medicine, Health Science Center, 3330 Hospital Drive N.W., Calgary, Alberta, T2N 4N1, Canada

INTRODUCTION

The cAMP dependent protein kinase (ATP:protein phospho-transferase, EC 2.7.1.37, PKA) is one of the major enzymes involved in the phosphorylation of proteins. PKA exists in two major forms referred to as type I and type II which differ in their regulatory subunits and are differently regulated and distributed among tissues (Taylor et al., 1990). It is well known that a cyclic nucleotide dependent protein kinase activity exists in bovine retinal rod outer segments (ROS) (Lee, R.H. et al., 1981, Farber, D.B. et al., 1979, Shinozawa, T., and Yoshizawa, T. 1986, Lolley, R.N. et al.,1977). It has also been shown that the addition of PKA or its activators to purified photoreceptor ROS preparations results in the specific phosphorylation of endogenous proteins (Farber, D.B. et al., 1979). In a previous study we have isolated and characterized protein kinase C from bovine ROS (Wolbring and Cook, 1991) and demonstrated specific phosphorylation of endogenous substrates by protein kinase C (unpublished results). We were also able to purify the catalytic subunit of PKA from bovine rod outer segments and to characterize the holoenzyme of PKA (Wolbring et al., submitted). We describe here the distribution of PKA subunits within the retina using light microscopie immunolabeling techniques.

RESULTS

Immunolocalisation of protein kinase A subunits in bovine retina:

By immunohistochemistry the question was investigated whether PKA is present in bovine retinal cells especially in the photoreceptor cells (Fig.1;). Foe this purpose we used antibodies directed against the catalytic subunit and the type I(RI) and typeII (RII) regulatory subunits of bovine heart and skeletal muscle PKA.
Very intensive immunostaining was observed in the outer and inner segments of bovine photoreceptors by using the anti-catalytic-subunit antibody. The same results was obtained by using the anti-RII-subunit antibody whereas no staining was observed in this cell layers with the anti-RI-subunit antibody. In the inner nuclear layer and in the ganglion cell layer strong to faint immunostaining was observed by using the anti-catalytic-subunit and the anti-RI-subunit antibody.
From the immunostaining the conclusion can be drawn that the highest amount of PKA is concentrated in the outer and inner segments of the photoreceptor cells with respect to the other retinal cell layers. Moreover from the immunostaining results it can be concluded that the PKA of the photoreceptor cells is exclusivly of the type II form. The latter conclusion was confirmed by western blotting. Only the anti catalytic-and anti RII-subunit antibodies reacted with the partially purified holoenzyme prepared from bovine rod outer segments (Wolbring et al., submitted).

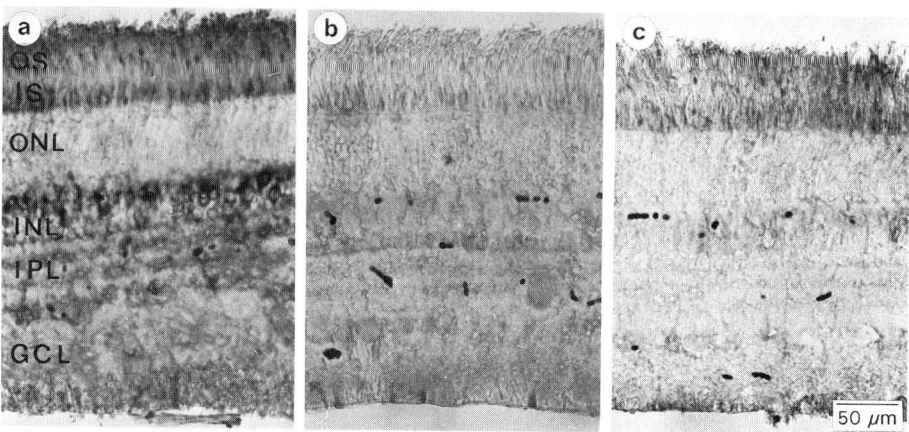

Fig. 1. Light microscopic immunoperoxidase staining of bovine retina with antibodies directed against different subunits of PKA. (a) anti catalytic subunit, (b) anti regulatory subunit type I (RI), (c) anti regulatory subunit type II (RII). The predominant binding sites of the antibodies against both the catalytic and the regulatory subunit type II are the outer (OS) and inner (IS) segments of the visual cells. ONL outer nuclear layer, INL inner nuclear layer, GCL ganglion cell layer.

DISCUSSION

We have shown that PKA of outer segments of bovine photoreceptor cells is of the type II form using specific antibodies against the catalytic subunit and against the type I and type II regulatory subunit of PKA. PKA is relativly abundant in the photoreceptor, as compared to other cell layers of the retina, and may therefore play a role in phototransduction. These antibodies were used in Western blotting (Wolbring et al., submitted) and immunohistochemical procedures. There are no substrates known to be specific for either the type I or type II form of the PKA. The difference so far known is that the activity of the type II form of PKA can be regulated by its state of phosphorylation (Taylor et al., 1990). The type I form is regulated by the Mg-ATP concentration and shows no autophosphorylation (Taylor et al., 1990).

Therefore it might well be that the phosphorylation state and therefore the activity of PKA of the photoreceptor cells is regulated by the state of illumination or by the activity of protein kinases and protein phosphatases.

REFERENCES

Farber, D.B. et al., (1979): Cyclic Nucleotide Dependent Protein Kinase and the Phosphorylation of Endogenous Proteins of Retinal Rod Outer Segments. *Biochemistry* 18, 370-378

Lee, R.H. et al,, (1981): Protein Kinases of Retinal Rod Outer Segment: Identification and Partial Characterization of Cyclic Nucleotide Dependent Protein Kinase and Rhodopsin Kinase. *Biochemistry 20*, 7532-7538

Lolley, R.N. et al.,(1977): Protein Phosphorylation in Rod Outer Segment from Bovine Retina: Cyclic Nucleotide-Activated Protein Kinase and its Endogenous Substrate. Biochim. Biophys. Res. Comm. 78 572-578

Shinozawa, T., and Yoshizawa, T. (1986): Cyclic Nucleotide-Dependent Phosphorylation of Proteins in Rod Outer Segments in Frog Retina. *J. Biol. Chem. 261*, 216-223

Taylor, S.S., et al., (1990): cAMP-Dependent Protein Kinase: Framework for a Diverse Family of Regulatory Enzymes. *Annu. Rev. Biochem. 59*, 971-1005

Wolbring, G., and Cook, N.J. (1991): Rapid Purification and Characterization of Protein Kinase C from Bovine Retinal Rod Outer Segment. *Eur. J. Biochem. 201*, 601-606

Author index
Index des auteurs

Abdulaev N.G., 299
Akaike T., 123
Alexiev U., 205
Amemiya T., 101
Amemiya Y., 37
Artemyer N.O., 361

Balashov S.P., 111, 115, 119
Barlow R.B. Jr, 283
Bauer U., 97
Baverstock J., 55
Bibikov S.I., 333
Birge R.R., 283
Blest A.D., 399
Bok. D., 407
Borgese F., 355
Boukra N., 303
Bousché O., 59
Braiman M.S., 233
Braslavsky S.E., 151
Brennan A.M.J., 275
Brown M., 399
Bruckert F., 379, 387
Büldt G., 97, 213

Carne A., 75
Cassim J.Y., 25
Catty P., 387
Chan T., 67
Chang C.W., 123
Chang M., 93
Chen A.H., 243
Chizhov I.V., 171
Chu K., 135
Cladera J., 33
Clausen J.A., 399
Colmenares L.U., 251
Converse S.C.A., 275
Conway S., 75
Cook N.J., 415
Courtois Y., 395
Crouch R.K., 111
Cugnoli C., 267, 271

Dancsházy Z., 175
Davies A., 55
DeCaluwé L.L.J., 59, 63
DeGrip W.J., 59, 63
Delaney A., 399
Dencher N.A., 97, 213, 217, 221
Dér A., 197, 343

Derguini F., 243
Deterre P., 383, 387
Diller R., 143
Doolittle F., 89
Drachev L.A., 159, 163, 167
Dracheva S.V., 163
Draheim J.E., 25
Dratz E.A., 21, 361
Ducreux G., 351
Dumler I., 347
Duñach M., 45, 295
Dyall-Smith M., 89

Ebrey T.G., 111, 115, 119
Eisfeld W., 139, 143
Engelhard M., 171, 317
Engelmann D.M., 3
Eremin S.V., 167
Etchebest C., 9

Fahmy K., 67
Fan X., 93
Faure J.P., 351, 355
Fedorovich I.B., 275
Feng Y., 111
Ferrando E., 325
Fiévet B., 355
Findlay J.B.C., 75, 267
Fioravanti R., 271
Foster R.G., 59
Franklin P., 243
Fujikawa K., 237
Fung B.D.K., 407

Garcia-Quintana D., 295
Garnovskaya M., 347
Garriga P., 295
Gärtner W., 79, 151
Gergely C., 193
Gerwert K., 127, 155
Gibson N.J., 25
Golisano O., 271
Goureau O., 395
Govardoskii V.I., 83
Govindjee R., 111, 115
Govorunova E.G., 343
Grant K.M., 275
Grishanin R.N., 333

Haase W., 415
Hamanaka T., 37

Hamm H.E., 361
Hara R., 255, 259
Hara T., 255, 259
Hara-Nishimura I., 255, 259
Hardie R., 403
Hargrave P.A., 51
Havelka W.A., 329
Heberle J., 213, 217, 221
Helgerson S.L., 21
Herring P.J., 411
Hess B., 171
Hessling B., 155
Heymann J.A.W., 329
Heyn M.P., 201, 205, 247
Hildebrandt V., 97
Hirayama J., 321
Höltje H.D., 213
Höltje M., 213
Holmes M., 89
Hu S., 243

Ihara K., 101
Ikonen M., 189
Imamoto Y., 321
Imasheva E.S., 119
Iwasa T., 287, 299

Janssen J.J.M., 59

Kahn T.W., 3
Kamekura M., 89
Kamo N., 237, 321
Karneyeva N.V., 119
Kaulen A.D., 159, 163, 167, 333
Keen J.N., 75, 267
Kelly L., 399
Keszthelyi L., 339, 343
Khitrina L.V., 167
Khodonov A.A., 167
Khorana H.G., 201, 205
Kibelbek J., 375
Kikkawa S., 287, 299
Kishigami A., 263
Kito Y., 37, 411
Koch K.W., 391
Komrakov A.Y., 159
Kondo M., 259
Kono M., 111, 115
Korf H.W., 59
Korolkov S., 347

Kozhevnikov N.M., 179
Krebs M.P., 201

Lambert C., 361
Langmack K., 55
Lavergne D., 351
Lavery R., 13
Le Coutre J., 127
Lee M., 67
Lee R.H., 407
Lemmetyinen H., 189
Lepoivre M., 395
Lévy D., 41
Lieberman B., 407
Litman B.J., 375
Litvin F.F., 119
Liu R.S.H., 251
Lohrmann R., 143, 147
Lott J.S., 75
Lukashev E., 111, 115

Maeda A., 131
Manyosa J., 295
Marti T., 205
Marwan W., 313, 333
Mascarelli F., 395
McGregor R.A., 75
Menguy L., 303
Menick D.R., 111
Michinomae M., 411
Mills J.S., 361
Minke B., 403
Mirshahi A., 351
Mirshahi M., 351, 355
Mitchell D.C., 375
Miyashita Y., 101
Mollaaghababa R., 201
Moltke S., 201
Monk P.D., 75
Montrone M., 313
Motais R., 355
Mourant J., 135
Mukohata Y., 101

Nakagawa M., 287, 299
Nakanishi K., 243
Narita K., 411
Nato A., 351, 355
Needleman R., 93
Neumann J.M., 41
Nina M., 17
Nishimura M., 255, 259
Nobes C., 55
Nullans G., 303

Oesterhelt D., 115, 313, 325, 329, 333

O'Gara E., 399
Okazaki H., 123
Oliver J., 355
Ormos P., 135
Oshida O., 123
Ostrovsky M.A., 275
Otomo J., 105
Otto H., 205
Ozaki K., 255

Padrós E., 29, 33, 225
Pages F., 383
Pardo L., 45
Partridge J.C., 411
Paulsen R., 369
Pepe I.M., 267, 271
Pfister C., 379, 383, 387
Plangger A., 369
Popot J.L., 9, 13
Pottinger J.D.D., 75
Pusch C., 143

Rando R.R., 365
Rarick H.M., 361
Razaghi A., 355
Renthal R.D., 21
Röhlich P., 83
Rohr M., 151
Rothschild K.J., 59
Roux B., 17
Ryba N.J.P., 75

Saibil H.R., 55
Sakmar T.P., 67
Samatey F.A., 9
Sasabe H., 105, 237
Scharf B., 317
Scherling S., 183
Scherrer P., 205
Schulenberg P., 151
Schwemer J., 277
Seidou M., 37, 411
Seigneuret M., 41
Sekiya N., 263
Sepulcre F., 225
Sharkov A.V., 171
Sharonov A., 189
Shichida Y., 291, 321
Shu-Hua L., 75
Siebert F., 317
Sigrist H., 183
Silvo B.R., 243
Sinclair A., 75
Sineshchekov O.A., 343
Skiba N.P., 361
Skulachev V.P., 229, 333

Smith J.C., 17
Soppa J., 115
Souvignier G., 155
Spengler F., 277
Spudich J.L., 309
Stockburger M., 139, 143, 147
Sugiyama Y., 101
Száraz S., 197
Szél A., 83

Taguchi T., 101
Takahashi T., 263
Tallent J.R., 283
Tateno M., 101
Terakita A., 255
Thiedemann G., 217
Tittor J., 115
Tkachenko N., 189
Tokaji Z., 175
Tokunaga F., 263
Tomioka H., 105, 237, 321
Torres J., 29
Tóth-Boconádi R., 197
Towner P., 79
Tsuda M., 287, 299
Tufféry P., 13

Ulrich A.S., 247
Urabe Y., 105
Urban P.F., 303

VanAalten D.M.F., 63
Vandewalle A., 355
Van Dorsselaer A., 303
VanOostrum J., 63
Van Veen T., 83
Váró G., 193
Venien-Bryan C., 55
Virmaux N., 303

Wakabayashi K., 37
Wallat I., 247
Walter T.J., 233
Wang J., 243
Watts A., 247
Wilkinson J.R., 55
Wolbring G., 415
Wrede P., 97

Yamane H., 407
Yasukawa T., 123
Yoshihara K., 263, 411
Yoshizawa T., 71, 321

Zaccaï G., 9

List and address of contributors
Liste et adresse des intervenants

Najmoutin G. Abdulaev, Shemyakin Institute of Bioorganic Chemistry, Russian Academy of Sciences, U1, Miklukko-Maklaya, 16/10, 117871 GSP Moscow V-437, Russie

Ulrike Alexiev, Biophysics Group, Freie Universität Berlin, Arnimallee 14, D-1000 Berlin 33, Allemagne

Sophia Arnis, Institut für Biophysik, Universität Freiburg, Albertstrasse 23, D-7800 Freiburg, Allemagne

Alexander S. Arseniev, Shemyakin Institute of Bioorganic Chemistry, Russian Academy of Sciences, U1-Miklukho-Maklaya 16/10, 117871 Moscow V-437, Russie

Igor Artamonov, Shemyakin Institute of Bioorganic Chemistry, Russian Academy of Sciences, U1, Miklukko-Maklaya 16/10, 117871 Moscow - V 437, Russie

George H. Atkinson, The University of Arizona, College of Arts and Sciences, Faculty of Science, Department of Chemistry, Tucson, Arizona 85721, USA

Sergei P. Balashov, Biology Faculty, Moscow State University, 119899 Moscow, Russie

Sergei I. Bibikov, A.N. Belozersky Institute of Physico-Chemical Biology, Moscow State University, 119899 Moscow, Russie

Robert R. Birge, Center for Science and Technology, Room 1-014, Syracuse, NY 13244-4100, USA

Roberto Bogomolni, University of California Santa Cruz, Department of Chemistry and Biochemistry, Santa Cruz, CA 95064, USA,

Mark Braiman, University of Virginia Health Sciences Center, Biochemistry Department, Box 440, Charlottesville, VA 22908, USA

Ch. Brauchle, Institut für Physikalische Chemie der Universität München, Sophienstrasse 11, D-8000 München 2, Allemagne

Franz Bruckert, CEN/Grenoble, DBMS/BMC, BP 85 X, 38041 Grenoble, France

Georg Büldt, Freie Universität Berlin, Department of Physics/Biophysics, Arnimallee 14, D-1000 Berlin 33, Allemagne

Robert Callender, Physics Department, City College of New York, Convent Avenue at 138th Street, New York, NY 10031, USA

Yi Cao, University of California, Department of Physiology and Biophysics, California College of Medicine, Irvine, CA 92717, USA

Joseph Y. Cassim, Department of Microbiology, The Ohio State University, 484 West 12th Avenue, Columbus, OH 43210-1292, USA

Marc Chabre, CNRS-IPMC, 660 route des Lucioles, Sophia Antipolis, 06560 Valbonne, France

Igor Chizhov, General Physics Institute, Russian Academy of Sciences, Vavilov Street 38, Moscow, Russie

Josep Cladera, Unitat de Biofisica, Departament de Bioquimica i de Biologia Molecular, Universitat Autònoma de Barcelona, 08193 Bellaterra/Barcelona, Espagne

Julia Clausen, Developmental Neurobiology Group, Research School of Biological Sciences, Australian National University, GPO Box 475, Canberra, ACT 2601, Australie

Yves Courtois, INSERM U.118, Unité de Recherches Gérontologiques, 29 rue Wilhem, 75016 Paris, France

Carlo Cugnoli, Institute of Cybernetics and Biophysics, CNR, Via Dodecaneso 33, Genova, Italie

Zsolt Dancsházy, National Health Institute, NHLBI, Laboratory of Cell Biology, NIH, Building 3, Bethesda, MD 20892, USA

Florence Davidson, Massachusetts Institute of Technology, Departments of Biology and Chemistry, Room 18-506, 77 Massachusetts Avenue, Cambridge, MA 02139, USA

Anthony Davies, Department of Crystallography, Birkbeck College London, Malet Street, London WCIE 7HX, Grande-Bretagne

Lieveke DeCaluwé, Department of Biochemistry, University of Nijmegen, Adelbertusplein 1, 6525 EK Nijmegen, Pays-Bas

W.J. DeGrip, Department of Biochemistry, University of Nijmegen, PO Box 9101, 6500 HB Nijmegen, Pays-Bas

John Delaney, The University of Arizona, College of Arts and Science, Faculty of Science, Department of Chemistry, Tucson, Arizona 85721, USA

Norbert Dencher, Hahn-Meitner-Institut, BENSC-N1, Glienickerstrasse 100, D-1000 Berlin 39, Allemagne

András Dér, Institute of Biophysics, Biological Research Centre of the Hungarian Academy of Sciences, Temesvàri Krt 62, P.O.B 521, H-6701 Szeged, Hongrie

Fadila Derguini, Columbia University, Chemistry Department, Harremeyer Hall, Box 662, New York, NY 10027, USA

Andrej K. Dioumaev, General Physics Institute, Vavilov Street 38, Moscow 117942, Russie

Helène Dollfus, Laboratoire de Génétique Humaine, INSERM U.12, Hôpital des Enfants Malades, 149 rue de Sèvres, 75743 Paris Cedex 15, France

Lei Drachev, A.N. Belozersky Laboratory, Moscow State University, Bldg A, Moscow 119 899, Russie

Inna Dumler, Sechenov Institute of Evolutionary Physiology and Biochemistry, 194223 St Petersburg, Russie

Mireia Duñach, Unitat de Biofisica, Departament de Bioquimica i Biologia Molecular, Facultat de Medicina, Universitat Autònoma de Barcelona, 08193 Bellaterra/Barcelona, Espagne

Mike Dyall-Smith, The University of Melbourne, Dept of Microbiology, Melbourne 3052, Australie

Thomas Ebrey, Dept of Physiology and Biophysics, University of Illinois, 156 Davenport Hall, 607 South Mathews Street, Urbana, IL 91801, USA

Wolf Eisfeld, Max-Planck Institut für Biophysik-Chemie, Abteilung Spektroskopie, Göttingen-Nikolausberg PF2841, D-3400 Göttingen, Allemagne

Martin Engelhard, Max-Planck Institut für Ernährungsphysiologie, Rheinlanddamn 201, D-4600 Dortmund 1, Allemagne

Donald M. Engelmann, Dept of Molecular Biophysics and Biochemistry, Yale University, 260 Whitney Avenue, New Haven, CT 06511, USA

Karim Fahmy, Howard Hughes Medical Institute, Rockefeller University, 1230 York Avenue, New York, NY 10021, USA

Irina Fedorovich, Institute of Chemical Physics, Russian Academy of Sciences, Kosygin, 4, 117334 Moscow, Russie

Elisa Ferrando, Max-Planck Institut für Biochemie, Am Klopferspitz 18a, D-8033 Martinsried/München, Allemagne

John Findlay, Dept of Biochemistry and Molecular Biology, University of Leeds, Leeds LS2 9JT, Grande-Bretagne

David Garcia-Quintana, Unitat de Biofisica, Departament de Bioquimica i de Biologia Molecular, Edifici M, Universitat Autònoma de Barcelona, 08193 Bellaterra/Barcelona, Espagne

Wolfgang Gärtner, Max-Planck Institut für Strahlenchemie, Stiftstrasse 34-36, D-4330 Mulheim/Ruhr, Allemagne

Yahaloma Gat, Weizmann Institute, Dept of Organic Chemistry, Rehovot 76100, Israël

Klaus Gerwert, Max-Planck Institut für Ernährungsphysiologie, Rheinlanddamm 201, D-4600 Dortmund 1, Allemagne

Olivier Goureau, INSERM U.118, Unité de Recherches Gérontologiques, 29, rue Wilhem, 75016 Paris, France

Rajni Govindjee, Dept of Physiology and Biophysics, University of Illinois, 156 Davenport Hall, 607 South Mathews Street, Urbana, IL 61801, USA

Elena Govorunova, Institute of Biophysics, Biological Research Centre, Temesvàri Krt., H-6701 Szeged, Hongrie

Bob Griffin, Massachussets Institute of Technology, Francis Bitter National Magnet Laboratory, Cambridge, MA 021 39, USA

Michel Groesbeek, Gorlaeus Laboratory, State University of Leiden, P.O. Box 9502, 2300 Ra Leiden, Pays-Bas

Toshiaki Hamanaka, Department of Biophysical Engineering, Faculty of Engineering Science, Osaka University, Machikaneyama, Toyonaka/Osaka 560, Japon

Heidi Hamm, Department of Physiology and Biophysics, University of Illinois, Chicago College of Medicine, Chicago IL 60612, USA

Reiko Hara, Department of Biology, Kinki University School of Medicine, Osaka-Sayama/Osaka 589, Japon

Tomiyuki Hara, Department of Biology, Kinki University School of Medicine, Osaka/Sayama/Osaka 589, Japon

Ikuko Hara-Nishimura, Department of Cell Biology, National Institute for Basic Biology, Myodaijicho, Okazaki 444, Japon

Paul A. Hargrave, Dept of Ophtalmology, University of Florida, College of Medicine, J. Hillis Miller Health Center, Box J 284, Gainesville, FL 32610-0284, USA

Thomas Haub, Freie Universität Berlin, FB Physik, Arnimallee 14, D-1000 Berlin 33, Allemagne

Wendy Havelka, Max-Planck Institut für Biochemie, Am Klopferspitz 18A, D-8033 Martinsried/München, Allemagne

Joachim Heberle, Hahn-Meitner Institut, BENSC, Glienicker Strasse 100, D-1000 Berlin 39, Allemagne

Sam Helgerson, Baxter Healthcare Corp., Hyland Division, 1710 Flower Avenue, Duarte, CA 91010, USA

Richard Henderson, MRC Laboratory of Molecular Biology, Hills Road, Cambridge CB2 2QH, Grande-Bretagne

Judith Herzfeld, Department of Chemistry, Brandeis University, Waltham, MA 02254-9110, USA

Benno Hess, Max-Planck Institut für medizinische Forschung, Postfach 103820, Jahnstrasse 29, D-6900 Heidelberg 1, Allemagne

Benedikt Hessling, Max-Planck Institut für Ernährungsphysiologie, Rheinlanddamm 201, D-4600 Dortmund 1, Allemagne

Jürgen Heymann, Max-Planck Institut für Biochemie, Am Klopferspitz 18A, D-8033 Martinsried/München, Allemagne

Maarten P. Heyn, Biophysics Group, Physics Dept, Freie Universität Berlin, Arnimallee 14, D-1000 Berlin 33, Allemagne

Volker Hildebrandt, Freie Universität Berlin, Department of Physics, Biophysics Group, Arnimallee 14, D-1000 Berlin 33, Allemagne

Astrid Hoffmann, Freie Universität Berlin, Dept of Physics/Biophysics, Arnimallee 14, D-1000 Berlin 33, Allemagne

Klaus P. Hofmann, Institut für Biophysik und Strahlenbiologie, Albert-Ludwigs-Universität Freiburg, Alberstrasse 23, D-7800 Freiburg, Allemagne

Marjo Ikonen, University of Helsinki, Department of Physical Chemistry, Meritullinkatu 1C, 00170 Helsinki, Finlande

Yasushi Imamoto, Department of Biophysics, Faculty of Science, Kyoto University, Kitashirakawa-Oiwake-cho, Sakyo-ku, Kyoto 606-01, Japon

Tatsuo Iwasa, Department of Life Science, Himeji Institute of Technology, Harima Science Garden City, Kamigori Ako/ Hyogo 678-12, Japon

Frank Jäger, Institut für Biophysik und Strahlenbiologie, Albertstrasse 23, D-7800 Freiburg, Allemagne

Stefan Jäger, Universität Freiburg, Institut für Biophysik, Albertstrasse 23, D-7800 Freiburg, Allemagne

Josseline Kaplan, Laboratoire de Génétique, INSERM U.12, Hôpital des Enfants Malades, 149 rue de Sèvres, 75015 Paris, France

Andrey Kaulen, Belozersky Institute of Physico-Chemical Biology, Moscow State University, Len.Gory, Moscow, Russie

Laszlo Keszthelyi, Institute of Biophysics, Biological Research Center, Temesvàri Krt 62, H-6701 Szeged, Hongrie

Gobind H. Khorana, Massachusetts Institute of Technology, Departments of Biology and Chemistry, 77 Massachusetts Avenue, Boston, MA 02139, USA

Yuji Kito, Department of Biology, Faculty of Science, Osaka University, Machikaneyama, Toyonaka/Osaka 560, Japon

David Kliger, University of California, 546 Arroyo Seco, Santa Cruz, CA 95060, USA

Karl-W. Koch, Institut für Biologische Informationsverarbeitung, Postfach 1913, D-5170 Julich, Allemagne

Nikolai Kozhevnikov, St Petersburg State Technical University, Dept of Experimental Physics, Polytechnicheskaya 29, 195251 St Petersburg, Russie

K.V. Lakshmi, Francis Bitter National Magnet Laboratory, MIT Brandeis University, 170 Albany St/NW14-5117, Cambridge, MA 02139, USA

Romy Lambrecht, Max-Planck Institut für Biochemistry, Department of Membrane Biochemistry, Am Klopferspitz, D-8033 Martinsried/München, Allemagne

Helmut Langer, Ruhr Universität Bochum, Tierphysiologie N.D5, Universitätstrasse 150, D-4630 Bochum, Allemagne

Janos Lanyi, University of California, Dept of Physiology and Biophysics, Irvine, CA 92717, USA

D. Lavergne, Laboratoire de Morphogenèse Végétale Expérimentale, Université Paris XI - URA 115 CNRS, 91405 Orsay Cedex, France

Johannes Le Coutre, Max-Planck Institut für Ernährungsphysiologie, Rheinlanddamm 201, D-4600 Dortmund 1, Allemagne

Rehwa Lee, Dept of Anatomy and Cell Biology, UCLA, School of Medicine, Los Angeles, CA 90024, USA

Daniel Lévy, Laboratoire de Biophysique cellulaire, URA 526 - Equipe ATIPE, Université Paris VII, 2, place Jussieu, 75251 Paris Cedex 05, France

Burton Litman, University of Virginia, Biochemistry Department, 1300 Jefferson Park Avenue, Jordan Hall, Room 6-7, Charlottesville, VA 22908, USA

Robert S.H. Liu, Department of Chemistry, University of Hawai, 2545 The Mail, Honolulu/Hawaii 96822, USA

Nurit Livnah, Weizmann Institute of Science, Dept of Organic Chemistry, Rehovot 76100, Israël

Ralf Lohrmann, Max-Planck Institut für Biophysik-Chemie, Abteilung Spektroskopie, Göttingen-Nikolausberg PF2841, D-3400 Göttingen, Allemagne

Johan Lugtenburg, Gorlaeus Laboratory, State University of Leiden, P.O. Box 9502, 2300 RA Leiden, Pays-Bas

Akio Maeda, Department of Biophysics, Faculty of Science, Kyoto University, Kitashirakawa-Oiwake-cho, Sakyo-ku, Kyoto 606-01, Japon

Joan Manyosa, Unitat de Biofisica, Departament de Bioquimica i de Biologia Molecular, Bellaterra/Barcelona 08193, Espagne

Wolfgang Marwan, Max-Planck Institut für Biochemie, Am Klopferspitz 18A, D-8033 Martinsried/München, Allemagne

Olivier Meyer, Centre d'Etudes Pharmaceutiques, 5 rue Jean-Baptiste Clément, 92296 Châtenay-Malabry, France

Baruch Minke, Department of Physiology, The Hebrew University, P.O.B. 1172, 91010 Jerusalem, Israel

Ali Mirshahi, Laboratoire de Morphogenèse Végétale Expérimentale, Université Paris XI - URA 115 CNRS, 91405 Orsay Cedex, France

Stephan Moltke, Freie Universität Berlin, FB Physik, Arnimallee 14, D-1000 Berlin 33, Allemagne

Yasuo Mukohata, Department of Biology, Faculty of Science, Nagoya University, Chikusa-ku, Nagoya 464-01, Japon

Koji Nakanishi, Dept of Chemistry, Columbia University, Havemeyer Hall, New York, NY 10027, USA

A. Nato, Laboratoire de Morphogenèse Végétale Expérimentale, Université Paris XI - URA 115 CNRS, 91405 Orsay Cedex, France

Richard Needleman, Wayne State University, Gordon H. Scott Hall of Basic Medical Sciences, 540 Post Canfield Avenue, Detroit, Michigan 48201, USA

Jean-Michel Neumann, DBCM/SBPM, Bâtiment 532, CE/Saclay, 91191 Gif-sur-Yvette Cedex, France

Mafalda Nina, DBCM/SBPM, Bâtiment 532, C.E. Saclay, 91191 Gif-sur-Yvette, France

Dieter Oesterhelt, Max-Planck Intitut für Biochemie, D-8033 Martinsried/München, Allemagne

Daniel D. Oprian, Brandeis University, Graduate Department of Biochemistry, Waltham, MA 02254-9110, USA

Pál Ormos, Institute of Biophysics, Biological Research Center, Temesvàri Krt 62, Szeged 6701, Hongrie

Jun Otomo, Frontier Research Program, Riken Institute, Hirosawa 2-1, Wako/Saitama 351-01, Japon

Harald Otto, Freie Universität Berlin, FB Physik, Arnimallee 14, D-1000 Berlin 33, Allemagne

Esteve Padrós, Unitat de Biofisica, Facultat de Medicina, Universitat Autònoma de Barcelona, 08193 Bellaterra/Barcelona, Espagne

Frédérique Pages, CEN/Grenoble, DBMS/BMC, BP 85 X, 38041 Grenoble Cedex, France

Krzvsztof Palczewski, R.S. Dow Neurological Sciences Institute, Good Samaritan Hospital and Medical Center, 1120 NW 20th Avenue, Portland, OR 97209-1595, USA

Reinhard Paulsen, Universität Karlsruhe, Zoologisches Institut I, Kornblumenstraße 13, D-7500 Karlsruhe 1, Allemagne

Isodoro Mario Pepe, Institute of Cybernetics and Biophysics, CNR, Via Dodecaneso 33, Genova, Italie

David Pepperberg, Dept of Ophthalmology and Visual Sciences, University of Illinois, 1855 West Taylor Street, Chicago, IL 60612, USA

Linda Peteanu, Department of Chemistry, University of California, Berkeley, CA 94720-9989, USA

Claude Pfister, CEN/Grenoble, DBMS/BMC, BP 85 X, 38041 Grenoble Cedex, France

Jean-Luc Popot, Institut de Biologie Physico-Chimique, 13, rue Pierre et Marie Curie, 75005 Paris, France

Robert Rando, Dept of Biological Chemistry and Molecular Pharmacology, Harvard Medical School, 250 Longwood Avenue, Boston, MA 02115, USA

George Rayfield, University of Oregon, Physics Department, Eugene, OR 97403, USA

Ahmad Razaghi, INSERM U.86, 15 rue de l'Ecole de Médecine, 75270 Paris Cedex 06, France

Kevin Ridge, Massachusetts Institute of Technology, Dept of Chemistry, Room 18-506, 77 Massachusetts Avenue, Cambridge, MA 02139, USA

Jean-Louis Rigaud, DBCM/SBE, Centre d'Études Nucléaires de Saclay, Bât. 532, 91191 Gif-sur-Yvette Cedex, France

Pàl Röhlich, Laboratory of Electron Microscopy, Semmelweis Medical School, Tuzolto u 58, P.O. Box 95, Budapest, Hongrie

Helen Saibil, Department of Crystallography, Birkbeck College, University of London, Malet Street, London WCIE 711X, Grande-Bretagne

Thomas P. Sakmar, Howard Hughes Medical Institute, The Rockefeller University, 1230 York Avenue, New York, NY 10021, USA

Fadel Samatey, Institut de Biologie Physico-Chimique, 13 rue Pierre et Marie Curie, 75005 Paris, France

Hans Jürgen Sass, Freie Universität Berlin, FB Physik, Abteilung Biophysik, Arnimallee 14, D-1000 Berlin 33, Allemagne

Tudor Savopol, "Carol Davila" Université de Médecine, Département de Recherche de Biophysique, Bucarest, Roumanie

Peter Scherrer, Freie Universität Berlin, Abteilung Biophysik, FB Physik, Arnimallee 14, D-1000 Berlin 33, Allemagne

Gerhard Schertler, MRC Laboratory of Molecular Biology, Hills Road, Cambridge CB2 2QH, Grande-Bretagne

Joachim Schwemer, Department Biophysics, University of Groningen, Westersingel 34, 9718 CM Groningen, Pays-Bas

Michel Seigneuret, Laboratoire de Biophysique Cellulaire, URA 526 - Equipe ATIPE, Université Paris VII, 2, place Jussieu, 75251 Paris Cedex 05, France

Francesc Sepulcre, Unitat de Biofisica, Departament de Bioquimica i de Biologia Molecular, Edifici M, Universitat Autònoma de Barcelona, 08193 Bellaterra/Barcelona, Espagne

Inna Severina, A.N. Belozersky Institute of Physico-Chemical Biology, Moscow State University, 119899 Moscow, Russie

Mordechai Sheves, Weizman Institute, Organic Chemistry Rehovot, 76100 Rehovot, Israël

Yoshinori Shichida, Department of Biophysics, Faculty of Science, Kyoto University, Kitashirakawa-Oiwake-cho, Sakyo-ku, Kyoto 606-01, Japon

Friedrich Siebert, Max-Planck Institut für Biophysik, Kennedyallee 70, D-6000 Frankfurt, Allemagne

Hans Sigrist, Institute of Biochemistry, University of Berne, Freiestrasse 3, CH-3012 Bern, Suisse

Oleg Sineshchekov, Moscow State University, 119899 Moscow, Russie

Vladimir P. Skulachev, A.N. Belozersky Laboratory, Molecular Biology and Bioorganic Chemistry, Moscow State University, Moscow 119 899, Russie

Jeremy Smith, DBCM/SBPM, Bâtiment 532, CE/Saclay, 91191 Gif-sur-Yvette Cedex, France

Elena Spudich, Department of Microbiology and Molecular Genetics, University of Texas, 6431 Fannin, Houston, TX 77030, USA

John Lee Spudich, The University of Texas, Health Science Center at Houston, Department of Microbiology, Medical School, PO Box 20708, Houston, TX 77030, USA

Gali Steinberg, Department of Organic Chemistry, Weizmann Institute, 76100 Rehovot, Israël

Manfred Stockburger, Max-Planck Institut für Biophysik-Chemie, Göttingen-Nikolausberg, D-3400 Göttingen, Allemagne

Walther Stoeckenius, Max-Planck Institut für Biophysik, Kennedy-Allee 70, D-6000 Frankfurt/Main 70, Allemagne

Sriram Subramanian, Department of Biological Chemistry, Johns Hopkins University, School of Medicine, 725 N Wolfe Street, Baltimore, MD 21205, USA

Gerd Thiedemann, Hahn-Meitner-Institut, BENSC-N1, Glienicker Strasse 100, D-1000 Berlin 39, Allemagne

Jörg Tittor, Max-Planck Institut für Biochemie, Am Klopferspitz, D-8033 Martinsried, Allemagne

Zsolt Tokaji, Biological Research Center, Institute of Biophysics, Temesvàri Krt 62, H-6701 Szeged, Hongrie

Hiroaki Tomioka, Frontier Research Program, Riken Institute, 2-1 Hirosawa, Wako 351-01, Japon

Jaume Torres, Unitat Biofisica, Departament Bioquimica i Biologia Molecular, Facultat de Medicina, Universitat Autònoma de Barcelona, 08193 Bellaterra/Barcelona, Espagne

Paul Towner, Biochemistry Department, University of Bath, Claverton Down, Bath, BA2 7AY, Grande-Bretagne

Motoyuki Tsuda, Dept of Life Science, Himeji Institute of Technology, Harima Science Garden City, Kamigori Ako/Hyogo 678-12, Japon

Pierre Tufféry, Institut de Biologie Physico-chimique, 13 rue Pierre et Marie Curie, 75005 Paris, France

George Turner, University of California, Department Biochemistry and Biophysics, San Francisco, CA 94143, USA

Anne Ulrich, Department of Biochemistry, University of Oxford, South Parks Road, Oxford OX1 3QU, Grande-Bretagne

Paul Urban, INSERM U.338, Centre de Neurochimie, 5 rue Blaise Pascal, 67084 Strasbourg, France

Daan VanAalten, Department of Biochemistry, University of Nijmegen, Adelbertusplein 1 - P.O. Box 9101, 6500 HB Nijmegen, Pays-Bas

György Váró, Institute of Biophysics, Biological Research Center of the Hungarian Academy of Sciences, Temesvàri Krt 62, H-6701 Szeged, Hongrie

Catherine Venien-Bryan, Department of Biochemistry, University of Oxford, South Parks Road, Oxford OX1 3QU, Grande-Bretagne

Guilhem Vidiella, Cie Bertin, 59 rue Pierre Curie, Z.I. des Gatines, 78373 Plaisir Cedex, France

Noëlle Virmaux, INSERM U.338, Centre de Neurochimie, 5 rue Blaise Pascal, 67084 Strasbourg, France

Timothy Walter, University of Virginia, Health Sciences Center, 1300 Jefferson Park Avenue, Charlottesville, VA 22908, USA

Olaf Weidlich, Institut für Biophysik-Strahlenbiologie, Albertstrasse 23, D-7800 Freiburg, Allemagne

Gregor Wolbring, Max-Planck Institut für Biophysik, Heinrich-Hoffmann Strasse 7, D-6000 Frankfurt/Main 71, Allemagne

E.K. Wolff, Universität Witten/Herdecke, Naturwissenschaftliche Fakultät, Stockumer Strasse 10, Postfach 6260 - D-6810 Witten, Allemagne

Tamio Yasukawa, Material Systems Engineering, Faculty of Technology, Tokyo University of Agriculture and Technology, 2-4 Nakamachi, Koganei, Tokyo 184, Japon

Kazuo Yoshihara, Suntory Institute for Bioorganic Research, 1-1 Wakayamadai, Shimamoto, Osaka 618, Japon

Tôru Yoshizawa, The University of Electro-Communications, 1-5-1 Chofuguoka, Chofu/Tokyo 182, Japon

Giuseppe Zaccaï, Institut Laue Langevin, B.P. 156 X, 38042 Grenoble Cedex, France

Laszlo Zimanyi, University of California, Dept of Physiology and Biophysics, Irvine, CA 92117, USA

Colloques INSERM
ISSN 0768-3154

Other *Colloques* published as co-editions by John Libbey Eurotext and INSERM

153 Hormones and Cell Regulation (11th European Symposium). *Hormones et Régulation Cellulaire (11e Symposium Européen).*
Edited by J. Nunez and J.E. Dumont.
ISBN : John Libbey Eurotext 0 86196 104 8
INSERM 2 85598 324 X

158 Biochemistry and Physiopathology of Platelet Membrane. *Biochimie et Physiopathologie de la Membrane Plaquettaire.*
Edited by G. Marguerie and R.F.A. Zwaal.
ISBN : John Libbey Eurotext 0 86196 114 5
INSERM 2 85598 345 2

162 The Inhibitors of Hematopoiesis. *Les Inhibiteurs de l'Hématopoïèse.*
Edited by A. Najman, M. Guignon, N.C. Gorin and J.Y. Mary.
ISBN : John Libbey Eurotext 0 86196 125 0
INSERM 2 85598 340 1

164 Liver Cells and Drugs. *Cellules Hépatiques et Médicaments.*
Edited by A. Guillouzo.
ISBN : John Libbey Eurotext 0 86196 128 5
INSERM 2 85598 341 X

165 Hormones and Cell Regulation (12th European Symposium). *Hormones et Régulation Cellulaire (12e Symposium Européen).*
Edited by J. Nunez, J.E. Dumont and E. Carafoli.
ISBN : John Libbey Eurotext 0 86196 133 1
INSERM 2 85598 347 9

167 Sleep Disorders and Respiration. *Les Evénements Respiratoires du Sommeil.*
Edited by P. Lévi-Valensi and D. Duron.
ISBN : John Libbey Eurotext 0 86196 127 7
INSERM 2 85598 344 4

169 Neo-Adjuvant Chemotherapy. *Chimiothérapie Néo-Adjuvante.*
Edited by C. Jacquillat, M. Weil, D. Khayat.
ISBN : John Libbey Eurotext 0 86196 150 1
INSERM 2 85598 349 5

171 Structure and Functions of the Cytoskeleton. *La Structure et les Fonctions du Cytosquelette.*
Edited by B.A.F. Rousset.
ISBN : John Libbey Eurotext 0 86196 149 8
INSERM 2 85598 351 7

Colloques INSERM
ISSN 0768-3154

172 The Langerhans Cell. *La Cellule de Langerhans.*
Edited by J. Thivolet, D. Schmitt.
ISBN : John Libbey Eurotext 0 86196 181 1
INSERM 2 85598 352 5

173 Cellular and Molecular Aspects of Glucuronidation. *Aspects Cellulaires et Moléculaires de la Glucuronoconjugaison.*
Edited by G. Siest, J. Magdalou, D. Burchell
ISBN : John Libbey Eurotext 0 86196 182 X
INSERM 2 85598 353 3

174 Second Forum on Peptides. *Deuxième Forum Peptides.*
Edited by A. Aubry, M. Marraud, B. Vitoux
ISBN : John Libbey Eurotext 0 86196 151 X
INSERM 2 85598 354 1

176 Hormones and Cell Regulation (13th European Symposium). *Hormones et Régulation Cellulaire (13e Symposium Européen).*
Edited by J. Nunez, J.E. Dumont, R. Denton
ISBN : John Libbey Eurotext 0 86196 183 8
INSERM 2 85598 356 8

179 Lymphokine Receptors Interactions. *Interactions Lymphokines-récepteurs.*
Edited by D. Fradelizi, J. Bertoglio
ISBN : John Libbey Eurotext 0 86196 148 X
INSERM 2 85598 359 2

191 Anticancer Drugs (1st International Interface of Clinical and Laboratory responses to anticancer drugs). *Médicaments anticancéreux (1re Confrontation internationale des réponses cliniques et expérimentales aux médicaments anticancéreux).*
Edited by H. Tapiero, J. Robert, T.J. Lampidis
ISBN : John Libbey Eurotext 0 86196 223 0
INSERM 2 85598 393 2

193 Living in the Cold (2nd International Symposium). *La Vie au Froid (2e Symposium International).*
Edited by A. Malan, B. Canguilhem
ISBN : John Libbey Eurotext 0 86196 234 9
INSERM 2 85598 395 9

Colloques INSERM
ISSN 0768-3154

194 Progress in Hepatitis B Immunization. *La Vaccination contre l'hépatite B.*
Edited by P. Coursaget, M.J. Tong
ISBN : John Libbey Eurotext 0 86196 249 4
INSERM 2 85598 396 7

196 Treatment Strategy in Hodgkin's Disease. *Stratégie dans la maladie de Hodgkin.*
Edited by P. Sommers, M. Henry-Amar,
J.H. Meezwaldt, P. Carde
ISBN : John Libbey Eurotext 0 86196 226 5
INSERM 2 85598 398 3

198 Hormones and Cell Regulation (14th European Symposium). *Hormones et Régulation Cellulaire (14ᵉ Symposium Européen).*
Edited by J. Nunez, J.E. Dumont
ISBN : John Libbey Eurotext 0 86196 229 X
INSERM 2 85598 400 9

199 Placental Communications : Biochemical, Morphological and Cellular Aspects. *Communications placentaires : aspects biochimique, morphologique et cellulaire.*
Edited by L. Cedard, E. Alsat, J.C. Challier,
G. Chaouat, A. Malassiné
ISBN : John Libbey Eurotext 0 86196 227 3
INSERM 2 85598 401 7

204 Pharmacologie Clinique : Actualités et Perspectives. (6ᵉ Rencontres Nationales de Pharmacologie clinique).
Edited by J.P. Boissel, C. Caulin, M. Teule
ISBN : John Libbey Eurotext 0 86196 225 7
INSERM 2 85598 454 8

205 Recent Trends in Clinical Pharmacology (6th National Meeting of Clinical Pharmacology).
Edited by J.P. Boissel, C. Caulin, M. Teule
ISBN : John Libbey Eurotext 0 86196 256 7
INSERM 2 85598 455 6

206 Platelet Immunology : Fundamental and Clinical Aspects. *Immunologie plaquettaire : aspects fondamentaux et cliniques.*
Edited by C. Kaplan-Gouet, N. Schlegel,
Ch. Salmon, J. McGregor
ISBN : John Libbey Eurotext 0 86196 285 0
INSERM 2 85598 439 4

Colloques INSERM
ISSN 0768-3154

207 Thyroperoxidase and Thyroid Autoimmunity. *Thyroperoxydase et auto-immunité thyroïdienne.*
Edited by P. Carayon, T. Ruf
ISBN : John Libbey Eurotext 0 86196 277 X
INSERM 2 85598 440 8

208 Vasopressin. *Vasopressine.*
Edited by S. Jard, R. Jamison
ISBN : John Libbey Eurotext 0 86196 288 5
INSERM 2 85598 441 6

210 Hormones and Cell Regulation (15th European Symposium). *Hormones et Régulation Cellulaire (15e Symposium Européen).*
Edited by J.E. Dumont, J. Nunez, R.J.B. King
ISBN : John Libbey Eurotext 0 86196 279 6
INSERM 2 85598 443 2

211 Medullary Thyroid Carcinoma. *Cancer Médullaire de la Thyroïde.*
Edited by C. Calmettes, J.M. Guliana
ISBN : John Libbey Eurotext 0 86196 287 7
INSERM 2 85598 440 0

212 Cellular and Molecular Biology of the Materno-Fetal Relationship. *Biologie cellulaire et moléculaire de la relation materno-fœtale.*
Edited by G. Chaouat, J. Mowbray
ISBN : John Libbey Eurotext 0 86196 909 1
INSERM 2 85598 445 9

215 Aldosterone. Fundamental Aspects. *Aspects fondamentaux.*
Edited by J.P. Bonvalet, N. Farman, M. Lombes, M.E. Rafestin-Oblin
ISBN : John Libbey Eurotext 0 86196 302 4
INSERM 2 85598 482 3

216 Cellular and Molecular Aspects of Cirrhosis. *Aspects cellulaires et moléculaires de la cirrhose.*
Edited by B. Clément, A. Guillouzo
ISBN : John Libbey Eurotext 0 86196 342 3
INSERM 2 85598 483 1

217 Sleep and Cardiorespiratory Control. *Sommeil et contrôle cardio-respiratoire.*
Edited by C. Gaultier, P. Escourrou, L. Curzi-Dascalora
ISBN : John Libbey Eurotext 0 86196 307 5
INSERM 2 85598 484 X

Colloques INSERM
ISSN 0768-3154

218 Genetic Hypertension. *Hypertension génétique.*
Edited by J. Sassard
ISBN : John Libbey Eurotext 0 86196 313 X
INSERM 2 85598 485 8

219 Human Gene Transfer. *Transfert de gènes chez l'homme.*
Edited by O. Cohen-Haguenauer, M. Boiron
ISBN : John Libbey Eurotext 0 86196 301 6
INSERM 2 85598 497 1

223 Mechanisms and Control of Emesis. *Mécanismes et contrôle du vomissement.*
Edited by A.L. Bianchi, L. Grélot, A.D. Miller, G.L. King
ISBN : John Libbey Eurotext 0 86196 363 6
INSERM 2 85598 511 0

224 High Pressure and Biotechnology. *Haute pression et biotechnologie.*
Edited by C. Balny, R. Hayashi, K. Heremans, P. Masson
ISBN : John Libbey Eurotext 0 86196 363 6
INSERM 2 85598 512 9

LOUIS-JEAN
avenue d'Embrun, 05003 GAP cedex
Tél. : 92.53.17.00
Dépot légal : 761 — Novembre 1992
Imprimé en France